# Nanoscale Biophysics of the Cell

Mohammad Ashrafuzzaman

# Nanoscale Biophysics of the Cell

Springer

Mohammad Ashrafuzzaman
Department of Biochemistry, College
of Science
King Saud University
Riyadh
Saudi Arabia

ISBN 978-3-030-08461-5 ISBN 978-3-319-77465-7 (eBook)
https://doi.org/10.1007/978-3-319-77465-7


Softcover re-print of the Hardcover 1st edition 2018


Printed on acid-free paper

This Springer imprint is published by the registered company Springer International Publishing AG part of Springer Nature
The registered company address is: Gewerbestrasse 11, 6330 Cham, Switzerland

# Preface

Understanding the cell using principles of physics is a fast expanding interdisciplinary area of modern scientific research. Cellular systems are largely composed of structures that work at the nanometer (nm) to micrometer (μm) dimensions. As such, physical states and functional expressions of cellular structures in biological systems are often better described by classical physics and quantum mechanics. Soft condensed matter physics has evolved to deal with precise aspects of the complex environment. Technology has also evolved to a point that we can even visualize and monitor the physical energetics-based fluctuations and dynamics of cellular systems, especially at a single molecule level, such as single lipid movement and flipping on cell surface, small drug transport through cellular boundaries, ion channel dynamics in membrane, and nucleic acid-based disorders, among others. Significant research data are available on the subject that may be congregated to explain crucial cellular processes where scale-dependent physics principles are applicable. Through 'Nanoscale Biophysics of the Cell,' we explore how to harness the abundant research findings to explain the cellular processes.

Biological cell research focuses mostly on understanding the cellular structural and functional aspects including especially the cell signaling pathways. The knowledge helps find the origins of cell-based diseases. Interdisciplinary approaches to the cell research ensure participation from multiple groups and individuals to the platform to create collaboration among medical professionals, pharma scientists, biomedical engineers, social scientists, biologists, chemists, and physicists. Four classes of professionals construct most of the perfect interdisciplinary research teams:

(i) Experts in the cell (biologists),
(ii) Experts in principles in scientific research (physicists, mathematicians, chemists, and biologists),

(iii) Experts with a thorough understanding of the delivery mechanisms into cellular compartments (pharmaceutical science people and biomedical engineers),
(iv) Experts to understand the socioeconomic impact of cell-based research inquiries.

Over the last three decades, collaborations among the four groups have increased dramatically due to increasing health risks and the emergence of complex medical issues. The reason for this mutual interest is obviously the modern biology's pre-eminent role on the front lines of scientific research targeting medical discoveries. The collaborating connections between the physical and biomedical sciences have been developing rapidly since especially the groundbreaking discoveries in molecular genetics. There is clearly a need for a continuing dialogue and a cross-fertilization between life sciences and physical sciences. Due to the inter-disciplinary nature of the life sciences, numerous new border areas are being created and developed. Therefore, disciplines such as mathematical biology, biophysics, computational biology, biostatistics, biological physics, theoretical biology, biological chemistry (and of course biochemistry), and biomedical engineering have been undergoing exponential growth. The ultimate goal of all these interdisciplinary scientific subjects and fields lies in understanding one thing 'life.'

What is life?

How does life work?

Why does life fail?

Probably, these are the questions scientists have always been after. The answers lie in using sophisticated technologies to address cellular aspects determining the very causes of life. Application of appropriate techniques that help not only to understand the cell but also the smaller nanoscale dimension structures inside the cell is required. The techniques have to address the cell-based structures that undergo local interactions and energetics-driven fluctuations at even femtosecond (fs) timescale. Principles of nanotechnology, imaging, femtochemistry, etc., help understand the smaller ordered or disordered individual and complex sections, interactions, and dynamics inside the cell. In the nm scale, we can now visualize, measure, physically handle, and manipulate the small-scale constructs of cells. We find various types of smaller building blocks performing versatile activities within cells, the traditionally considered building block of life. Of course, the cell is still the ultimate kingdom where most of the life's processes are found. But within the cell, there are important sections or tiny worlds where crucial processes originate.

Distinguishable diversity present in biological systems is a simple result of the multitude of possible combinations of the finite number of structural elements. The physiological function of biological systems follows from the complexity of the specific organization of various molecular systems where individual molecule's functions are mostly governed by fundamental laws of physics. The principle of complexity begetting function is familiar to physicists and has often been referred to as an emergent phenomenon. It is a characteristic feature of atomic systems to display new emergent properties as they become more complex. Especially, certain biological aspects of molecules at nm scale rely on resultant physical actions that are governed by specific physics laws. For example, in structures of biomolecules like proteins, enzymes localized fluctuations, displacements, disorders are occasionally found to happen at low dimensions. Although at a global scale, such small-scale regulations in structures may mean almost nothing in eyes of biologists. But in physicists' careful eyes, these nm scale perturbations may seem to be just the initiation of unknown major changes in the systems' biological expressions. In such small-scale changes, some fundamental physics principles, such as harmonic and anharmonic oscillation energetics, can be utilized to understand the mechanical condition and predict the future trends. This type of fundamental mechanism-based understanding in low-scale biological systems may appear quite important, especially when things come to get dealt for understanding the initiation of major disease conditions. Functional fundamental physics rules are hidden in mainly low-dimension biological systems that dictate the normal functioning of building blocks making whole living organisms. Some of this low-scale biological system-based physics will be pinpointed or discovered and utilized in this book along with my earlier book 'Membrane Biophysics' published by Springer-Verlag in 2012. Current experimental techniques are sophisticated enough to measure parameters related to structural and functional properties of biomolecules at a global scale as well as, in many cases, at the nm scale building block levels. These experiments aid our understanding of low-scale structure functioning physics principles, develop analytical expressions to connect various parameters explaining the physics, perform the numerical computations to address the role of these parameters, and compare between many sources of experimental data.

The book is focused on a detailed description of the diverse mechanisms and phenomena associated with structure and function of the cell and its fundamental units inside, especially those at nm scale. There are nine chapters presenting the fundamentally relevant materials.

It has been possible for me to take this big project due to my training and active research in condensed matter physics including dealing especially with soft biological systems at a single molecular level. My decadelong experience in teaching biophysics related courses at both the undergraduate and graduate levels has also played a crucial role in understanding the importance of accumulating current relevant information about this subject in a book. I have added enough novel scientific ideas and models into this information and presented them in a professional manner, systematically, so that the interested readers find them quite understandable and entertaining. It is a hope that this monograph will be found of use as source of valuable information and conceptual inspiration to both students of biophysics, biology, biomedical engineering, pharmaceutical science, medical science, and expert researchers. We have avoided going into the details on calculations and presented only the necessary forms of physical formulas to make the presentations quite understandable for all. A bulk of experimental data are also avoided from presentation, and only the summary of them are mentioned. Discussion on nanoscales of cell could be extended to covering many other areas as this is a large field now. But we limited us within most important aspects that we found necessary for addressing the important physics active in cell biology at low dimensions.

Riyadh, Saudi Arabia
January 2018

Mohammad Ashrafuzzaman

# Acknowledgements

I am thankful to Profs. Jack Tuszynski and Michael Houghton of Alberta University and Dr. Chih-Yuan Tseng of MDT Canada Inc. for many insightful discussions on various aspects that helped develop some of the ideas incorporated in this book. Jack Tuszynski read the chapters and provided important suggestions regarding formatting of this book. I am very much grateful to him. Writing of this book would be impossible without using a bulk of experimental and theoretical data from publications of various authors (all are quoted in references). It is a great pleasure to thank all of them for their contributions in the field. Hundreds of discussions with colleagues, academic friends, and research group members at both King Saud University and Alberta University helped me shape ideas while writing this book during last five years. Editorial assistance and encouragement provided by the staff members at Springer, especially Dr. Angela Lahee, are thankfully acknowledged. I am thankful to my parents whose encouragements were always with me. I especially value the emotional support given by my wife Anwara and two sons Imtihan and Yakin Ahmeds. Anwara has been always on trips between my Canada residence and Saudi workstation to provide much needed social supports in my academic activities.

Riyadh, Saudi Arabia
January 2018

Mohammad Ashrafuzzaman

# Contents

# Chapter 1
# General Introduction and Nanoscale View of the Cell

Interdisciplinary scientific initiatives that focus at understanding of the cell are enormous. It started considerably as early as in seventeenth century when an eminent physicist Robert Hooke discovered the cell. The general or microscopic concepts about cells suggest that a collection of varieties of them construct a body. A living system can be subdivided into various functional sections and its smallest unit is cell what contains even smaller structures and hosts a lot of physical natures, trends, and techniques. Cell's appearance and function are ensured due to the biological expressions of thousands of genes, as well as the ongoing activities of inner molecular structures following physical and chemical formalisms.

Attempts at applying physics principles to living systems containing varieties of cells can be traced to the early creators of modern science. Galileo applied physical principles to analyze the structure of animal bones, Newton applied optics to the perception of color, Volta and Cavendish studied animal electricity, and Lavoisier demonstrated that the respiration process is an oxidative chemical reaction. Robert Mayer got inspired by physiological studies to formulate the first law of thermodynamics. Another important area of application of physics to physiology is hydrodynamics. Blood flow was analyzed by Poiseuille, and air flow in the lungs has been described consistently with aerodynamics. Hermann von Helmholtz laid the foundations for the fundamental theories of vision and hearing. The list of physicists making huge impacts on biology and physiology is very long. A few of the well-known figures who have crossed the rapidly thinning current boundary between physics and biology have just been mentioned. Delbrück, Kendrew, von Bekesy, Crick, Meselson, Hartline, Gamow, Schrödinger, Hodgkin, Huxley, Fröhlich, Davydov, Cooper, and Szent-György have undoubtedly pushed the frontier of the life sciences in the direction of exact quantitative analysis.

The tools to observe small things like cells and organisms have mostly been developed using fundamental physics principles and engineering techniques. The applications of these tools with their ever-enriching sophistications in ever-growing number of biological fields are beyond imagination. Hard-core physics principles are now utilized in exploring deep biological problems too. For example,



M. Ashrafuzzaman, *Nanoscale Biophysics of the Cell*,
https://doi.org/10.1007/978-3-319-77465-7_1

techniques like X-ray, crystallography are regularly used in exploration of the biological systems. Physical modeling of the DNA double helix structures by Huxley and Hodgkin based on X-ray crystallography data of Franklin was an inspiration for subsequent studies aiming at understanding tuned biological aspects of cell using pinpointed application of physics techniques even at nanometer (nm)-scale dimensions. Another physical technique nuclear magnetic resonance (NMR) has found its enormous applications in understanding biological cell's molecular processes. NMR has been used to study biological cells since the early days (almost seven decades ago) of its development on the shoulders of giants like Stern, Rabi, Bloch, Ernst, Wüthrich, Lauterbur. Novel discoveries on understanding the structures of various proteins inside cells, membranes are made regularly these days. We can now also track the low nm resolution (as close as a lipid cross section) cell surface dynamics based on the mosaic model of Springer and Nicholson, discovered about 5 decades ago with subsequent developments. The twenty-first century scientific advancement in optical microscopy by Betzig, Hell and Moerner, and many others ensures observing the cellular system's nanoworld what was once considered impossible as in 1873, the microscopist Ernst Abbe stipulated a physical limit for the maximum resolution of traditional optical microscopy—it could never become better than 0.2 micrometers (μms). We visualize the pathways of individual molecules movements inside living cells. We can see how molecules create synapses between nerve cells in the brain; we can track proteins involved in diseases as they aggregate; we follow individual proteins in fertilized eggs while dividing into embryos. Optical microscopy technique is now so powerful that it lets us observe nm resolution structural formation, fluctuations, displacements, movements, etc., ahead of understanding the underlying mechanisms.

Concepts of many-body interactions, charge particle interactions, stochastic principles, principles of electrostatics and electrodynamics, etc., are found quite applicable to addressing various phenomena in cellular systems. Classical mechanics principles have long been used in understanding mechanisms behind cellular systems' functions. Besides, quantum mechanics principles have recently been found important in exploring cell dynamics. Computational techniques are popularly used now to address all kinds of molecular properties, from a molecule's stability and movement to its reactions with other molecules. Since early 90s, the computing power and algorithms began catching up with theory to enable the application of calculations to relevant biomolecules like proteins, drugs, and materials, thanks to Karplus, Warshel, and Levitt among others whose groundbreaking initiatives made this possible. Computer visualization of the simulated behavior, real-life movement of complex entities like drugs, proteins, large molecular systems like liposomes, membrane monolayers and bilayers, ion channels help understand the nm resolution active physics principles involved in constructing the biological systems.

The cell exploration has now systematically been taken to a nanodimensional world what is popularly known as 'nanotechnology.' Femtosecond (fs) chemistry address of cell reaction research is also within our understanding realm. We can now regulate the cell constituents' structure and functions at nm-dimension resolution. Both slow and fast kinetics of them can be manipulated. That is why we can no longer just believe totally that the cell is the building block of life. Specialized building blocks that are smaller than and within the cell have been discovered. Of course, the cell is the kingdom where most of the life's processes are collectively found but cell also hosts pretty important sections or many smaller worlds where most of the cellular processes might be distinguishably originated at or controlled from. We can now put our fingerprints at the genetics, related genes or codes in nucleic acids and often can act at the DNA structures to create pinpointed changes and repairs at low nm dimensions. Nanoscale biophysics is a subject that aims to deal with this kind of exploration. Cell's internal nanostructures are understood using biophysics techniques what has become quite popular nowadays. The physical understanding of the nanostructure geometry, function, and their alteration/ modulation helps address important questions about life. Nanoscale biophysics of the cell and nanotechnology of the cell are complementary to each other. Nanoscale biophysics helps us understand biological processes, while nanotechnology helps manipulate the nanostructures of the cell as required to repair especially disease states of the structures. This book will address these aspects quite in detail.

The cell's nanoscale world and the related biology fields will be scrutinized through the analysis of existing and novel experimental and theoretical data along with the techniques of their production, methods of interpretations, and understanding of their applications. Structural and functional mechanism perspectives of both normal cell conditions and perturbed conditions under the influences of diseases are the focal points here. A lot of specific cell-based structural components have been taken into consideration as examples for explaining the biophysics aspects of mainly the nm-dimension cell components.

Following the address on the defining aspects of cell (see also Ashrafuzzaman and Tuszynski 2012a), the book starts with the search of a descriptive definition of life. With the development of technology over centuries, we have been able to revise our knowledge on the cell by going deep into its finest structures. Consequently, life's definition, once used to be made up using biological perspectives, now requires considerations of physical principles and phenomena for clarifications. Cell hosts varieties of life processes, such as nonstop dynamics, diffusion, and transport aspects what need to be explained using low-dimension scale functioning physics principles. Considerable existing rules and models, as well as novel understanding and analytical expressions, are required. New physics is yet to emerge to help understand many unexplored areas. I have accumulated here a considerable amount of conceptual approaches what readers may find as their mind opening and thought creating for further explorations.

Biological cells are enriched with molecular machines (see Wilson 1925; Taylor 2001) which account for most of the molecular-level solid-state structures that regulate cellular functions. Their pre-creation stage molecular mechanisms, states, and energetics of physical existence and physiological function aspects are inspected in detail. The book contains rigorous discussion where molecular machines have been scrutinized in regard to their physical roles in creating cell-based natural nanotechnology. The principles of this nanotechnology along with artificial biophysics techniques can be utilized to regulate the functions of cellular machines. These principles may also indirectly help us design agents for repairing the cellular disorders.

Understanding of the states of cell-based diseases (as a result of cellular disorders including altered structures of cellular machineries) and drug effects has been historically made mainly using chemical, biochemical, biological, and medical techniques. Careful biophysical analysis reveals that the physiological disorders in mostly cell-based structures occur at nm-scale resolution, so do the effects of most of the drugs. The use of biophysics principles in analyzing and understanding these small-scale structural and functional issues is new. The ideas are therefore not always well organized. I have made an attempt to explain the structural and functional disorders of specific ion channels in various cancer cells. This analysis will help us understand how the rise of disorders in the structures and functions of cell-based general molecular machines participates in altering the physiology of cells.

Quantum mechanics application in cell is quite new, but it offers some promising tools to initiate understanding crucial cellular processes. Quantum mechanical calculations and other related aspects are briefly provided to mainly pinpoint the future trends in cell research.

As mentioned earlier, physics principles besides biological and chemical properties are naturally in built that drive physical structures into functioning. Interaction, force, and energy principles are found to determine the regulation of the statics, dynamics, and technological architectures (specifically nanotechnology) of the cell. Determination of correct expressions of parameters related to universal mechanisms, e.g., energetics, entropy, that dictate various cellular processes, is often not any easy task. But these tasks appear relatively easier with the application of physics laws, using mostly classical and occasionally quantum mechanical calculations that are relevant in the complex biological systems. Disorders in disease conditions can also be addressed through understanding the perturbations in structures and functions of related cellular components using physics principles. Such perturbations require the change of order parameters ensuring the necessary free energy changes that cause the transitions of structural states from one energy trap (often is referred to as a harmonic potential well) to other ones. Special attention has been given on explaining the ion channel aberrations in cancer cells and other disease conditions. The readers will get a summary of huge information on all of these topics. The knowledge will create catalysts for future studies of the cell in regard to its treatment at the nanoscale components where various technologies are available.

Subsequent chapters of the book will address specific areas covered in fundamental and medically relevant aspects of the nanoscale view of the cell. The explorations will be done on mainly the applications of specific modern physics principles and latest technologies to address the cellular biology. We have avoided presenting complicated equations, formulas, and their derivations as the book is primarily written for students, researchers, faculties, and biomedical scientists working in the interdisciplinary areas of science. We have instead presented a lot of physical models and sketches to explain complicated biological aspects in a simplistic manner so that with minimal physics knowledge and interests readers with nonphysics backgrounds face no issues to understand most of the aspects covered in this book. Before going to the advanced stages, the readers may still require understanding of the necessary basic physics, chemistry, and technology that a cell hosts in its various low-dimension compartments. Therefore, I wish to provide the basics of a few introductory areas involved in low-dimension biology and biophysics of the cell here.

## 1.1 What is a Cell?

The basic unit of a living organism is called cell. Cell is the primary unit which provides the basis of the biological structures and functions. Cells are the building blocks of life. It means that life perhaps originates at the cell. Alternatively, the collective actions of cellular activities are responsible for life.

From the commonly accepted definition of a cell presented above, it is clear that the cell holds two unique defining features: (i) structural aspect that actually helps make a cell and (ii) mechanism aspect that qualifies the cell. The latter follows from the former—the mechanism aspect is often the result or the collective expression of the structural aspect. From the perspective of structural aspect, we observe plant and animal cells to differ. From the perspective of mechanism aspect, we observe cells from various organs within the same body to differ. Besides, cells of the same organ may also differ from both structural and mechanism perspectives while the organ falls into any disease state. Moreover, the dynamic nature of various structural components of a cell makes it functionally quite variable. The disease-affected cells may not show the characteristics of a normal cell. The same cell being affected by different kinds of diseases may also behave differently. A neuron makes the type of a cell which is primarily for communicating information throughout the body. All the mentioned contradictions make our task quite hard when we attempt to provide a unique definition for cell. Cells from different organs in normal and disease states may host organ- and state-specific structures and characteristics that are subject to be continued fluctuations. But the general definition of a cell emerging from the first paragraph of this section is perhaps still traditionally the best.

## 1.2 Optophysics to Discover Cells

Optophysics techniques helped discovering the μm-dimension objects cells. In 1590, the microscope was invented by two lens grinders Hans Lippershey and Zacharias Janssen in the Netherlands who placed two lenses in a tube. People discovered a way to see tiny objects through a microscope. During fifteenth and sixteenth centuries, a revolution happened in the field of microscope and telescope. People came to know ways to observe many objects that were either too small or too far to see. After bringing lens in the pathway of sighting, things appeared big that were tiny ones or closer that were distant ones. The modern foundation was made by many scientists including Robert Hooke, Galileo Galilei, Newton, and many other icons. The origin even dates back to eleventh century when an Arab scientist Ibn Al-Haytham first discovered the concept of a camera and published his 'Book of Optics.' Ibn Al-Haytham was the first to experiment to prove that we see as a result of light from objects traveling in a straight line into our eyes. Ibn Al-Haytham and his student Kamal Ad-Din observed the camera obscura phenomenon which is defined as 'dark chamber.' A cardboard with a tiny pinhole on it allowed the light to pass, and thus, concept of a camera (popularly referred as pinhole camera) emerged. Haytham's famous optical diagram and eye diagram from perspectives of light transmission and reflection both made the background to understand the proof of principles in optophysics. In today's pinhole cameras, lenses are added and thus the optical diagrams of modern cameras have further evolved in complexity and have become more powerful.

During the seventeenth century, a British scientist Robert Hooke who was primarily a physicist developed his own microscope to study various objects and published his results in Micrographia in 1667 (Gutenberg eBook 1964). He viewed a thin cutting of cork through his microscope and found empty spaces contained by walls (see Fig. 1.1). He called them 'cells.' More precisely, he saw plant cells. Within a decade, in 1678, a Dutch scientist Antony van Leeuwenhoek, one of the earliest microbiologists, reported that he had observed little animals—bacteria and protozoa through his simple microscope. This discovery was confirmed by plant cell discoverer Hooke through his compound microscope. Within a short time, humankind found the smallest unit of a plant and little 'animal' called bacteria. Both of these discoveries happened, thanks to the technological developments in optophysics. People learned to use machine-made eyes to observe things. These discoveries created opportunities for scientists to try to understand if there were similarities between smaller organisms and larger ones.

The discovery of cell and bacteria initiated the exploration of things at μm scale. Both of these discoveries were just observational. Nothing was known about the activities of these smallest units. Now we just do not see the cells, we go beyond to see the constituents of the cells. A cell is perhaps the smallest building block holding almost complete criteria of a living body. But the cell's constituents have independent structures and functions. They are very different within. The cellular constituents are often orders of magnitude smaller than the μm-scale cell. With the development of microscopic techniques, it is no longer hard to explore smaller scales than those at cell dimensions. The cellular constituent observation-based understanding of cells found the beginning of its in-depth understanding with the discovery of nm-scale detailed DNA double helical structure by Watson and Crick in 1953 based on the famous X-ray photograph of DNA produced by Franklin (Watson and Crick 1953). Within a year of this groundbreaking discovery, another great scientist Linus Pauling was awarded a Nobel Prize in chemistry for his famous works on the structure of molecules and crystals, including a clear picture on chemical bonds. The nanoscale-level first-ever complete understanding of

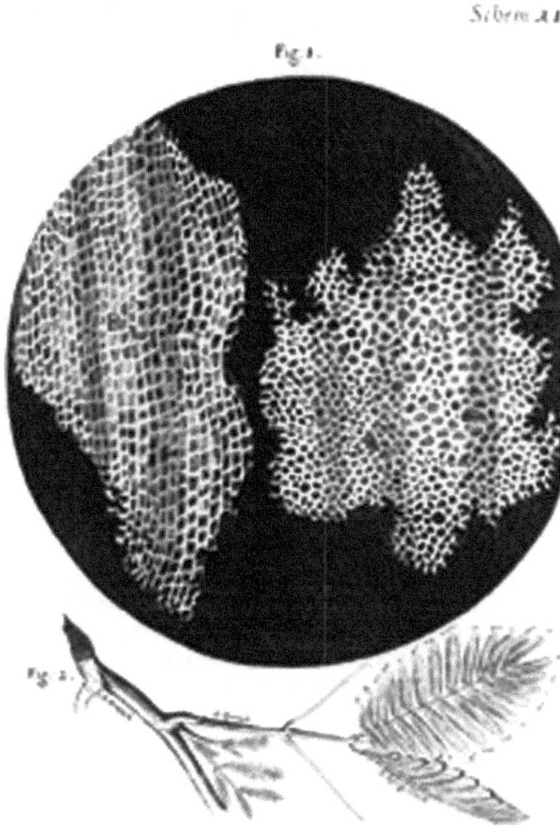

**Fig. 1.1** Hooke's cork cells. To understand the original details, interested readers can read from The Project Gutenberg eBook, Micrographia by Robert Hooke, Printed by Jo. Martyn, and Ja. Allestry, Printers to the Royal Society, London (1964). The Project Gutenberg eBook has granted automated permission to reuse the contents in the book. Other sources can be the following links: http://www.nobelprize.org/educational/physics/microscopes/timeline/index.html. Hooke: http://www.science-of-aging.com/timelines/hooke-history-cell-discovery.php, http://www.scribd.com/doc/69985133/24-Robert-Hooke-Micrographia-1665

biomolecule structures happened about 100 years after the first isolation of nucleic acid in cell by a Swiss physician Friedrich Miescher. But the nanoscale structure understanding led to the creation of a huge platform for biomolecule structure studies which subsequently changed the primitive view about the cell. The modern-day concept of the cell emerged on various structural studies about different compartments within cell interior and exterior regions. A cell's generally accepted model is shown in Fig. 1.2. We shall not provide details on cell components here as there is quite a large amount of literature available on these topics.

## 1.3 Cell Imaging

How a cell looks like under microscope is known. In biological research, microscopic images are taken regularly. We also took many such images recently to demonstrate biophysical aspects of the cell. We specifically aimed at understanding the functional physics at cellular compartments with dimensions as low as nm- to µm-scale range. We shall explain a few example cases in this chapter.

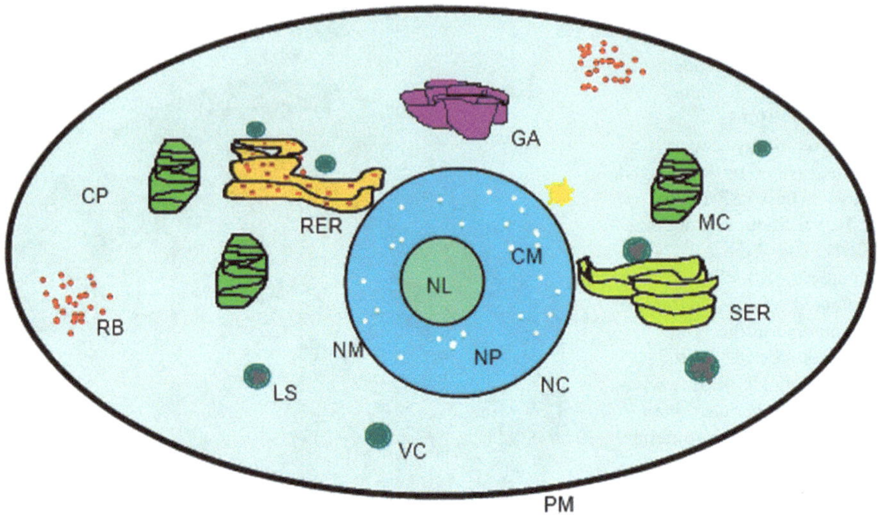


**Fig. 1.2** Schematic diagram of a cell showing different parts (cell's model structure). In no way, any component schematized here represents the true structure observed in biological cell. PM: plasma membrane, CP: cytoplasm, VC: vacuole, LS: lysosome, RB: ribosome, MC: mitochondrion, GA: Golgi apparatus, RER: rough endoplasmic reticulum, SER: smooth endoplasmic reticulum, NC: nucleus, NM: nuclear membrane, NP: nucleoplasm, CM: chromatin, NL: nucleolus. The constituents shown here are found in an animal cell. In plant cell in addition to all these chloroplasts with photosynthesis ability exist. A plant cell (not an animal cell) also consists of a cell wall surrounding the plasma membrane which provides tensile strength and protection against mechanical and osmotic stress. Special credits are due for Imtihan Ahmed of Computing Science Department, University of Alberta, Canada, who helped to draw this model diagram

### *1.3.1 Cell's General View Under Microscope*

Figure 1.3 shows how cells look like under microscope. A549 cell line (that are the adenocarcinomic human alveolar basal epithelial cells) are imaged under microscope.

In Fig. 1.3, we observe simple cell images suggesting cells to be just like spheres with no constituent structures. Here we observe the cell surface and that gives us an impression of how cells look like. To observe cell constituents we need to go deeper. For that, we need to use the power of technological developments and perform high-resolution microscopy imaging. Sometimes we take help of external agents (e.g., nanoparticles) without or with fluorescence tags attached and let them migrate to cell surface or cell-based constituent's structural vicinity. Through them, we can observe certain regions quite clearly. Here, we briefly show two independent studies that had constructed cell surface images using two independent techniques for general demonstration purposes.

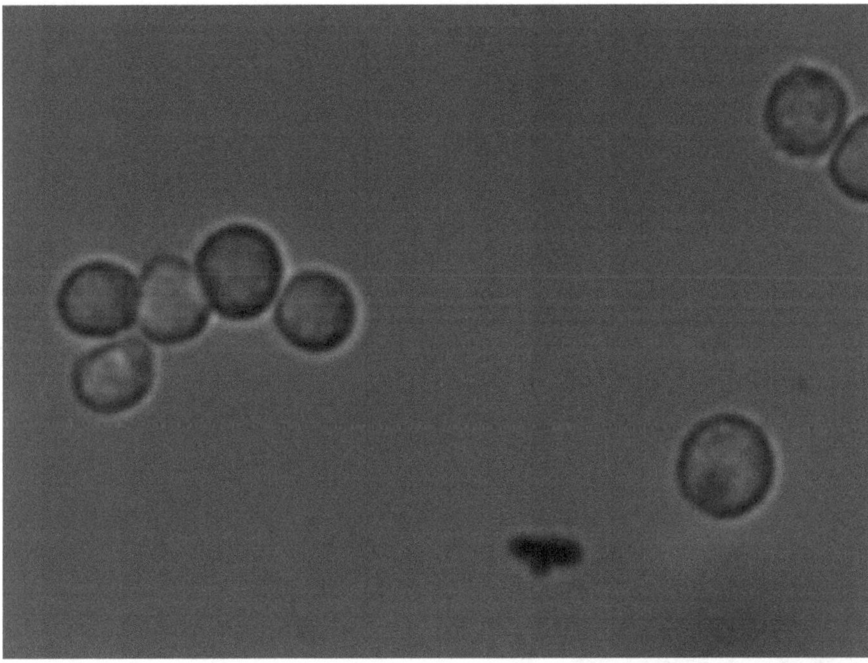

**Fig. 1.3** Cells are imaged under microscope. We incubated 1 mL cells in cell culture media Dmem F12 at 37oC. A549 was supplied by ATCC (http://www.atcc.org/products/all/CCL-185.aspx). We then added 5 μL cell into 5 μL trypan blue to a glass slide (Fisher). We then took image [under microscope Axiovert 200 M (Zeiss)] of the plate of which a section is presented here. This imaging was performed by Ashrafuzzaman with the help of coworkers at the Department of Oncology, University of Alberta, Canada, during summer, 2014 (yet to be published)

### *1.3.2 Atomic Force Microscopy to Observe Cell Surface Landscapes*

Atomic force microscopy (AFM) is a powerful imaging technique that is frequently used to track the low nm-dimension structural changes, surface landscapes, etc. Here we shall provide a few images that we recently captured for demonstrating primarily the statistical cell adsorption mechanisms of chemotherapy drugs (CDs), e.g., colchicines. For details, see Ashrafuzzaman et al. (2016). We just present a few figures to demonstrate how AFM emerged to help us investigate deep into low-dimension scale structural components of the cell.

**Independent Cell and Cell Targeted Drug Characterization**

In Fig. 1.4a, an inverted optical microscope image is shown where position of the probe and distribution of cells at a large scale is clearly seen. The area of the image is (700 × 520) $\mu m^2$. A large-scale AFM topography image of the cell distribution is shown in Fig. 1.4b. The average size of the cell is approximately 1.2 μm. The

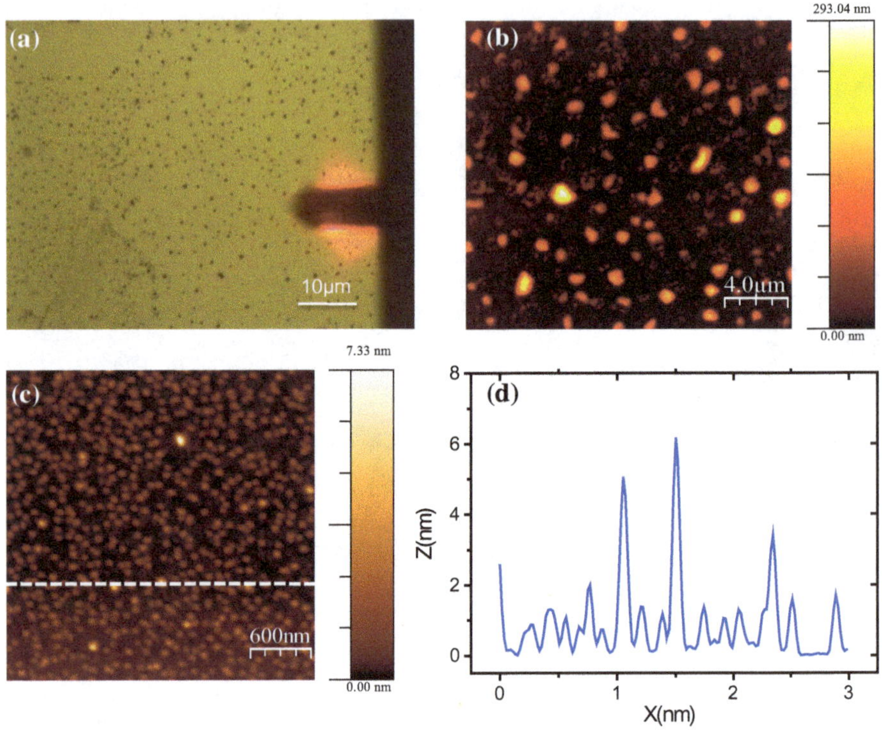


**Fig. 1.4** Cell and colchicine TM-AFM image on $SiO_2$ substrate. **a** Optical microscope image at large scale. **b** AFM height image of distribution of cells at a large scale. The height of the cells is presented in color scale 0 (dark fields)—293.04 nm (light fields). **c** AFM height image of distribution of colchicine (1 μM) molecules. The height of the colchicine molecules is presented in color scale 0 (dark fields)—7.33 nm (light fields). (**d**) A height profile recorded along the dashed line in Fig. 1.4c. For details, see Ashrafuzzaman et al. (2016)

shape of all cells is not identical. This is expected because the shape of the cell on surface depends on the range of interactions between the cells and the surface. The shape of the cell can also be deformed due to tip–sample interactions during scanning the probe over the surface.

In Fig. 1.4b, AFM height image of the distribution of cells at a large scale is presented. The heights of the cells are shown in color scale 0 (dark fields)—293.04 nm (light fields). Thus, we can even detect small geometrical fluctuations here in color code. This type of information is very important to understand cell surface mechanical properties.

In Fig. 1.4c, AFM height images of distribution of colchicine (1 μM) molecules are shown. Colchicine molecules are on average with 1 nm dimension. The heights of the colchicine molecules are presented in color scale 0 (dark fields)—7.33 nm (light fields).

Finally in Fig. 1.4d, a height profile that was recorded along the dashed line in Fig. 1.4c is presented. This provides the linear scale conversion for color scale.

The images in Fig. 1.4 demonstrate the cell surface structures at nm-scale resolution. This kind of inspection is always crucial while dealing with nanoparticle adsorption in the cell surface. This imaging helps quantify the cell surface adsorption as well as understand the modulation of the cell surface geometry what happens at nm scale as a result of the surface adsorption of drugs or general nanoparticles.

### *1.3.3 Confocal Imaging to Observe Cell Substances*

Confocal imaging is a powerful technique that helps to image high-resolution small structural components in cell. Here, we present one of our captured confocal microscope images of the A549 cell line illustrating deep inside into cell structures, e.g., tubulin. See Fig. 1.5. This imaging was performed by Ashrafuzzaman and coworkers at the Department of Oncology, University of Alberta, Canada during summer, 2012.

In Fig. 1.5, we observe cellular constituents, e.g., microtubules. This has been possible due to the use of a high-resolution confocal microscope imaging. This technology helps us see in the nanoscale-order resolution, as low as 1 nm or even fraction of nanoscale-dimension cellular constituents. This developed setting helps us model the μm-dimension cell's structural concept almost accurately at nm resolution. The rise of disease conditions and corresponding low-resolution structural changes are also monitored using imaging techniques. Drug effects through direct

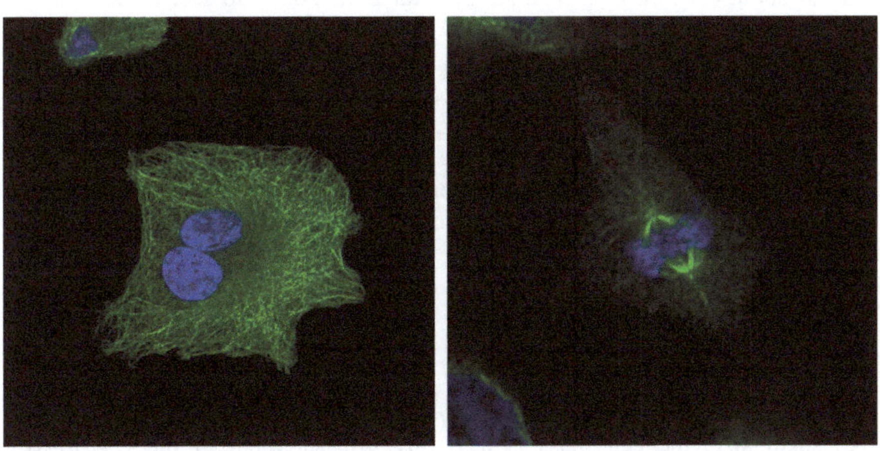

**Fig. 1.5** Confocal imaging to observe cell interior structures. Confocal microscope image of the A549 cell line illustrating deep into cell structure like tubulin (Ashrafuzzaman, unpublished). The tubular polymers grow in μm scale while the diameter in nm scale

drug binding at the structural sites are also monitored. Yet, we lack in understanding the role of many cell compartments in detail unless we perform various chemical analysis on the components and perform studies on the components' dynamics. Various constituents are not well structured or the structures are not mutually distinguishable to offer understandable images. That depends mainly on the kind of structures or indirectly the state of the structures, namely solid state, liquid state. That leaves the cell studies still in not so under control condition.

The nanoscale-dimension structural studies through the application of mainly optophysics techniques and varieties of chemical analysis of the cell constituents led to understanding the versatile biological natures of individual structures and compartments most of what rely on corresponding physical properties. We can now brand the type of state of every structure and often of every compartment which is a collection of unitary, binary or tertiary structures. Based on their structural state classification, we can also understand state-specific mechanisms that are governed by fundamental laws of physics. We can also extend our understanding by correlating the physical structural states and related mechanisms with the corresponding biological expressions. The definition of the cell underlies in the collective actions what we often brand as resultant of the mechanisms contained in all these physical structures making the known biological shape of a cell.

## 1.4 Cell Measurements

Cell is small and it has even smaller different distinguishable sections. Types of measurements on cell can therefore be global, localized, compartmental, etc. Unaided eyes can see some big cells as unaided human eyes can see objects as small as about 0.1 mm long. Human eyes have the ability to see, without or with magnifying glasses, an amoeba proteus, a human egg, and a paramecium. Light microscopes, electron microscopes, and other imaging devices aid us with high resolution to go beyond general look at cells to the individual components. AFM imaging, confocal imaging, etc., provide most of the structural information deep inside cells at nm-order resolution of the dimensions or beyond. Besides imaging based mostly geometry related measurements, there are various other types of measurements we are able to perform on cells. This section will be dedicated to explaining a few example cases.

### *1.4.1 Cell Mass*

Like small size, the cell mass is also tiny. Measurement of the cell mass is sometimes considered important. Recently, Grover et al. (2011) published a nice article where a technique was used to measure the single-cell mass, volume, and density. Here, approximately 500 cells per hour with a density precision of

0.001 g $mL^{-1}$ were utilized for measurement. A microfluidic mass sensor was used to measure the density of single living cells. Each cell was weighed in two fluids of different densities.

Cell density measurements generally rely on buoyant force to transport cells to a location where the surrounding fluid density most closely equals the density of the cell. But during centrifugation, the lengthy exposure to concentrated solutions of the substance used to construct the density gradient may inadvertently affect the density and viability of the cells. Here, a clear buoyant force dependence exists. The hypothesis adopted in Grover et al. (2011) that the ideal method for measuring a single-cell density would not rely on buoyant force to slowly push the cell, but would measure the buoyant force on the cell directly, in any desired fluid. Then, use the information to calculate density, mass, and volume of the cell. Such a method was used around 250 B.C. by Archimedes of Syracuse, who was asked by his king to determine whether a crown was made of pure gold. We all know the story. Archimedes measured the density of the crown by hanging it in a fluid. He used a balance to measure the submerged crown's buoyant mass $m_b$ using the following equation:

$$m_b = m\left(1 - \frac{\rho_f}{\rho}\right) \tag{1.1}$$

Here, $m$ is the mass of the crown, $\rho_f$ is the fluid density and $\rho$ is the density of the crown. The buoyant mass of the crown in two fluids of different known densities, for example, water and air, was measured. Archimedes could then calculate the density of the crown and compare it to the known density of pure gold. The same illustration was used for the cell measurements (see Fig. 1.6).

The intrinsic cell-to-cell variation in density was reported nearly 100-fold smaller than the mass or volume variation. As a result, changes in cell density can be measured which is indicative of cellular processes that would be otherwise undetectable by mass or volume measurements. Four examples were reported: identifying *Plasmodium falciparum* malaria-infected erythrocytes in a culture, distinguishing transfused blood cells from a patient's own blood, identifying irreversibly sickled cells in a sickle-cell patient, and identifying leukemia cells in early stages of responding to a drug treatment (Grover et al. 2011). For details on techniques, interested readers may learn directly from Grover et al. (2011). Results are shown in Fig. 1.7.

Beads with known density and size were measured, see Fig. 1.7a: the bead volume, mass, and density distributions for a sample of 5.0-µm-diameter polystyrene beads. Measurements were then performed on a cell sample, *P. falciparum* malaria-infected erythrocytes (see Fig. 1.7c). *P. falciparum* causes significant physical changes in infected erythrocytes (Mrema et al. 1979; Mills et al. 2007).

Following identical techniques, mass and density measurements of live and dead gram-negative and gram-positive bacteria were recently measured (see Lewis et al. 2014). Buoyant mass measurements were performed for gram-negative *E. coli* and gram-positive *L. innocua* bacterial cells. Buoyant mass of a growing population of

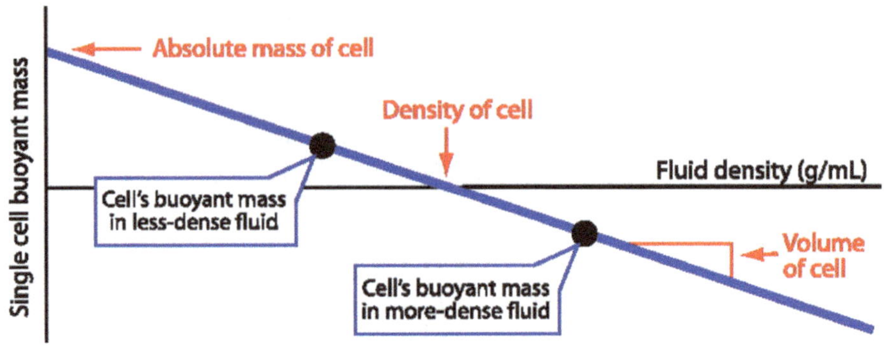


**Fig. 1.6** The method of Archimedes was applied to measure single-cell mass, volume, and density. By weighing a cell in two fluids of different densities and plotting the linear relationship between buoyant mass and fluid density, the absolute mass, volume, and density of the cell can be determined from the *y*-intercept, slope, and *x*-intercept, respectively. Presented with due permission from Grover et al. (2011)

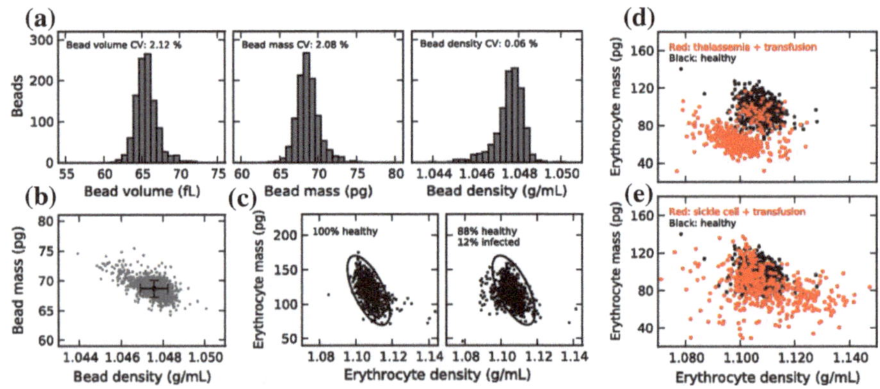


**Fig. 1.7** **a** Bead mass, volume, and density distributions for a population of 5.0-µm-diameter polystyrene beads ($n$ = 1069). **b** A scatter plot of bead mass versus density. Here, error bars are ± one standard deviation of the mean. **c** Erythrocyte mass versus density for 690 cells from a healthy erythrocyte culture (*Left*) and an equal number of cells from a culture containing approximately 12% *P. falciparum*-infected erythrocytes (*Right*). **d** Single erythrocyte mass versus density for an individual with suspected thalassemia trait who also received a transfusion of normal (nonthalassemic) blood 4 d prior to collection (red; $n$ = 502 cells) compared to a random nonthalassemic, nontransfused individual (black; $n$ = 502 cells). The patient's own erythrocytes (red) are offset from a normal patient's erythrocytes (black), except for a small number of normal erythrocytes the thalassemic patient received during the transfusion (red points clustered on black points). *E. Erythrocytes* from an individual with sickle-cell anemia who also received a blood transfusion 35 d before collection (red; $n$ = 502 cells) compared to the same nontransfused individual as in D (black; $n$ = 502 cells). The widening of the distribution of erythrocyte densities in sickle-cell anemia reported is consistent with other studies (Rodgers et al. 1985), with the more dense cells likely representing irreversibly sickled cells (Bertles and Milner 1968). For details, see Grover et al. (2011)

bacterial cells was monitored in real time. The buoyant mass (found in femtogram (fg) order) data were used to determine average density and total mass of live and dead cells. Buoyant masses of live and dead cells with and without exposure to hydrogen peroxide stress were measured. Live cells generally were found to have larger buoyant masses than dead cells.

## *1.4.2 Electrical Measurements*

Besides observing cells and their components at low nm-dimension resolution, we do various physical measurements to demonstrate cells' transport properties, electrical conditions, mechanical properties, etc. These measurements distinguish the intercompartment and intracompartment biological and chemical conditions and provide information on comparative criteria. Capacitance measurements of a cell's compartment, current measurements across cell-based physiological barriers, etc., using established techniques provide crucial low-dimension scale biological and chemical information on cell constructs. We have discussed, quite in detail, the electrical, capacitive and related issues in regard to the cell membrane in our previous book 'Membrane Biophysics' (Ashrafuzzaman and Tuszynski 2012a). We shall avoid repeating all those details here. Although we shall brief a few of those that concern the electrical measurements at the nm-dimension scales. An important field 'Electrophysiology' has evolved based on mainly the electrical conductance properties of the cell. Due to the lack of space, we wish not to detail on the electrophysiology here but readers will find their application-based descriptions in several places in subsequent chapters.

### 1.4.2.1 Membrane Potential

Membrane potential is the electrical potential difference between the interior and exterior of a cell. If the potential of the region just outside the membrane is $V_o$ and the potential of the region just inside the cell near the membrane is $V_i$ the membrane potential of the cell is $V_i - V_o$. Using traditional definition of electrical potential, we can also define the membrane potential to be the energy required to transfer unit charge from the exterior to the interior of a cell crossing through the cell membrane. For example, if transfer of $Q$ coulomb charge from exterior to interior of a cell requires energy of $W$ joules the potential difference, hence the membrane potential of the cell will be $W/Q$ volts.

Both cellular interior and exterior regions exist with electrical conditions represented by electrical potentials. Electrical potential of both regions depends mainly on what are the constituents making the regions. The fluids on both sides of the mainly lipid membrane contain high concentrations of various ions—both cations and anions. Among the cations sodium ($Na^+$), potassium ($K^+$) and calcium ($Ca^{+2}$) are mentionable especially while chloride ($Cl^-$) is the important anion. Although both cations and anions exist in both interior and exterior regions of a cell the concentrations of sodium and chloride ions in the exterior is higher than that in the interior. Similarly, potassium ions exist in higher concentration in the interior region than the exterior region of a cell. The interior region importantly exists with the giant presence of protein anions.

Due to the differences in charge types and concentrations between intracellular and extracellular regions, they exist with different potential conditions; as a result, membrane exists in an electric field which accounts for the membrane potential. The membrane therefore plays a role of the cell's battery by providing a continuous source of electrical energy originated from the potential imbalance between intracellular and extracellular regions. This naturally inbuilt source of electrical energy plays important roles in regulating many cellular processes like transmitting signals between different parts of a cell, exciting ion channels across the membrane, etc. As the concentrations of each of the ion types present in the intracellular and extracellular regions are different there always exists a concentration gradient for each ion across the cellular membrane. This gradient causes a tendency for the ions to cross through the membrane. Potassium ion tries to move from intracellular to the extracellular region while sodium and chloride try to flow in the opposite direction. The natural tendency of the movement of charges across membrane causes changes in the membrane resting potential. Similarly, the changes in membrane resting potential due to natural reasons or due to artificial stimulus drive charges across the membrane. These dynamical processes are keys to normal functioning of cellular systems.

#### 1.4.2.2 Resting Potential

Resting potential is simply the potential of a membrane's interior region in absence of any excitation. This is in fact the membrane potential of nonexcitable cell or the membrane potential of an excitable cell in absence of any excitations. Besides the uneven distribution of other charges as explained earlier about 10 times higher $Na^+$ concentration on the outside and 20 times higher $K^+$ concentration on the inside of a membrane cause altogether the huge charge density gradient. As a result, in condition of rest or in absence of any excitation, the cell membrane is polarized maintaining an effective negative interior charge which accounts for a negative interior resting potential of the order of about −70 mV (see Fig. 1.8).

The chemical gradients across the membrane in rest cause resting potential to get built (see Fig. 1.8b). The approximately −70 mV potential gradient is built across the small-scale membrane thickness 3–7 nm, at a very high rate of potential gradient build up $\sim 10^7$ V/m.

ATP powered ion pumps or ion transporters play crucial roles. In an animal cell plasma membrane sodium–potassium pumps ($Na^+$ or $K^+$-ATPase) help to build sodium and potassium gradients across the membrane. The resting potential may also be altered due to the changes of acidic environments across the cells. For example, in cancer cell membrane, due to extra acidic conditions, the resting potential may alter considerably.

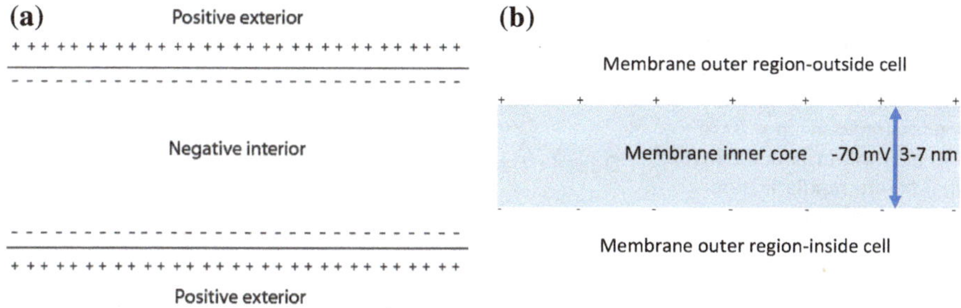


**Fig. 1.8** **a** About −70 mV potential in the membrane interior region relative to the membrane exterior region is a general electrical condition found due to a resultant negatively charged interior and positively charged exterior membrane regions. **b** A model diagram is presented here to demonstrate the opposite polarities between two sides of the cell membrane. The cellular interior region is negatively polarized, while exterior region is positively polarized. A potential difference (∼−70 mV) between cellular interior and exterior regions is maintained within a distance (membrane thickness) ∼ 3–7 nm

### 1.4.2.3 Resting Potential and Neuron Membrane

Brains communicate with other parts of the body through neuron cells. The resting potentials in neuron membranes help transfer the messages in the form of electrical pulses. The membrane of a neuron is about 8 nm thick containing two thick layers of fat molecules embedding larger protein molecules. In the polarized state, membrane effectively maintains the −70 mV resting potential due to uneven concentrations of anions and cations on both sides. In the polarized state, the membrane is permeable to $K^+$ ions but does not allow big $Na^+$ ions to cross through it. With sending information along the axon toward the axon terminal, a nerve impulse occurs. This way the transmission of information from one neuron to other neurons or cells happens. The nerve impulse or action potential is created by depolarizing current. The passage of electrical currents happens due to the movements of sodium and potassium ions across the membrane.

### 1.4.2.4 Action Potential

An action potential is an event which lasts for a short time and in it the electrical cell membrane potential rapidly rises and then falls following a special type of trajectory. The pattern of a typical action potential has been sketched here (see Fig. 1.9).

In several types of excitable cells like neurons, muscle cells, endocrine, action potentials occur. An action potential occurs during the time when a neuron sends information down an axon which goes away from the cell body. These potentials are caused by an exchange of ions across the neuron membrane. Any stimulus first causes sodium channels to open, with it sodium ions move into the neuron that leads neuron to experience depolarization. Potassium channels usually take a longer

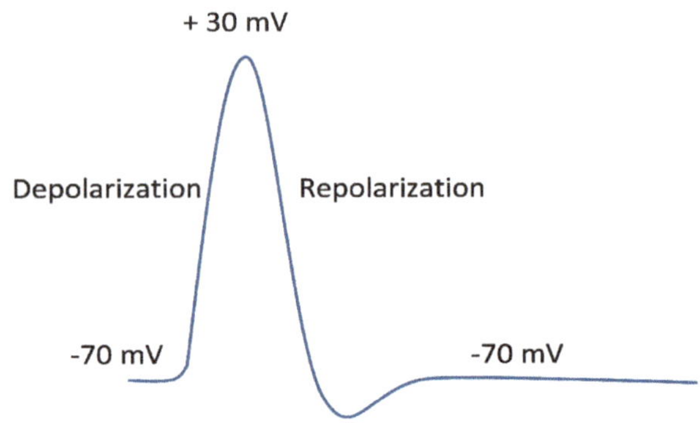


**Fig. 1.9** A schematic diagram of action potential illustrating depolarization and repolarization. Membrane potential goes from −70 to +30 mV within about 1 ms (ms) before repolarization forces the trend to reverse and finally the resting potential goes back to −70 mV after briefly experiencing a hyperpolarized state

time to open but when these channels do open, potassium starts moving out of the cells which reverses the depolarization. Consequently, sodium channels start to close. At this repolarization phase, the action potential goes past the −70 mV, a state referred as hyperpolarization. The ion concentrations across the cell gradually go back to the resting level and cell returns to the usual resting potential −70 mV.

Similar to the concepts discussed here, depolarization across a plasma membrane causes the creation of action potential. Certain external stimuli reduce the charge across the plasma membrane. Stimulus may originate from various kinds of sources. Mechanical stimuli like stretching, sound waves activate mechanically gated sodium channels across the membrane. Certain neurotransmitters like acetylcholines open ligand-gated sodium channels. Certain electrical shock may also stimulate and causes depolarization. The favorable diffusion of sodium ions into the cell locally reduces the membrane's resting potential. If the reduction is considerable, e.g., if the potential reduces to the threshold voltage (in mammalian neurons, about −50 mV), an action potential is generated in the cell. This kind of action potentials usually last for less than 1 ms. Action potentials generated by voltage-gated calcium channels may last much longer, of the order of 100 ms or more. The action potential is very much organ-specific and appears with lot of complexity because in different parts the stimuli appear from different sources with different nature. For example, in some types of neurons a long burst of rapidly emitted sodium spikes appear due to the slow calcium spike induced driving force whereas in cardiac muscle cells muscle contraction takes place due to the rapid onset of a calcium spike provoked from an initial fast sodium spike.

#### 1.4.2.5 Membrane as a Capacitor

A cell membrane separates charges on both sides of it. The inner core of a membrane exists with low dielectric state while the outside with high dielectric state. Parsegian wrote a nice article almost 5 decades ago to demonstrate the dielectric aspects (Parsegian 1969). The membrane generally acts as an insulator with conducting media on both sides.

From simple electrostatic analysis, we know the capacitance of a body is defined as the amount of charge needed across it to be stored to create unit potential difference between two terminals. If a potential gradient $V$ can hold a charge $Q$ across a capacitor the capacitance $C$ can be defined as

$$C = Q/V \tag{1.2}$$

A cell membrane structure is equivalent to a model structure where a relatively low dielectric medium is surrounded by two conducting media on both sides (intracellular and extracellular regions). This biological arrangement makes a membrane equivalent to a capacitor.

To calculate the membrane capacitance we need to use the principles of standard electrostatics. Here membrane is comparable to the insulator medium. The capacitancc of a cell membrane can thus be defined as

$$C_m = \kappa\varepsilon_0/d \tag{1.3}$$

$\kappa$ is the dielectric constant for the membrane's inner core and $\varepsilon_0$ is the permittivity of free space. $d$ is the membrane thickness. A low dielectric medium (inner layer) exists in between two conducting media (outside membrane). Depending on the variations in the values of $\kappa\varepsilon_0/d$ in various types of cells the values of capacitance of the corresponding membranes vary. But the value is often found to be of the order of 1.0 μF/cm$^2$ in normal cells. However, the value may vary substantially in cells under disease conditions.

In model lipid bilayer membrane, under control conditions (Ashrafuzzaman et al. 2006; Hwang et al. 2003; Lundbæk et al. 2005), the membrane capacitance can be measured quite easily using an electrophysiology set up and is found little different. Our own measurements of capacitance across dioleoylphosphatidylcholine (DOPC)/n-decane lipid bilayer provided the value 3.72 ± 0.24 nF cm$^{-2}$ (mean ± S.D., number of repeats $n$ = 11) (see Ashrafuzzaman et al. 2006). The capacitance was found so robust in this model lipid bilayer membrane that even due to the effects of amphipathic molecules the value did not change much. We incubated the model DOPC membrane with an antifusion peptide Z–Gly–Phe at 800 μM and measured the value at 3.94 ± 0.38 nF cm$^{-2}$ (mean ± S.D., $n$ = 10) (Ashrafuzzaman et al. 2006), this value was not much different than the modified value including the effects of other bilayer-active amphiphiles (Hwang et al. 2003; Lundbæk et al. 2005).

Most importantly, it is mentionable that the cholesterol level, phospholipids and glycolipids, membrane proteins, hydrocarbons, etc., together are responsible for certain value of biological cell membrane capacitance. Unlike animal cytoplasmic membrane, bacteria (prokaryotes) does not have cholesterol which may account for a considerable effect on membrane's electrical condition and hence the capacitance.

Understanding the capacitive effect of membrane helps understand the electrical properties of membrane through a model often referred as Electrical Circuit Model of the Cell Membrane (see Fig. 1.10).

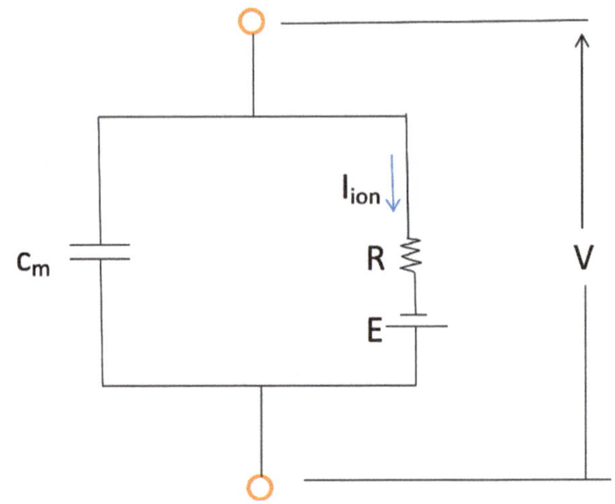


**Fig. 1.10** Cell membrane showing the capacitive effects. A parallel capacitor and resistor combination is the schematic form representing the electrical equivalent circuit

Here, the membrane is assumed to appear with a capacitor in parallel to a resistor. The not necessarily ohmic resistance acts against the flow of ions across the membrane which is represented by ion current $I_{\text{ion}}$. The capacitive current is $C_m \mathrm{d}V/\mathrm{d}t$. The capacitive current and the ion current together conserve the current flow between the inside and outside of the membrane. Therefore,

$$C_m \mathrm{d}V/\mathrm{d}t + I_{\text{ion}} = 0 \tag{1.4}$$

The theoretical calculation of $I_{\text{ion}}$ is a long-standing challenge. The following Goldman–Hodgkin–Katz (GHK) current equation is one such expression for $I_{\text{ion}}$ across membrane:

$$I_{\text{ion}} = D/L\left(z^2F^2/RT\right)V\left([N]_{\text{in}} - [N]_{\text{out}}\exp\{-zFV/RT\}\right)/\left(1 - \exp\{-zFV/RT\}\right) \tag{1.5}$$

Here, $D$ is the Einstein's diffusion constant, $L$ is the membrane thickness, $[N]_{\text{in}}$ and $[N]_{\text{out}}$ are ion concentrations inside and outside the cell across the membrane. $z$ is the valence of the ionic species, $F$ is the Faraday constant, $R$ is the universal gas constant and $T$ is the absolute temperature in Kelvin. Due to limited scopes, we wish not to detail on the calculations (related to GHK equation) here. However, the readers will find them in a lot of focused articles and textbooks [e.g., see Ashrafuzzaman and Tuszynski (2012a), Sterratt (2014)].

#### 1.4.2.6 Excitability and the State of Membrane Potential

Neurons, muscle cells, and similar ones use their membrane potentials as signals. The operation of nervous system, muscle contraction, etc., depend on the generation and propagation of electrical signals and membrane potentials in these cases mainly serve the purposes. Influences of membrane potentials can be regulated by controlling certain cellular processes like the ionic current carrying ion channels across membranes. The ion channel transport of ionic currents happens across a geometric length scale which is limited by the nanoscale dimension of membranes. The central dogma of this nanoscale cell signaling requires not a detailed but a brief analysis in the current book.

Electrical cell signaling is very much cell type specific. To understand better, the cells are grouped into two categories, namely nonexcitable cells and excitable cells. Nonexcitable cells maintain stable equilibrium potentials. If an external applied current perturbs the membrane potential of a nonexcitable cell, the withdrawal of the current ensures the potential to be back to its equilibrium state. Epithelial cells, photoreceptors, etc., are found to fall into the nonexcitable cell category.

Research findings suggest that nonexcitable cells are found not to generate all action potentials in response to depolarizing stimuli due to a lack of voltage-gated $Na^+$ or $Ca^{2+}$ channels (Rink and Jacob 1989; Fewtrell 1993; Clapham 1995; Berridge 1997). Consequently, membrane potential changes are proposed to influence the localized (intracellular) concentration of $Ca^{2+}$ ions ($[Ca^{2+}]_i$), responses mainly by altering the driving force for $Ca^{2+}$ entry through ligand-gated or second messenger-operated channels. However, in a subsequent study (Mahaut-Smith et al. 1999) it was reported that, during stimulation of a nonexcitable cell (e.g., metabotropic purinoceptors), membrane depolarization evokes an increase in $Ca^{2+}$ concentration in the interior cellular regions, primarily due to release of $Ca^{2+}$ from intracellular stores. Although depolarization in nonexcitable cells was found to result in a decrease in $[Ca^{2+}]_i$, hyperpolarization causes an increase of $[Ca^{2+}]_i$ during activation of mast cells, lymphocytes and related cell lines (Prenner et al. 1988; Lewis and Cahalan 1989; Demaurex et al. 1992). All these investigative research works suggest that the electrogenic influences in nonexcitable cells may also originate from various organ-specific mechanisms.

On the other hand in excitable cells a strong applied current causes the membrane potentials to go through a large excursion, called 'action potential,' before eventually returning to rest. Most neurons, cardiac cells, smooth and skeletal muscle cells, secretory cells, etc., fall into the excitable cell category. We shall mainly address the aspects of electrical signal propagation, mathematical modeling, etc., in excitable cells. It is a century-old problem and has reached to a very high level of understanding following few groundbreaking discoveries like the work of Alan Hodgkin and Andrew Huxley, who developed the first quantitative model of the propagation of an electrical signal along a squid giant axon in 1952. Hodgkin–Huxley theory is applicable for not only on electrophysiology but also on applied mathematics through appropriate modifications. The creation of a new field in mathematics 'the study of excitable systems' has been possible, thanks to the remarkable simplification and extension of the Hodgkin–Huxley theory. We shall explain all the details here. Hodgkin–Huxley model of the action potential which is a quantitative model has been presented in Fig. 1.11.

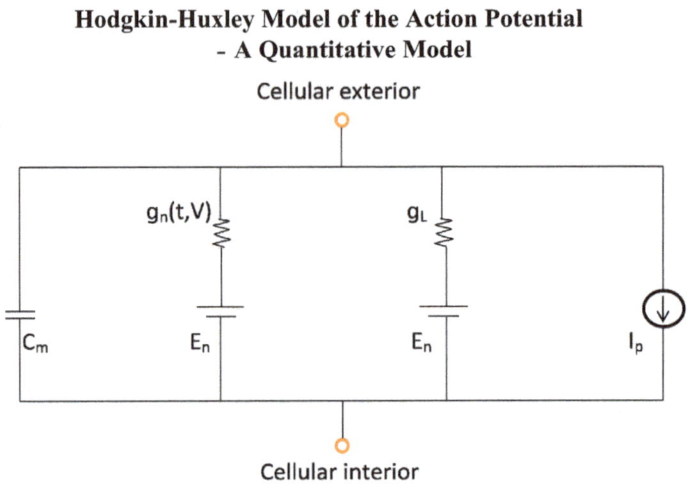


**Fig. 1.11** Capacitor resistor representation of the cell membrane. This is a more detailed form of representation in circuit model than that presented earlier in this chapter. For convenience, we shall use letter g for the symbol g in the text

The cell membrane has both biochemical and biophysical characteristics. The biophysical characteristic (some specific aspects) of a cell membrane is represented by the well-accepted Hodgkin–Huxley model. The lipid bilayer is represented as a capacitor. The low dielectric ($\sim$2) hydrophobic region inside the membrane relative to the outside hydrophilic regions with high dielectric values ($\sim$80) (Parsegian 1969) makes the cell membrane as almost a perfect capacitor. Voltage-gated and leak ion channels are represented by nonlinear ($g_n$) and linear ($g_L$) conductance, respectively. The electrochemical gradients driving the flow of ions are represented by batteries ($E_n$ and $E_L$), and ion pumps and exchangers are represented by current sources ($I_p$). The values of the batteries are determined from the Nernst potential of specific ionic species.

In an idealized cell with a small portion of the membrane blown up into an idealized circuit, we can present the Hodgkin–Huxley model for calculating the membrane current ($I_m$) by the following current equation:

$$I_m = C_m \mathrm{d}V/\mathrm{d}t + I_K + I_{Na} + I_L \qquad (1.6)$$

Here, $V$ is the membrane voltage, $I_K$ and $I_{Na}$ are the potassium and sodium currents, respectively, and $I_L$ is the sum of all leakage currents due to the flow of other ions moving passively through the membrane.

The charge stored in the capacitive membrane $q_m = C_m E$ where $E$ is the voltage across the capacitor which is comparable to the transmembrane potential.

Earlier in this chapter, we have explained how current across the membrane conserves which is possible due that the cell membrane being modeled as a capacitor in parallel with ionic currents. Now if ionic currents are considered to be depending on both transmembrane voltage $V$ and time $t$ the membrane capacitive current follows.

$$C_m \mathrm{d}V/\mathrm{d}t + I_{\mathrm{ion}}(V,t) = 0 \tag{1.7}$$

In the Hodgkin–Huxley theory besides the main currents, sodium and potassium ion currents across the membrane, all other small currents are together considered as leakage current $I_{\mathrm{L}}$.

In the squid giant axon, the I–V curves of open $Na^+$ and $K^+$ channels are approximated by linear equations. Therefore, the membrane capacitive current equation becomes

$$C_m \mathrm{d}V/\mathrm{d}t = -g_{\mathrm{Na}}(V - V_{\mathrm{Na}}) - g_{\mathrm{K}}(V - V_{\mathrm{K}}) - g_{\mathrm{L}}(V - V_{\mathrm{L}}) + I_{\mathrm{ext}} \tag{1.8}$$

Here, $I_{\mathrm{ext}}$ is the external applied current. $g_{\mathrm{Na}}$, $g_{\mathrm{K}}$, and $g_{\mathrm{L}}$ represent conductance (reciprocal of Ohmic resistance) for $Na^+$ and $K^+$ ions and other ions responsible for leakage currents, respectively. $V_{\mathrm{Na}}$, $V_{\mathrm{K}}$, and $V_{\mathrm{L}}$ are membrane resting potentials corresponding to $Na^+$, $K^+$ ions and other leakage ions across the membrane. The previous first-order ordinary differential equation can be rewritten as a more general form of equation representing capacitive current of membrane like the following:

$$C_m \mathrm{d}V/\mathrm{d}t = -g_{\mathrm{eff}}\left(V - V_{\mathrm{eq}}\right) + I_{\mathrm{ext}} \tag{1.9}$$

Here the grand conductance across the membrane is

$$g_{\mathrm{eff}} = g_{\mathrm{Na}} + g_{\mathrm{K}} + g_{\mathrm{L}} \tag{1.10}$$

And the membrane resting potential is

$$V_{\mathrm{eq}} = (g_{\mathrm{Na}} V_{\mathrm{Na}} + g_{\mathrm{K}} V_{\mathrm{K}} + g_{\mathrm{L}} V_{\mathrm{L}})/g_{\mathrm{eff}} \tag{1.11}$$

Specifically, in voltage-gated ion channels, the channel conductance $g_{\mathrm{i}}$ is a function of both time and voltage ($g_{\mathrm{n}}(t, V)$), while in leak channels $g_{\mathrm{L}}$ is a constant ($g_{\mathrm{L}}$).

The above description can be summarized as like as described in the original paper of Hodgkin–Huxley on a quantitative description of membrane current (Hodgkin and Huxley 1952). The electrical behavior of a membrane may be represented by a network as shown in previous Fig. 1.11. Here Current can be carried through the membrane either by charging the membrane capacitor or by the movement ions $Na^+$, $K^+$, etc., through the corresponding resistances in parallel. The ionic current corresponding to a specific ion is proportional to the difference between the membrane potential and the equilibrium potential for the specific ion. Here the proportional constant is the Ohmic conductance specific to the corresponding ion.

*Voltage and Time-Dependent Conductance in the Hodgkin–Huxley Model*

As explained earlier the total membrane current $I_m$ can be subdivided into mainly two categories which are capacitive current and ionic currents. Thus in normal condition, the following equation is valid:

$$I_m = C_m \mathrm{d}V/\mathrm{d}t + g_{\mathrm{Na}}(V - V_{\mathrm{Na}}) + g_{\mathrm{K}}(V - V_{\mathrm{K}}) + g_{\mathrm{L}}(V - V_{\mathrm{L}}) \tag{1.12}$$

The equation here gives the values of the membrane capacitance to be independent of the magnitude and sign of $V$ and is little affected by the time course of $V$ (Table 1 of Hodgkin et al. (1952), Measurement of current–voltage relation). Evidence of capacitive current and ionic currents to be parallel are well established in the study by Hodgkin et al. (1952), (Measurement of current–voltage relation). A major reservation, however, in the earlier equation is that it takes no account of dielectric loss in the membrane. But as the capacitive surge was found to be reasonably close to that calculated for a perfect condenser (Hodgkin et al. (1952), Measurement of current–voltage relation) it was then predicted that the mentioned dielectric condition inside membrane would not change the pattern of the equation dramatically. This has so far been found with modern approaches considering other constituents inside membranes not unrealistic.

*The Potassium Conductance*

Hodgkin–Huxley investigated the time dependence of ionic conductance too using the experimental investigation considering, for example, the case of potassium ion conduction. From the famous 1952 paper (Hodgkin and Huxley 1952), A quantitative description of membrane current) it is clearly suggestive that the rise of potassium conductance associated with depolarization of a potential is followed by the fall of conductance associated with repolarization to the resting potential. Here the nonlinear rise (depolarizing effect) of conductance $g_{\mathrm{K}}$ is found to mathematically match with $(1 - \exp(-t))^4$ while the fall with $\exp(-4t)$. These two different fourth-order forms explain nicely of the marked infexion for the rise while a simple exponential for the fall for $g_{\mathrm{K}}$. Identical mathematical assumption using other order forms would also be possible but there often would perhaps be needed other term, e.g., a term representing inactivation would be necessary in the case of third power in the mentioned representing exponential power series expansion.

Following a detailed analysis in (Hodgkin and Huxley (1952), A quantitative description of membrane current) the generalized form of $g_{\mathrm{K}}$ stands at

$$g_{\mathrm{K}} = \left\{ g_{\mathrm{K}\infty}^{1/4} - \left[ g_{\mathrm{K}\infty}^{1/4} - g_{\mathrm{K0}}^{1/4} \right] \exp(-t/\tau_n) \right\}^4 \tag{1.13}$$

$g_{\mathrm{K}\infty}$ is the value which the conductance finally attains and $g_{\mathrm{K0}}$ is the conductance at $t = 0$. $\tau_n$ is the inverse of the sum of the rate constants addressing the timescale of the resultant net inward flow of ions. The proposed equation best fits with the experimental results as presented in that paper (Hodgkin and Huxley 1952).

*The Sodium Conductance*

The transient change in sodium conductance $g_{Na}$ was addressed considering two variables, both of which obey first-order equations. Following few formal assumptions (see Hodgkin and Huxley 1952) $g_{Na}$ was found to well fit with experimental observations by the following equation:

$$g_{Na} = g'_{Na}[1 - \exp(-t/\tau_m)]^3 \exp(-t/\tau_h) \tag{1.14}$$

Here $g'_{Na}$ is the value which the sodium conductance would attain in the case of consideration of the proportion of the inactivating molecules on the outside boundary of the membrane. $\tau_m$ and $\tau_h$ are the inverse of the net transfer rate constants toward inside and outside directions, respectively.

Detailed analysis of the rate constants and other related aspects of membrane conductance addressed in Hodgkin–Huxley models are not only interesting but also very important. But we wish to offer an invitation for the readers to understand details going beyond our book and by reading all of the groundbreaking papers published by this pair independently and with others in the early 50s. The readers may also go through a few more models which were subsequently built on the basis of Hodgkin–Huxley models to perform better qualitative analysis and understand of the various aspects of the Hodgkin–Huxley models [see explanations and references in 'Membrane Biophysics' (Ashrafuzzaman and Tuszynski 2012a)].

## 1.5 Small Scales of Cell and Physics of Life

By 'small scales of cell,' we consider structures with their structure specific processes and functions that exist in small dimensions, work within small impact region and do small independent activities that may have collective big impacts as global cellular processes. Analysis on historical developments of cellular research as progressed with the developments of observation/microscopic techniques follows from Fig. 1.12.

Started with a simple microscope observation of the cell, slow but consistent technological progresses pushed the cell research all way down to quantum mechanical handling (see Chap. 9) of the cellular processes. We are now more interested to enrich our knowledge at the molecular level. This has become possible by developing various techniques, e.g., theoretical, experimental and computational methods to handle a single biomolecule. A flowchart presented in Fig. 1.13 describes a schematic of the length scale of biological molecules and complexes in the context of larger macroscopic length-scale entities where our modern-day science has been successfully pushed to Leake (2013).

At or beyond (lower than) nm-scale cell, world is now within our scientific reach. Classical and quantum mechanical techniques are fast capturing the hardly

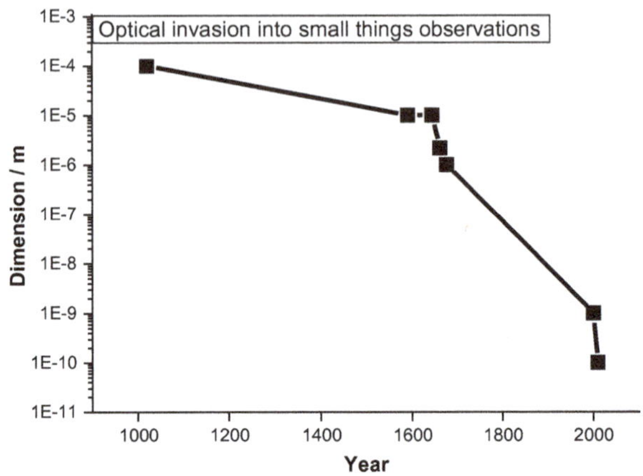


**Fig. 1.12** Observable object dimension versus Era plot—schematic sketch based on closest possible available information. The extreme right/bottom point represents quantum mechanical capacity at the hydrogen bond scale level. The quantum mechanical aspect is yet to develop substantially. We have discussed this briefly in Chap. 9. All other points represent the development in observation as progressed with the discovery of better microscopic techniques, starting with naked eye view which progressed all way down to nm-scale confocal imaging resolution. The middle four points over sixteenth century represent the observation scale based on discovery of magnified optical instrumentations by Hans Lippershey, Giambattista Odierna, Robert Hooke, and Antony Van Leeuwenhoek, respectively (explained earlier)

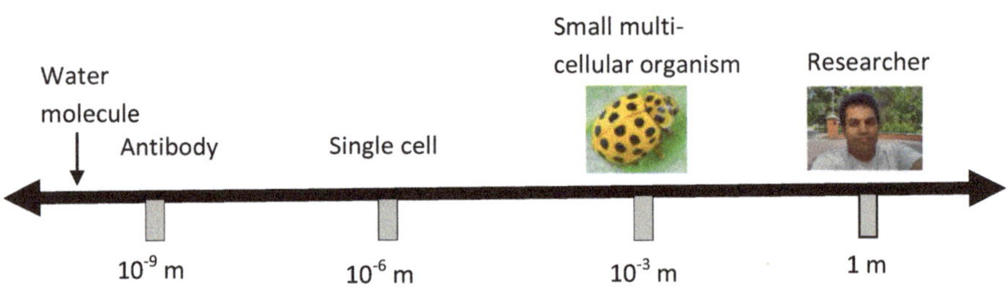


**Fig. 1.13** A schematic of the length scale of biological molecules and complexes in the context of larger macroscopic length-scale entities. As sketched in Fig. 1.12 'Observable object dimension versus Era plot' the quantum mechanical treatment falls here beyond 1 nm scale which is in the zone of water molecule in this figure. This figure is reconstructed in light of one presented in Leake (2013) due to unavailability of the copyright permission to reproduce it free

known or even unknown cellular realms. The study of fine cellular structures and understanding of their tuned functions perhaps hold some remarkable parameters, characters, clues that can provide a scientifically correct (or at least close to the correct) interpretation of long argued definition of life.

Physicist Erwin Schrödinger wrote a brief essay entitled 'What is Life?', asking: 'How can the events in space and time which take place within the spatial boundary of a living organism be accounted for by physics and chemistry?' The 70 years following this seminal work have seen enormous developments in our understanding of biology on the cellular and molecular scale. Physics has been found to play various key roles in solving many central problems through the development and application of new physical science techniques, biophysical analysis, and rigorous intellectual insight. The single-molecule bioscience investigations involve robust methods for understanding molecular-level details and thus often help us assess the role of cellular compartments and structures. Thus, we achieve general understanding on most of the criteria that altogether help define life. Beside general physical macroscopic and microscopic observation and measurement techniques, we have obtained enough knowledge to handle the small scales of the cell even using the powerful approaches of quantum mechanics. The future is perhaps more glorious when we shall have the power to use wave function to address small scales of the cell at lower than nm dimensions using advanced quantum mechanical calculations. This will then help us address not just the structural transitions but also intrastructure fluctuations, energetics, electron tunneling, etc., and thus, the hidden clues of life will start evolving. There is a strong possibility that the uncertainty principle, one of the most important quantum mechanical formulas, may be found generally applicable (yet undiscovered to my knowledge) to address the ultra-low cellular structure or substructure-based physics. Then, life may even require another novel theoretical definition or explanation.

In biological explorations, a lot about life have already been known or often guessed. One such latest realization has been made by latest Nobel Laureate in Physiology or Medicine Yoshinori Ohsumi who thinks that life is related to both 'autophagy' (termed for the degradation process of cytoplasmic constituents in the cellular lysosome/vacuole) and synthesis of cellular constructs mainly proteins. Cell-based degradation is a fundamental function, just as essential to the functioning of life as synthesis (for details, see Yoshinori 2014). The synthesis has been modestly addressed using physics principles (see, e.g., Basu and Chowdhury 2007; Caniparoli and Lombardo 2014), but valid physics principles are yet to be applied to explain the autophagy to discover its role in defining life. However, as autophagy has got attention now more than ever before, attempts have also been started by many scientists from various disciplines besides biology. One study was recently performed by a Korean group who attempted, using the computer-aided analysis, to provide some insights into how interactions of subcellular components such as genes and protein modules/complexes regulate autophagy and then impact on the dynamic behaviors of living cells as a whole (Han et al. 2015). Future developments to address the low scale biological processes of the cell including synthesis, autophagy, etc., using laws of physics (statistical, classical and quantum mechanical, etc.) and various computational modelings may provide further insights into our understanding of life.

## 1.6 Synthesis of Cell

The synthesis of the cellular components is now a reality. Computational understanding of cell-based structures in silico, calculation of various parameters mimicking the values as close as those observed in the biological environment and simulating them in computer are now quite popular because of their simplistic nature. The in vitro synthesis of the simplest forms of life that exist in minimal cells in wet laboratory is a challenging aspect. Synthetic biology field has got success by dealing with this subject. Here, often information from in silico studies that match generally with in vitro assay studies is utilized.

### *1.6.1 Creation of Quasi-Cellular System and Detection of System Constructs*

Synthetic biology deals with the de novo creation of living organisms in laboratory. Thus, it has a profound impact in basic science and in applied biotechnology (Szostak et al. 2001; Forster and church 2006; Stano et al. 2011). The first stage toward creating living organism is represented by the construction of quasi-cellular systems which are often referred to as 'minimal cells' that are liposome-based compartments. These minimal cells contain the minimal but sufficient number of biomolecules. These biomolecules interact with each other within liposome and emulate some of the fundamental properties of natural/living systems. Semi-synthetic approach is used for the construction of minimal cells (Luisi et al. 2006). Figure 1.14 demonstrates such a construction of a molecular species in solution.

Quasi-cellular systems that are able to produce proteins directly from DNA can be obtained by encapsulating the cell-free transcription/translation system PURESYSTEM™(PS) in liposomes (Calviello et al. 2013). It is possible to detect the intravesicle protein production using DNA encoding for green fluorescent protein (GFP) and monitoring the fluorescence emission over time. The entrapment of solutes in small-volume liposomes is a fundamental and open problem. Stochastic simulation is a valuable tool in the study of biochemical reaction at nanoscale range. Quick Direct-method Controlled (QDC), a stochastic simulation software based on the well-known Gillespie's SSA algorithm, was used recently and a suitable model describing the PS reactions network was developed (Calviello et al. 2013). It predicts, from inner species concentrations, the resulting fluorescence signal that is observable experimentally.

By varying the initial concentrations of the three main molecules involved in the PS: DNA, enzymes, consumables, Calviello et al. were able to stochastically simulate the time course of GFP production (Calviello et al. 2013). The sigmoid fit of the GFP production curves allowed to extract three quantitative parameters which are significantly dependent on the various initial states. 575 nm diameter

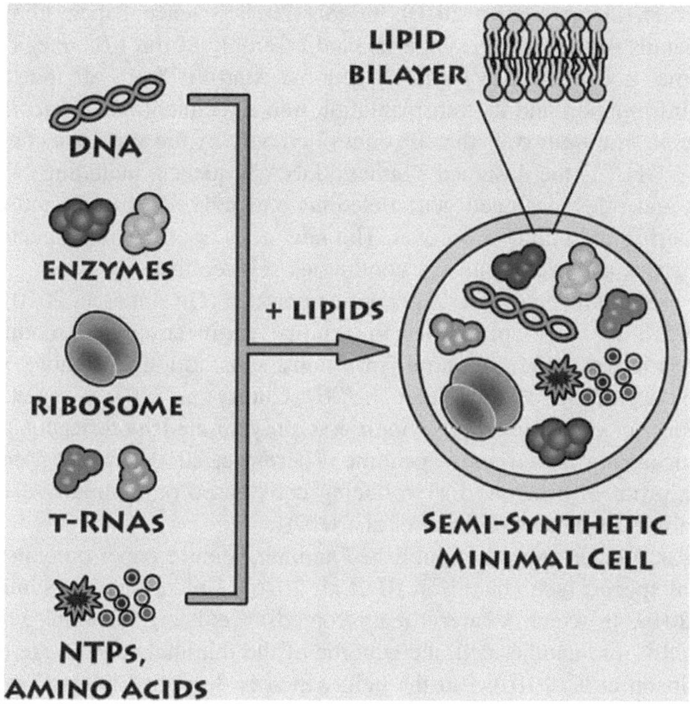


**Fig. 1.14** Construction of a semi-synthetic cell. These cells are composed of the minimal number of genes, enzymes, ribosomes, tRNAs, and low molecular weight compounds that are encapsulated within a synthetic compartment as in the case of lipid vesicles. The resulting construct is similar to living cells and displays minimal living properties (self-maintenance, self-reproduction, and possibility to evolve) is generally designed on the basis of the minimal number of functions required and on the minimal complexity of the biochemical elements needed for its construction. Copied from Chiarabelli et al. (2009)

liposome vesicles were used for extended studies. The stochastic behavior of the PS was detailed. Thus, an experimental approach was introduced which could aim at recording the GFP production kinetics in very small microemulsion droplets or liposomes, and inferring, by using the simulation as a reverse engineering procedure, the internal solutes distribution, and shed light on the still unknown forces driving the entrapment phenomenon. For details, readers are encouraged to go through the referred papers. We just wished here to familiarize the interested readers with the topic 'synthesis of quasi-living cellular systems.'

### *1.6.2 Practical Creation and Commercial Breakthrough*

Genomics entrepreneur Craig Venter and his group got success to create a synthetic cell that contains the smallest genome of any known, independent organism

(Gibson et al. 2010; Venter 2010). In this 2010 Science paper, Gibson et al. reported details of the design, synthesis, and assembly of the 1.08–mega–base pair Mycoplasma mycoides JCVI-syn1.0 genome starting from digitized genome sequence information and its transplantation into a recipient *M. capricolum* cell to create new *M. mycoides* cells that are controlled only by the synthetic chromosome. Cell's only DNA is the designed synthetic DNA sequence, including 'watermark' sequences and other designed gene deletions and polymorphisms, and mutations acquired during the building process. The new cells appear with expected phenotypic properties and are capable of continuous self-replication.

Venter group utilized synthetic genomic approach (Gibson et al. 2010) stands in sharp contrast to other approaches to genome engineering that modify natural genomes via introduction of multiple insertions, substitutions, deletions, etc. (Itaya et al. 2005; Itaya 1995; Mizoguchi et al. 2007; Chun et al. 2007; Wang et al. 2009). Venter group created synthetic genome has only limited modifications from the naturally occurring *M. mycoides* genome. Therefore, 2010 work of Venter group provides a proof of principle for producing cells based on computer-designed (in silico) genome sequences (Gibson et al. 2010).

This year, Venter group has published another Science paper on constitution of an artificial species (see Hutchison III et al. 2016). Unlike the first synthetic cells made in 2010, in which Venter's team copied an existing bacterial genome and transplanted it into another cell, the genome of the minimal cells is like nothing in nature (Gibson et al. 2010). But the cell, which is described in this year's paper, constitutes a brand new, artificial species (Hutchison III et al. 2016). We have to wait to see the reaction to this scientific development.

Functioning with 473 genes, the cell is a milestone in Venter team's 20-year quest to reduce life to its bare essentials and, by extension, to design life from scratch (Callaway 2016). When the JCVI-syn1.0 cells were unveiled in 2010, many considered it as the dawn of synthetic life. However, the organism's genome was found to be built by copying an existing plan and not through design—and its bloated genome of more than 1 million DNA bases was anything but minimal. The syn3.0's genome was in fact designed by trial and error, rather than being based on a fundamental understanding of how to build a functioning genome.

## 1.7 Physical States of Cell

Biological cells are generally considered to fall into the category of soft matter. When we see a cell we in fact see its outside look that means we see the cell surface or membrane. Beyond the membrane there exist various things as shown in Fig. 1.2 that addresses the schematic diagram on cellular components. These cellular constituents fall into different physical structure categories. Solid, liquid, gas are the major physical states in nature. Between solid and liquid, there exists another state called plasma state. All these states appear with distinguishable physical properties, certain kinds of interparticle interactions, certain types of shapes and sizes, etc. The mechanical and electrical properties of various states are different. In a cell, we generally find all of these states but gas. Of course, gas vesicles (containing gas) are used by Archaea, bacteria and planktonic microorganisms, possibly to control vertical migration by regulating the gas content and thereby buoyancy, or possibly to position the cell for maximum solar light harvesting. Some of these mentioned cellular structures permanently fall within a state while others temporarily or to be more specific many structures experience transitions between states. Sometimes some of the building blocks of any structure fall into a physical state class but the collective structure may not necessarily fall into the same class rather they are often found to fall into a different state. In this section, we shall learn about all these structures, in brief (Ashrafuzzaman 2015a, b).

Cell constituents can roughly be classified into three major classes as said earlier. If we think simple, we understand it quite easily. Let us hypothetically consider that we disintegrate all constituents from each other and allow them to freely float in a container or test tube. We would then discover the cell constituents as like as shown in Fig. 1.15. Here, three layers are shown which are separated mainly following the molecular weights of the components creating a complete cell. Those which make solid-state structures naturally fall at the bottom due to heavy gravity. Plasma state creating constituents take the position above solid-state components and the liquid-state components go above the plasma-state components. The three distinctive state classes may also consist of various further distinctive subclasses. Because not all constituents creating a class solid, plasma or liquid have identical molecular weights. That is why, probably in realistic consideration the above mentioned Fig. 1.15 may even appear with further classifications within a state class and the schematic diagrams take the form shown in Fig. 1.16.

### *1.7.1 Solid-State Structure*

Solid generally stands for something rigid in structure. Its general shape, size, and structure are visibly nonchangeable at a certain thermodynamic condition. That means general topology is deterministic. But the molecular-level structure may show vibration and even localized displacements although often keeping the outlook not so much changed. A protein, for example, is generally a rigid structure biomolecule in which the atomistic arrangements follow laws that are applicable to solid-state structures. That brings a protein to a solid-state structure class in biological system. Proteins usually appear with certain type of structures like α-helix, β-sheet, etc. Within such structures, the constituent atomic arrangements are found to be formatted considering most of the deterministic properties that are applicable to this level in solid bodies. For example, the interatomic interaction, bonding, etc., often appear to create certain permanent type of rigidity, elasticity, etc., in the geometric topology of the structure. With the change of thermodynamic condition the structures undergo structural transitions globally but in atomistic physical coordination the interaction, bonding, etc., may remain unchanged as long as the

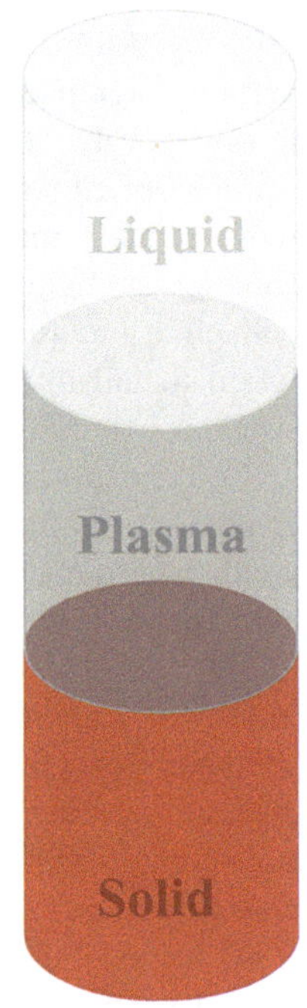


**Fig. 1.15** Different phases of a cell are shown here. Different colors are dedicated to representing different phases. If solid-, liquid-, and plasma-state structures get separated from each other and the whole constituents are placed in a test tube, the structures are expected to get separated due to their differences in molecular weights. Solid occupies the bottom, liquid the top, and plasma the middle. Column height for each phase is shown here arbitrarily

thermodynamic perturbation does not push the structure out to a totally different thermodynamic phase. For example, within crystalline phase the modest thermodynamic fluctuations may sometimes force the biomolecules to transform between varieties of crystal structure forms within the solid-state structure class.

### *1.7.2 Liquid-State Structure*

Cytoplasm and nucleoplasm are liquids. They provide support to cell's internal structures playing the role of a medium for their suspension. They hold various nonliquid substances but the liquids pose to have most of the liquid-state characteristics. Although it is generally a gel type liquid, its main part is cytosol consisting of huge amount of water, ions, etc. (Goodsell 1991).

### *1.7.3 Plasma-State Structure*

Cell membrane consists of lipids. Mitochondrial and nuclear membrane both behave as barrier, but the barrier properties are subject to perturbation. The structure

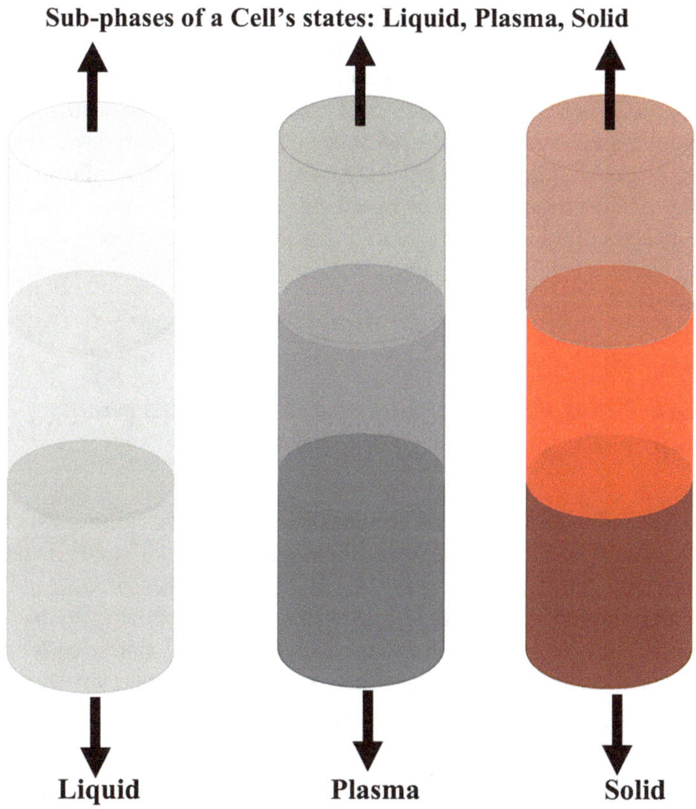


**Fig. 1.16** Different phases and sub-phases of a cell are shown here. Different colors are dedicated to representing different phases and sub-phases. Three sub-phases of solid-, liquid-, and plasma-state structures are schematically shown to be separated within their grand phase states. Up and down arrows suggest for possibility of more sub-phases within the grand phase states. The sub-phases within any grand phase state are separated from each other due to constituents within a phase state class having different molecular weights. Column height for each phase or sub-phase is shown here arbitrarily

of membrane follows some average geometric properties but within there is found a lot of dynamic nature. Membrane therefore is not a true liquid, nor a solid. Rather it takes the properties of a plasma state or we often refer it as a liquid crystal structure (Ashrafuzzaman and Tuszynski 2012b). The famous fluid mosaic membrane model was the first such successful cell's plasma-state demonstration (Singer and Nicolson 1972). The readers are also encouraged to read about subsequent developments that are explained in various other studies [huge information has been presented in Ashrafuzzaman and Tuszynski (2012a), Nicolson (2014)].

The above mentioned three different kinds of structures are usually found in cell constituents and we can simply say these three structures together make a cell. Most of the disease states therefore occur in mainly these three states. Physical properties of those states are quite explainable or understandable using various physics laws, e.g., Hooke's law for mechanical properties, Coulomb law for electrostatic interaction related properties, laws of general diffusion for the states of liquid density imbalances, etc. Therefore, application of various physics laws to understanding cell functioning and addressing disease states is inevitable. Similarly, the discovery of drugs to target cellular structures which are nothing but certain physical states requires the consideration of physics-based mechanisms. The most important aspect of the cell-based communication occurs through general and controlled cell transport among various cell compartments. We have addressed on this issue in Chap. 6 quite rigorously.

### *1.7.4 Cell Fluid Chemical States and the Consequent Influences on States of the Structures*

Cell's interior and exterior regions are both occupied with various structures. But most of these structures are hosted by liquid substance like water in which electrolytes are contained. Concentrations of ions often are measured in molarity and acidity or alkalinity of water-soluble substances (pH, 'potential of hydrogen'), etc., control many cell-based mechanisms, creation of the cell states, cells' normal or abnormal existence, etc. These chemical properties are found to exert important influences on biophysical states of cells.

One of the most important aspects of a body fluid compartment is that it is often specific with respect to ionic compositions. Various ions exist in a specific compartment and this is valid for cell compartments as well. Cellular exterior fluid, cellular interior fluid, mitochondrial interior fluid, etc., are different with respect to the presence of various ionic components. The ionic components account mainly as sodium, potassium, chloride, phosphorous, etc. Various proteins, peptides, etc., also exist in charged forms. Water is the main component in most of these fluids. All these ions' presence causes certainly molarity for them. Measuring their concentrations in cell fluids is an ongoing issue, started long ago in biological, clinical, physiological, chemical, etc., fields. For example, about half a decade ago Beilin et al. (1966a) did such a detailed investigation on red blood cells (RBCs) of healthy human adults to measure the sodium, potassium, and water contents in RBCs. Other investigations addressing on similar aspects are mentionable in references (Beilin et al. 1966b; Valberg et al. 1965; Singer et al. 1964; Kessler et al. 1961). These earlier investigations along with modern measurements we find that sodium concentration in RBCs is generally measured to be of the order of 6–8 mEq per Kg cells with modest difference between female (lower) and male (Beilin et al. 1966a). Age-dependent variation is negligible. Red blood cell potassium concentration in women ~92 mEq per kg cells while in men ~88 mEq per kg cells (Beilin et al. 1966a). The same report also suggested that about ninety percent of the variation of water contents of the red blood cells had been accounted for due to variation in sodium, potassium, and hemoglobin contents of the cells. In both animal and plant cells, ions need to have a continuous supply for sustainable existence of the functional cells. Maintaining certain concentration or specific values of molarity for certain ions need to be ensured for healthy cells. For example, an important ingredient phosphorous needed to maintain key certain proportion of specific molecules such as nucleic acids, phospholipids, and ATP (Schachtman et al. 1998) and this is involved in controlling key enzyme reactions and in the regulation of metabolic pathways (Theodorou and Plaxton 1993).

Normal behaving cells, abnormal acting cells, cells that undergo transitions, etc., contain specific ions with specific molarity in their physiological environments. Measurements of such molarities are therefore often considered even important clinical measurements nowadays for various diseases like cancer, allergy, etc., where cell-based ions face a change in their concentrations as a result of the rise of disease conditions. Often, the specific ions certain molarity maintenance may therefore be considered as a standard therapy for certain disease states. Molarities are found to regulate protein conformations, ion channel functions and activities of even other tuned molecular machines active in cells. Thus, molarity may affect certain criteria of the cell states liquid, plasma, solid that are explained earlier. We shall learn about problem-based issues in subsequent chapters.

## 1.8 Femtochemistry to Watch Cell-Based Molecular Conditions and Interactions in Real Time

Cellular compartments coexist or get separated by distance as small as in the order of nm. Dealing at this nm scale means we do nanotechnology. The cell membrane and other intracellular structures fall within this limit. Most of the interactions or interatom bonding mechanisms work at a distance that scales in low nm or angstrom ($10^{-10}$ m) order. For example, cell-based capacitive effects, intracompartment conductance effects (cell membrane and other intracellular compartments, etc.), interatomic bonding effects, etc., may be mentionable. These mechanisms can be indirectly addressed using various classical techniques including electrophysiology record of currents, electrical capacitive measurements, inter- and intracompartmental diffusion measurements, microscopy imaging, etc. Most of the mechanisms can be detected at timescale as low as microsecond ($10^{-6}$ s). But recently, a new field 'femtochemistry' got developed as a result of groundbreaking experiments of famous Arab scientist Ahmed Zewail (1946–2016) who got Nobel Prize in chemistry in 1999. It was the first time ever possible with rapid laser technique to study in slow motion how atoms in a molecule move during a chemical reaction down to tens of femtoseconds (fs). 1 fs = $10^{-15}$ s. Zewail's femtochemistry enables us to perform molecular filming using fs time-resolved laser spectroscopy and understand why certain chemical reactions take place but not others.

Femtosecond laser spectroscopy has been found as a powerful technique to study biological molecules for decades in femtochemistry and femtobiology fields. Recently, the femtochemistry technique was fused with molecular biology and cell biology techniques to advance our understanding of various human diseases, notably cancer, and how their treatments work. This potential new field has evolved with a new name called 'femtomedicine' (FMD) (see Elsevier News 2015) which can deal with nuclear structural damage related aspects. Here we shall focus at explaining a few cases, in brief.

### *1.8.1 DNA Damage Detection Using Femtochemistry Approaches*

Origins of many cell-based diseases are now quite known to be at DNA. Understanding DNA damage and repairing the damages are both at the prime focus of advanced biological cell-based recent research in chemistry, biochemistry, biophysics, bioinformatics, molecular medicine, etc. Conventional microscopic techniques including confocal microscopy explained earlier in this chapter are unable to address these small-scale high-resolution issues inside cells. We need to go beyond. The DNA damage is recently found possible to be studied using a sort of molecular filming technique called femtosecond time-resolved laser spectroscopy. The fs laser spectroscopy technique is like a high-speed camera, which uses two pulses of light: one pulse to start a reaction and the other to monitor the way the molecules react. This technique lets researchers watch how molecules interact in real time. This investigation helps researchers to understand the molecular reactions in cells. It also helps reveal how cells become cancerous as a result of malfunctioning in cell-based various reactions.

DNA damage is the initial and crucial step in the development of cancer. FMD approach helps us to go back to the very beginning to find out what causes DNA damage in the first place, then mutation, then cancer (Elsevier News 2015).

FMD leads to the discovery of a reductive damaging mechanism in DNA (Wang et al. 2009; Lu 2010a, b; Nguyen et al. 2011). For example, Nguyen et al. recently performed a direct observation of ultrafast-electron-transfer reactions that unravels high effectiveness of reductive DNA damage (Nguyen et al. 2011). Oxidative DNA damage is a well-known mechanism, but the relative role of reductive DNA damage is unknown. The prehydrated electron ($e^-_{pre}$), a novel species of electrons in water, is a fascinating species due to its fundamental importance in chemistry, biology, and the environment. The $e^-_{pre}$ is an ideal agent to observe reductive DNA damage. Nguyen and co-researchers report both the first in situ femtosecond time-resolved laser spectroscopy measurements of ultrafast-electron-transfer (UET) reactions of $e^-_{pre}$ with various scavengers ($KNO_3$, isopropanol, and dimethyl sulfoxide) and the first gel electrophoresis measurements of DNA strand breaks induced by $e^-_{pre}$ and $OH^{\bullet}$ radicals co-produced by two-UV-photon photolysis of water. The femtosecond generation of radicals and subsequent DNA damage mechanisms are schematized in Fig. 1.17. The investigators strikingly found that *the yield of reductive DNA*

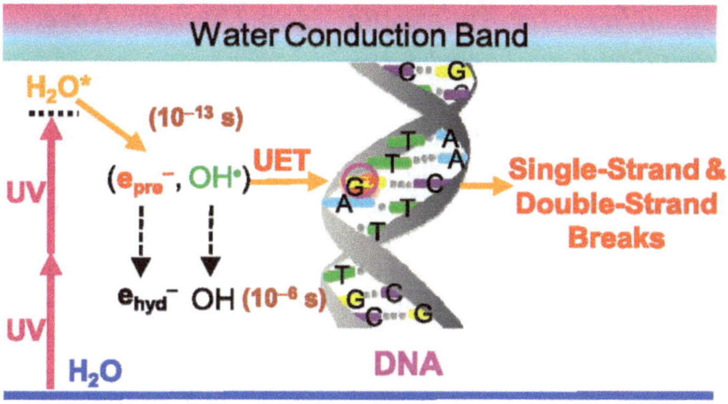


**Fig. 1.17** Two photon photolysis of water. Two-photon excitation of water readily generates the main radicals ($e^-_{pre}$ and $OH^•$) of radiolysis of water within $10^{-14}$ s after irradiation at a UV wavelength of 266–380 nm. The attacks of radicals can then cause single-strand and double-strand breaks of the nucleic acid DNA

*strand breaks induced by each* $e^-_{pre}$ *is twice the yield of oxidative DNA strand breaks induced by each* $OH^•$ *radical.*

The FMD study results (for details, see Nguyen et al. 2011) not only unravel the long-standing mystery about the relative role of radicals in inducing DNA damage under ionizing radiation, but also challenge the conventional notion that oxidative damage is the main pathway for DNA damage. The success of this investigation also shows the potential of femtomedicine which is a field in femtochemistry as a new transdisciplinary frontier and the broad significance of UET reactions of $e^-_{pre}$ in many processes in chemistry, physics, biology, and the environment.

Recently, the reductive damaging mechanism in human lung and skin normal cells was further demonstrated, which causes serious reductive DNA damage and genetic mutations and might be related to the pathology of diseases, especially cancer (Lu et al. 2013). Contrast differences are observed in the effects of an antioxidant (EGCG) on human normal and abnormal (cancer) cells: It caused DNA damage, mutations, and cell death more effectively in normal cells than in cancer cells. These results provided a compelling explanation for the confirmed observation in clinical trials that increases of some cancers such as lung and skin cancers are associated with antioxidants [The Alpha-Tocopherol, Beta Carotene (ATBC) Cancer Prevention Study Group, 1994 (Albanes et al. 1996; Omenn et al. 1996)]. DeNicola et al. also recently showed evidence in mice that several oncogenes actively induce transcription of Nrf2—the transcription factor that mainly regulates physiological antioxidant pathways—that is required for tumor initiation (DeNicola et al. 2011).

The thorough knowledge on DNA damage helps us design promising, efficient, economical, and rational approaches for discovering compounds as drugs. This has great pharmaceutical importance.

### 1.8.2 Femtochemistry Approaches Help Discover Novel Compounds to Target Cells

A molecular mechanism-based anticancer drug discovery program using FMD approaches helped Lu et al. to find a previously unknown class of nonplatinum-based halogenated molecules (called FMD compounds) as potent antitumor agents for the effective treatment of cancers (Lu et al. 2015). These compounds generally comprise an aromatic ring (rather than a platinum coordinating ion), coupled to two $NH_2$ groups as electron transfer promoter and one or more halogen atoms (Cl, Br or I), such as 4,5-dichloro/dibromo/diiodo-1,2-diaminobenzene (benzenediamine/phenylenediamine) and 4(3)-chloro/bromo/iodo-1,2-diaminobenzene (benzenediamine/phenylenediamine) (shortened as FMD-nX-DABs or $B(NH_2)_2X_n$ with X = Cl, Br or I, and $n = 1, 2$).

The in vitro and in vivo studies of the FMD compounds for targeted chemotherapy of cervical, breast, ovarian, and lung cancers report that these FMD agents led to DNA damage, cell cycle arrest in the S phase, and apoptosis in cancer cells. It also reported that such a FMD compound caused an increase of reduced glutathione (GSH, an endogenous antioxidant) levels in human normal cells, while it largely depleted GSH in cancer cells. The FMD agents exhibited no or little toxicity toward normal cells/tissues, while causing significant cytotoxicity against cancer cells, as well as suppression and delay in tumor growth in mouse xenograft models of cervical, ovarian, breast, and lung cancers. The results are not presented here, but interested readers may enrich their knowledge from Lu et al. (2015).

As said here, the FMD approaches helped Lu et al. to find FMD compounds through understanding the DNA damages in cells (Lu et al. 2015). Other recent studies of FMD in cancer have led to the discoveries of a reductive damaging mechanism in DNA (Wang et al. 2009; Nguyen et al. 2011), and living cells (Lu et al. 2013), and of the molecular mechanisms of existing anticancer agents (Lu 2007; Lu et al. 2007; Wang et al. 2006; Wang and Lu 2007, 2010). These investigations have offered unique opportunities to develop new effective drugs for targeted chemotherapy of cancer. The finding of a reductive DNA-damaging mechanism might be an important step for effective prevention and therapy of cancer.

FMD techniques have led to the finding of a previously unknown family of nonplatinum-based anticancer FMD compounds as potent antitumor agents for natural targeted chemotherapy without the need of using an antibody to bind to cancer cells (Lu et al. 2015). Upon entering normal and abnormal (cancer) cells, the discovered FMD compounds, which are highly oxidizing agents, have effects just opposite to those induced by antioxidants and are capable of generating the anticancer effects on cancer cells while having minimal toxicity toward normal cells. They are highly reactive with weakly bound electrons rich in the more reduced intracellular or intranuclear environment of cancer cells. In contrast, the FMD compounds are much less toxic toward normal cells due to the lack of a reduced intracellular or intranuclear environment in normal cells. FMD compounds have no or low systemic or acute toxicity in the body. They are effective anticancer agents that can preferentially kill cancer cells and are, therefore, useful for natural targeted chemotherapy of cancer and potentially other chemotherapy treatable disorders. Thus, FMD may eventually get established as a popular cell reaction monitoring technique that may even help discovering novel compounds as drugs to act in the cell-based damages especially to repair DNA damages.

## References

Ashrafuzzaman, Md., Tuszynski, J., Membrane Biophysics, Springer (Heidelberg), 2012a, ISSN 1618-7210, ISBN 978-3-642-16104-9 ISBN 978-3-642-16105-6 (eBook), https://doi.org/10.1007/978-3-642-16105-6.

Wilson, E.B. The Cell in Development and Heredity. 3rd ed. New York: The Macmillan Company; 1925.

Taylor, E.W. E.B. Wilson Lecture: The Cell as Molecular Machine. Mol Biol Cell. 2001 Feb; 12 (2): 251–254.

The Project Gutenberg eBook, Micrographia by Robert Hooke, Printed by Jo. Martyn, and Ja. Allestry, Printers to the Royal Society, London (1964).

J.D. Watson and F.H.C. Crick. 1953. Molecular structure of nucleic acids. A structure for deoxyribose nucleic acid. Nature 171, 737–738.

Md. Ashrafuzzaman, Z. Khan, M. Alanazi, M.S. Alam. 2016. Cell surface binding and lipid interactions behind chemotherapy drug induced ion pore formation in membranes. Submitted.

William H. Grover, Andrea K. Bryan, Monica Diez-Silva, Subra Suresh, John M. Higgins, and Scott R. Manalis. Measuring single-cell density. PNAS, 2011, vol. 108, 10992–10996.

Mrema JE, Campbell GH, Miranda R, Jaramillo AL, Rieckmann KH (1979) Concentration and separation of erythrocytes infected with Plasmodium falciparum by gradient centrifugation. Bull World Health Organ 57:133–138.

Mills JP, et al. (2007) Effect of plasmodial RESA protein on deformability of human red blood cells harboring Plasmodium falciparum. Proc Natl Acad Sci USA 104:9213–9217.

Rodgers GP, Schechter AN, Noguchi CT (1985) Cell heterogeneity in sickle cell disease: Quantitation of the erythrocyte density profile. J Lab Clin Med 106:30–37.

Bertles JF, Milner PF (1968) Irreversibly sickled erythrocytes: A consequence of the heterogeneous distribution of hemoglobin types in sickle-cell anemia. J Clin Invest 47:1731–1741.

Christina L. Lewis, Caelli C. Craig and Andre G. Senecal. Mass and Density Measurements of Live and Dead Gram-Negative and Gram-Positive Bacterial Populations. Appl. Environ. Microbiol. 2014, 80 no. 12 3622–3631.

Parsegian, A. Energy of an Ion crossing a Low dielectric Membrane: solutions to four relevant electrostatic problems. Nature 221, 844–846 (1969).

Md Ashrafuzzaman, M A Lampson, D V. Greathouse, R E. Koeppe II, and O S Andersen. Manipulating lipid bilayer material properties using biologically active amphipathic molecules. J. Phys.: Condens. Matter 18 (2006) S1235–S1255.

Hwang T C, Koeppe R E II and Andersen O S. 2003. Biochemistry 42 13646–58.

Lundbæk J A, Birn P, Tape S E, Toombes G E, Sogaard R, Koeppe R E II, Gruner S M, Hansen A J and Andersen O S. 2005. Mol. Pharmacol. 68 680–9.

Sterratt, D.C. 2014. Goldman-Hodgkin-Katz Equations. Encyclopedia of Computational Neuroscience. Springer, New York. pp 1–3.

Rink, T.J., Jacob, R.: Calcium oscillations in non-excitable cells. Trends Neurosci. 12, 43–46 (1989). PMID: 2469208, https://doi.org/10.1016/0166-2236(89)90133-1.

Fewtrell, C.: $Ca^{2+}$ oscillations in non-excitable cells. Annu. Rev. Physiol. 55, 427–454 (1993).
Clapham, D.E.: Calcium signaling. Cell 80, 259–268 (1995).
Berridge, M.J.: Elementary and global aspects of calcium signalling. J. Physiol. 499, 290–306 (1997).
Mahaut-Smith, M.P., Hussain, J.F., Mason, M.J.: De-polarization-evoked $Ca^{2+}$ release in a non-excitable cell, the rat megakaryocyte. J. Physiol. 515, 385–390 (1999).
Penner, R., Matthews, G., Neher, E. Regulation of calcium influx by second messengers in rat mast cells. Nature 334, 499–504 (1988).
Lewis, R.S., Cahalan, M.D. Mitogen-induced oscillations of cytosolic $Ca^{2+}$ and transmembrane $Ca^{2+}$ current in human leukemic T cells. Cell Regul. 1, 99–112 (1989).
Demaurex, N., Schlegel,W., Varnai, P., Mayr, G., Lew, D.P., Krause, K.H. Regulation of $Ca^{2+}$ influx in myeloid cells. Role of plasma membrane potential, inositol phosphates, cytosolic free $[Ca^{2+}]$, and filling state of intracellular $Ca^{2+}$ stores. J. Clin. Invest. 90, 830–839 (1992).
Hodgkin, A.L., Huxley, A.F. A quantitative description of membrane current and its application to conduction and excitation in nerve. J. Physiol. 117 (4), 500–544 (1952). PMID: 12991237.
Hodgkin, A.L., Huxley, A.F., Katz, B.: Measurements of current-voltage relations in the membrane of the giant axon of Loligo. J. Physiol. 116 (4), 424–448 (1952). PMID: 14946713.
Mark C. Leake. The physics of life: one molecule at a time. Philos Trans R Soc Lond B Biol Sci. 2013 Feb 5; 368(1611): 20120248.
Y. Ohsumi. Historical landmarks of autophagy research. Cell Research (2014) 24:9–23.
A. Basu and D. Chowdhury. Modeling protein synthesis from a physicist's perspective: A toy model. Am. J. Phys. 75, 931 (2007).
L. Caniparoli, P. Lombardo. Nonequilibrium stochastic model for tRNA binding time statistics. Phys. Rev. E 89, 012712 (2014).
Han K, Kim J, Choi M. Autophagy mediates phase transitions from cell death to life. Heliyon. 2015 Sep 26; 1(1):e00027. https://doi.org/10.1016/j.heliyon.2015.e00027.
Szostak JW, Bartel DP, Luisi PL. Synthesizing life. Nature. 2001; 409 (6818):387–390.
Forster A, Church G. Towards synthesis of a minimal cell. Molecular systems biology. 2006; 2 (9140d519-fc79-e312-0a8a-bfb7cec33a4e):45.
Stano P, Carrara P, Kuruma Y, Pereira de Souza T, Luisi PL. Compartmentalized reactions as a case of soft-matter biotechnology: synthesis of proteins and nucleic acids inside lipid vesicles. Journal of Materials Chemistry. 2011; 21(47):18887–18902.
Luisi P, Ferri F, Stano P. Approaches to semi-synthetic minimal cells: a review. Naturwissenschaften. 2006; 93(e975186c-6d35-e5a8-7236-239c98c5ea6f):1–14.
Chiarabelli C, Stano P, Luisi PL. Chemical approaches to synthetic biology. Current opinion in biotechnology. 2009; 20(4):492–497.
Lorenzo Calviello, Pasquale Stano, Fabio Mavelli, Pier Luigi Luisi, and Roberto Marangoni. Quasi-cellular systems: stochastic simulation analysis at nanoscale range. BMC Bioinformatics. 2013; 14(Suppl 7): S7.
Daniel G. Gibson, John I. Glass, Carole Lartigue, Vladimir N. Noskov, Ray-Yuan Chuang, Mikkel A. Algire, Gwynedd A. Benders, Michael G. Montague, Li Ma, Monzia M. Moodie, Chuck Merryman, Sanjay Vashee, Radha Krishnakumar, Nacyra Assad-Garcia, Cynthia Andrews-Pfannkoch, Evgeniya A. Denisova, Lei Young, Zhi-Qing Qi, Thomas H. Segall-Shapiro, Christopher H. Calvey, Prashanth P. Parmar, Clyde A. Hutchison III, Hamilton O. Smith, J. Craig Venter. Creation of a Bacterial Cell Controlled by a Chemically Synthesized Genome. Science, 2010, 329: 52–56.
M. Itaya, K. Tsuge, M. Koizumi, K. Fujita. Combining two genomes in one cell: stable cloning of the Synechocystis PCC6803 genome in the Bacillus subtilis 168 genome. Proc. Natl. Acad. Sci. U.S.A. 102, 15971 (2005). https://doi.org/10.1073/pnas.0503868102 pmid:16236728.
M. Itaya. An estimation of minimal genome size required for life. FEBS Lett. 362, 257 (1995). https://doi.org/10.1016/0014-5793(95)00233-y pmid:7729508.

lyde A. Hutchison III, Ray-Yuan Chuang, Vladimir N. Noskov, Nacyra Assad-Garcia, Thomas J. Deerinck, Mark H. Ellisman, John Gill, Krishna Kannan, Bogumil J. Karas, Li Ma, James F. Pelletier, Zhi-Qing Qi, R. Alexander Richter, Elizabeth A. Strychalski, Lijie Sun, Yo Suzuki, Billyana Tsvetanova, Kim S. Wise, Hamilton O. Smith, John I. Glass, Chuck Merryman, Daniel G. Gibson, J. Craig Venter. Design and synthesis of a minimal bacterial genome. Science, 2016, 351:aad6253. https://doi.org/10.1126/science.aad6253.

Ewen Callaway. 'Minimal' cell raises stakes in race to harness synthetic life. Craig Venter's creation comes as CRISPR gene-editing methods provide alternative ways to tinker with life's building blocks. Nature, 2016, 531: 557–558.

J. Craig Venter. Multiple personal genomes await. Nature, 2010, 464: 676–677.

Md. Ashrafuzzaman. 2015a. Phenomenology and energetics of diffusion across cell phase states. Saudi J. Biol. Sci., 22, 666–673.

Md. Ashrafuzzaman. 2015b. Diffusion across cell phase states. Biomedical Sciences Today, 1:e4.

Goodsell, D.S. Inside a living cell. Trends Biochem. Sci., 1991, 16 (6): 203–6.

Ashrafuzzaman, M., and J. A. Tuszynski. 2012b. Regulation of channel function due to coupling with a lipid bilayer, J. Comput. Theor. Nanosci. 9: 564–570.

G.L. Nicolson. 2014. The Fluid—Mosaic Model of Membrane Structure: Still relevant to understanding the structure, function and dynamics of biological membranes after more than 40 years. Biochimica et Biophysica Acta (BBA) – Biomembranes, 1838, 1451–1466.

L. J. BEILIN, G. J. KNIGHT, A. D. MUNRO-FAURE, AND J. ANDERSON. The Sodium, Potassium, and Water Contents of Red Blood Cells of Healthy Human Adults. Journal of Clinical Investigation, Vol. 45, No. 11, 1966a.

Beilin, L. J., G. J. Knight, A. D. Munro-Faure, and J. Anderson. The measurement of sodium concentration in human red blood cells. J. Gen. Physiol. 1966b, 50, 61.

Valberg, L. S., J. M. Holt, E. Paulson, and J. Szivek. Spectro-chemical analysis of sodium, potassium, calcium, magnesium, copper, and zinc in normal human erythrocytes. J. dlin. Invest. 1965, 44, 379.

Singer, M. M., H. R. Hoff, S. Fisch, and A. C. De-Graff. Red blood cell potassium. Therapeutic implications. J. Amer. med. Ass. 1964, 187, 24.

Kessler, E., M. R. Levy, and R. L. Allen, Jr. Red cell electrolytes in patients with edema. J. Lab. dlin. Med. 1961, 57, 32.

Daniel P. Schachtman, Robert J. Reid and S.M. Ayling. Phosphorus Uptake by Plants: From Soil to Cell. Plant Physiology, February 1998, vol. 116 no. 2 447–453.

Theodorou ME, Plaxton WC (1993) Metabolic adaptations of plant respiration to nutritional phosphate deprivation. Plant Physiol 101:339–344.

Elsevier News. Media contact, Nienke Swankhuisen, Elsevier. 2015. Scientists discover new molecules that kill cancer cells and protect healthy cells. Website: https://www.elsevier.com/about/press-releases/research-and-journals/scientists-discover-new-molecules-that-kill-cancer-cells-and-protect-healthy-cells.

C.R. Wang, J. Nguyen, Q.B. Lu. Bond breaks of nucleotides by dissociative electron transfer of nonequilibrium prehydrated electrons: a new molecular mechanism for reductive DNA damage. J. Am. Chem. Soc., 131 (2009), pp. 11320–11322.

Lu QB. 2010a. Cosmic-ray-driven electron-induced reactions of halogenated molecules adsorbed on ice surfaces: Implications for atmospheric ozone depletion and global climate change. Phys Rep 487:141–167.

Lu QB. 2010b. Effects and applications of ultrashort-lived prehydrated electrons in radiation biology and radiotherapy of cancer. Mutat Res 704:190–199.

J. Nguyen, Y. Ma, T. Luo, R.G. Bristow, D.A. Jaffray, Q.B. Lu. Direct observation of ultrafast electron transfer reactions unravels high effectiveness of reductive DNA damage. Proc. Natl. Acad. Sci. U. S. A., 108 (2011), pp. 11778–11783.

L.Y. Lu, N. Ou, Q.B. Lu. Antioxidant induces DNA damage, cell death and mutagenicity in human lung and skin normal cells. Sci. Rep., 3 (2013), p. 3169 (1–11).

Albanes D, Heinonen OP, Taylor PR, Virtamo J, Edwards BK, Rautalahti M, Hartman AM, Palmgren J, Freedman LS, Haapakoski J, Barrett MJ, Pietinen P, Malila N, Tala E, Liippo K, Salomaa ER, Tangrea JA, Teppo L, Askin FB, Taskinen E, Erozan Y, Greenwald P, Huttunen JK. Alpha-tocopherol and beta-carotene supplements and lung cancer incidence in the alpha-tocopherol, beta-carotene cancer prevention study: effects of base-line characteristics and study compliance. J. Natl. Cancer Inst., 88 (1996), pp. 1560–1570.

Omenn GS, Goodman GE, Thornquist MD, Balmes J, Cullen MR, Glass A, Keogh JP, Meyskens FL, Valanis B, Williams JH, Barnhart S, Hammar S. Effects of a combination of beta-carotene and vitamin A on lung cancer and cardiovascular disease. N. Engl. J. Med., 334 (1996), pp. 1150–1155.

DeNicola GM, Karreth FA, Humpton TJ, Gopinathan A, Wei C, Frese K, Mangal D, Yu KH, Yeo CJ, Calhoun ES, Scrimieri F, Winter JM, Hruban RH, Iacobuzio-Donahue C, Kern SE, Blair IA, Tuveson DA. Oncogene-induced Nrf2 transcription promotes ROS detoxification and tumorigenesis. Nature. 2011 Jul 6; 475(7354):106–9.

Qing-Bin Lu, Qin-Rong Zhang, Ning Ou, Chun-Rong Wang, Jenny Warrington. In Vitro and In Vivo Studies of Non-Platinum-Based Halogenated Compounds as Potent Antitumor Agents for Natural Targeted Chemotherapy of Cancers. EBioMedicine, Volume 2, Issue 6, June 2015, Pages 544–553.

Q.B. Lu. Molecular reaction mechanisms of combination treatments of low-dose cisplatin with radiotherapy and photodynamic therapy. J. Med. Chem., 50 (2007), pp. 2601–2604.

Q.B. Lu, S. Kalantari, C.R. Wang. Electron transfer reaction mechanism of cisplatin with dna at the molecular level. Mol. Pharm., 4 (2007), pp. 624–628.

C.R. Wang, A. Hu, Q.B. Lu. Direct observation of the transition state of ultrafast electron transfer reaction of a radiosensitizing drug bromodeoxyuridine. J. Chem. Phys., 124 (2006), p. 241102 (1–4).

C.R. Wang, Q.B. Lu. Real-time observation of a molecular reaction mechanism of aqueous 5-halo-2′-deoxyuridines under UV/ionizing radiation. Angew. Chem. Int. Ed., 46 (2007), pp. 6316–6321.

C.R. Wang, Q.B. Lu. Molecular mechanism of the dna sequence selectivity of 5-halo-2′-deoxyuridines as potential radiosensitizers. J. Am. Chem. Soc., 132 (2010), pp. 14710–14713.

Hiroshi Mizoguchi, Hideo Mori, Tatsuro Fujio, (2007) Escherichia coli minimum genome factory. Biotechnology and Applied Biochemistry 46 (3):157.

Cheryl D. Chun, Oliver W. Liu, Hiten D. Madhani, (2007) A Link between Virulence and Homeostatic Responses to Hypoxia during Infection by the Human Fungal Pathogen Cryptococcus neoformans. PLoS Pathogens 3 (2):e22.

S. J. Singer, G. L. Nicolson, (1972) The Fluid Mosaic Model of the Structure of Cell Membranes. Science 175 (4023):720–731.

# Chapter 2
# Cell Surface Dynamics

Cell surface materials are quite dynamic. The timescales of dynamics often fall at the order of millisecond to microsecond or even lower ranges. Both slow and fast dynamics of cell surface are known to follow general physics rules. Cell contains biological components having mostly chemical and physical properties. Cell surface structure can better be explained using its physical properties, specifically mechanical and electrical properties. These properties generally serve the causes of specific surface dynamics. This chapter will focus at an in-depth analysis of the cell surface dynamics.

## 2.1 Topology of the Cell Surface

About five decades ago, it was demonstrated that different structural elements of plasma membranes are mobile in the plane of the cell surface. These studies led to the discovery of the universally accepted fluid-mosaic membrane model (F-MMM) on the structure of the cell membranes (Taylor et al. 1971; Singer and Nicolson 1972). Thermodynamic principles of the organization of membrane lipids and proteins were found keys to the fluid nature of the membrane. Besides in hundreds of references, we also explained the model quite in detail in our previous book 'Membrane Biophysics' (see Ashrafuzzaman and Tuszynski 2012a). After over 40 years, this original cell membrane model remains relevant for describing the basic nanostructures of a variety of intracellular and cellular membranes of plant and animal cells and lower forms of life (Nicolson 2014). In this chapter, we shall provide a brief description of the cell surface topology.

Cell surface inspection has had a long history. Various techniques, most of which are based on microscopic inspections, have been applied to observe the physical status of the cell surface and collect information on surface topology. After careful processing of this information, the ultimate goals among scientists are to reach at acceptable conclusions to present models that best describe cell surface



M. Ashrafuzzaman, *Nanoscale Biophysics of the Cell*,
https://doi.org/10.1007/978-3-319-77465-7_2

structures and related functions. We shall provide a historical details (see also Ashrafuzzaman and Tuszynski 2012a) and analyze a few focused cases to present the most acceptable models (e.g., see Singer and Nicolson 1972; Nicolson 1976; Nicolson et al. 1977; Jacobson et al. 1995; Escribá et al. 2008; Nicolson 2014).

### 2.1.1 Basic Fluid-Mosaic Membrane Model

The original nanoscale model for F-MMM that was proposed in 1972 is schematized in Fig. 2.1 (for details see Singer and Nicolson 1972). The model mainly considers the following:

(i) The cell membrane is a two-dimensional (2D) plane;
(ii) Phospholipids help make a viscous or fluid bilayer;
(iii) Integral proteins are mobile in the oriented 2D solution.

F-MMM depicted biological membranes as a matrix made up of a mostly fluid bilayer of various phospholipids with mobile integral globular membrane proteins (MPs) and glycoproteins that were intercalated into the fluid lipid bilayer membrane.

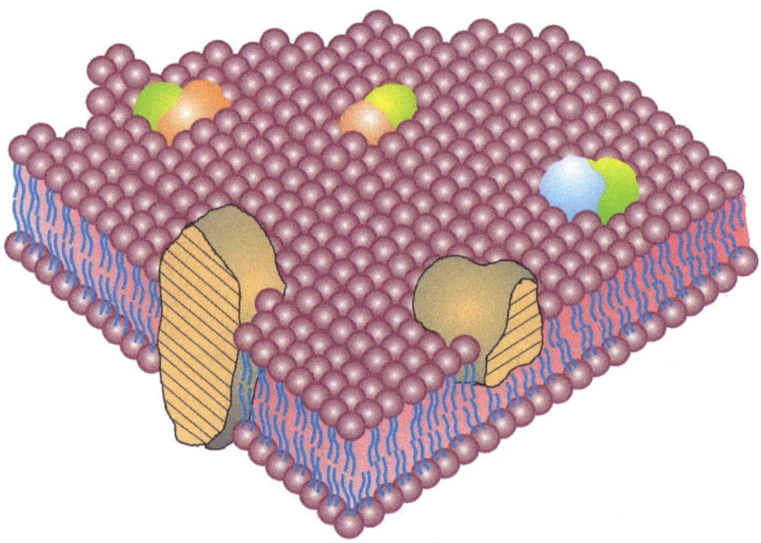

**Fig. 2.1** The F-MMM of biological membrane structure that was originally proposed in 1972. In this sub-micro- or nanosized cross-sectional structural view of a cell membrane, the solid bodies with stippled cut surfaces represent integral globular MPs, which at intermediate range are randomly distributed in the membrane plane. At short range, some integral MPs form certain integral protein complexes. The figure represents integral proteins in a fluid bilayer lipid phase, and it does not contain other membrane-associated structures or membrane domains of different compositions or phases. Copied from Singer and Nicolson (1972)

The F-MMM discovery was based on a rich history of explorations of biological membranes. During the time of discovery of the F-MMM, there were other half a century old to latest models, such as the Unit Membrane Model of Robertson (1959, 1960, 1981) or Danielli–Davson Tri-Layer (protein–lipid–protein) Model of membrane structure (Danielli and Davson 1935). The Danielli–Davson Tri-Layer membrane model was based on another lipid bilayer proposal of Gorter and Grendel (1925), with added unfolded sheets of proteins on either side of a lipid bilayer. Some transmembrane protein components were added later to reconcile observations on intramembranous particles like those found by Pinto da Silva and Branton, who successfully freeze fractured cell membranes with surface-bound ferritin markers (Silva and Branton 1970) and the discovery of transmembrane proteins (Stoeckenius and Engelman 1969). The basic unit model has remained a tri-layer structure with most proteins present in extended beta configurational forms, bound to the lipid bilayer by forces like electrostatic and other hydrophilic ones (Robertson 1981). There was another alternative to the tri-layer membrane structure models proposed based on repeating sub-units of lipoproteins without a supporting matrix of lipid bilayer (see Benson 1966; Green et al. 1987). The earlier membrane models, prior to the discovery of the F-MMM, did not take into account the ability of components in membranes to rapidly move laterally and dynamically and change their topographic distributions. This aspect was added in the F-MMM (Singer and Nicolson 1972). F-MMM thus successfully cleared certain limitations in both the Unit Membrane (Robertson 1959, 1960, 1981) and Sub-unit Membrane (Benson 1966; Green et al. 1987) models to explain existing data on the structures of biological membranes (Gorter and Grendel 1925; Danielli and Davson 1935; Robertson 1959, 1960, 1981; Stoeckenius and Engelman 1969; Singer and Nicolson 1972).

The original F-MMM model (Singer and Nicolson 1972) does not require extensive revisions beyond the versions published in 1976 (see Fig. 2.2) (Nicolson 1976; Nicolson et al. 1977). In 1976, elaborations of the original F-MMM (Singer and Nicolson 1972) were made by including several novel features (Nicolson 1976), as follows:

(i) Interactions of extracellular matrix and membrane-associated cytoskeletal components with cell membranes,
(ii) The potential influence of the mentioned interactions on the mobility and distribution of transmembrane glycoproteins, and
(iii) The possibility that less mobile lipid–protein or lipid–lipid domains might exist in membranes as frozen or semi-frozen islands in a sea of fluid phospholipids.

In the revised 1976 models, the cell membranes were observed to be much less homogeneous and little more complicated than the original F-MMM depiction (Nicolson 1976). The 1976 models contained additional information, not included in the original model, e.g., as explained earlier, protein and lipid aggregations and segregation into domains, cytoskeletal interactions, and extracellular matrix interactions. However, the revised F-MMM cartoons (Nicolson 1976) still contain all of the basic elements of the original F-MMM (Singer and Nicolson 1972).

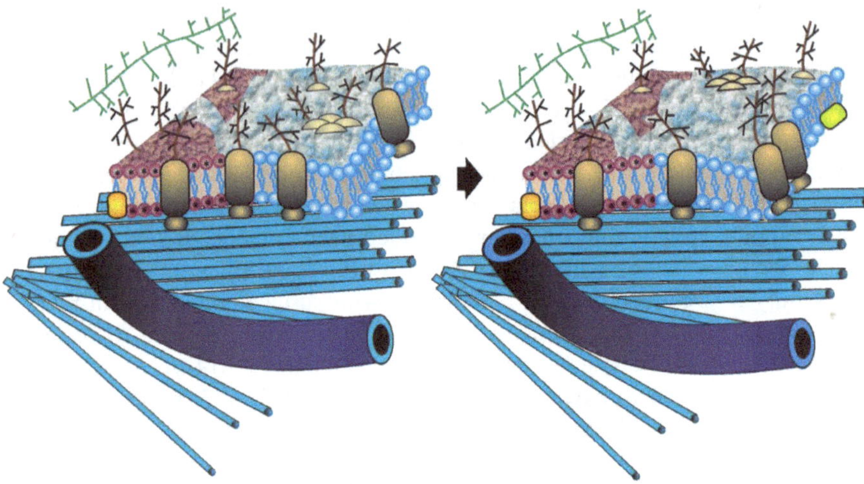

**Fig. 2.2** A modified version of the schematic F-MMM of biological membrane structure, as proposed originally in 1976. In this version, different snapshots in time are represented by the two left and right panels. Some integral membrane glycoproteins are relatively free to diffuse laterally in the plane of the membrane within a fluid lipid region or domain, whereas others are 'anchored' or relatively impeded by a cytoskeletal assemblage or an ordered or even solid lipid phase. In this arrangement, an integral membrane glycoprotein complex is also being displaced by membrane-associated cytoskeletal components in an energy-dependent process. This figure suggests some possible integral MP and lipid mobility restraint mechanisms, but it does not accurately present the sizes or structures of integral MPs, lipids, cytoskeletal structures, polysaccharides, sub-micro- or nanosized domains or membrane-associated cytoskeletal structures or their specific or nonspecific crowding in the membrane. Copied from Nicolson (1976)

Beside the intracellular and extracellular influences on the dynamics of the plasma membrane, the packing of components into very compact structures and domains is found to contribute to maximizing the mosaic nature of certain membranes. For example, cell junctions, adhesion sites, lipid rafts, mitochondrial inner membranes, viruses, and other compact membranous structures possess limited lateral macromobility of specific membrane components while still exhibiting the basic microstructure of the F-MMM.

The thermodynamic considerations included important ingredients in F-MMM. Here, the concept of hydrophobic interactions and their importance in the thermodynamics of protein structure are addressed using the concept incorporated in Kauzmann (1959), Singer (1992). The basis of biological membrane stability lies in two fundamental thermodynamic properties, as follows:

(i) the propensity of hydrophobic structures: to self-associate and exclude water interactions, which happens via driving entropically by water exclusion, and
(ii) the propensity of hydrophilic structures: to interact with the aqueous environment.

Major membrane components lipids, mainly phospholipids, are observed to self-assemble with their hydrophobic tails excluding water to form bilayers due to the physical energy provided by the van der Waals forces and the effects of hydrophobic interactions (Singer 1971). MPs, especially the integral or intrinsic proteins, interact with membrane lipids due to mainly hydrophobic forces and much less due to hydrophilic interactions between the lipid headgroups and hydrophilic groups on proteins. This concept did not preclude or diminish the significance of hydrophilic interactions between membrane lipids and integral membrane globular proteins. It simply described the relative importance of hydrophobic interactions in determining the basic microstructure of cell membranes. Recently, we have added some novel concepts in the MP–lipid interaction scenario by discovering the screened Coulomb interactions (SCIs) among charge distributions in the complex. This model calculation detects charge property-based interactions in a MP–lipid complex as major contributors for stabilizing the complex (see details in Ashrafuzzaman and Tuszynski 2012a, b).

### 2.1.2 Latest Status of the Fluid-Mosaic Membrane Models

Nicolson recently presented his analysis on huge amounts of data, hypothesis, and models of membrane structures and functions that emerged since the discovery of the original models by Singer and himself (Singer and Nicolson 1972). Subsequent developments over the span of four decades ensured inclusion of a few more important features related to membrane structure and function (all important studies are quoted in Nicolson 2014), such as:

(i) Transmembrane interactions with membrane structures exist and influence membrane dynamics. This hypothesis helps us explain the controls over membrane structure, component mobility, and importantly function.
(ii) The recent discoveries of lipid rafts and specialized membrane domains, membrane-associated 'fences' and membrane 'fenceposts,' etc. They are possibly involved in controlling and restraining membrane component distribution.

As said, during last five decades, there have been tremendous amount of articles published in the filed addressing membrane dynamics. A lot of novel data and hypotheses have been published that helped put forward various concepts and models. This is why, a revision on the F-MMM, original (Singer and Nicolson 1972) or revised (Nicolson 1976), is required. Fortunately, one of the two original authors Nicolson is still there to deal with this, so did he write an article recently (Nicolson 2014). Here, he again attempted to consider all valid data and ideas and put forward a revised F-MMM as presented in Fig. 2.3a. Nicolson considered the data published mainly during 1972–2014, presented an analysis, and cleared some of the misconceptions inherent in the 1970s era models. We also encourage readers to try to understand the membrane surface mechanical properties through our model diagram sketch (by Imtihan Ahmed and Mohammad Ashrafuzzaman), as presented in Fig. 2.3b.

In the revised model as seen in Fig. 2.3, different integral proteins, glycoproteins, lipids, and oligosaccharides are represented by different colors, and where the membrane has been peeled-up to view the inner membrane surface, cytoskeletal fencing is apparent that restricts the lateral diffusion of some, but not all transmembrane glycoproteins. Other important lateral diffusion restriction mechanisms are also represented, such as lipid domains, integral membrane glycoprotein

(a)

(b)

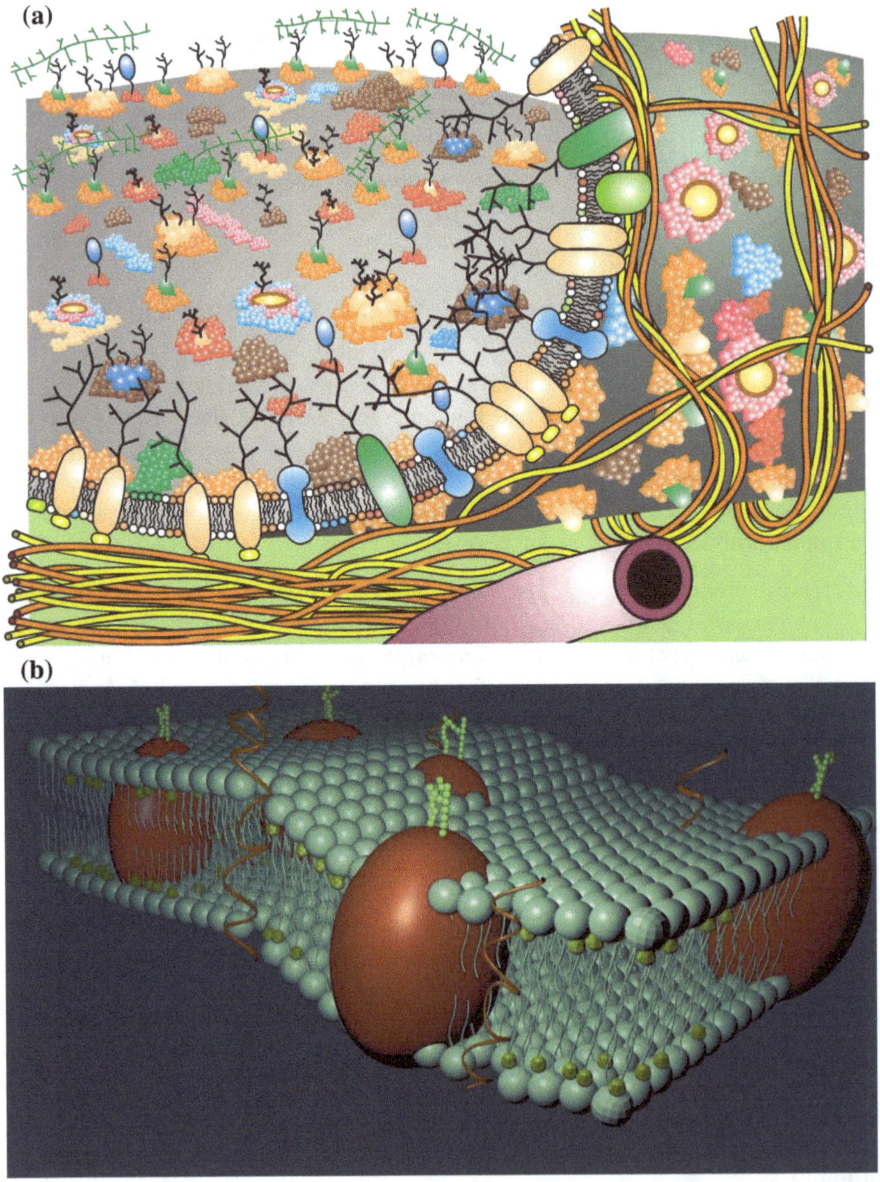

◂**Fig. 2.3** **a** An updated F-MMM that contains information on membrane domain structures and membrane-associated cytoskeletal and extracellular structures (see Nicolson 2014) **b** Focused and simplified membrane structure has been presented here. This diagram was sketched by Imtihan Ahmed and Mohammad Ashrafuzzaman to include in book 'Membrane Biophysics' (Ashrafuzzaman and Tuszynski 2012b). The main purpose of this model diagram is to show the three-dimensional look of the membrane showing that the membrane bilayer surface is not as smooth as is thought to be. The mechanical properties (especially the membrane elasticity) of the membrane surface are visible through change of bilayer thickness at random locations which may evolve due to mainly the mismatches between various integral MPs coupled to bilayer across the bilayer thickness. Two lipid monolayers are the primary components of a membrane. Near lipid headgroups (light blue color) are seen the cholesterol, residing in the hydrophobic region. Globular proteins (in red) are shown to reside across the membrane. Alpha helical proteins have both hydrophilic and hydrophobic parts. Membrane bending, membrane thickness change, etc., are also schematically diagrammed here which are consistent with the real membrane surrounding a cell (for detailed analysis, see Ashrafuzzaman and Tuszynski 2012a). For simplicity, we did not schematize the presence of different ion channels, or other complicated MP structures, membrane domain structures, and membrane-associated cytoskeletal and extracellular structures, as shown in Fig. 2.3a

complex formation (seen in the membrane cutaway), polysaccharide–glycoprotein associations (at the far top left), and direct or indirect attachment of inner surface membrane domains to cytoskeletal elements (at lower left). Although Fig. 2.3a suggests some possible integral MP and lipid mobility restraint mechanisms, it does not accurately present the sizes or structures of integral MPs, cytoskeletal structures, polysaccharides, lipids, sub-micro- or nanosized domains or membrane-associated cytoskeletal structures, or their crowding in the membrane.

In the revised model in Fig. 2.3, inclusion of the following three rather important information is especially mentionable, such as:

(i) Integral MP lateral movements (Jacobson et al. 1995) (see Fig. 2.4);
(ii) Lipid domains around integral MPs and glycoproteins (Escribá et al. 2008) (see Fig. 2.5);
(iii) Membrane surface appears unsmooth due to mechanical properties of the lipid bilayer membrane. The bilayer thickness change occurs as a result of the mismatch caused by various integral MPs while locally coupling with the bilayer. The locations where lipid bilayer does not contain integral proteins may show equilibrium bilayer thickness (Ashrafuzzaman and Tuszynski 2012a).

In model of Jacobson et al. (1995), integral MP lateral movements are described as follows:

(A) Transient confinement by obstacle clusters,
(B) Transient confinement by the cytoskeleton (fences),
(C) Directed motion by direct attachment to the cytoskeleton, and
(D) Free, random diffusion in the membrane plane.

In the 2008 cartoon presented by Escribá et al. (see Edidin 2003; Escribá et al. 2008 and see Fig. 2.5), cell membrane structure shows specialized lipid sub-

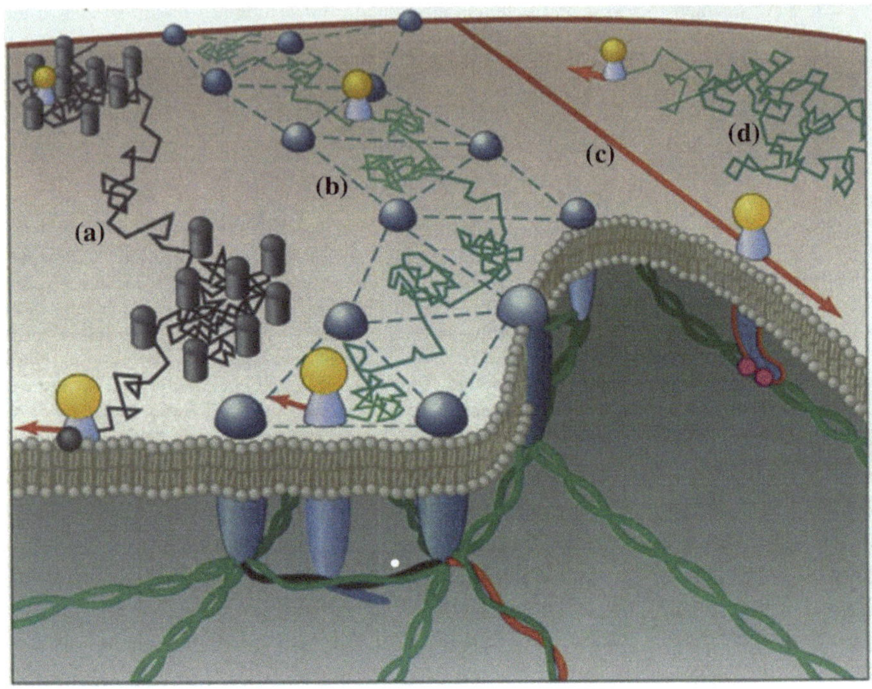


**Fig. 2.4** Modes of integral MP lateral mobility at the cell surface as envisioned by Jacobson and colleagues in 1995. Copied from Jacobson et al. (1995)

microdomains around integral MPs and glycoproteins. Here, severe heterogeneity in topographic distribution and asymmetry of different membrane lipids have been demonstrated. The limited mobility of lipids in specialized domains or rafts and in islands or in boundary lipids around integral MPs has resulted in updated proposals on membrane structural models that limit the fraction of completely free diffusing lipids and proteins in biomembranes (Edidin 2003; Garcia-Sáez and Schwille 2010; Lindblom and Orädd 2009; Verbe et al. 2003; Bagatolli et al. 2010; Mouritsen and Bloom 1984; Escribá et al. 2008; Somerharju et al. 2009).

Lateral mobility of integral MPs has been linked to various extracellular restrictions. They are, for example, binding to extracellular matrix, the formation of specialized membrane nanodomains, such as lipid rafts, and the formation of large supramolecular protein complexes or domains where high protein concentrations decrease lateral diffusion rates (Jacobson et al. 1987; Neumann et al. 2010a, b; Kusumi et al. 2004; Lenaz 1987). Membrane-associated cytoskeletal and peripheral MP barriers to free lateral movements of integral MPs have also been linked to inner membrane surface corrals where MPs can move freely within a corral or skeletal fencework, but they only rarely cross over to adjacent corrals (Jacobson et al. 1987, 1995; Edidin et al. 1991; Kusumi et al. 2012; Sheetz 1983; Radhkrishnan et al. 2010). Membrane-associated cytoskeletal systems have been

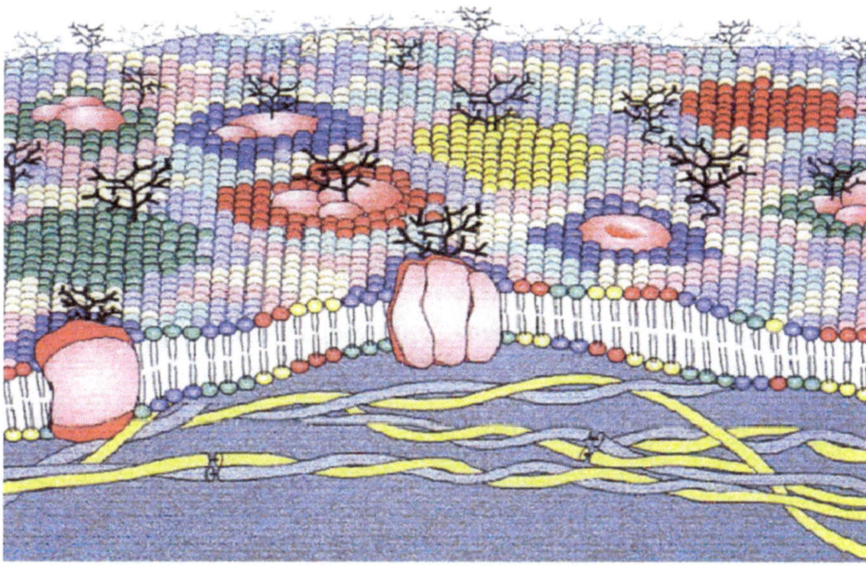

**Fig. 2.5** A schematic illustration of a modification of the fluid-mosaic membrane model, as envisioned by Escribá and colleagues in 2008. Different lipids are indicated in various colors forming specialized domains around integral MPs and gylcoproteins as well as being asymmetrically distributed across the membrane. Copied from Escribá et al. (2008)

known for some time to be dynamically associated with immobilization or directed movements of transmembrane integral proteins. However, this type of domain restriction by cytoskeletal fencework is a relatively new concept (Edidin et al. 1991; Zhang et al. 1993; Jacobson et al. 1995; Verbe et al. 2003; Kusumi et al. 2011, 2012). In addition, membrane integral protein components themselves may be involved in the dynamic formation of nanosized domains by initiating cis-interactions at the extracellular surface resulting in macromolecular complexes (Liu et al. 2012).

Most of the membrane dynamics explained here are especially observed to happen on a quite small scale, in nanometer (nm) dimensions. The lipid domains, rafts, etc., are observed to be geometrically stable at low nm scale dimension. Similarly, the dynamics of MPs happens at various low nm scale dimensions. The revised model is capable of explaining membrane arrangement and dynamics at as low as the nm scale dimensions.

The revised model, presented in Fig. 2.3, incorporates recent information on membrane domains, lipid rafts, and cytoskeletal fencing that were unknown in the 1970s. The literature abounds with membrane models, but it should be clear that for basic membrane structure at the nanostructural level the F-MMM has been severely tested and found worthy, if not completely accurate at higher levels of organization where crowding and specialized domain formation are important structural considerations.

## 2.2 Dynamics of Membrane Domains

Membrane domains randomly exist on various locations of cell membrane. The physical pictures of diffusion of plasma membrane lipids and proteins hint for the presence of membrane domains quite clearly. Some studies point to the presence of small, dynamic domains, and others provide clear picture of slowly diffusing, heavily clustered rafts.

### *2.2.1 Membrane Domains and Lipid Rafts—General Aspects and Roles in Cell Signaling*

A major membrane biology challenge has been to characterize the various cellular structures and mechanisms that impede free diffusion and dynamics in cell membranes and determine the consequences that membrane compartmentalization has on cellular biology. Understanding the membrane domain formation, membrane domain dynamics, status of lipid rafts, etc., may help raise some crucial scientific and medically important information that ultimately concerns the general cell surface dynamics and functions. In this chapter, we shall address on this issue in brief.

In cellular membrane, there exists a dynamic boundary with the environment, and the biological membranes are constituted by a specific set of lipids and proteins whose organization influences all cellular processes (Grecco et al. 2011). The pioneering fluid-mosaic model of Singer and Nicolson (1972) suggests that all membrane constituents such as lipids and MPs are continuously laterally mobile, diffuse freely, and distribute randomly. The large variety of different lipid species in biological membranes shows distinct physicochemical properties (Cronan 2003; van Meer et al. 2008; Coskun and Simons 2011; Ashrafuzzaman and Tuszynski 2012a). The existence of diverse lipid species importantly results in their lateral segregation into various membrane microdomains, because they tend to coalescence due simply to their physicochemical affinities (Lingwood and Simons 2010; Simons and Sampaio 2011a, b). This heterogeneous organization of membrane lipids into discrete microdomains leads to a diverse distribution of embedded MPs, which appears to be essential for the protein functionality (Neumann et al. 2010a, b; Simons and Sampaio 2011a, b; Kraft 2013) in cellular membranes. The lateral separation of various membrane lipids and proteins is now referred to as membrane domains (Schuck and Simons 2004a, b; Owen et al. 2010; Spira et al. 2012).

The lipid raft concept was introduced as early as three decades ago while explaining the generation of the glycolipid-rich apical membrane of epithelial cells (Simons and Van Meer 1988a). The proposed existence of specific lipid rafts or general membrane rafts emerged eventually as interesting concept in membrane organization (Simons and Ikonen 1997a, b). Eukaryotic cell membranes organize a variety of proteins related to signal transduction and membrane trafficking into various microdomains or rafts that are enriched in specific types of lipids such as cholesterol or sphingolipids (Simons and Ikonen 1997a, b).

The increase in raft lipids, ergosterol, and sphingolipids is observed to promote a raft coalescence process, induced by clustering of components (e.g., by lectins) of raft. This could lead to selective raft protein and lipid segregation in *trans*-Golgi network (TGN) membranes (see Fig. 2.6) (Simons and Sampaio 2011a, b).

In Fig. 2.6, it is observed that before clustering, proteins associate with rafts (red) to various extents (1). Clustering is induced, for example, via binding of a dimerizing protein (green) to a transmembrane raft protein (2). Scaffolded raft-associated proteins coalesce into a raft cluster. Clustered raft domain growth beyond a critical size induces budding (3). A transport container consisting of raft components finally pinches off from the parent membrane by fission at the boundaries of the domain. Additional protein machinery facilitates and regulates the budding process (4).

An energetic penalty is induced due to the immiscibility of the two liquid phases in the membrane that helps membrane bending to promote. The energetic penalty appears with increased effects in the case of lipid rafts, because of increased thickness and order of the raft domain compared to the more disordered vicinity (Schuck and Simons 2004a, b; Klemm et al. 2009). The bending energy can be calculated using the SCI protocol, explained in our book 'Membrane Biophysics' (see Ashrafuzzaman and Tuszynski 2012a). To avoid presenting too details, we shall not explain these theoretical calculations here. The sorting of proteins and lipids is fine-tuned by specific sorting, aided by accessory proteins that bind to raft cargo, such as the Ast1p protein that facilitates the delivery of Pma1p, the proton ATPase to the cell surface (Bagnat et al. 2001). Protein machinery involved in bending and release would also be required to bud the membrane domain into a transport vesicle, leading to regulated protein and lipid sorting at the exit from the TGN membrane.

Large raft membrane is reported in apical membrane of epithelial cells. Measuring long-range diffusion of several MPs by FRAP in the apical membrane of MDCK cells as compared with the same protein in the PM of a fibroblast, the possible conclusion was put forward that the apical membrane behaved as a percolating (continuous) phase at room temperature. Here, raft proteins freely diffused, whereas nonraft proteins dispersed into isolated domains (Meder et al. 2006). Also, the apical brush border membrane of small intestinal cells was described to behave as a large superraft domain stabilized by galectin-4 and another lectin, intelectin (Danielsen and Hansen 2008). These studies suggest for the existence of a wide distribution of sizes of lipid rafts.

Lipid rafts harbor specific proteins. One such raft-associated protein is commonly referred to as reggie or flotillin (see Morrow and Parton 2005; Babuke and Tikkanen 2007a, b; Otto and Nichols 2011; Stuermer 2011; Zhao et al. 2011), as example. The flotillin proteins are membrane-bound chaperones that are found to localize at lipid rafts. Here, they may recruit the proteins that need to be localized in lipid rafts to active and facilitate their interaction and oligomerization (Morrow and Parton 2005; Babuke and Tikkanen 2007a, b; Otto and Nichols 2011; Stuermer 2011; Zhao et al. 2011).

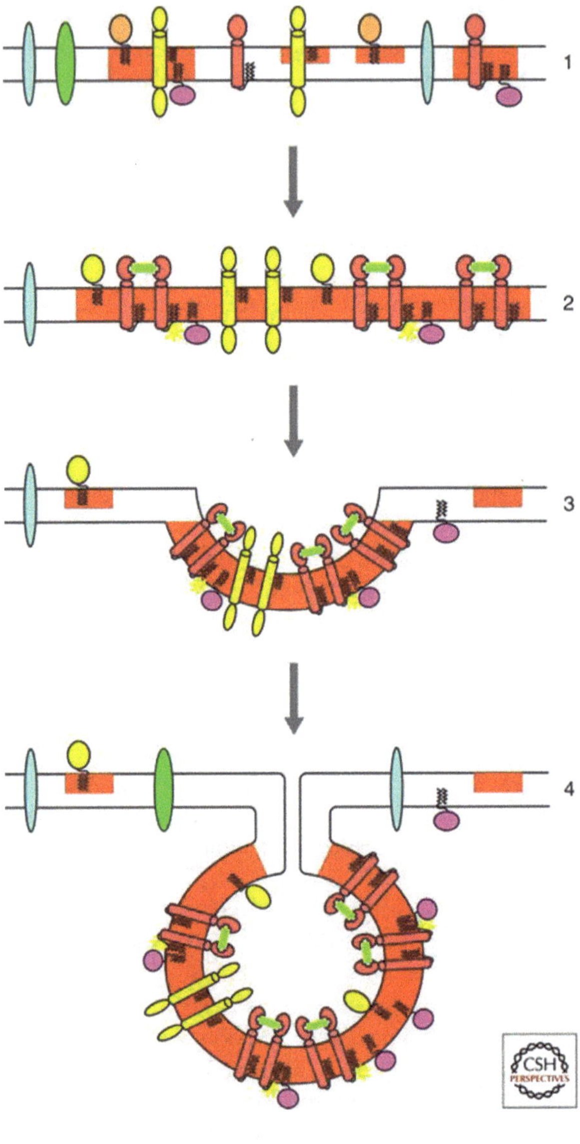


**Fig. 2.6** Clustering of rafts and domain-induced budding

Flotillin proteins are thus considered to be important components of lipid rafts which are found to play a central role in the organization of lipid rafts (Babuke and Tikkanen 2007a, b; Stuermer 2011; Zhao et al. 2011). Flotillin proteins are therefore considered bona fide markers of the subcellular localization of lipid rafts (Morrow and Parton 2005; Babuke and Tikkanen 2007a, b; Otto and Nichols 2011; Stuermer 2011; Zhao et al. 2011). The activity of flotillin is important for the correct functionality of numerous raft-associated cellular processes, including membrane sorting, trafficking, cell polarization, and signal transduction. Consequently, the perturbation of the activity of flotillin causes serious defects in signal transduction and membrane trafficking (Morrow and Parton 2005; Babuke and Tikkanen 2007a, b; Otto and Nichols 2011; Stuermer 2011; Zhao et al. 2011), which seems to be related to the occurrence of severe diseases such as Alzheimer's disease and Parkinson's disease (Michel and Bakovic 2007).

The lipid raft association of reggie proteins and their joint role in cell signaling are now quite evident. In cell signaling, this kind of membrane domain association of proteins plays crucial roles. Reggie proteins were originally discovered with suggestion that they would be plasma MPs accessible from the extracellular milieu. It thus suggests that the proteins contain transmembrane domains (Schroeder et al. 1994; Schulte et al. 1997). Subsequent findings show that reggie proteins do not traverse the membrane but rather associate with specifically the lipid bilayer by means of fatty acid modifications and hydrophobic amino acid stretches. This has been suggested to form a loop within the membrane without traversing it, analogously to the caveolin proteins (for details, see Bauer and Pelkmans 2006).

Saltiel and co-workers described a novel, PI3 kinase-independent insulin signaling pathway that proceeds through membrane rafts (Baumann et al. 2000). It was shown that in adipocytes, a constitutive, insulin-independent complex was formed between c-Cbl, a protooncogene, and a multifunctional adaptor protein named Cbl-associated protein or CAP (see Fig. 2.7). This complex was recruited into membrane rafts after insulin stimulation, which resulted in glucose uptake by the glucose transporter GLUT4. Recruitment of the CAP–Cbl complex into rafts was suggested to occur by means of a direct molecular interaction of CAP with reggie-2/flotillin-1.

### *2.2.2 Domains and Rafts in Simulated Membranes Containing Cholesterol*

In living cells, with possibility of having thousands of unique lipid species (Meer 2005), it is quite difficult to study various roles and effects of mixing various lipids. Model lipid systems have been used extensively to understand lipid mixing effects (Bennett and Tieleman 2013). In vitro constructed membrane systems exhibit a rich phase behavior for simple binary and ternary lipid mixtures containing cholesterol, including regions of liquid–liquid phase coexistence (Veatch and Keller 2005; Feigenson 2009). Cholesterol is observed to induce phase separation in ternary lipid mixtures with saturated and unsaturated lipids (Veatch and Keller 2005). In Fig. 2.8, an example of a ternary phase diagram of a giant unilamellar vesicle is shown. The triangle edges are the composition of each respective lipid. The point in the middle of the triangle is, therefore, a 1:1:1 ratio of the three lipids. Liquid-ordered–liquid-disordered phase coexistence is shown in the shaded region in the middle of the diagram. Tie-lines determine the composition of each phase. The ordered phase is rich in palmitoyl-sphingomyelin (PSM) and cholesterol, while the disordered phase is rich in dioleoyl-PC (DOPC).

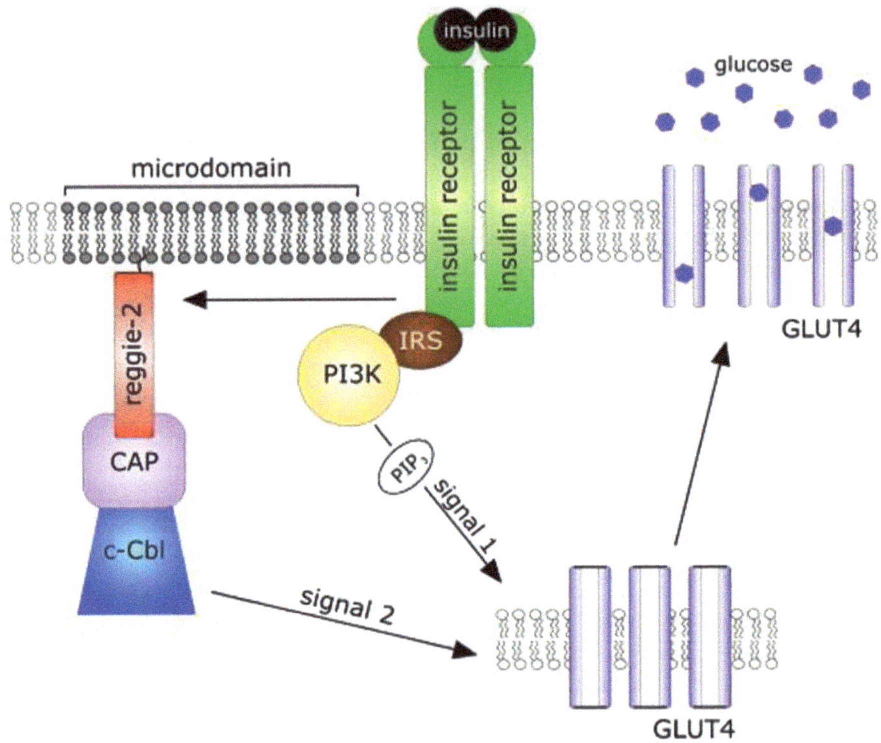


**Fig. 2.7** Formation of membrane microdomain and the role of reggie-2 in insulin signaling have been schematically demonstrated. Two signaling cascades are shown. (i) In adipocytes and muscle cells, insulin receptor activation triggers the classic signaling cascade through phosphatidylinositol 3-kinase, PI3 K. (ii) A second signaling route proceeds through membrane rafts and involves a ternary complex c-Cbl/CAP/reggie-2. Both pathways mediate the recruitment of the glucose transporter GLUT4 to the plasma membrane and uptake of glucose into the cell. Here, the role of membrane microdomain/rafts is especially mentionable (Babuke and Tikkanen 2007)

Large domains can easily be observed in model membranes. But in cell membranes, which may limit the applicability of in vitro phase behavior measurements to biological lipid rafts, the visibility of such domains may be limited. In giant plasma membrane vesicles, large domains have been visible at low temperatures, but not at higher temperatures, and fluorophore partitioning suggested liquid-ordered–liquid-disordered phase coexistence (Baumgart et al. 2007). It has been shown that near the critical point in the phase diagram, compositional fluctuations in model membranes result in transient domains with a correlation length 20–50 nm (Veatch et al. 2008). Using sub-diffraction limit stimulated emission depletion fluorescence spectroscopy, cholesterol-meditated trapping of lipids associated to rafts was observed in live cells (Eggeling et al. 2009). Recently, it was shown that the partitioning of fluorescently labeled lipid analogs between liquid-ordered and liquid-disordered model membranes was not correlated with their possible trapped

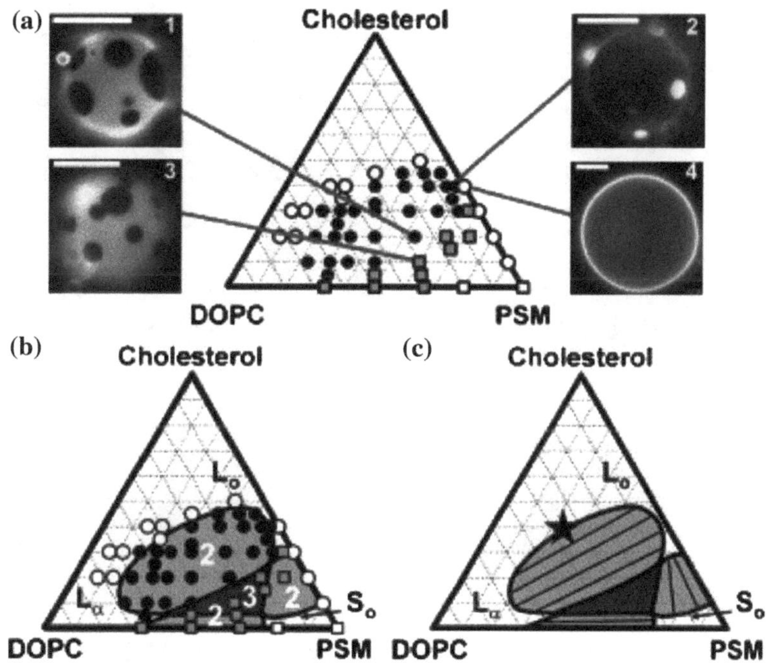


**Fig. 2.8** **a** Phases observed by fluorescence microscopy of GUVs containing mixtures of DOPC, PSM, and cholesterol at 25 °C. White symbols denote that membranes are in one uniform phase, either liquid (circles) or solid (squares). Black circles denote coexisting liquid phases, and gray squares denote coexisting solid and liquid phases. **b** Fluorescence microscopy data are consistent with a speculative underlying phase diagram that includes a region of three-phase coexistence. **c** Speculative location of tie-lines and a miscibility critical point (star). Copied from Veatch and Keller (2005)

diffusion in live cell membranes (Sezgin et al. 2012). Knowledge is still quite limited to conclude on how lipids behave—how they diffuse laterally in the membrane and self-associate. More mechanistic studies are required.

Coarse-grained (CG) models have been used to directly observe domain formation in model membrane systems. Using molecular details, one can simulate bilayers large enough and for long enough to directly observe domain formation in ternary lipid mixtures. Figure 2.9 illustrates examples of domain formation using CG models under various conditions. Bennett and Tielman summarized a few simulated data in a 2013 review (Bennett and Tielman 2013) to reconstruct this Fig. 2.9.

Phase separation of ternary lipid mixtures of lipid bilayers and small vesicles was shown to occur. They matched experimentally observed liquid-ordered domains enriched with high dipalmitoyl-PC (DPPC) and cholesterol contents and liquid-disordered domains enriched with high polyunsaturated DLiPC content and low cholesterol (Risselada and Marrink 2008). Interleaflet domain coupling was

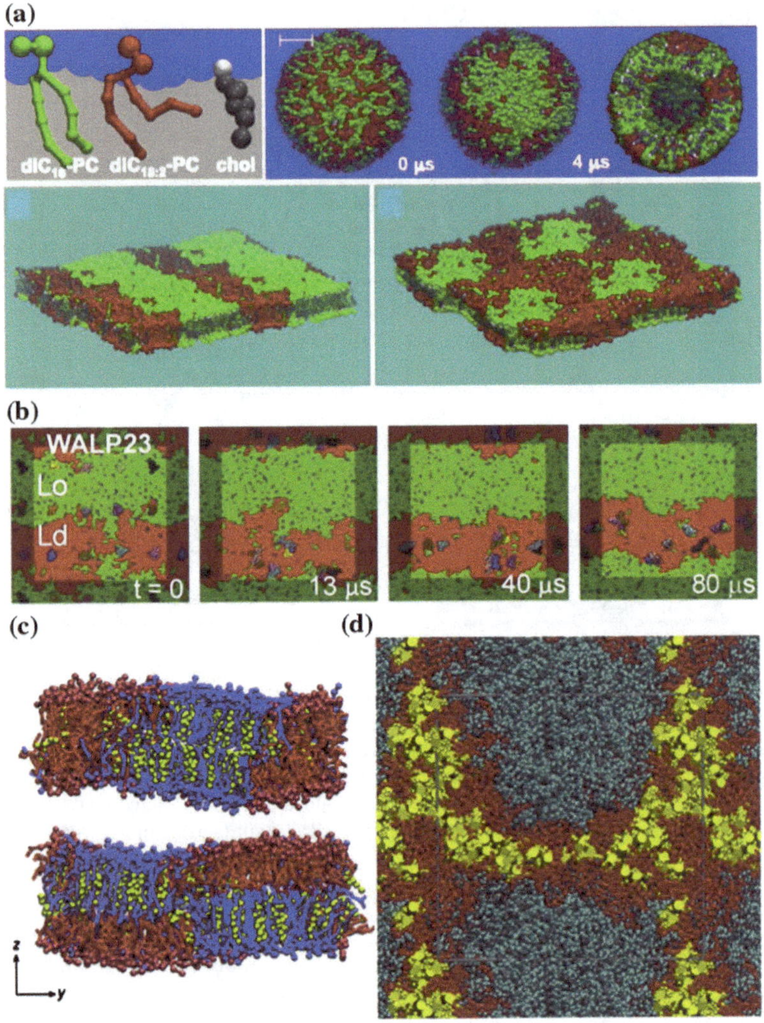


**Fig. 2.9** Rafts in simulated membranes, including a lipid mixture in vesicles and bilayers [(**a**), from Risselada and Marrink (2008)], partitioning of WALP23 in different phases [(**b**), from Baoukina et al. (2013)], an asymmetric bilayer [(**c**), from Perlmutter and Sachs (2011)], and a peptide-containing phase-separated membranes [(**d**), from Domanski et al. (2012)]

observed. It was linked with elastic theory, predicting a small surface tension-driven bilayer registration (Risselada and Marrink 2008). Fast cholesterol flip-flop in the disordered phase and slow flip-flop in the ordered phase were also observed. Later simulation of asymmetric ternary lipid mixtures (Perlmutter and Sachs 2011) reported that in the DLiPC:DPPC:CHOL symmetric bilayers liquid-ordered–liquid-disordered coexisting domains were directly observed. CHOL or chol stands for cholesterol. Domain formation in one leaflet was observed to be induced ordering in the opposite leaflet composed of pure unsaturated lipids. This increased the local lipid curvature. With longer saturated lipids' domain, antiregistration was detected (Perlmutter and Sachs 2011). Domain registration in bilayers composed of mixtures of single-chain polymers was studied by Pantano et al. (2011) using CG model of Shinoda and co-workers (Shinoda et al. 2007). Identical lipid tails but with either a charged or neutral polar headgroup was used (Pantano et al. 2011). Clustering of the charged headgroups cross-linked with an ion was found to cause domain formation. Thickness mismatch and local bilayer curvature were also found to register domains.

Following the simulations of monolayer mixtures that showed no clear domain formation (Baoukina et al. 2008; Baoukina and Tieleman 2011), another study investigated the properties of two types of large monolayers (80 nm × 80 nm, ca. 9000 lipids) on a timescale of 25 microsecond (μs). The first, mixture of DPPC, POPG, and DOPC (3:1:1), showed a transition from a homogeneous liquid-expanded (LE) to a mixed liquid-condensed (LC) state upon quenching from 310 to 290 K. The second, mixture of DPPC, DOPC, and cholesterol (5:3:4), showed a transition from liquid-disordered (Ld) to liquid-ordered (Lo). Figure 2.10 shows snapshots of the process of domain formation for the cholesterol-containing monolayer.

About a decade ago, Day and Kenworthy wrote a review to provide a brief summary of the classes of domains that had been characterized (Day and Kenworthy 2009). Here, the main focus was to identify the properties, especially the dynamic characteristics of lipid rafts in cells through measurements of the diffusion of lipids and proteins. Subsequently, besides many articles, another

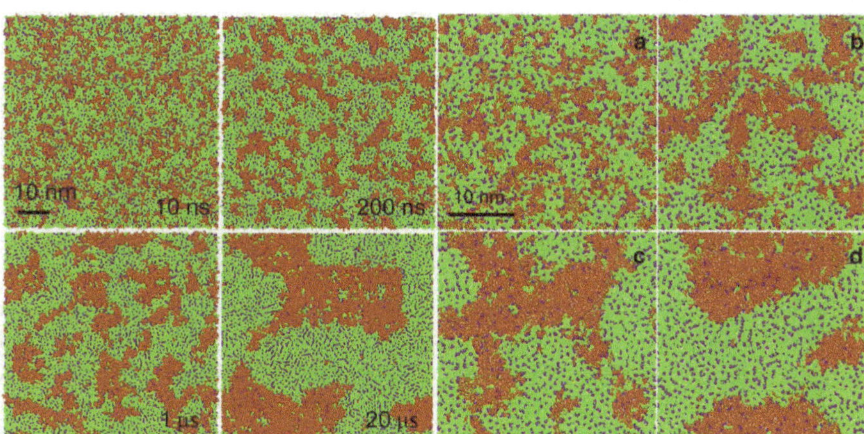


**Fig. 2.10** Transformation into Lo and Ld phases in a DPPC:DOPC:cholesterol 5:3:4 mixture in a bilayer and a monolayer at 290 K. In all panels, DPPC is shown in green, DOPC in orange, and cholesterol in purple. The left four panels are simulation snapshots of a monolayer at 30 mN/m. The right four panels (a, b, c, and d) are snapshots of a bilayer at 10 ns, 500 ns, 2 μs, and 11 μs, respectively. The left four panels are reproduced from Baoukina et al. (2012). The right four panels are reproduced from Baoukina et al. (2013). Taken from Bennett and Tieleman (2013)

important review may be especially mentionable which was written by Simons and Sampaio (2011a, b). Here, they summarized the emerging principles of membrane architecture with special emphasis on lipid organization and domain formation (Simons and Sampio 2011a, b). The effects of composition of lipids and cholesterol on domain formation have been addressed. Simulated membranes, vesicles, may be good platforms to address this aspect.

As described earlier in F-MMM, the basic structure of cell membrane is the lipid bilayer, composed of two apposing leaflets, forming a 2D liquid. This bilayer poses to have fascinating properties that are naturally inbuilt to perform the functions cells require. To coordinate the various phenomenological functions, the lipid bilayer has evolved the propensity to segregate its constituents laterally. This capability is based on both fast and slow dynamic liquid–liquid immiscibility and underlies the raft concept of membrane sub-compartmentalization. The cell signaling has some crucial roles drawn from membrane domains what have been explained here. The formation of membrane domain, lipid rafts and status of their membrane surface localization, static or dynamic nature, etc., require thorough inspection to understand their correct roles. Here, we shall focus on this issue quite in detail.

### *2.2.3 Membrane Domains and Lipid Rafts—Diffusion and Dynamics*

The interest on understanding of the dynamic nature of agents within the plane of the cell membrane started about five decades ago. Immediately after the ground-breaking paper of Frye and Edidin in 1970 that demonstrated that membrane antigens are not static entities, but rather can move within the cell membrane plane (Frye and Edidin 1970), many laboratories started characterizing the movement of membrane components. Over the decades, a variety of methods have been developed to probe the diffusion of lipid and protein components in cell membranes. Among them, the most widely used approaches are fluorescence correlation spectroscopy (FCS), fluorescence recovery after photobleaching (FRAP), and single-particle tracking (SPT) (Chen et al. 2006; Kenworthy 2007; Marguet et al. 2006; Bacia and Schwille 2007; Kusumi et al. 2005; Lommerse et al. 2004; Lagerholm et al. 2005; Jacobson et al. 2007; Kenworthy 2005). Although the measurements obtained using these techniques have been successful to reveal various features of cellular membranes, many are still not fully understood due to lack of enough evidences. The diffusion of molecules in simplified model membranes is as much as 50 times faster than that in rather complex live cells (Kusumi et al. 2004). Diffusion measurements in cell membranes also clearly suggest that the membranes are not too simple as are thought or they appear to be in most of the model membrane representations. Instead, they are heterogeneous, consisting of multiple classes of domains with various geometries and biophysical criteria.

Early measurements on diffusion of proteins and lipids pointed to the presence of membrane domains prior to the development of the hypothesis on lipid rafts (Jacobson et al. 1995; Edidin 1992; Kusumi and Sako 1996; Sheetz 1995). Early FRAP studies detected a role of the cytoskeleton in slowing protein diffusion (Sheetz et al. 1980; Wu et al. 1982) and also revealed evidence for heterogeneities that impede free diffusion (Edidin 1992, 1997; Yechiel and Edidin 1987; Edidin and Stroynowski 1991). In another study, examining the dependence of diffusion on bleach spot size (Yechiel and Edidin 1987), the mobility of proteins and fluorescent lipid analogs showed a decrease in mobile fraction ($M_f$) with increasing spot size. The diffusion coefficient was also reported heterogeneous across the cell surface. This suggests for the presence of a mixture of micrometer (μm)-size protein-rich and protein-poor domains (Yechiel and Edidin 1987). The ability of proteins to sense these domains depends on whether they are attached to the membrane by a transmembrane domain or GPI-anchor (Edidin and Stroynowski 1991). The barrier location usually lies 2–3 nm beneath the surface of the plasma membrane (Edidin et al. 1994, 1997).

Single-particle tracking (SPT) experiments demonstrated that molecules undergo periods in which they appear to be transiently confined (Simson et al. 1995). Furthermore, transient interactions of proteins with coated pits could also be detected by FRAP (Fire et al. 1991). Thus, there is ample evidence that the diffusion of molecules in cell membranes is affected by interactions with membrane components especially due to the heterogeneity in membrane structure.

Figure 2.11 demonstrates the diffusion-based biophysical methods used to study the membrane domains. Here, a protein or lipid of interest is labeled with a fluorescent tag (red molecules in Fig. 2.11). A focused laser spot is then used; the fluorophores in a small patch of membrane, referred to as the bleach region of interest (ROI) (yellow circle, Fig. 2.11), are irreversibly bleached with a brief pulse at high intensity (black molecules, Fig. 2.11).

Figure 2.11, top panel. Investigations on fluorescence recovery after photobleaching (FRAP) were performed. Using low-intensity laser excitation, the subsequent lateral diffusion of unbleached fluorophores from the surrounding membrane into the ROI, and corresponding movement of bleached molecules out of the ROI, can be monitored. By plotting the change in fluorescence intensity in the ROI versus time (right panel) and then fitting the curve to an appropriate equation, an average rate of diffusion (D) can be calculated. A second kinetic parameter, the mobile fraction ($M_f$), can also be acquired from a FRAP experiment. This value is a percentage of the recovered fluorescence in the ROI compared to the fluorescence intensity lost during bleaching. Typically, bleach ROIs in the μm range are used, and diffusion coefficients obtained range from $\sim$0.01 to 1 $\mu m^2\ s^{-1}$ for MPs and fluorescent lipid analogs. You can read from Chen et al. (2006), Kenworthy (2007) for further information.

Figure 2.11, middle panel. FCS measures the photon fluctuations arising from fluorescent molecules (red molecules) contained within a very small 3D excitation volume ($\sim$1 fmol) over time. In the case of MP or lipid, this three-dimensional area is further reduced to a transverse plane of the imaging region at the laser beam

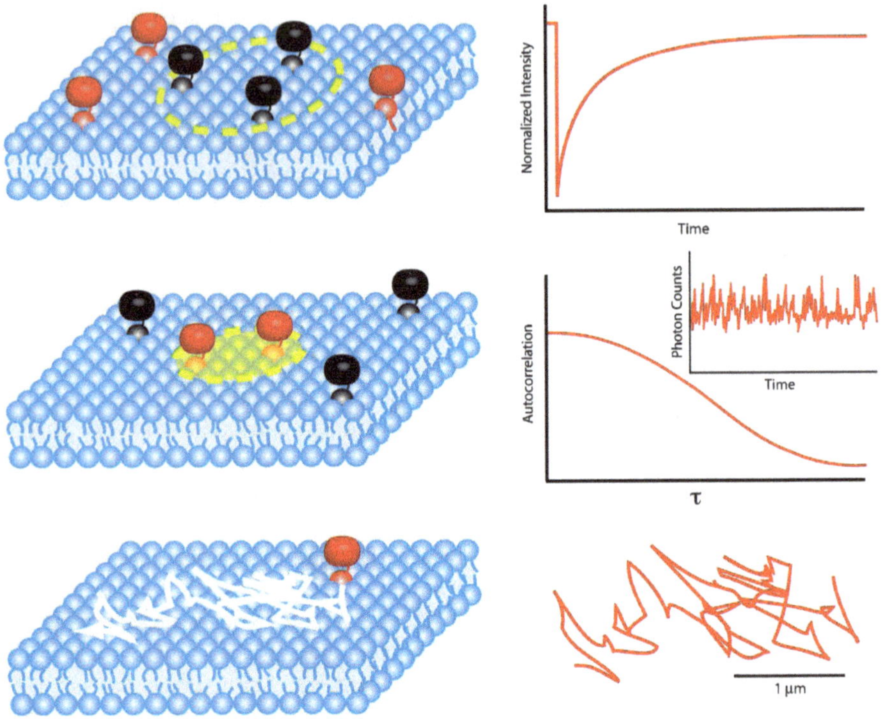


**Fig. 2.11** Diffusion-based biophysical methods used to study the membrane domains. Top. Fluorescence recovery after photobleaching (FRAP). Middle. FCS measures the fluctuations of photons arising from fluorescent molecules (red molecules) contained within a very small three-dimensional (3D) excitation volume ($\sim 1$ fmol) over time. Bottom. In single-particle tracking (SPT), a molecule of interest is tagged (genetically with a fluorescent protein or with antibodies conjugated to latex microspheres, colloidal gold, quantum dots, or traditional fluorophores) and then imaged as it diffuses in the membrane. Figure is copied from Day and Kenworthy (2009)

waist (yellow circle), often 0.2–1 μm in radius. By constant illumination at the excitation wavelength of the fluorophore in a fixed beam waist size, changes in emitted photons can be measured as a function of time (see inset, right panel). These intensity traces are used to calculate an autocorrelation curve (see right panel). Through fitting the autocorrelation curve with the appropriate models, a diffusion coefficient (D) can be calculated. Note that FCS typically can only measure D's > 0.1 $\mu m^2\ s^{-1}$ and thus is not useful for analyzing slowly diffusing MPs. For recent reviews of this approach, see Marguet et al. (2006), Bacia and Schwille (2007).

Figure 2.11, bottom panel. Using single SPT, a tagged molecule of interest is tracked down. Using a fast camera and computer-assisted analysis, the location of the molecule can be measured with high spatial accuracy ($\leq 10$ nm) and high temporal resolution [from 30 frames/s up to 50,000 frames/s (Umemura et al. 2008)] (right panel). The resulting trajectories are then used to generate plots of the mean squared displacement of the diffusing molecule versus time, which can in turn be classified into different modes of diffusion (ex. free diffusion versus confined diffusion). See Chen et al. (2006) and Kusumi et al. (2005) for more information about this technique.

### 2.2.4 *Size and Stability of Domains and Rafts on Detection of Diffusion and Dynamics*

We have addressed the detection of dynamics of the domains using example assays. However, the field of lipid raft is still at a technical impasse because the physical tools to study biological membranes as a liquid that is ordered in space and time are still being developed or poorly understood (Jacobson et al. 2007). This leads to a disconnection between the concepts of membrane domains and lipid rafts as derived from biochemical and biophysical assays and their existence in the living cell. Most of the studies can deal with lipid domains and rafts using synthetic membranes. Current computer simulation modeling can also not deal with too large system run for too long timescale so that they can address the energetically equilibrium condition properly. To be biologically significant, domains should appear with a minimum size of a few protein diameters and with a minimum lifetime corresponding to a short enzyme turnover time of ~ microseconds (Jacobson et al. 2007). Traditional and advanced imaging techniques can detect domains that are stable and of sufficient size.

In Jacobson et al. (2007), a nice analysis has been presented. Lipid and protein reporters imaged at the light and electron microscope level suggest for the existence of stable domains approaching to a micrometer (μm) dimension, or greater, in various cells (see Fig. 2.12). Microdomains of detergent-resistant transporters are reported to show stability in growing yeast cells for more than 10 min (Malinska et al. 2003). The exchange time of individual proteins was not determined. Microdomains in smooth muscle cells (Schutz et al. 2000) and macrophages Gaus et al. (2003) are reported to be stable for tens of seconds. In muscle, they were reported as cohesive entities that rotate as units. Individual lipid-probe molecules showed a domain residence time of about 10 ms (Schutz et al. 2000). Although larger domains can exist in various cells, the stabilizing factors are largely unknown (Vereb et al. 2003). Occasionally, proteins from the four-transmembrane domain tetraspanin superfamily may organize special microdomains as in microvilli (Roper et al. 2000), and HIV may induce tetraspanin-enriched domains as export portals in infected cells (Nydegger et al. 2006). Caveolae may also form μm-sized clusters in some cells (Rothberg et al. 1992).

When domains are small and transient, the biophysical challenges of measuring them are quite hard. Most of the lipid dimension scale dynamics of cell membrane and domains are quite fast, of the order of μs, and happen within nanoscale dimension (Ashrafuzzaman and Tuszynski 2012a). Computer modeling may help

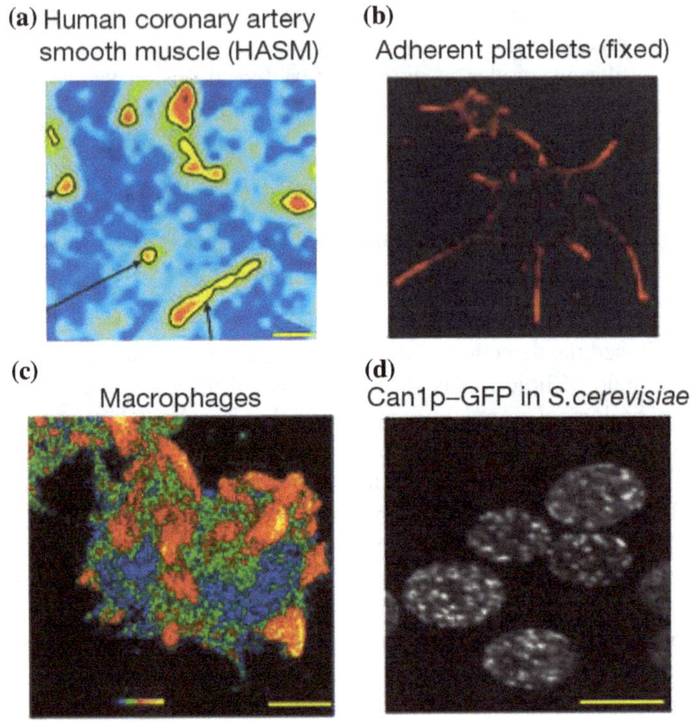


**Fig. 2.12** Lipid and protein domains in cell membranes. **a** Single domains, enriched in the fluorescent lipid analogue DMPECy5, imaged in human coronary artery smooth muscle cells (Schutz et al. 2000). **b** Cholesterol-rich domains in platelet processes (Heijnen et al. 2003) visualized using a noncytolytic derivative of biotinylated perfringolysin O followed by indirect immunofluorescence microscopy with an antibiotin monoclonal antibody (reprinted with permission from the International Society on Thrombosis and Haemostasis). **c** Lipid domains with greater relative order than the bulk membrane, visualized in living macrophages with the fluorescent probe, Laurdan (6-lauroyl-2-dimethylaminonaphthalene), where the warmer pseudocolors represent more ordered regions 60. **d** Domains formed by the proton–argenine symporter transporter (Can1p–GFP) in Saccharomyces cerevisiae (Malinska et al. 2003). The scale bars represent 1 μm in **a**, and 5 μm in **c** and **d**. Copied from Jacobson et al. (2007)

us understand this low dimension issues over the time when more and more computational power will be incorporated in supercomputing systems. Clearly, not just existing imaging techniques can deal with this low-dimensional dynamics, rather application of correct simulation program may not only help us address the dynamical aspects of membrane domains, rafts, and components therein but also help us address their energetics and statistical distribution including dynamics within their structures. This information may also be helpful in designing agents that bind to structures that are dynamic and distributed on the cell surface, following localized energetics determining their statistical mechanical probability in the local environment (Ashrafuzzaman et al. 2013).

## 2.3 Bacterial Membrane Domains and Lipid Rafts: Distribution and Dynamics

Membrane domains have been long recognized in eukaryotic cells only. For instance, polarized epithelial cells show a lateral membrane organization to distinguish a basolateral and an apical membrane macrodomain. It shows different lipid and protein compositions and is being specialized in different roles (Schuck and Simons 2004a, b; Simons and van Meer 1988; van Meer and Simons 1988). Neurons also have membrane domains with different lipids and proteins, which are catalogued according to their role in synapsis (Simons and Toomre 2000; Tsui-Pierchala et al. 2002). The existence of membrane domains is not an exclusive feature of eukaryotic cells. Membrane domains are now evident in bacteria and archaea.

Membrane organization is particularly important in unicellular organisms, as it represents the boundary between the organism and the environment, therefore, orchestrates many cellular processes that are essential for life of the system, such as cell division or signal transduction (Barak and Muchova 2013; Barak et al. 2008; Matsumoto et al. 2006; Muchova et al. 2010). The use of specific lipid dyes [e.g., nonyl-acridine orange (NAO)] helped to demonstrate the presence of cardiolipin-enriched domains at the cell polar regions and at the division septum in Escherichia coli and Bacillus subtilis bacterial cells (Kawai et al. 2004; Mileykovskaya and Dowhan 2000, 2009; Rosch et al. 2007). The specificity of this dye for the detection of cardiolipin has been questioned recently (Oliver et al. 2014). But the localization pattern of NAO-enriched domains in E. coli and B. subtilis suggests a lateral organization of lipids in bacterial membrane that may be correlated with cell division and morphogenesis (Kawai et al. 2004; Mileykovskaya and Dowhan 2000, 2009; Rosch et al. 2007).

The existence of lipid rafts has been traditionally associated with eukaryotic cells, because their assembly depends on the presence of cholesterol, which is absent from the membranes of most bacteria and archaea. Thus, the assembly of lipid rafts in eukaryotes has been considered a fundamental step during the evolution of cellular complexity, suggesting that bacteria and archaea were too simple to require such a sophisticated organization of their signaling networks and membrane-associated protein complexes. However, bacteria also show a variety of membrane-associated sensory complexes, such as the ones involved in bacterial chemotaxis, which organize into large clusters that integrate and amplify stimuli before transmitting the signal to downstream proteins (Sourjik 2004; Bray et al. 1998). This demonstrates that bacteria are also able to organize their signal transduction systems into signaling platforms of a certain complexity.

In addition to the above-mentioned findings, it was recently shown that bacteria are able to organize many signal transduction cascades and protein transport into functional membrane microdomains (FMMs) constituted by specific lipids; i.e., bacterial membranes contain lipid rafts similar to those found in eukaryotic cells (Simons and Ikonen 1997a, b). The assembly of FMMs involves the biosynthesis of

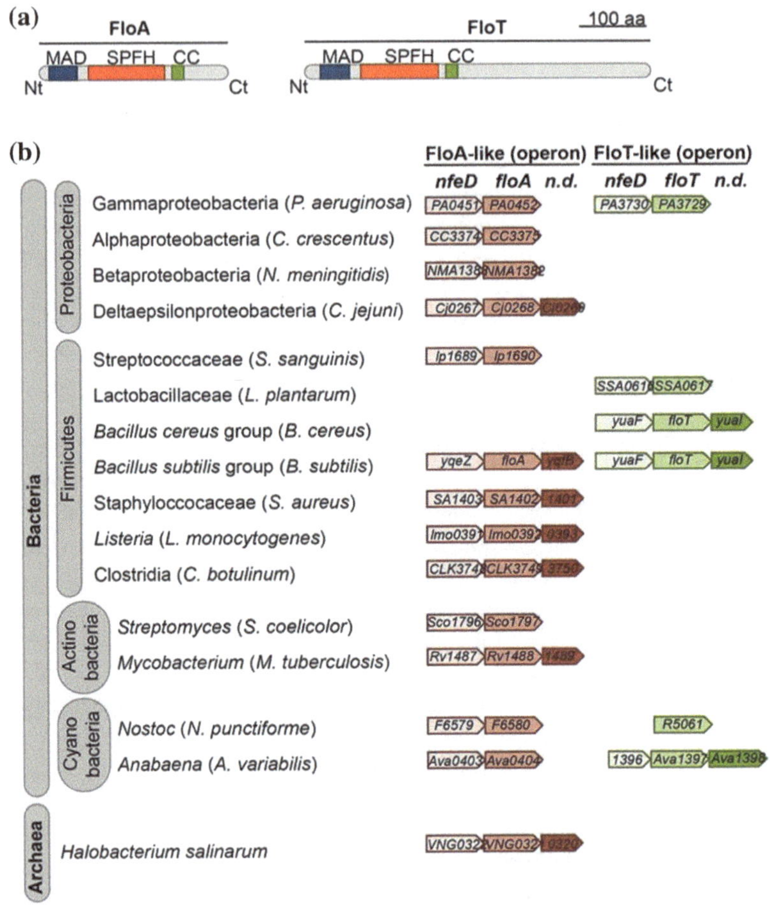


**Fig. 2.13** Taxonomic distribution of FloA and FloT in bacteria. **a** Schematic representation of the molecular structures of two different flotillin proteins, FloA and FloT, from the model organism Bacillus subtilis. MAD represents a membrane-anchoring region (whether this is a transmembrane helix or a hairpin loop has not yet been experimentally addressed). SPFH is a typical protein domain of flotillin proteins, and CC represents a coil-coiled region that localizes at the C-terminal regions of these two proteins. **b** Distribution of the FloA and FloT operons in bacteria. The first gene of the operon codes for an NfeD-like protein. The second gene is the flotillin-encoding gene. The third gene, coding for a protein of unknown function, is less well conserved and is absent from various species. Shown is the architecture of the operons from bacterial species from different phyla as a reference. The operons contain provisional gene names given by genome annotation. S. sanguinis, Streptococcus sanguinis; C. botulinum, Clostridium botulinum; n.d., not determined. Copied from Bramkampa and Lopez (2015)

polyisoprenoid lipids in the membrane and their colocalization with flotillin-like proteins, which are also present in bacteria (Tavernarakis et al. 1999). Bacterial flotillins seem to play a role similar to the one played by eukaryotic flotillins, acting as protein scaffolds in recruiting proteins that need to be localized in lipid rafts to promote interactions and oligomerization (Good et al. 2011; Langhorst et al. 2005). Similar to eukaryotic flotillin proteins, flotillins in bacteria play an essential role in organizing and maintaining the correct architecture of the FMMs. The discovery of FMMs in bacterial membranes led many laboratories in the last decades to experimentally test critical aspects of this discovery. The recent review of Bramkampa and Lopez (2015) is a nice resource that compiled the available information about the existence and biological role of bacterial lipid rafts.

Current state of knowledge, suggesting for the possibility of existence of FMMs in bacteria, seems to be universal, given that almost all bacterial species harbor at least one flotillin-like protein-encoding gene in their genome. An overview of the presence of flotillin genes in bacteria is shown in Fig. 2.13. We have learnt the role of flotillin in domain formation earlier in this chapter.

## References

R.B. Taylor, W.P.H. Duffus, M.C. Raff, S. de Petris. 1971. Redistribution and Pinocytosis of Lymphocyte Surface Immunoglobulin Molecules Induced by Anti-Immunoglobulin Antibody. Nature new biology 233, 225–229.

S.J. Singer, G.L. Nicolson. The Fluid Mosaic Model of the structure of cell membranes. Science, 175 (1972), pp. 720–731.

Md Ashrafuzzaman and J. Tuszynski. 2012a. Membrane Biophysics, Springer-Verlag, Berlin, Heidelberg. ISBN: 978-3-642-16104-9 (Print), 978-3-642-16105-6 (Online).

G.L. Nicolson. 2014. The Fluid—Mosaic Model of Membrane Structure: Still relevant to understanding the structure, function and dynamics of biological membranes after more than 40 years. Biochimica et Biophysica Acta (BBA)—Biomembranes, 1838, 1451–1466.

G.L. Nicolson. Transmembrane control of the receptors on normal and tumor cells. I. Cytoplasmic influence over cell surface components. Biochim. Biophys. Acta, 457 (1976), pp. 57–108.

G.L. Nicolson, T. Ji, G. Poste. The dynamics of cell membrane organization. G. Poste, G.L. Nicolson (Eds.), Dynamic Aspects of Cell Surface Organization, Elsevier, New York (1977), pp. 1–73.

K. Jacobson, E.D. Sheets, R. Simson. Revisiting the fluid mosaic model of membranes. Science, 268 (1995), pp. 1441–1442.

P.V. Escribá, J.M. Gonzáles-Ros, F.M. Goñi, P.K.J. Kinnunen, L. Vigh, L. Sánchez-Magraner, A. M. Fernández, X. Busquets, I. Horváth, G. Barceló-Coblijn. Membranes: a meeting point for lipids, proteins and therapies. J. Cell. Mol. Med., 12 (2008), pp. 829–875.

J.D. Robertson. The ultrastructure of cell membranes and their derivatives. Biochem. Soc. Symp., 16 (1959), pp. 3–43.

J.D. Robertson. The molecular structure and contact relationships of cell membranes. Prog. Biophys. Biophys. Chem., 10 (1960), pp. 343–418.

J.D. Robertson. Membrane structure. J. Cell Biol., 91 (1981), pp. 191s–204s.

J.F. Danielli, H. Davson. A contribution to the theory of permeability of thin films. J. Cell. Comp. Physiol., 5 (1935), pp. 495–508.

E. Gorter, F. Grendel. On bimolecular layers of lipoids on the chromocytes of the blood. J. Exp. Med., 41 (1925), pp. 439–443.

P. Pinto da Silva, D. Branton. Membrane splitting in freeze-etching. Covalently bound ferritin as a membrane marker. J. Cell Biol., 45 (1970), pp. 598–605.

W. Stoeckenius, D.M. Engelman. Current models for the structure of biological membranes. J. Cell Biol., 42 (1969), pp. 613–646.

A.A. Benson. On the orientation of lipids in chloroplast and cell membranes. J. Am. Oil Chem. Soc., 43 (1966), pp. 265–270.
D.E. Green, D.W. Allmann, E. Bachmann, H. Baum, K. Kopaczyk, E.F. Korman, S. Lipton, D.H. MacLennan, D.G. McConnell, J.F. Perdue, J.S. Rieske, A. Tzagoloff. Formation of membranes by repeating units. Arch. Biochem. Biophys., 119 (1987), pp. 312–335.
W. Kauzmann. Some factors in the interpretation of protein denaturing. Adv. Protein Chem., 14 (1959), pp. 1–63.
S.J. Singer. The structure and function of membranes—a personal memoir. J. Membr. Biol., 129 (1992), pp. 3–12.
S.J. Singer. The molecular organization of membranes. L.I. Rothfield (Ed.), Structure and Function of Biological Membranes, Academic Press, New York (1971), pp. 145–222.
Ashrafuzzaman, Md.; Tuszynski, J. Regulation of Channel Function Due to Coupling with a Lipid Bilayer. Journal of Computational and Theoretical Nanoscience, Volume 9, Number 4, April 2012b, pp. 564–570.
M. Edidin. Lipids on the frontier: a quarter century of cell-membrane bilayers. Nat. Rev. Mol. Cell Biol., 4 (2003), pp. 414–418.
A.J. Garcia-Sáez, P. Schwille. Surface analysis of membrane dynamics. Biochim. Biophys. Acta, 1798 (2010), pp. 766–776.
G. Lindblom, G. Orädd. Lipid lateral diffusion and membrane heterogeneity. Biochim. Biophys. Acta, 1788 (2009), pp. 234–244.
G. Verbe, J. Szöllosi, J. Matkó, P. Nagy, T. Farkas, L. Vigh, L. Mátyus, T.A. Waldmann, S. Damjanovich. Dynamic, yet structured: the cell membrane three decades after the Singer–Nicolson model. Proc. Natl. Acad. Sci. U. S. A., 100 (2003), pp. 8053–8058.
L.A. Bagatolli, J.H. Ipsen, A.C. Simonsen, O.G. Mouritsen. An outlook on the organization of lipids in membranes: searching for a realistic connection with the organization of biological membranes. Prog. Lipid Res., 49 (2010), pp. 378–389.
O.G. Mouritsen, M. Bloom. Mattress model of lipid–protein interactions in membranes. Biophys. J., 46 (1984), pp. 141–153.
P. Somerharju, J.A. Virtanen, K.H. Cheng, M. Hermansson. The superlattice model of lateral organization of membranes and its implications on membrane lipid homeostasis. Biochim. Biophys. Acta, 1788 (2009), pp. 12–23.
K. Jacobson, A. Ishihara, R. Inman. Lateral diffusion of proteins in membranes. Annu. Rev. Physiol., 49 (1987), pp. 163–175.
A.K. Neumann, M.S. Itano, K. Jacobson. Understanding lipid rafts and other related membrane domains, F1000. Biol. Reprod., 2 (2010a), pp. 31–36.
A. Kusumi, I. Koyama-Honda, K. Suzuki. Molecular dynamics and interactions for creation of stimulation-induced stabilized rafts from small unstable steady-state rafts. Traffic, 5 (2004), pp. 213–230.
G. Lenaz. Lipid fluidity and membrane protein dynamics. Biosci. Rep., 7 (1987), pp. 823–837.
M. Edidin, S.C. Kuo, M.P. Sheetz. Lateral movements of membrane glycoproteins restricted by dynamic cytoplasmic barriers. Science, 254 (1991), pp. 1379–1382.
A. Kusumi, T.K. Fujiwara, R. Chadda, M. Xie, T.A. Tsunoyama, Z. Kalay, R.S. Kasai, K.G. Suzuki. Dynamic organizing principals of the plasma membrane that regulate signal transduction: commemorating the fortieth anniversary of Singer and Nicolson's fluid-mosaic model. Annu. Rev. Cell Dev. Biol., 28 (2012), pp. 215–250.
M.P. Sheetz. Membrane skeletal dynamics: role in modulation of red cell deformability, mobility of transmembrane proteins, and shape Semin. Hematol., 20 (1983), pp. 175–188.
K. Radhkrishnan, A. Halasz, D. Vlachos, J.S. Edwards. Quantitative understanding of cell signaling: the importance of membrane organization. Curr. Opin. Biotechnol., 21 (2010), pp. 677–682.
F. Zhang, G.M. Lee, K. Jacobson. Protein lateral mobility as a reflection of membrane microstructure. Bioessays, 15 (1993), pp. 579–588.
A. Kusumi, K.G. Suzuki, R.S. Kasai, K. Ritchie, T.K. Fujiwara. Hierarchical mesoscale domain organization of the plasma membrane. Trends Biochem. Sci., 36 (2011), pp. 604–615.

P. Liu, X. Wang, M.S. Itano, A.K. Neumann, K. Jacobson, N.L. Thompson. The formation and stability of DC-SIGN microdomains require its extracellular moiety. Traffic, 13 (2012), pp. 715–726.

Grecco HE, Schmick M, Bastiaens PI. 2011. Signaling from the living plasma membrane. Cell 144:897–909.

Cronan JE. 2003. Bacterial membrane lipids: where do we stand? Annu Rev Microbiol 57:203–224. https://doi.org/10.1146/annurev.micro.57.030502.090851.

van Meer G, Voelker DR, Feigenson GW. 2008. Membrane lipids: where they are and how they behave. Nat Rev Mol Cell Biol 9:112–124. https://doi.org/10.1038/nrm2330. CrossRefMedline.

Coskun U, Simons K. 2011. Cell membranes: the lipid perspective. Structure 19:1543–1548. https://doi.org/10.1016/j.str.2011.10.010.

Lingwood D, Simons K. 2010. Lipid rafts as a membrane-organizing principle. Science 327:46–50. https://doi.org/10.1126/science.1174621. Abstract/FREE Full Text.

Neumann AK, Itano MS, Jacobson K. 2010b. Understanding lipid rafts and other related membrane domains. F1000 Biol Rep 2:31. https://doi.org/10.3410/b2-31.

Simons K, Sampaio JL. 2011a. Membrane organization and lipid rafts. Cold Spring Harb Perspect Biol 3:a004697. https://doi.org/10.1101/cshperspect.a004697.

Kraft ML. 2013. Plasma membrane organization and function: moving past lipid rafts. Mol Biol Cell 24:2765–2768.

Schuck S, Simons K. 2004a. Polarized sorting in epithelial cells: Raft clustering and the biogenesis of the apical membrane. J Cell Sci 117: 5955–5964.

Owen DM, Rentero C, Rossy J, Magenau A, Williamson D, Rodriguez M, Gaus K. 2010. PALM imaging and cluster analysis of protein heterogeneity at the cell surface. J Biophotonics 3:446–454. https://doi.org/10.1002/jbio.200900089. CrossRefMedline.

Spira F, Mueller NS, Beck G, von Olshausen P, Beig J, Wedlich-Soldner R. 2012. Patchwork organization of the yeast plasma membrane into numerous coexisting domains. Nat Cell Biol 14:640–648. https://doi.org/10.1038/ncb2487.

Simons K, Van Meer G. 1988a. Lipid sorting in epithelial cells. Biochemistry 27: 6197–6202.

Simons K, Ikonen E. 1997a. Functional rafts in cell membranes. Nature 387:569–572.

Klemm RW, Ejsing CS, Surma MA, Kaiser HJ, Gerl MJ, Sampaio JL, de Robillard Q, Ferguson C, Proszynski TJ, Shevchenko A, et al. 2009. Segregation of sphingolipids and sterols during formation of secretory vesicles at the trans-Golgi network. J Cell Biol 185: 601–612.

Bagnat M, Chang A, Simons K. 2001. Plasma membrane proton ATPase Pma1p requires raft association for surface delivery in yeast. Mol Biol Cell 12: 4129–4138.

Meder D, Moreno MJ, Verkade P, Vaz WL, Simons K. 2006. Phase coexistence and connectivity in the apical membrane of polarized epithelial cells. Proc Natl Acad Sci 103: 329–334.

Danielsen E, Hansen G. 2008. Lipid raft organization and function in the small intestinal brush border. J Physiol Biochem 64: 377–382.

Morrow IC, Parton RG. 2005. Flotillins and the PHB domain protein family: rafts, worms and anaesthetics. Traffic 6:725–740.

Babuke T, Tikkanen R. 2007. Dissecting the molecular function of reggie/flotillin proteins. Eur J Cell Biol 86:525–532.

Otto GP, Nichols BJ. 2011. The roles of flotillin microdomains—endocytosis and beyond. J Cell Sci 124:3933–3940.

Stuermer CA. 2011. Reggie/flotillin and the targeted delivery of cargo. J Neurochem 116:708–713.

Zhao F, Zhang J, Liu YS, Li L, He YL. 2011. Research advances on flotillins. Virol J 8:479. https://doi.org/10.1186/1743-422x-8-479.

Michel V, Bakovic M. 2007. Lipid rafts in health and disease. Biol Cell 99:129–140. https://doi.org/10.1042/bc20060051.

W.T. Schroeder, S. Stewart-Galetka, S. Mandavilli, D.A. Parry, L. Goldsmith, M. Duvic. Cloning and characterization of a novel epidermal cell surface antigen (ESA). J. Biol. Chem., 269 (1994), pp. 19983–19991.

T. Schulte, K.A. Paschke, U. Laessing, F. Lottspeich, C.A. Stuermer. Reggie-1 and reggie-2, two cell surface proteins expressed by retinal ganglion cells during axon regeneration. Development, 124 (1997), pp. 577–587.

M. Bauer, L. Pelkmans. A new paradigm for membrane-organizing and -shaping scaffolds. FEBS Lett., 580 (2006), pp. 5559–5564.

C.A. Baumann, V. Ribon, M. Kanzaki, D.C. Thurmond, S. Mora, S. Shigematsu, P.E. Bickel, J.E. Pessin, A.R. Saltiel. CAP defines a second signalling pathway required for insulin-stimulated glucose transport. Nature, 407 (2000), pp. 202–207.

Tanja Babuke, Ritva Tikkanen. 2007. Dissecting the molecular function of reggie/flotillin proteins. European Journal of Cell Biology 86, Pages 525–532.

G. van Meer. Cellular lipidomics. EMBO J., 24 (2005), pp. 3159–3165.

W.F. Drew Bennett, D. Peter Tieleman. Computer simulations of lipid membrane domains. Biochimica et Biophysica Acta (BBA)—Biomembranes. Volume 1828, Issue 8, August 2013, Pages 1765–1776.

S.L. Veatch, S.L. Keller. Seeing spots: complex phase behavior in simple membranes. Biochim. Biophys. Acta Biomembr., 1746 (2005), pp. 172–185.

G.W. Feigenson. Phase diagrams and lipid domains in multicomponent lipid bilayer mixtures Biochim. Biophys. Acta Biomembr., 1788 (2009), pp. 47–52.

T. Baumgart, A.T. Hammond, P. Sengupta, S.T. Hess, D.A. Holowka, B.A. Baird, W.W. Webb. Large-scale fluid/fluid phase separation of proteins and lipids in giant plasma membrane vesicles. Proc. Natl. Acad. Sci. U. S. A., 104 (2007), pp. 3165–3170.

S.L. Veatch, P. Cicuta, P. Sengupta, A. Honerkamp-Smith, D. Holowka, B. Baird. Critical fluctuations in plasma membrane vesicles. ACS Chem. Biol., 3 (2008), pp. 287–293.

C. Eggeling, C. Ringemann, R. Medda, G. Schwarzmann, K. Sandhoff, S. Polyakova, V.N. Belov, B. Hein, C. von Middendorff, A. Schonle, S.W. Hell. Direct observation of the nanoscale dynamics of membrane lipids in a living cell. Nature, 457 (2009), pp. 1159–1162.

E. Sezgin, I. Levental, M. Grzybek, G.n. Schwarzmann, V. Mueller, A. Honigmann, V.N. Belov, C. Eggeling, ú. Coskun, K. Simons, P. Schwille. Partitioning, diffusion, and ligand binding of raft lipid analogs in model and cellular plasma membranes. Biochim. Biophys. Acta Biomembr., 1818 (2012), pp. 1777–1784.

H.J. Risselada, S.J. Marrink. The molecular face of lipid rafts in model membranes. Proc. Natl. Acad. Sci. U. S. A., 105 (2008), pp. 17367–17372.

S. Baoukina, E. Mendez-Villuendas, W.F.D. Bennett, D.P. Tieleman. Computer simulations of the phase separation in model membranes. Faraday Discuss., 161 (2013), pp. 63–75.

J.D. Perlmutter, J.N. Sachs. Interleaflet interaction and asymmetry in phase separated lipid bilayers: molecular dynamics simulations. J. Am. Chem. Soc., 133 (2011), pp. 6563–6577.

J. Domanski, S.J. Marrink, L.V. Schafer. Transmembrane helices can induce domain formation in crowded model membranes. Biochim. Biophys. Acta Biomembr., 1818 (2012), pp. 984–994.

W. Shinoda, R. DeVane, M.L. Klein. Multi-property fitting and parameterization of a coarse grained model for aqueous surfactants. Mol. Simul., 33 (2007), pp. 27–36.

D.A. Pantano, P.B. Moore, M.L. Klein, D.E. Discher. Raft registration across bilayers in a molecularly detailed model. Soft Matter, 7 (2011), pp. 8182–8191.

S. Baoukina, L. Monticelli, H.J. Risselada, S.J. Marrink, D.P. Tieleman. The molecular mechanism of lipid monolayer collapse. Proc. Natl. Acad. Sci. U. S. A., 105 (2008), pp. 10803–10808.

S. Baoukina, D.P. Tieleman. Lung surfactant protein SP-B promotes formation of bilayer reservoirs from monolayer and lipid transfer between the interface and subphase Biophys. J., 100 (2011), pp. 1678–1687.

S. Baoukina, E. Mendez-Villuendas, D.P. Tieleman. Molecular view of phase coexistence in lipid monolayers. J. Am. Chem. Soc., 134 (2012), pp. 17543–17553.

Charles A. Day, Anne K. Kenworthy. Tracking microdomain dynamics in cell membranes. Biochimica et Biophysica Acta (BBA)—Biomembranes. Volume 1788, Issue 1, January 2009, Pages 245–253.

Kai Simons and Julio L. Sampaio. Membrane Organization and Lipid Rafts. Cold Spring Harb Perspect Biol 2011b; 3:a004697.

L.D. Frye, M. Edidin. The rapid intermixing of cell surface antigens after formation of mouse-human heterokaryons. J. Cell. Sci., 7 (1970), pp. 319–335.

Y. Chen, B.C. Lagerholm, B. Yang, K. Jacobson. Methods to measure the lateral diffusion of membrane lipids and proteins. Methods, 39 (2006), pp. 147–153.

A.K. Kenworthy. Fluorescence recovery after photobleaching studies of lipid rafts. T. McIntosh (Ed.), Lipid Rafts, Humana Press, Towata, NJ (2007).

D. Marguet, P.F. Lenne, H. Rigneault, H.T. He. Dynamics in the plasma membrane: how to combine fluidity and order. EMBO J., 25 (2006), pp. 3446–3457.

K. Bacia, P. Schwille. Fluorescence correlation spectroscopy. Methods Mol. Biol., 398 (2007), pp. 73–84.

A. Kusumi, H. Ike, C. Nakada, K. Murase, T. Fujiwara. Single-molecule tracking of membrane molecules: plasma membrane compartmentalization and dynamic assembly of raft-philic signaling molecules. Semin. Immunol., 17 (2005), pp. 3–21.

P.H. Lommerse, H.P. Spaink, T. Schmidt. In vivo plasma membrane organization: results of biophysical approaches. Biochim. Biophys. Acta, 1664 (2004), pp. 119–131.

B.C. Lagerholm, G.E. Weinreb, K. Jacobson, N.L. Thompson. Detecting microdomains in intact cell membranes. Annu. Rev. Phys. Chem., 56 (2005), pp. 309–336.

K. Jacobson, O.G. Mouritsen, R.G. Anderson. Lipid rafts: at a crossroad between cell biology and physics. Nat. Cell Biol., 9 (2007), pp. 7–14.

A.K. Kenworthy. Fleeting glimpses of lipid rafts: how biophysics is being used to track them J. Investig. Med., 53 (2005), pp. 312–317.

M. Edidin. Patches, posts and fences: proteins and plasma membrane domains. Trends. Cell. Biol., 2 (1992), pp. 376–380.

A. Kusumi, Y. Sako. Cell surface organization by the membrane skeleton. Curr. Opin. Cell Biol., 8 (1996), pp. 566–574.

M.P. Sheetz. Cellular plasma membrane domains. Mol. Membr. Biol., 12 (1995), pp. 89–91.

M.P. Sheetz, M. Schindler, D.E. Koppel. Lateral mobility of integral membrane proteins is increased in spherocytic erythrocytes. Nature, 285 (1980), pp. 510–511.

E.S. Wu, D.W. Tank, W.W. Webb. Unconstrained lateral diffusion of concanavalin A receptors on bulbous lymphocytes. Proc. Natl. Acad. Sci. U. S. A., 79 (1982), pp. 4962–4966.

E. Yechiel, M. Edidin. Micrometer-scale domains in fibroblast plasma membranes. J. Cell Biol., 105 (1987), pp. 755–760.

M. Edidin, I. Stroynowski. Differences between the lateral organization of conventional and inositol phospholipid-anchored membrane proteins. A further definition of micrometer scale membrane domains. J. Cell Biol., 112 (1991), pp. 1143–1150.

M. Edidin. Lipid microdomains in cell surface membranes. Curr. Opin. Struct. Biol., 7 (1997), pp. 528–532.

M. Edidin, M.C. Zuniga, M.P. Sheetz. Truncation mutants define and locate cytoplasmic barriers to lateral mobility of membrane glycoproteins. Proc. Natl. Acad. Sci. U. S. A., 91 (1994), pp. 3378–3382.

R. Simson, E.D. Sheets, K. Jacobson. Detection of temporary lateral confinement of membrane proteins using single-particle tracking analysis. Biophys. J., 69 (1995), pp. 989–993.

E. Fire, D.E. Zwart, M.G. Roth, Y.I. Henis. Evidence from lateral mobility studies for dynamic interactions of a mutant influenza hemagglutinin with coated pits. J. Cell Biol., 115 (1991), pp. 1585–1594.

Y.M. Umemura, M. Vrljic, S.Y. Nishimura, T.K. Fujiwara, K.G. Suzuki, A. Kusumi. Both MHC class II and its GPI-anchored form undergo hop diffusion as observed by single-molecule tracking. Biophys. J., 95 (2008), pp. 435–450.

Malinska, K., Malinsky, J., Opekarova, M. & Tanner, W. Visualization of protein compartmentation within the plasma membrane of living yeast cells. Mol. Biol. Cell 14, 4427–4436 (2003).

Schutz, G. J., Kada, G., Pastushenko, V. P. & Schindler, H. Properties of lipid microdomains in a muscle cell membrane visualized by single molecule microscopy. EMBO J. 19, 892–901 (2000).

Gaus, K. et al. Visualizing lipid structure and raft domains in living cells with two-photon microscopy. Proc. Natl Acad. Sci. USA 100, 15554–15559 (2003).

Vereb, G. et al. Dynamic, yet structured: The cell membrane the decades after the Singer-Nicolson model. Proc. Natl Acad. Sci. USA 100, 8053–8058 (2003).

Roper, K., Corbeil, D. & Huttner, W. B. Retention of prominin in microvilli reveals distinct cholesterol-based lipid micro-domains in the apical plasma membrane. Nature Cell Biol. 2, 582–592 (2000).

Nydegger, S., Khurana, S., Krementsov, D. N., Foti, M. & Thali, M. Mapping of tetraspanin-enriched microdomains that can function as gateways for HIV-1. J. Cell Biol. 173, 795–807 (2006).

Rothberg, K. G. et al. Caveolin, a protein component of caveolae membrane coats. Cell 68, 673–682 (1992).

Heijnen, H. F. G. et al. Concentration of rafts in platelet filopodia correlates with recruitment of c-Src and CD63 to these domains. J. Thromb. Haemos. 1, 1161–1173 (2003).

Ashrafuzzaman M, Tseng CY, Kapty J, Mercer JR, Tuszynski JA. A computationally designed DNA aptamer template with specific binding to phosphatidylserine. Nucleic Acid Ther. 2013 Dec;23(6):418–26. https://doi.org/10.1089/nat.2013.0415. Epub 2013 Nov 26.

Schuck S, Simons K. 2004b. Polarized sorting in epithelial cells: raft clustering and the biogenesis of the apical membrane. J Cell Sci 117:5955–5964. https://doi.org/10.1242/jcs.01596.

Simons K, van Meer G. 1988. Lipid sorting in epithelial cells. Biochemistry 27:6197–6202. https://doi.org/10.1021/bi00417a001.

van Meer G, Simons K. 1988. Lipid polarity and sorting in epithelial cells. J Cell Biochem 36:51–58. https://doi.org/10.1002/jcb.240360106.

Simons K, Toomre D. 2000. Lipid rafts and signal transduction. Nat Rev Mol Cell Biol 1:31–39. https://doi.org/10.1038/35036052.

Tsui-Pierchala BA, Encinas M, Milbrandt J, Johnson EM, Jr. 2002. Lipid rafts in neuronal signaling and function. Trends Neurosci 25:412–417. https://doi.org/10.1016/s0166-2236(02)02215-4.

Barak I, Muchova K. 2013. The role of lipid domains in bacterial cell processes. Int J Mol Sci 14:4050–4065. https://doi.org/10.3390/ijms14024050.

Barak I, Muchova K, Wilkinson AJ, O'Toole PJ, Pavlendova N. 2008. Lipid spirals in Bacillus subtilis and their role in cell division. Mol Microbiol 68:1315–1327. https://doi.org/10.1111/j.1365-2958.2008.06236.x.

Matsumoto K, Kusaka J, Nishibori A, Hara H. 2006. Lipid domains in bacterial membranes. Mol Microbiol 61:1110–1117. https://doi.org/10.1111/j.1365-2958.2006.05317.x.

Muchova K, Jamroskovic J, Barak I. 2010. Lipid domains in Bacillus subtilis anucleate cells. Res Microbiol 161:783–790. https://doi.org/10.1016/j.resmic.20 https://doi.org/10.07.006.

Kawai F, Shoda M, Harashima R, Sadaie Y, Hara H, Matsumoto K. 2004. Cardiolipin domains in Bacillus subtilis Marburg membranes. J Bacteriol 186:1475–1483. https://doi.org/10.1128/jb.186.5.1475-1483.

Mileykovskaya E, Dowhan W. 2000. Visualization of phospholipid domains in Escherichia coli by using the cardiolipin-specific fluorescent dye 10-N-nonyl acridine orange. J Bacteriol 182:1172–1175. https://doi.org/10.1128/jb.182.4.1172-1175.2000.

Mileykovskaya E, Dowhan W. 2009. Cardiolipin membrane domains in prokaryotes and eukaryotes. Biochim Biophys Acta 1788:2084–2091. https://doi.org/10.1016/j.bbamem.2009.04.003.

Rosch JW, Hsu FF, Caparon MG. 2007. Anionic lipids enriched at the ExPortal of Streptococcus pyogenes. J Bacteriol 189:801–806. https://doi.org/10.1128/jb.01549-06.

Oliver PM, Crooks JA, Leidl M, Yoon EJ, Saghatelian A, Weibel DB. 2014. Localization of anionic phospholipids in Escherichia coli cells. J Bacteriol 196:3386–3398.

Sourjik V. 2004. Receptor clustering and signal processing in E. coli chemotaxis. Trends Microbiol 12:569–576. https://doi.org/10.1016/j.tim.2004.10.003.

Bray D, Levin MD, Morton-Firth CJ. 1998. Receptor clustering as a cellular mechanism to control sensitivity. Nature 393:85–88. https://doi.org/10.1038/30018.

Simons K, Ikonen E. 1997b. Functional rafts in cell membranes. Nature 387:569–572. https://doi.org/10.1038/42408.

Tavernarakis N, Driscoll M, Kyrpides NC. 1999. The SPFH domain: implicated in regulating targeted protein turnover in stomatins and other membrane-associated proteins. Trends Biochem Sci 24:425–427. https://doi.org/10.1016/s0968-0004(99)01467-x.

Good MC, Zalatan JG, Lim WA. 2011. Scaffold proteins: hubs for controlling the flow of cellular information. Science 332:680–686. https://doi.org/10.1126/science.1198701.

Langhorst MF, Reuter A, Stuermer CA. 2005. Scaffolding microdomains and beyond: the function of reggie/flotillin proteins. Cell Mol Life Sci 62:2228–2240.

Marc Bramkampa and Daniel Lopez. 2015. Exploring the Existence of Lipid Rafts in Bacteria. Microbiol. Mol. Biol. Rev., vol. 79: 81–100.

# Chapter 3
# Cell Surface Diffusion and Adsorption

Diffusion of adsorbates on cell surface is an important natural process. Once materials reach at the surface of the cell, they fall on a dynamic two-dimensional (2D) plane. The dynamical nature of the constituents of the plane has been explained in detail from a mainly thermodynamic point of view in Chap. 2. The underlying mechanisms of the cell surface diffusion appear with various types of explanations. Biochemical and biological analyses of surface assay experiments provide observation-based explanations on the surface diffusion mechanisms. General physics rules on diffusion and adsorption have been tried to explain the cell surface diffusion and adsorption mechanisms theoretically. Most of the explanations appear with no conclusive and universal models that may explain the general surface diffusion of adsorbates and various other natural materials on and across the quite dynamic plane of the surface of the biological cell. In this chapter, we shall try to address this crucial issue considering some of our self-developed strategies as well as available information from various important studies.

## 3.1 Introduction

Both experimental observations and theoretical models may help understand biological cell surface diffusion. A latest study demonstrates a self-consistent analysis of diffusing colloidal probe interactions due to conservative and nonconservative forces to characterize important biophysical cell surface properties that are involved with surface diffusion (Duncan et al. 2016). Duncan et al. have determined here the 2D trajectories of micron-sized polyethylene glycol (PEG)-coated silica colloids relative to adherent epithelial breast cancer cell perimeters to allow measurement of the position-dependent diffusivities and interaction potentials. Various types of nanoparticles (NPs) without or with drugs attached act on cell surface to diffuse across and/or interact with surface constructs, e.g., membrane-based lipids, proteins, while used in biological applications especially for medical purposes (see,



M. Ashrafuzzaman, *Nanoscale Biophysics of the Cell*,
https://doi.org/10.1007/978-3-319-77465-7_3

e.g., Nel et al. 2009; Wang and Wang 2014). A study shows that cellular uptake of gold NPs depends on the sedimentation and diffusion velocities of the NPs and is independent of size, shape, density, surface coating, and initial NP concentration (Cho et al. 2011). Another recent study emphasizes how relevant and important it is to evaluate and monitor the size and agglomeration state of NPs in the biological medium, in order to interpret correctly the results of the in vitro toxicological assays as regards to the measurements of the cellular uptake of the NPs (Halamoda-Kenzaoui et al. 2015). All these mentioned investigations (including many other available) altogether suggest that we know some of the quantitative features of the surface diffusion and cell surface absorption of surface adsorbates. But we still need a clearer understanding of the underlying mechanisms. More than two decades ago, a theoretical analysis was made to describe the surface diffusion of membrane particles from a spherical shell onto a thin cylindrical process (Berk et al. 1992). Here, provision was made for different rates of diffusion within the two morphologically distinct regions. A recent review article focuses on the physiochemical mechanisms underlying NP uptake into cells (Zhang et al. 2015). When NPs approach to the close vicinity of a cell, the interactions between the NPs and the cell membrane generate forces from different origins. This leads to the membrane surface diffusion and wrapping of the NPs, followed by statistical cellular uptake. This article discusses how the kinetics, energetics, and forces are related to NPs' cell surface interactions and dependent on the size, shape, and stiffness of NPs, the biomechanical properties of the cell membrane, as well as the local environment of the cells (Zhang et al. 2015). In this chapter, we aim at understanding the general features and mechanisms behind materials' (e.g., NPs, viruses, drugs, peptides) cell surface diffusion that may lead to statistical cell surface quantitative adsorption, rejection, and/or cellular uptake.

## 3.2 Diffusion of Cell Surface Adsorbates to Target Random Locations on Surface Lattice Structures

Understanding the general interaction phenomena between surface adsorbates with cell surface, especially the cell membrane, is still poorly addressed. Molecular details of the interaction mechanisms and pathways are even less known. Using conventional experimental observations, we obtain ideas and quantitative measurements but those are phenomenological address requiring in-depth analysis using numerical methods. There are general diffusion formulas applied on cell surface but due to the complex dynamical character of the liquid–fluid interface of the cell surface the correct formulas are yet to be discovered. In this section, we shall attempt to address the cell surface diffusion and possible surface adsorption using a novel approach utilizing a set of general theoretical formulas and strategic computer simulation studies on the NP-cell surface interactions in order to measure the probability of the cell membrane absorption of NPs. My laboratory member Noor Halder (who recently left for the University of Western Australia for starting his Ph.D. studies) and I have been working on this project (still unfinished but enough conclusions have been found). These studies intensify our understanding of the NP–membrane interaction mechanisms and provide a design guideline for other NPs. Here we attempt to explain the mechanisms of correlation between dynamic cell surface adsorbates and dynamic cell surface lattice structures. Here, dynamic cell surface lattice structure means we consider the mobile lattice structure constructed by the continuously diffusive cell surface constituents lipids, membrane proteins (MPs), etc.

Beside general theoretical treatment of the problem, a computational study has been adopted to identify the key characteristics of diffusive NPs through addressing their binding to the cell surface considering different surface properties. We have developed a strategy to calculate trapping probability of NPs inside a cell surface-based defect as a result of NPs' random diffusion on cell surface. We have successfully developed a universal algorithm to help visualize the NP movement and interaction on cell surface. Our algorithm now contains some basic features but eventually we wish to extend the algorithm to cover most of the aspects of NP movement, interactions with the cell surface targets, and delivery beyond the cell surface boundary what is known as the cell membrane.

### 3.2.1 General Background

With the rapid development of nanotechnology, NPs now have received intense interest due to their potential applications in medical diagnostics and therapeutics (Allen and Cullis 2004; Connor et al. 2005; Gao et al. 2005; Ghosh et al. 2008; Meyers et al. 2006; Shubayev et al. 2009; Geng et al. 2007; Paciotti et al. 2004). Though this technology provides a new route to use NP as a transport tool for carrying drugs, it poses the potential undesirable effects, including the negative impacts on human health, namely nanotoxicity (Ai et al. 2011; Curtis et al. 2006).

A cell membrane hosts dynamic lipids and MPs safeguarding the cell's interior substances and manages different bioprocesses, such as communication among cells (Yue and Zhang 2012). In the field of biomedicine and biochemistry, it is crucial to understand the interaction between cell membrane and NPs targeting the cells. Different molecular mechanisms of drug effects on membranes have been identified. Some theoretical analysis of NP diffusion across cell surface and other structural cell boundaries has been presented recently in two of my publications (Ashrafuzzaman 2015a, b). These papers produce some considerable suggestions on quantitative distribution with statistical nature of active cellular particles by measuring their mobility in the cell environment.

In recent years, numerous experimental studies have been performed to address the interactions between NPs and cell membrane experimentally (e.g., see Meyers et al. 2006; Lapotko et al. 2007), theoretically (Ginzburg and Balijepalli 2007) and using simulations (Yang and Ma 2010). Several mechanisms for the NP–membrane interaction have been proposed based on experimental aspects. A number of studies based on effects of NP size for interaction with the cell membrane are discussed. In Lu et al. (2009), it is shown that the cell interaction highly depends upon NP size. There exist other experimental studies on the interactions between cell membrane and NPs (Connor et al. 2005; Davda and Labhasetwar 2002; Qhobosheane et al. 2001; Nativo et al. 2008; Pan et al. 2007).

Although various experimental searches available, there is a lack of supporting computational results. Li et al. investigated the interactions of graphene and few-layer graphene (FLG) microsheets with three cell types and with model lipid bilayers by combining coarse-grained molecular dynamic (MD), all-atom MD, analytical modeling, confocal fluorescence imaging, and electron microscopic imaging (Li et al. 2013). The simulations of Li et al. (2013) reveal direct bilayer penetration that begins with localized piercing at sharp corners or at protrusions along graphene edges followed by propagation along the edge to achieve full penetration. For a small graphene flake, however, the Brownian motion and entropic driving forces in the near-membrane region first position the flake orthogonal to the bilayer plane, which then leads to spontaneous corner piercing. All-atom-steered MD simulations track the free energy evolution during corner piercing and reveal only a small energy barrier, comparable to the Boltzmann's energy $k_BT$, where $k_B$ is the Boltzmann constant and $T$ is the absolute temperature. The bottom-line information we may draw from these investigations is that the graphene experiences general principles of surface interactions, surface motion, and statistical mechanical manner while penetrating into the cell's interior through membrane.

The structures of NP complexes on the cell surface as a result of surface diffusion and interactions can be understood quite in detail by computer simulation and modelling approaches. It is very hard and time consuming for getting a conclusion due to variation of NP characteristics. Even simple questions such as the effects of size variation of the NP complexes or surface charge changes can be difficult to answer experimentally. Some precise control over building such process is necessary in order to answer such minute queries (Chapel and Ersatile 2012). Computer modelling of this system may help considering the influence of different parameters such as charge of NPs, building block properties, and different other physiological conditions. The computer modelling may also be able to provide more detailed information about the structures of NPs and the cell surface. In this study, the use of computer simulation to explore the multiparameter space for complex nanostructures is described where experiments cannot give classical answers.

The subsequent section contains brief discussions about some of the computational methods that are used and have been developed to model the structure of individual NPs and their interactions with the surroundings by describing both theoretical basis and computational approaches. As a prototypical system, we consider the problem of the interactions of the cell surface with different-sized NPs.

Before going to the computational analysis, we need to know some of the aspects of the targeted cell surface adsorbate NPs and cell surface diffusion of what we wish to analyze using computational technique. Section 3.2.2 will focus on this issue.

### 3.2.2 Nanoparticle-Based Drug Delivery Systems

NP-based systems can be used to deliver therapeutic entities such as small molecule drugs, peptides, proteins, and nucleic acids, either alone or in combinations (Gindy and Prud'homme 2009; Ruoslahti et al. 2010). Their properties could be attributed to their sizes, NPs surface properties, cell death, drug pharmacokinetics, etc. NPs can enter into the cell surface more efficiently than larger micromolecules and therefore could be used as effective transport and delivery systems. For therapeutic applications, drugs can either be integrated in the matrix of the particle or attached to the particle surface (Suri et al. 2007). A drug-targeting system should be able to control the fate of a drug entering the biological environment. Nanosystems with different compositions and biological properties have been extensively investigated for drug and gene delivery applications. It is also important to understand the roadblocks of drugs such as stability of therapeutic agents in the living cell environment. There are various types of NPs which can be used as drug carriers in medicines. These NPs can be polymeric NPs, metal NPs such as gold, silver, magnetic NPs, lipid NPs, dendrimers, silica NPs, liposomes, etc.

To achieve efficient drug delivery, it is important to understand the interactions of NPs with the biological environment of the cell, targeting cell surface receptors, drug release, multiple drug administration, and molecular mechanisms of the cell that concern interactions with receiving drugs, etc.

**Important Characteristics for Drug Delivery using NPs**

(i) *Particle Size*. Particle size and size distribution are the most important characteristics of NPs. Size has important role to determine the ability (from geometric perspective) of drug delivery system. In addition, size may influence drug loading, drug release, and stability of NPs (Sing and Lillard 2009).

(ii) *Surface Properties of NPs or Surface Charge*. The surface properties of NPs play important roles in effective drug delivery. The zeta potential (means electrokinetic potential) of an NP is commonly used to characterize the surface charge property of NPs (Couvreur et al. 2002). It reflects the electrical potential of particles and is influenced by both the composition of the particle and the medium in which it is dispersed. NPs with a zeta potential above $\pm 30$ mV have been shown to be stable in suspension, as the surface charge prevents aggregation of the particles (Sing and Lillard 2009). The zeta potential also can be used to determine whether a charged active material is encapsulated within the center of the NP or on the surface.

(iii) *Drug Loading*. Drug loading capacity of a nanodelivery system should be high. In general, there are two methods which are associated with drug loading, i.e., incorporation method and absorption method. The incorporation method requires the drug to be incorporated at the time of NP formulation. The absorption methods call for absorption of the drug after NP formation (Sing and Lillard 2009).

(iv) *Drug Release*. Drug release can occur by diffusion or erosion. If the diffusion of the drug is faster than erosion, then the mechanism of release of drug is largely controlled by a diffusion process. In general, the drug release rate depends on five criteria as follows:

(1) Drug solubility;
(2) Drug diffusion through the NP matrix;
(3) Desorption of the surface bound or adsorbed drug;
(4) NP matrix erosion or degradation;
(5) The combination of erosion and diffusion processes.

(v) *Targeted Drug Delivery*. Targeted drug delivery is of two types. The actively achieved targeted delivery requires the therapeutic agent to be achieved by conjugating the therapeutic agent or carrier system to a tissue or cell-specific ligand. Passive targeting is achieved by incorporating the therapeutic agent into a macromolecule or NP that passively reaches the target organ (Sing and Lillard 2009).

### *3.2.3 Computer Simulation Addressing NP Diffusion on Cell Surface*

As discussed earlier, we are working on developing a universal in silico simulation-based understanding of the cell surface adsorbates' general diffusion on the surface of the cell and their possible cell interactions. In this section, we shall provide a brief explanation on how we perform the simulations using various computational approaches and develop necessary algorithms.

#### 3.2.3.1 Computer Simulation Techniques

Here, different computational techniques will be described. Molecular approaches are commonly used in studies relating to the formation and evolution of individual NPs. Classical computer simulation based on different classical molecular approaches includes Monte Carlo (MC), MD, and Brownian dynamics (BD) methods. Some of those methods are used in equilibrium and other used in nonequilibrium states. The easiest method to implement is perhaps the MC method.

(a) **Brownian Dynamics**. In cases where various species, such as dispersions of NPs, polymer chains, polyelectrolytes and individual molecules such as salt particles, the timescale characterizing the motion of each species can differ by several orders of magnitude. This type of behavior is due to different characteristics of such elements in one solution. To simulate the dynamics of such systems when the fast motion of the small species is not relevant approaches, in such case, the BD method can be used (Satoh 2011). In BD, the fast motion of the smaller particles is not considered and the impact on the large objects is represented by a combination of random and frictional forces. Here, Newton's law of motion is replaced by Langevin equations of motion. It can be used to describe the motion of molecules in reality. It is a simplified version of Langevin dynamics and corresponds to the limit where no average acceleration takes place.

The original Langevin equation describes Brownian motion, and the apparently random movement of a particle in a fluid due to collisions with the molecules of the fluid is as follows:

$$m\frac{\mathrm{d}^2\mathbf{x}}{\mathrm{d}t^2} = -\lambda\frac{\mathrm{d}\mathbf{x}}{\mathrm{d}t} + \boldsymbol{\eta}(t). \tag{3.1}$$

The degree of freedom of interest here is the position x of the particle, and m denotes the particle's mass. The force acting on the particle is written as a sum of a viscous force proportional to the particle's velocity (Stokes' law) and a noise term $\eta(t)$ (the name given in physical contexts to terms in stochastic differential equations which are stochastic processes) representing the effect of the collisions with the molecules of the fluid. The force $\eta(t)$ has a Gaussian probability distribution with correlation function as follows:

$$\langle \eta_i(t)\eta_j(t')\rangle = 2\lambda k_\mathrm{B} T \delta_{i,j}\delta(t - t'), \tag{3.2}$$

where as said earlier, $k_\mathrm{B}$ is Boltzmann constant and $T$ is the temperature.

In general, the BD simulation technique can be described as a microscopic method in which explicit solvent molecules are replaced instead by a stochastic force. The technique takes advantage of the fact that there is a large separation in timescales between the rapid motions of solvent molecules and the more sluggish motion of polymers or colloids.

(b) **Molecular Dynamics**. MD is a computer simulation method for studying the physical movements of atoms and molecules in the context of simulation studies. The method was originally conceived within theoretical physics in the late 1950s (Alder and Wainwright 1959) but is applied today mostly in chemical physics, materials science, and the modeling of biomolecules. MD method is quite complex in nature as a molecular system consists of vast number of particles of various properties, but it can simulate such system which has both equilibrium and nonequilibrium states. This is a well-established method which is based on Newton's laws of motion (Nedyalkova et al. 2012). An MD simulation can experience conformational and momentum changes such that different parts of the phase space accessible to the molecules can be explored. MD can calculate various structural and dynamic properties by providing several mechanisms for controlling temperature and pressure of a system. MD can realistically describe the motion of small molecules across bilayer cell membrane channels. It can also describe different lipid–protein interaction in a bilayer lipid membrane. The most time-consuming part of MD simulation is the calculation of the force on each atom. The main difficulty with such an approach is that it cannot be used to model the behavior of realistic processes containing NPs (Stoll 2014).

(c) **Monte Carlo**. Monte Carlo (MC) simulation is a broad class of computational algorithms that rely on repeated random sampling to obtain numerical results. It is used in many areas of mathematics, physics, chemistry, etc. It contains one random variable and uses a sort of random path via a statistical sampling of multiple replicates. These types of simulation methods are mainly used in three distinct problem classes, namely optimization, numerical integration, and generating draws from a probability distribution. MC is based on Markov chain properties, where successive trials depend on the immediate predecessor. This approach is successfully used to describe building processes between polymers and the complexion process between NPs and polyelectrolyte chains.

#### 3.2.3.2 Proposed Conditions for Developing the Computational Algorithms

An NP can move on a cell surface in horizontal, vertical as well as diagonal directions. The movements of such NPs are based on various properties like surface charges, size of NP, and its shape. In order to travel a particular distance from an initial NP location, it is necessary to acquire minimum energy to travel the distance. In the proposed algorithm, we evaluate a probability function which calculates the probability to interact with a defect which is situated at a unit or multiple of a unit distance. Let us consider the distance between NP and defect/channel (CH) is $d$. Here and onward, we shall consider CH for generally representing either defect of channel that can trap NP randomly moving on cell membrane surface. This distance $d$ has a lower cutoff value, e.g., $d_0$, below which the NP gets trapped inside CH. Above $d_0$, the NP may either move freely or fall under physical interactions with CH following conventional long-range interactions like Coulomb, van der Waals interactions (Ashrafuzzaman and Tuszynski 2012a, b; Ashrafuzzaman et al. 2014).

Let us assume '$a$' being the lattice constant of the lipid membrane surface (see Fig. 3.1). We can consider '$a$' to be the lipid head-group dimension where $a^2$ 0.6 nm$^2$ (Mannock et al. 1992, 1994). The probability of interaction of NP and CH (in case when an NP and a CH randomly move on the surface of the cell) with unit trapping distance and its multiples is not same. We may calculate the probability of interactions due to three different trapping distances with the value of unit distance $d_0 = a$, $2a$, and $3a$ (for example) for relative comparison and visualization in graphical representations. While $d_0 = a$, it may be arbitrarily denoted as step I (SI) hopping interaction, $d_0 = 2a$, as step II (SII) hopping interaction, and $d_0 = 3a$, as step III (SIII) hopping interactions, respectively.

There are two distances between NP and CH, qualitatively different, namely trapping distance, $d \leq d_0$ at which NP gets trapped inside CH and beyond this distance; i.e., for $d > d_0$, NP does not interact with CH. In Fig. 3.1, the NP movement is shown in simplest way. An NP can move on cell surface in different

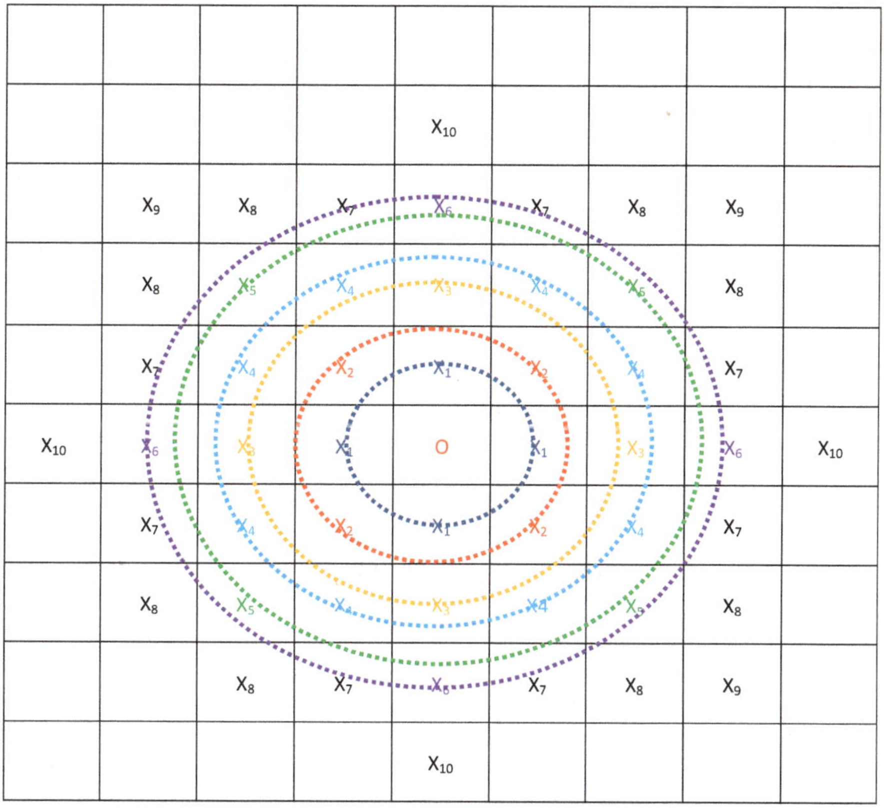


**Fig. 3.1** Relative arrangement of NP and CH. The small circle 'O' in the middle represents the initial location of NP. All other cells represent the possible random positions of CH which is the possible target of the NP to interact. For simplicity, we consider CH to be fixed at a location $X_i$ ($X_i$ is the center of $i$th lattice constant), while O is the initial location of NP. $i = 1, 2, 3, \ldots$, etc., represent all possible locations $X_1, X_2, X_3, \ldots$, etc., for CH at a diagonal distance $x_1, x_2, x_3, \ldots$, etc. Here, we consider that both NP and CH can only be stably located at the center of any cell, O and $X_i$. We have modelled this lattice structure (square lattice is an arbitrary choice) equivalent to consideration of a pattern for the arrangement of lipids on a cell surface (see Chap. 2 for details)

directions. The movement can be in four-way directions, four diagonal in case of square lattice structure as shown in Fig. 3.1. In initial condition, let us assume the location of NP is in the middle of the diagram. It is shown as 'O'. The surrounding middle points of squares are the possible locations of different defects/channel (CH). An NP can move randomly on cell surface and can interact with channel once it travels the minimum distance between past position and nearest channel position. In Fig. 3.1, the minimum distance to interact nearest channel is assumed $d_0$. If NP gets energy to travel distance $d$, such that $d > d_0$, there is a maximum chance to interact with nearest channel. However, NP and CH interaction depends on many other biological and physical parameters too.

**Nearest Neighbor Locations**. As shown in Fig. 3.1, the possible nearest neighbor location of the channel from NP's initial location 'O' is at $X_1$. The number of possibilities of $X_1$ is counted as four due that there are four possibilities for NP to jump to the center of any nearest neighbor cell. This means that if NP, at time $t_0$, is in position 'O', after time $t_1$ it can reach at any of four different positions. The distance travelled between $t_0$ and $t_1$ is $d$, such that $d > d_0$.

**Next-Nearest-Neighbor Locations and Beyond**. The channel position is mentioned as $X_2$. It will be covered if minimum distance $\sqrt{2}d_0$ is travelled by NP. $X_2$ is situated in the nearest diagonal of the starting point of NP.

Similarly, location $X_3$ can be reached in four ways for NP by travelling a minimum distance of $2d_0$. It can be noted that, in case of covering $X_4$ position, there are a totally eight ways. To reach at these positions, NP needs to travel a minimum distance of $\sqrt{5}d_0$. In this way, the distances and the number of ways the NP can travel to CH can be calculated. The figure gives an idea about relative arrangement of NP and CH, while all the squares represent their random locations. All the possible distances are shown in Table 3.1.

Table 3.2 presents the relation of actual distance and number of possible positions for an NP for hopping to. Within exactly a unit distance $d_0$ ($x_1$ in Fig. 3.1), an NP can be available in four possible lattice locations. Similarly, the NP can be available in four possible locations within one unit diagonal distance $x_2 = \sqrt{2}d_0$ which covers the possibility for it to hope to the next neighboring cells. Values in columns A (Table 3.3), and B and C (Table 3.4) represent frequency ($f_{Si}$) for hoping through unit distance, $d_0 = a$ (SI), and $d_0 = 2a$ (SII) and $d_0 = 3a$ (SIII), respectively, where we have considered, for simplicity, the possibility of holding realistically the value of $d_0$ to be only integer multiple of $a$ (see Fig. 3.1). Here, $f_{Si}$ ($i$ = I, II, III, etc.) is $f_{SI}$ for hopping condition SI, $f_{SII}$ for hopping condition SII, and $f_{SIII}$ for hopping condition SIII, respectively. These are valid approximations for the cases in which we consider NP to move only through cells that means it walks at a step equivalent to unit cell dimension or in multiples.

**Table 3.1** Representation of distance abbreviation ($x_i$ for distance of $X_i$ from O) and its actual distance representation in terms of unit distance $d_0$

| Distance abbreviation, $x_i$ | Actual distance |
|---|---|
| $x_1$ | $d_0$ |
| $x_2$ | $\sqrt{2}d_0$ |
| $x_3$ | $2d_0$ |
| $x_4$ | $\sqrt{5}d_0$ |
| $x_5$ | $2\sqrt{2}d_0$ |
| $x_6$ | $3d_0$ |
| $x_7$ | $\sqrt{10}d_0$ |
| $x_8$ | $\sqrt{13}d_0$ |
| $x_9$ | $3\sqrt{2}d_0$ |
| $x_{10}$ | $4d_0$ |
| etc. | etc. |

**Table 3.2** Representation of number of possible position with respect to possible distance

| Actual hopping step distance of NP in unit of $d_0$ | Number of possible positions of CH, $f$ |
|---|---|
| 1.00 | 4 |
| 1.41 | 4 |
| 2.00 | 4 |
| 2.24 | 8 |
| 2.83 | 4 |
| 3.00 | 4 |
| 3.16 | 8 |
| 3.61 | 8 |
| 4.00 | 4 |
| 4.12 | 8 |
| 4.24 | 4 |
| 4.47 | 8 |
| 5.00 | 12 |
| 5.10 | 8 |
| 5.39 | 8 |
| 5.66 | 4 |
| 5.83 | 8 |
| 6.00 | 4 |
| 6.08 | 8 |
| 6.32 | 8 |
| 6.40 | 8 |
| 6.71 | 8 |
| 7.00 | 4 |
| etc. | etc. |

**Table 3.3** Number of CH approaching possibilities for NP at a hopping step (walking step) equal to the unit lattice constant $a$

| Hopping step distance of NP in unit of $d_0 = a$ | $f_{SI}$ (A) |
|---|---|
| 0 | 0 |
| 1 | 8 |
| 2 | 12 |
| 3 | 16 |
| 4 | 32 |
| 5 | 28 |
| 6 | 40 |
| 7 | 40 |
| 8 | 48 |
| 9 | 68 |
| 10 | 56 |
| etc. | etc. |

**Table 3.4** Number of CH approaching possibilities for NP at a hopping step equal to the integer multiple of unit lattice constant *a*

| Hopping step distance of NP in unit of $d_0 = 2a$ | $f_{SII}$ (B) | Hopping step distance of NP in unit of $d_0 = 3a$ | $f_{SIII}$ (C) |
|---|---|---|---|
| 0 | 0 | 0 | 0 |
| 2 | 20 | 3 | 36 |
| 4 | 48 | 6 | 100 |
| 6 | 68 | 9 | 156 |
| 8 | 88 | 12 | 196 |
| 10 | 124 | 15 | 260 |
| 12 | 140 | 18 | 336 |
| 14 | 176 | 21 | 372 |
| 16 | 196 | 24 | 428 |
| 18 | 224 | 27 | 492 |
| 20 | 228 | 30 | 556 |
| etc. | etc. | etc. | etc. |

Here, we have considered for possibilities for hopping step unit distance of 2 and 3 times of the unit lattice constant *a*

In Table 3.1, we have presented the case for an $n \times n$ array where $n$ is a finite number. We may consider $n$ = 10, 20, 30, 40, 50, …, etc. When the value of $n$ is small, we may be able to calculate the values of both columns in Table 3.1 manually, as are done here (see Tables 3.1, 3.2, 3.3, and 3.4). When the value of $n$ is too high which is in fact the case for biological cell surface lattices, we cannot address this physical NP–CH communications manually. Therefore, we are in process of developing a program using MATLAB (http://www.mathworks.com/products/matlab/) and incorporate these preliminary results into the program to further revise it to constructing an analytical expression for numerical computations. A final program will be developed to address the scenario for all realistic finite values of $n$.

In Fig. 3.1, the number of possible positions where an NP can move with respect to different distances from center is shown. An NP may have four possible destinations if it travels $d_0$ distance from center or its initial location. In the next step, it can cover its four diagonals on the next move. To cover the diagonal distance for first time, the NP must have the energy for travelling distance $\sqrt{2}d_0$. Similarly, there are also four different ways to travel nearby $2d_0$ distance in the next term. If an NP has energy to travel distance equal to $\sqrt{5}d_0$, NP can interact with CH with a total number of eight different possibilities. In this way, the number of possible positions covered by an NP with respect to distance is described in Table 3.2.

Generally, Fig. 3.2a, b can be presented as a graphical representation of Tables 3.2 and 3.3, respectively. The horizontal axis represents the possible distance of any channel with respect to initial NP position. The vertical axis is the number of possible channel locations (as mentioned as ‘number of available CH’ in Fig. 3.2) where an NP may hope to.

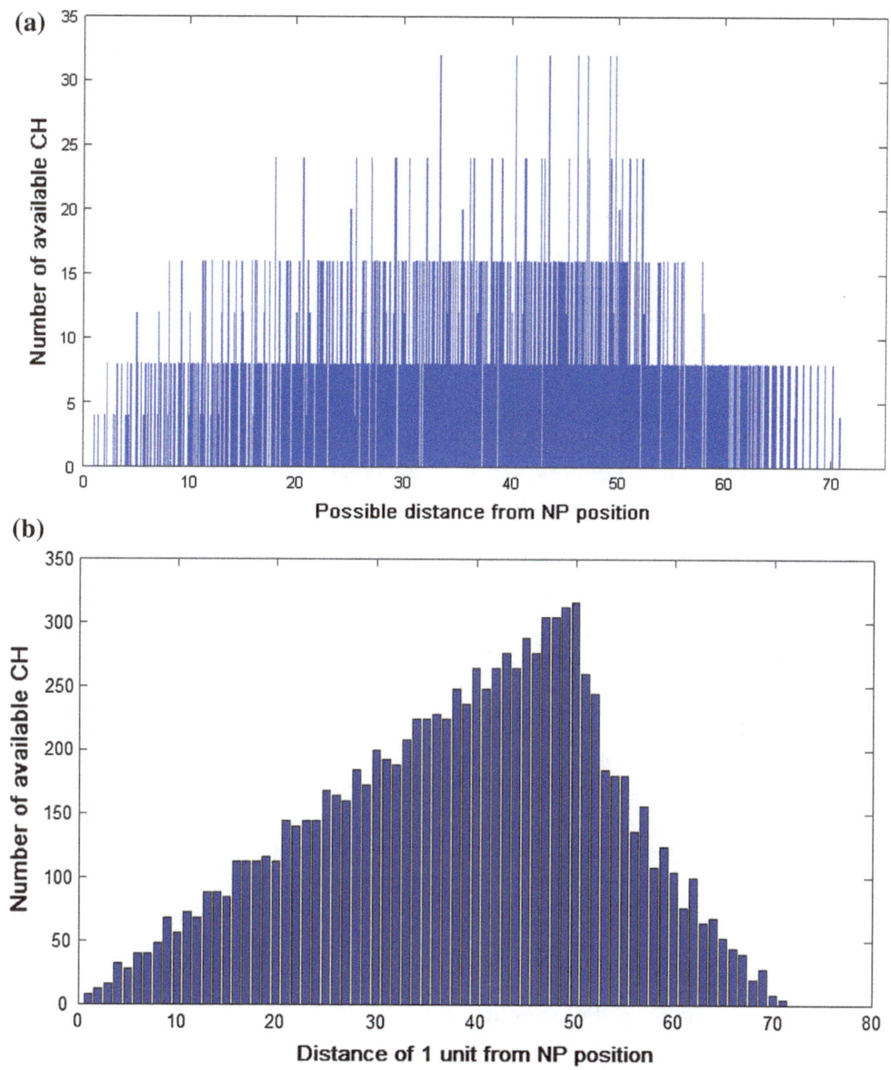


**Fig. 3.2** **a** This plots the frequency/possibility ($f$) for NP to reach at channel locations as a function of actual channel distance from NP's initial location as shown in Table 3.2 for a 50 × 50 lattice. **b** This plots the frequency/possibility ($f_{SI}$) for NP to reach at channel locations as a function of channel distance from NP's initial location as shown in Table 3.3 for a 50 × 50 lattice

Let us assume NP is situated in the center of the cell surface with structure having lattice numbers $50j \times 50j$ where $j = 1, 2, 3, \ldots$, etc., an integer number. The 2D $X - Y$ $(x, y)$ plane cell surface structure is an assumption which is a valid approximation for a small section on the surface of a cell. For $j = 1$, the maximum $X$ and $Y$ directional $X -$ Y plane distances that to be covered by an NP [assumed to be at the center (0,0)] are both $50d_0$s, where $d_0$ is the minimum $X$ and $Y$ directional distance (unit of the hopping step distance), and NP needs to hope, for reaching at the farthest located CHs in the $X$ and $Y$ directions on the $X - Y$ plane. In this case, an NP can also travel another further maximum distance $50\sqrt{2}d_0$ along the diagonal direction of the $X - Y$ plane. The number probability (plotted as the 'number of available CH' as they are proportional to each other) for NP to interact a CH situated at $50d_0$ distance is found to be maximum (see Fig. 3.2b) for $j = 1$. It can also be noted that the number of possible NP–CH interactions (see vertical axis values of Fig. 3.2b) suddenly drops down while the distance goes beyond $50d_0$. Generally, as seen in Fig. 3.2, the number of NP–CH interactions (frequency) increases steadily, while the lattice space dimension ($x$ or $y$) increases. At a certain distance (say $50d_0$), the frequency is maximum and after that it decreases drastically.

In Fig. 3.2, the number of possible moves of NP covering corresponding possible distances is plotted. In this case, any NP acquires possibility to visit a nearby CH, while it obtains sufficient energy and satisfies other various related physiological conditions. An NP can also be approximated to virtually travel a unit distance (see Fig. 3.2b) rather than the actual distance (see Fig. 3.2a). In such cases, the frequency of traveling such unit distance will be different from considering the actual distances. For step distance 1 (in unit of the lattice constant $a$), the frequencies which correspond for displacements in between two consecutive steps but fall lower than the corresponding midpoints between the two consecutive lattice centers will be added to the lowest distance. For example, $1.41a$ distance has four possibilities to cover exact distance, so it will be added to lower step distance 1 (in unit $a$) (see the comparing contents in Tables 3.2 and 3.3). Conversely, the displacement $2.83a$ for NP has four possibilities; hence, the frequency for this displacement will be added to the next upper distance class, i.e., to the distance 3 (in unit $a$), and so on.

Similarly, in hopping step distances 2 and 3, the frequencies $f_{\mathrm{SII}}$ and $f_{\mathrm{SIII}}$, respectively, are shown in Table 3.4. These two frequencies for step II and III, and the frequency for step I are plotted in Fig. 3.3 for mutual comparisons. Figure 3.3a plots bar diagrams but in Fig. 3.3b are plotted the step diagrams. But for theoretical understanding of the trend, we would require average linear diagrams as shown in Fig. 3.4 corresponding to the diagrams in Fig. 3.3.

Figures 3.1, 3.2, and 3.3 present the plots of $N_{\mathrm{p}}$ for values of $d$ considered for actual distance and the distance following unit distance 1 (SI), unit distance 2 (SII), and unit distance 3 (SIII). By using Tables 3.3 and 3.4, the plots in Figs. 3.1, 3.2, and 3.3 are drawn. The red, yellow, and blue color bars are representing the frequency step I, II, and III, respectively. Figures 3.2 and 3.3 are representation in stair and line plot, respectively.

Form those plots, it is quite understandable that in the middle of the propagation the value of frequency (option) is the highest. From the initial point to highest peak, the propagation happens on a straight line and after reaching the peak it steps down suddenly and then maintains a slope (see Fig. 3.4).

So far, we analyzed the data for $j = 1$ that is considering a 2D cell surface section with lattice dimension $50j \times 50j = 50 \times 50$. For bigger surface sections, e.g., for $j = 2, 4, 8, 16, 32$, etc., i.e., for the cell surface sections $100 \times 100$,

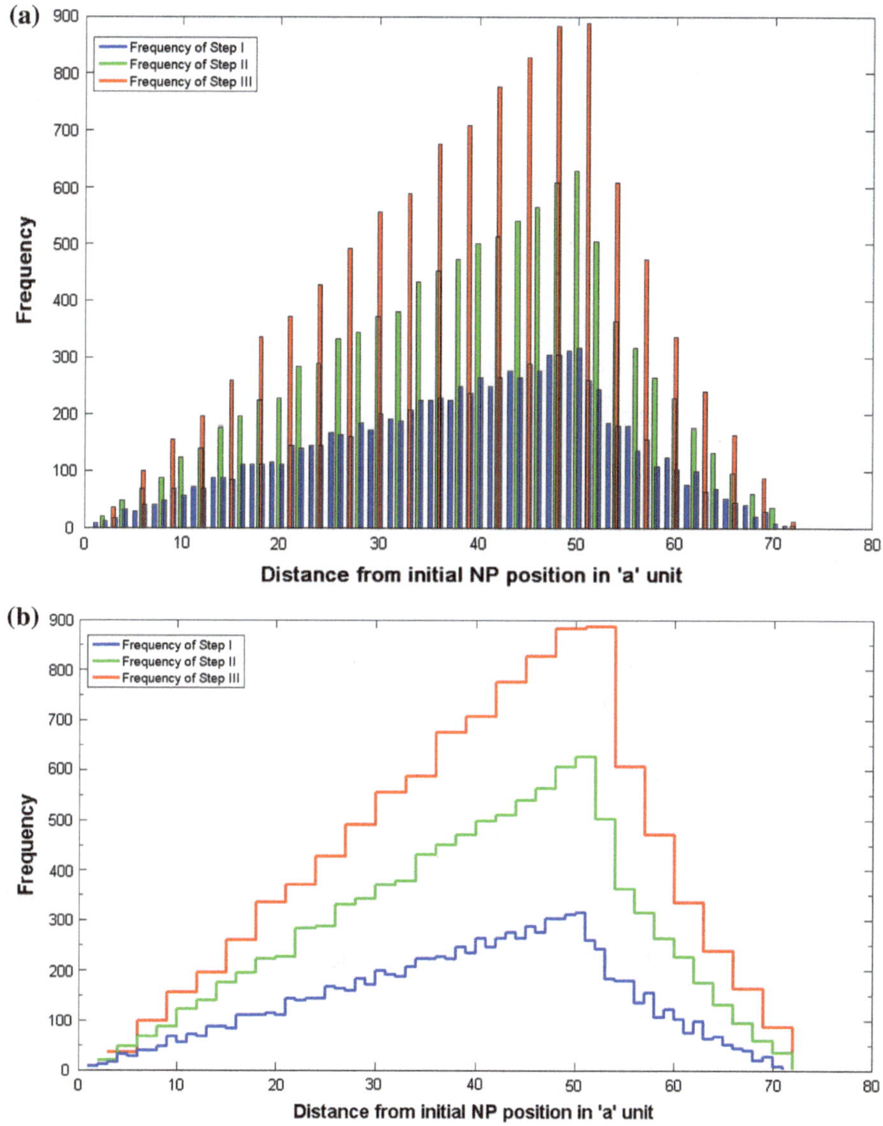


**Fig. 3.3** **a** This plots the frequency/possibility ($f_{\mathrm{Si}}$) (here i=I, II, III) for NP to reach at channel locations as a function of channel distance from NP's initial location as shown in Tables 3.3 and 3.4 for a 50 × 50 lattice. Here and later as before, step I (SI), step II (SII), and step III (SIII). Three frequencies plotted here are $f_{\mathrm{SI}}$, $f_{\mathrm{SI}}$, and $f_{\mathrm{SI}}$. **b** This plots the frequency/possibility ($f_{\mathrm{Si}}$) for NP to reach at channel locations as a function of channel distance from NP's initial location as shown in Tables 3.3 and 3.4 for a 50 × 50 lattice

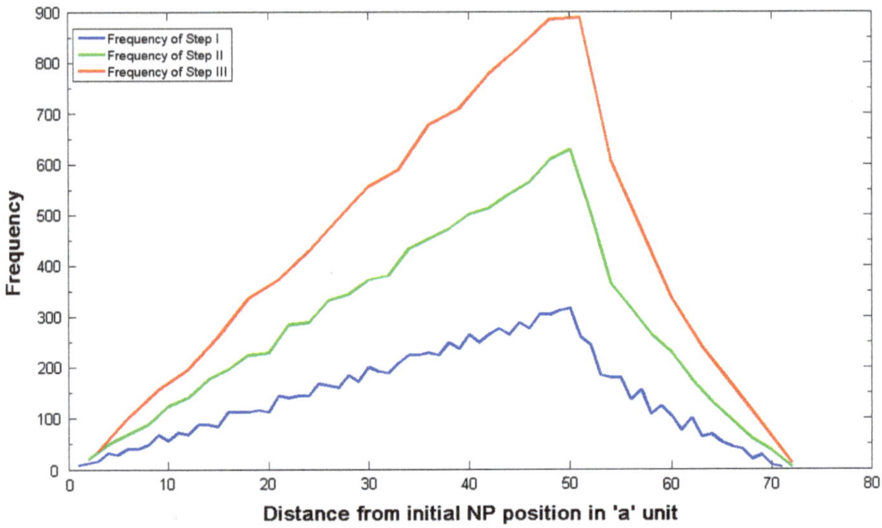


**Fig. 3.4** Here, the linear plots show the comparable patterns of the frequencies for theoretical analysis

200 × 200, 400 × 400, 800 × 800, 1600 × 1600, etc., respectively, we may plot the frequencies as like as of 50 × 50 (see Fig. 3.5).

We could alternately plot Fig. 3.5 as in Fig. 3.6. Figure 3.6 helps compare the lattice dimension dependence for any specific hopping step class SI, SII, or SIII.

The behavior of each subplot is quite same. After reaching a certain peak point, the curves fall fast before slowing down to linearly progress downward to finally touch the *x*-axis. We may try to find ad hoc theoretical function(s) for these plots to develop analytical forms for the plotted frequencies. Once we obtain appropriate analytical forms, we may be able to succeed performing the research for any lattice (small to large *XY* sections) of the cell surface lattices as long as we may consider that as 2D. The 2D representation would serve most of the purposes required for addressing interactions of independent NPs, drugs, peptides, MPs, etc., with the cell surface ingredients, e.g., lipids and various other proteins, channels, as these interactions occur mostly in narrow localized regions.

#### 3.2.3.3 Theoretical Function Representation: Generalized Formulation of the Frequency Graph

In this section, a generalized formula is targeted to be derived from the frequency graph. By analyzing the patterns of the graphs (see Figs. 3.5 and 3.6), it can be assumed that there exists some uniform perceptual structure in the diagrams. In Fig. 3.7, a model structure is shown which generalizes the frequency graphs shown in Figs. 3.5 and 3.6.

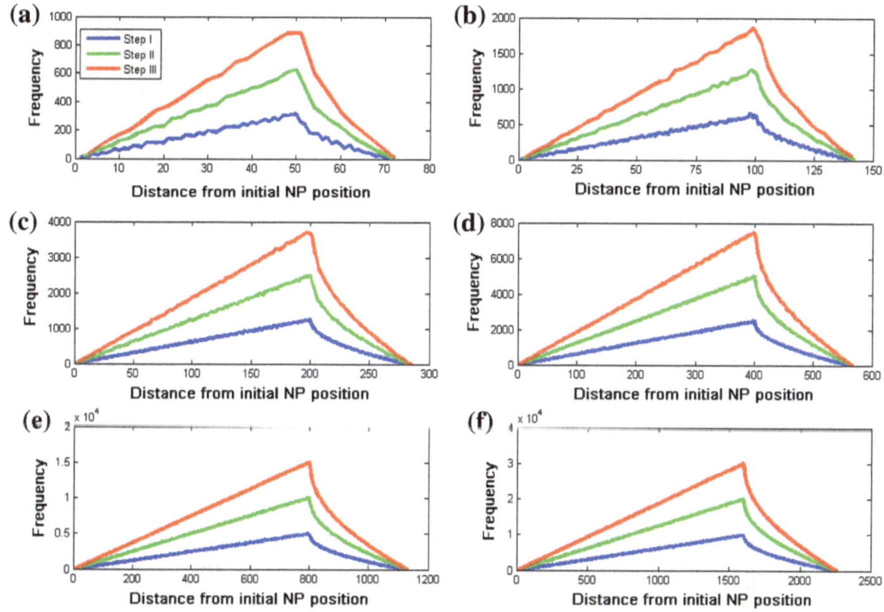


**Fig. 3.5** Frequency plots for comparable graphical interpretations with respect to various cell surface dimensions of XY lattice sections. **a** 50 × 50, **b** 100 × 100, **c** 200 × 200, **d** 400 × 400, **e** 800 × 800, **f** 1600 × 1600

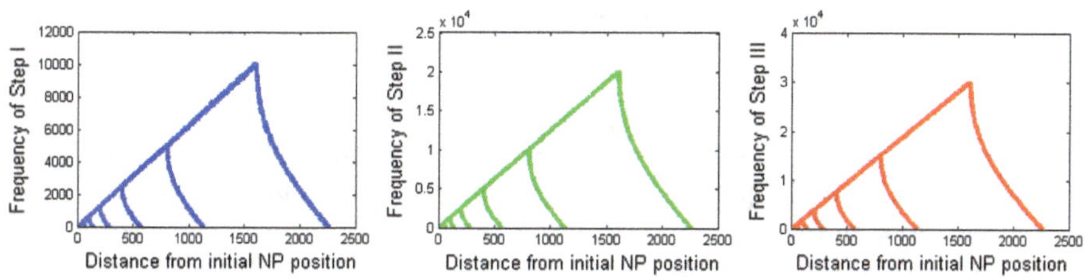


**Fig. 3.6** Frequency plots for comparable graphical interpretations with respect to various cell surface dimensions of XY lattice sections. In every graph (left to right), plots are for lattice size a, b, c, d, e, and f (see Fig. 3.5), i.e., a: 50 × 50, b: 100 × 100, c: 200 × 200, d: 400 × 400, e: 800 × 800, f: 1600 × 1600. Left panel for SI, middle panel for SII, right panel for SIII, respectively

The base of the structure AE represents the distance from initial NP position where the point $E$ is the maximum distance of the specific cell surface lattice section. Let us assume (as before) the dimension of the cell surface is $n \times n$. In such case, the value of $E$ (i.e., distance from center to the maximum reachable point in $XY$ plane of the lattice space with dimension $n \times n$) will be $\sqrt{2}n$ (explained in detail earlier). However, the $x$-axis value of point $K$ is $n$. In order to formulize a general equation, the curve is divided into three parts. The first part is the straight

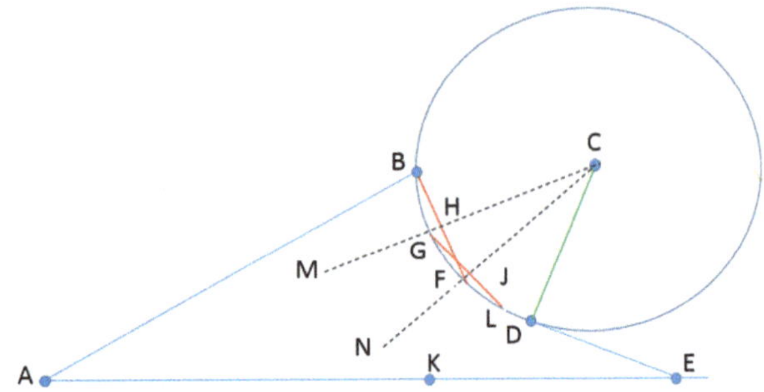


**Fig. 3.7** Dissection of frequency graphs presented in Figs. 3.5 and 3.6

line AB, the second part is the arc of curvature BGFD, and the third part is the straight line DE which is inclined on AE.

These three parts of equation are derived here.

(i) **Equation of AB.**
The line AB is inclined on base AE with a particular angle, and it is extended through the origin *A* (0, 0). The *x*-axis value of end point *B* of AB is *n*, whereas the *y*-axis value varies on different steps of frequency (see Figs. 3.5 and 3.6). It is noted that the *y*-axis values for frequency of SII and SIII are almost 2 and 3 times, respectively, of that of SI. This confirms the slope of line AB for SII (let us assume $m_2$); in particular, dimension of cell surface section is double of $m_1$ (where $m_1$ is the slope of AB for frequency plot for SI). Similarly, for frequency plot at SIII, the slope $m_3$ will be three times $m_1$.

In general, the equation of a line with slope *m* is defined as

$$Y = mX + C, \tag{3.3}$$

where *m* is the slope of the straight line and *C* is the intercept with *Y*-axis.

However, in case of equation of straight line AB, the value of *C* = 0.

And the slope *m* is

$$m = \frac{\text{value of } Y \text{ component of point } B}{\text{value of } X \text{ component of point } B} = \frac{\text{value of } Y \text{ component of point } B}{\text{surface dimention}(n)} \tag{3.4}$$

After calculating $m_1$, $m_2$, and $m_3$, the equation of AB for SI, SII, and SIII can be defined as: $Y_1 = m_1X_1$, $Y_2 = m_2X_2$, and $Y_3 = m_3X_3$, respectively.

(ii) **Equation of Arc BGFD.**
The arc BGFD is the part of a circle. In order to calculate the arc, it is necessary to identify the center, *C* of the circle. The center *C* will be at some point where perpendicular bisectors of the arc BGFD meet. More than three perpendicular bisectors are needed to identify the center *C*. In Fig. 3.7, two bisectors (CM and CN) of line BF and GL are shown. These two bisectors intersect BF and GL at *H* and *J,* respectively. However, they converge themselves at point *C* which is assumed to be the center of the circle. Hence, in this section, the point *B* and center *C* are identified.

(iii) **Equation of DE.**
We can use identical strategy as used in derivation of equation for AB.

#### 3.2.3.4 Probability of NP–CH Interactions

So far, we have manually analyzed frequency (see Figs. 3.5 and 3.6) and developed analytical form (Sect. 3.2.3.3) for general calculation of $f_{\mathrm{Si}}$ on a cell surface section. While an NP experiences diffusion on a cell surface to interact with a CH staying at an unknown location inside a unit cell within a 2D lattice space with dimension $n \times n$, we have developed these results. $f_{\mathrm{Si}}$ is a number of possibilities whose inverse correspondingly appears as proportional to the probability ($p_{\mathrm{Si}}$) of meeting between NP and CH. With this probability function $p_{\mathrm{Si}}$, one can easily add (through normalization) all biological information of the cell surface; e.g., parameters related to the dynamics of both NP and CH such as their mobilities (inverse of the cell surface frictions) can be used a condensed matter theoretical calculations in Ashrafuzzaman and Beck (2004a, b), thermodynamic conditions, etc., to finally calculate the probability of interactions between dynamic NP and CH on a cell surface. These are what we have incorporated while developing algorithms (briefed in Sect. 3.2.4). We shall avoid detailing them here due to strategic reasons.

The above-explained calculations require modifications for the case where more than one NP and CH are available on the 2D lattice space with dimension $n \times n$.

### *3.2.4 Developing Algorithms*

We wish to keep this part mostly confidential as long as we do not finish getting a patent. But a general strategy can be discussed here.

Here is a sample strategy provided on how we may obtain a program package for computational address of the NP dynamics on the cell surface to interact with CH and possibly experience a probability to diffuse beyond the cell membrane.

```
Initialize: Surface_dimension, NP_location
Surface_x_axis_value ← Surface_dimension(x axis)
Step 1: Calculate each surface point w.r.t. NP_location
Step 2: Create DistancePointMatrix where first column contain distance
 from NP and second
column contains number of points available in particular distance
for outer = 1 to Surface_x_axis_value
       temp ← -1
       for inner = 0 to outer
               temp ← temp + 1
               arrayIndex1 = arrayIndex1 + 1
```

```
            //calculate distance from NP_location
            arrDist(arrayIndex1) ← sqrt(outer^2 + temp^2);
    if (temp == 0) || (temp == outer)
      arrNoOfPoint(arrayIndex1) ← 4;
    else
      arrNoOfPoint(arrayIndex1) ← 8;
    end if
      end for
end for
//Create distance-number-of-point matrix
DistancePointMatrix = [arrDist', arrNoOfPoint'];
Step 3: Find the indexes which has same distances.
Step 4: Add distance wise number of points which have multiple but
same distanced-value
Step 5: Remove the whole rows whose indexes are repeated and among
which is the maximum index
Step 6: Present the reduced matrix
```

### 3.2.5 *Concluding Remarks*

The algorithm we developed helps us address the general movement of NPs on cell surface and their occasional approach to the cell surface-based interaction sites caused by the presence of MPs, specialized lipid structural defects, etc. As we can address the cell surface diffusion of NPs or general cell surface adsorbates theoretically and computationally using our developed advanced algorithm, we can utilize these techniques to address the general drug diffusion on cell surface and their possible cell surface adsorption. Both of the cell surface diffusion and cellular adsorption follow general physics formalisms (as explained earlier) including statistical mechanical rules on surface interactions of the agents moving randomly on the surface of the cell. Therefore, we now easily understand two important aspects regarding drug effects, namely the cell membrane-based drug adsorption which causes the cell membrane-based drug cytotoxicity and drugs' cellular uptake probability in a quantitative measurements, computationally. We shall provide more details in our publications and patents (in preparation).

## 3.3 Cell Surface Movement of Viruses and Pharmaceutical Nanoparticles and Their Cell Entry Mechanisms

NPs are now trackable during their uptake into living cells using highly sensitive fluorescence microscopes. In this chapter, we have addressed spectroscopic tracking of cell surface adsorbates/materials while moving on the surface of the cell.

Trajectories of the motion of NPs can be tracked down. Analysis of these trajectories helps us distinguish between random movement and active transport. Physical parameters such as the average transport velocity, mean free path, diffusion coefficient can be determined. As explained earlier in this chapter, pharmaceutical NPs, viruses, various drugs, etc., diffuse through cell surface randomly and interact with the cell surface constituents, as a consequence may experience statistical probability of cell surface pinning and occasionally cellular uptake. Here, 'statistical probability' is used to refer to the NP interactions with the cell surface interaction sites that require energetic binding costing free energy in a thermodynamic environment where cells exist. Therefore, an analysis of the trajectories provides important information regarding the history of interactions, uptake pathway and location of pharmaceutical NPs, drugs, viruses, etc.

Recent studies provide some in-depth analysis on single-particle tracking (SPT) and determination of the mean-squared displacement (Ruthardt et al. 2011). Advances in SPT include millisecond (ms) alternating-laser excitation for removal of spectral cross talk, alternating wide field (WF), and total internal reflection fluorescence (TIRF) microscopy for sensitive experiments at the plasma membrane and three-dimensional (3D) tracking strategies. The prospects of single-cell microscopy techniques and their impact on unraveling the interactions of NPs and viruses with cells have been quite addressed to date (Ruthardt et al. 2011; Shevchuk et al. 2008; Byrne et al. 2015).

Viruses, known as gene transfer particles, are also valuable tools for investigating the possible entry pathways into cells (Oh and Park 2014; Pelkmans et al. 2001). Multicolor fluorescence microscopy on living or fixed cells has been used as the main optical technique to study the cell entry of viruses and pharmaceutical NPs (Lakadamyali et al. 2003; Rust et al. 2004; Ewers et al. 2005; Payne 2007; Brandenburg et al. 2007). As a regular cell assay practice, compartments of the endocytic pathway in living cells are visualized by fluorescently labeled markers (e.g., dye-coupled transferrin, cholera toxin B, or dextrans) or by cellular expression of marker proteins fused to fluorescent proteins (e.g., clathrin–green fluorescent protein, caveolin–green fluorescent protein) (Watson et al. 2005). The entry of fluorescent-labeled viruses or nonviral vectors can be followed in the living cell. Then, colocalization analysis with cellular compartments is used to analyze the viral entry route. Specific entry pathways can be blocked by using chemical inhibitors or small-interfering RNA (iRNA) and test for internalization of gene carriers. Various studies of NP uptake reveal that all tested gene carriers enter cells by endocytosis (Rémy-Kristensen et al. 2001; Kopatz et al. 2004; Rejman et al. 2005), even the ones connected to cell-penetrating peptides (Rinne et al. 2007; Lundin et al. 2008; Mäe et al. 2009). Besides having the cell membrane-penetrating ability, majority of viruses exploit the cellular endocytotic machinery for productive infection (Helenius et al. 1980; Marsh et al. 1997; Döhner et al. 2005). The internalization pathway is not generalized; rather, it follows specificity in regard to type of cell, virus, or gene carrier (Rejman et al. 2005; von Gersdorff et al. 2006; Marsh and Helenius 2006). Use of multiple endocytosis pathways, blocking of specific pathways that enhances the internalization through alternative pathways, etc., are mechanisms quite addressed now. Real-time tracking of viruses and NPs may help us understand the cell environment localization, interaction, internalization, compartmentalization, etc. High spatial and temporal resolution spectroscopic techniques are required. Powerful microscopies (often confocal microscopes) are needed for imaging.

### *3.3.1 Single-Particle Tracking as Cell Internalization Assessment*

Microscopy, such as wide field (WF) microscopy, is typically used for tracking particle in SPT technique. WF is a highly sensitive fluorescence microscopy technique popularly used in cell research. Single gene carriers' internalization can be continuously followed from their initial cell attachment point on the cell membrane through the various steps of the uptake process and intracellular trafficking pathways. Imaging single gene carriers that are labeled with only a few fluorophores, at high temporal resolution over several minutes, requires methods of ultrasensitive detection (schematic representation is shown in various publications, e.g., Ruthardt et al. 2011).

Individual fluorescent particles can be identified as bright spots on a dark background in the WF microscopic images. Every frame in the movie represents the position of the particles at a certain time. The trajectory of the particle is obtained by extracting the particle's *X*- and *Y*-coordinates. Particles for tracking are selected with consideration of their intensity, size, and shape. *X*- and *Y*-coordinates are obtained by fitting a 2D Gaussian function to the particle's intensity profile (see Fig. 3.8a). Finally, the particle coordinates are used for calculating the corresponding trajectories (with nanometer (nm) level precision) based on a nearest-neighbor algorithm (Godinez et al. 2009; Sbalzarini and Koumoutsakos 2005). For details, see Fig. 3.8.

Trajectories are analyzed for understanding particle's motional behavior. The particle motion provides information on the surroundings and interactions of the particle with the cell compartments. There are a few parameters to address the motion. Mean-squared displacement (MSD) is such an important one (Saxton and Jacobson 1997) (see Fig. 3.8b). The MSD $<r^2>$ represents for the average of the squared distances between a particle's start and end positions over all time lags of certain length $\Delta t$ within one trajectory. From evaluation of MSD plot, information about the mode of motion can be obtained (see Fig. 3.8c). The mode of motion helps us interpret in a biological context and create conclusions on the location, and environment of the tracked particle can be drawn quite correctly.

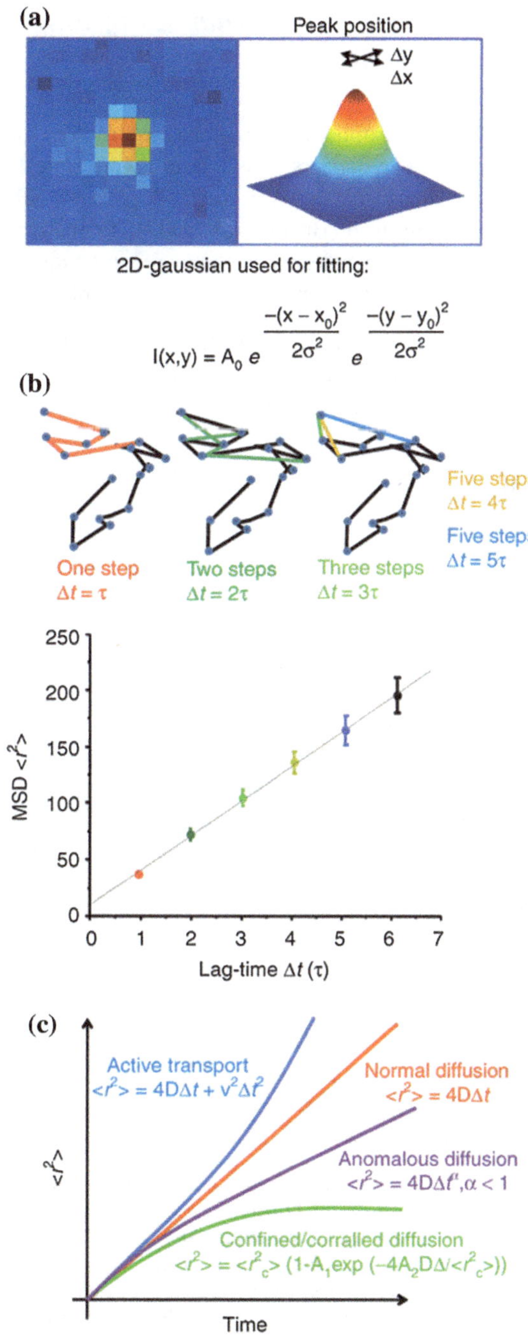


**Fig. 3.8** Tracking and MSD. **a** The *x*- and *y*-coordinates of the particle at a given time point are derived from the central position of its diffraction-limited intensity profile by, for example, fitting it to a 2D Gaussian function. Hence, a positional accuracy far below the optical resolution is obtainable. **b** MSD plots. Upper panel. A schematic trajectory showing the distance travelled between consecutive frames (blue circles in trajectory). The average of all steps within the trajectory for each time lag $\Delta t$, with $\Delta t = \tau$ (red), $\Delta t = 2\tau$ (dark green), and so on (where $\tau$ = acquisition time interval from frame to frame) gives a point in the MSD plot. **c** The time dependence of the MSD allows the classification of the type of diffusional behavior. A linear plot indicates normal diffusion and can be described by $\langle r^2\rangle = 4D\Delta t$ (D, diffusion coefficient) for a 2D analysis and $\langle r^2\rangle = 6D\Delta t$ in three dimensions. A quadratic dependence of $\langle r^2\rangle$ on $\Delta t$ indicates directed motion and can be fitted by $\langle r^2\rangle = v^2\Delta t^2 + 4D\Delta t$ ($v$, mean velocity). When the MSD asymptotically approaches a maximum value for larger $\Delta t$, the system is undergoing confined or corralled diffusion with $\langle r^2\rangle = \langle r_c^2\rangle\,[1 - A_1\exp(-4A_2D\Delta t/\langle r_c^2\rangle)]$. Anomalous diffusion has the form $\langle r^2\rangle = 4D\Delta t^{\alpha}$ and $\alpha < 1$ (sub-diffusive). For theoretical details, see Ruthardt et al. (2011)

### 3.3.2 *Kinetics and Dynamics of Single Nanoparticles in Living Cells*

The living cell uptake kinetics and internalization dynamics in real time for single polymeric gene carriers (polyplexes) were investigated a decade ago. In this study, sensitive fluorescence WF microscopy and SPT with a temporal resolution of 300 ms were used (Bruin et al. 2007). Trajectories showing the full internalization process of epidermal growth factor receptor-targeted polyplexes were obtained. The trajectories included three sectarian dynamics, namely (see Fig. 3.9)

(i) The cell surface dynamics: attachment to the cell membrane;
(ii) The cellular uptake: cell internalization through the cell surface;
(iii) The intracellular dynamics: active transport toward the nucleus.

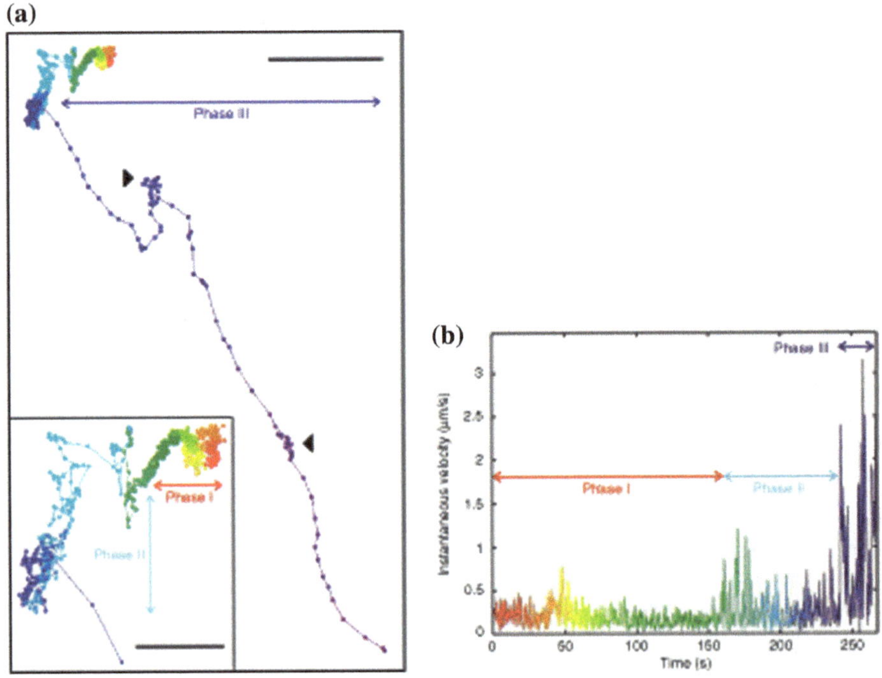


**Fig. 3.9 Trajectory of an epidermal growth factor (EGF)+ particle. a** Trajectory of an EGF+ particle attached to a HuH7 cell. The particle was tracked for 4 h and 30 min at a frame rate of 300 ms. The trajectory starts immediately after the polyplex attaches to the cell membrane. The inset shows an enlargement of the first part of the trajectory. The color of the trajectory is shown as changing from red to blue with progressing time. Phase I was observed until $t$ = 160 s and phase II until $t$ = 240 s, followed by phase III. Scale bar: 3 µm; scale bar inset: 1 µm. **b** Instantaneous velocity plot of the trajectory presented in **a**. The instantaneous velocity of the EGF+ particle is plotted versus time for each frame. The color coding corresponds to matching time points in the trajectory. Phase I: slow directed motion, phase II: diffusion, phase III: directed motion along microtubules

Three phases of motion were reported in Bruin et al. (2007). Phase I shows slow directed motion, phase II consists either of normal, anomalous, and confined diffusion or a combination of them, and phase III shows fast, active transport in the cytoplasm. Phase I occurs shortly after attachment to the cell membrane where the polyplexes show slow directed transport with typical velocities of ~0.015 μm/s. This transport is attributed to the movement of the underlying actin cytoskeleton mediated by the epidermal growth factor (EGF) receptor and linker proteins as revealed by dual-color microscopy. The polyplexes are thus considered to be transported by the retrograde actin flow within the cell. A similar behavior is found for untargeted polyplexes where the cell surface-attached polyplexes colocalize with actin fibers of the underlying cortical actin network (Bausinger et al. 2006). SPT of the membrane-bound polyplexes follows various motional behaviors including 2D free diffusion, anomalous diffusion as well as complete immobilization. Similarly, magnetic polyplexes are reported to display the three phases of internalization, independent of the applied magnetic field (Sauer et al. 2009).

Most particles are found quenched during phase I, so they are still accessible to trypan blue and attached to the extracellular membrane (see Fig. 3.10a). Tracking of unquenched particles (see Fig. 3.10b) reveals diffusive or directed motion at a later time point (see Fig. 3.10c, d). Particles that exhibit normal and confined diffusion (phase II) or directed motion (phase III) are observed not quenched. Thus, internalization is predicted to occur during phase I. As endocytic vesicles containing the particles escape from the actin network, particle diffusion and internalization can be observed (phase II).

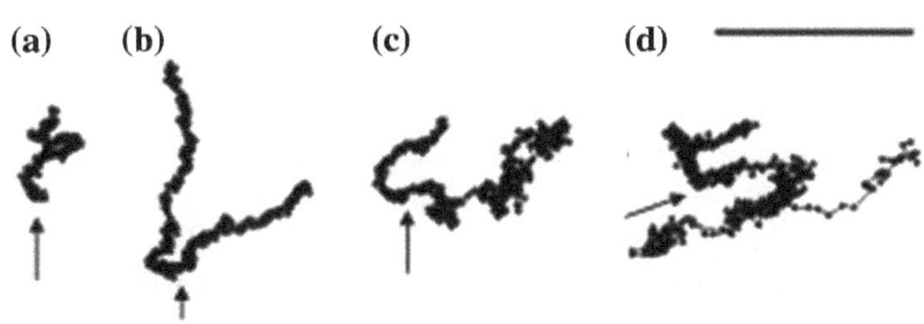


**Fig. 3.10 Trajectories of epidermal growth factor (EGF)+ particles during quenching experiments**. The trajectories were recorded at a frame rate of 500 ms. Trajectories starting shortly after the time point of addition to the cell are shown. The arrow indicates the time point of quenching. **a** Trajectory of a particle that was quenched during phase I, indicating extracellular attachment. **b** Trajectory of an unquenched EGF+ particle displaying phase I behavior at addition of the quencher. **c** Trajectory of a particle showing phase I behavior at addition of trypan blue and diffusion at a later time point. **d** Trajectory of a particle showing phase I at addition of trypan blue. The particle was not quenched; it started diffusion (phase II) shortly after addition of the quencher and directed motion (phase III) at a later time point. Scale bar: 2 μm

### 3.3.3 Mechanisms of Viral Entry into Cells

The metastable conformation of viruses that are quasi-inert macromolecular assemblies changes during the cell entry processes. Membrane rupture or fusion and genome uncoating may occur. Adjusting to local environment is another mechanism viruses adopt with. Virions tune their physical properties and thus enhance penetration and uncoating. Influenza virus, for example, softens at low pH to uncoat (Greber 2016). The physical properties such as stiffness and pressure of adenovirus control uncoating and membrane penetration. The cell membrane mechanical properties' perturbation may be important processes helping viral entry into cells.

Enveloped viruses that contain a protein shell with specific nucleic acid genomics and lipid envelope deliver their genomes into the cellular interior region by fusion of their lipid membranes with a host cell membrane, e.g., an endosomal membrane (Mercer et al. 2010). Nonenveloped viruses either directly penetrate a host membrane or deliver a subviral particle containing the genome into the cytosol (Suomalainen and Greber 2013). Membrane-penetrating peptides of both enveloped and nonenveloped viruses mask themselves usually in a glycoprotein of the virion membrane. These peptides are unmasked by cellular cues such as proteases, metal ions, and reducing agents (Suomalainen and Greber 2013), and some motor proteins mediated mechanical properties that act on the virus during entry (Yamauchi and Greber 2016).

Cells are known to sense, generate, and bear mechanical forces. Cells can convert these forces into particular responses, in processes of mechanotransduction (Bao and Suresh 2003). Unlike cells or organelles, which adjust their stiffness to resist tension to respond to external forces, viruses are not commonly known to adjust their mechanical properties under any biological forces, but rather to break or rupture. Such 'inertia' offers opportunities for the virus to use mechanical events to break the capsid open and release the viral genome for transcriptional activation during entry into cells (Greber 2016). Cellular cues thus are able to modulate the physical properties of viruses and affect viral entry into cells and infectivity.

The article written by Greber earlier this year (Greber 2016) has attempted to integrate physical properties of single virus particles with mechanical or chemical cues from host cells. Last year, Greber group (Luisoni et al. 2015) wrote another paper to demonstrate on the mechanics of virus and host cell to explain the cell membrane penetration and the cell entry. Here, they found that during the cell entry, nonenveloped viruses undergo partial uncoating to expose membrane lytic proteins for gaining access to the cytoplasm. The viral membrane lytic protein-VI is found to trigger a positive feedback loop involving lysosomal secretion, endocytosis, and lipid modification and thereby enhances the membrane disruption activity of the virus and infection. Adenovirus uses membrane piercing to induce and hijack cellular wound removal processes that facilitate further membrane disruption and infection. Certain lipids like ceramide in the membrane are also found to favor membrane binding of the lytic proteins and virus rupture of endosomes. A flowchart has been put forward to explain the mechanism in brief (see Fig. 3.11).

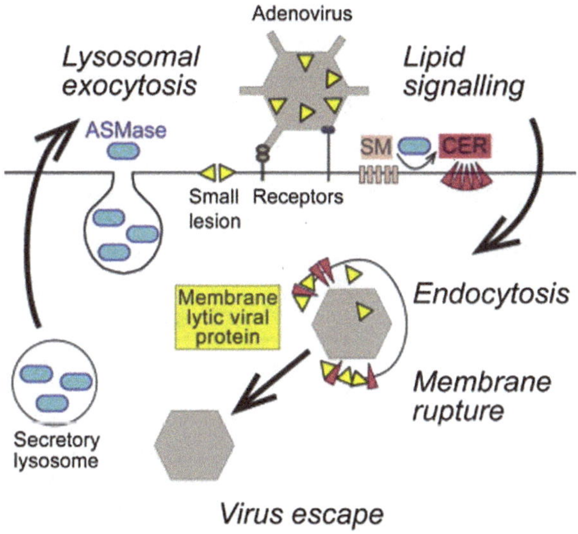


**Fig. 3.11** Schematic representation of the summary of lipid signaling-enhanced virus penetration of endosomal membranes. (1) Limited human adenovirus species (HAdV) uncoating leads to exposure of the membrane lytic protein-VI and influx of $Ca^{2+}$ through small lesions of the plasma membrane. (2) $Ca^{2+}$ influx triggers lysosomal exocytosis and delivery of ASMase to the outer surface of the plasma membrane. (3) Acid sphingomyelinase (ASMase) converts sphingomyelin (SM) to ceramide (Cer). (4) Cer increase accelerates virus endocytosis. (5) Cer increasingly recruits protein-VI, and this enhances the destabilization of the endosomal membrane until enough protein-VI has been released. (6) Endosomal membrane breaks, and virus is released to the cytosol. For details, see Luisoni et al. (2015)

### *3.3.4 Tracking the Viral Entry into Cells*

A three-stage entry process has been detected for the cell entry of influenza viruses (Ruthardt et al. 2011). This is slightly different than that explained earlier for the polyplexes (Lakadamyali et al. 2003). Shortly after addition of the viruses to the cells, at stage I, virus particles were found to show a slow actin-dependent movement in the cell periphery. At stage II, a short unidirectional movement toward the perinuclear region was found. This is followed by intermittent and bidirectional movement in the nuclear periphery in stage III (see Fig. 3.12). The comparatively rapid stage II movement was followed by the initial acidification of the endosomes. Consequently, virus fusion occurred at the end of stage III (Ruthardt et al. 2011).

The cell entry process follows a specific pathway with a number of cellular factors involved which has been simplistically schematized in the model presented shown in Fig. 3.13 (Luo 2012). The influenza virus is adsorbed at the cell plasma membrane by endocytosis (Bottcher et al. 1999). The virus particle then quickly enters coated vesicles (endosomes) inside the cell. The virus particles are internally transported inside the endosomes to locations near the nucleus (Lakadamyali et al. 2003). There are now a lot

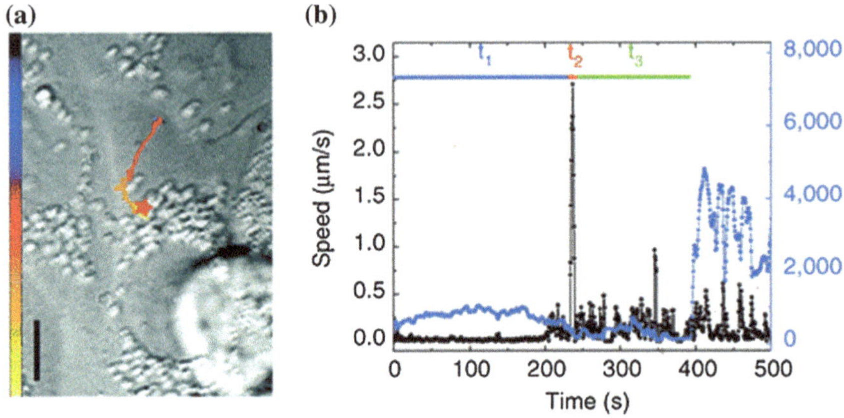


**Fig. 3.12** **Entry of influenza virus**. **a** The trajectory of a DiD-labeled virus inside a cell. The color of the trajectory codes time with the colored bar indicating a uniform time axis from 0 s (black) to 500 s (yellow). The red star indicates the fusion site. Bar = 10 μm. **b** The instantaneous velocity (black) and DiD fluorescence intensity (blue) during the trajectory of the virus. $t1$, $t2$, and $t3$ are the durations of stages I, II, and III, respectively. Stage I starts with contact of the virus with the cell, stage II consists of a rapid unidirectional translocation from the cell periphery to the perinuclear region, and stage III is the period between stage II and fusion and shows intermittent and bidirectional movement in the nuclear periphery. Fusion of the virus with the acidified endosome always occurs after the rapid movement in stage II. For details see, Ruthardt et al. (2011)

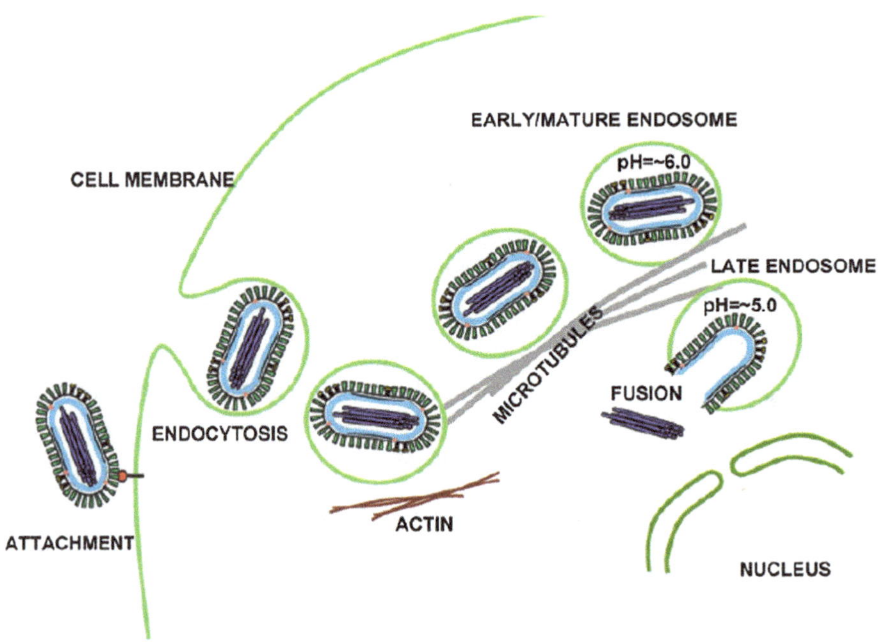


**Fig. 3.13** Model for the entry pathway of influenza virus

of animated videos explaining the viral entry. One such is found in the link [Link] that helps understand the viral entry into the cell quite easily.

Confocal microscopy can be used to track the viral particles 3D entry into cells. Figure 3.14 shows the assembly of HIV-1 particles on the dorsal cell surface. A 3D trajectory of a nascent assembly site from the beginning of assembly to release is shown in Fig. 3.14a. A 2D trajectory projection, the fluorescence intensity during the assembly process, and the instantaneous velocity are shown in Fig. 3.14b, c. The phase I is shown in red, phase II in yellow, and phase III in cyan. The residual gag molecules' fluorescence in the plasma membrane made it possible to reconstruct the cell membrane, shown as the brown surface in Fig. 3.14a. Using the spinning disk confocal microscope measurements, the total assembly time of HIV-1 was found to be $\sim$30 min.

## 3.4 Cell-Penetrating Peptides and Cell Uptake Mechanisms

Cell-penetrating peptides (CPPs) provide an effective strategy for intracellular drug delivery. CPPs' diffusion into cells is therefore an important aspect. Peptide-based delivery has potency to increase the uptake of drugs in tumor cells. The efficacy of the treatment thus is increased sometimes by several folds in conventional small molecular drugs or oligonucleotide-based therapeutics. One relevant clinical application of cell-penetrating peptides that are found effective is cancer therapeutics. We shall familiarize CPPs, their diffusion on cell surface, and the consequent cell uptake mechanisms in this chapter.

### *3.4.1 Cell-Penetrating Peptides*

The definition of the CPPs lies in the name itself. That means the peptides which have ability to penetrate into the cellular interior region fall in this category. CPPs are a class of diverse peptides, typically with 5–30 amino acids, that unlike most peptides can cross the cellular membrane. CPPs also often are referred to as protein transduction domains (PTDs), Trojan peptides, or membrane translocating sequences (MTS) (Järver and Langel 2006). Besides independent cell membrane effects, these peptides have been found to show great potential in the field of drug delivery. CPPs can deliver a wide range of biomolecules such as proteins, peptides, oligonucleotides (ON), and NPs to a variety of cell types and to different cellular compartments, both in vivo and in vitro. CPPs vary greatly in size, molecular weight, amino acid sequence, and charge. But most of them share the common feature via their ability to rapidly translocate the plasma membrane and enable delivery to the cytoplasm or nucleus (Lindgren et al. 2000). The mechanism of the cellular internalization of hydrophilic macromolecules through the classical

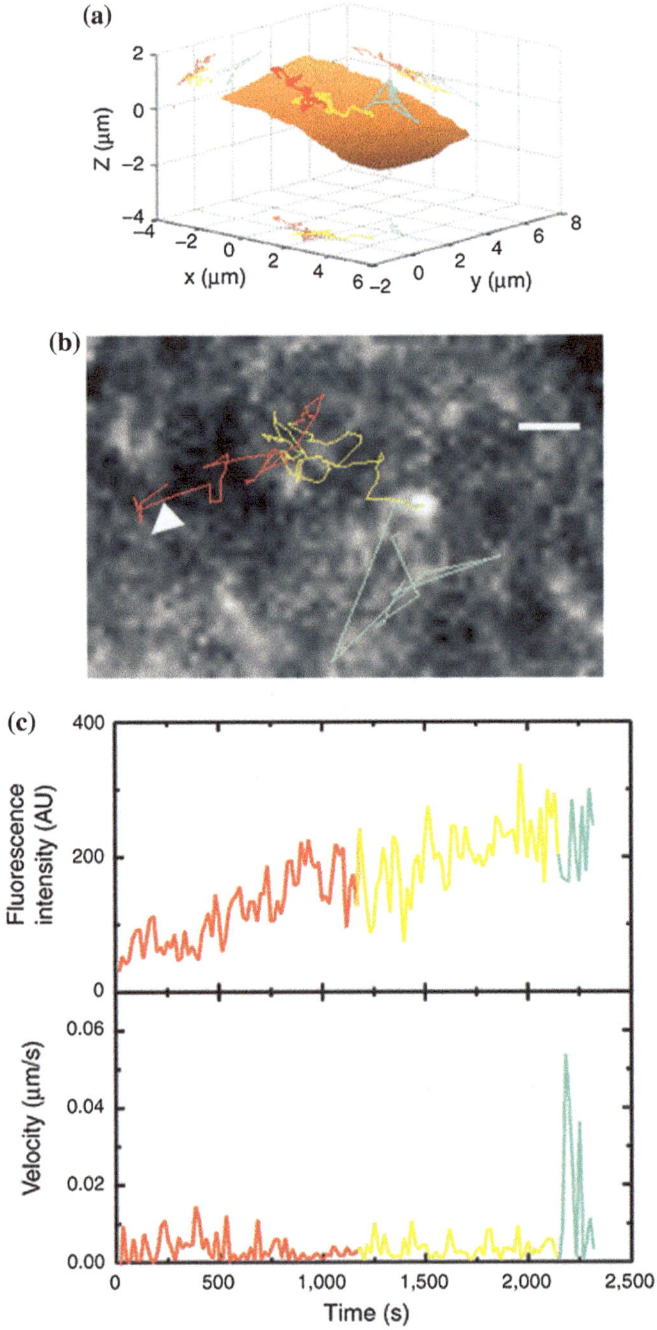
(a)
Z (µm)
x (µm)
y (µm)
(b)
(c)
Fluorescence intensity (AU)
Velocity (µm/s)
Time (s)

◂**Fig. 3.14** 3D tracking of HIV-1 assembly and release. **a** The trajectory of an HIV-1 particle observed using a spinning disk confocal microscope is shown here. The plasma membrane is shown in brown. The trajectory is color-coded (phase I, red; phase II, yellow; phase III cyan). **b** A 2D projection of the trajectory superimposed on a single confocal image of the assembly site. Bar = 1 μm. **c** The fluorescence intensity (upper panel) and instantaneous velocity (low panel) during assembly and release of the HIV-1 particle tracked in panel (**a**). For details, see Ruthardt et al. (2011)

endocytosis pathway is quite demonstrated. During last almost two decades, some specialized peptides have been demonstrated to translocate across the plasma membrane of eukaryotic cells by a seemingly energy-independent pathway. These peptides are found helpful in the intracellular delivery of macromolecules with molecular weights several times greater than their own. The cellular delivery using these cell-penetrating peptides offers advantages over conventional techniques because it is efficient for a range of cell types, can be applied to cells en masse, and has potential therapeutic applications (Lindgren et al. 2000).

### 3.4.2 Probability Functions Related to CPP Diffusion

For peptide internalization, we may construct a model flowchart as shown in Fig. 3.15. Here, the definition of various probability functions related to CPP diffusion is as follows:

$P_{SA}$ is the cell surface adsorption probability
$p_{FB}$ is the feedback to the cellular exterior region probability
$p_{CI}$ is the cellular internalization probability
$p_{MA}$ is the permanent membrane adsorption probability.

For the case of damage free flow of peptides across the cell membrane from extracellular region into intracellular region, the number of peptides is conserved, so

$$P_{SA} = p_{MA} + p_{FB} + p_{CI} \tag{3.5}$$

In case of localized cell membrane diffusion of peptides, we can consider equal probability for membrane internalized peptides to be released on both sides of the membrane that is to extracellular (with feedback probability $p_{FB}$) and intracellular (cellular internalization probability $p_{CI}$) regions. This approximation is considerably correct as locally membrane appears to be a planar entity having identical hydrophilic environments (in a broader sense) on both cellular exterior and interior regions. Therefore, we can consider the following approximation:

$$p_{FB} \approx p_{CI} = 1/2(P_{SA} - p_{MA}) \tag{3.6}$$

Earlier in this chapter, we have provided statistical mechanical analysis on the free diffusion of NPs and their possible trapping in any defects or interaction sites

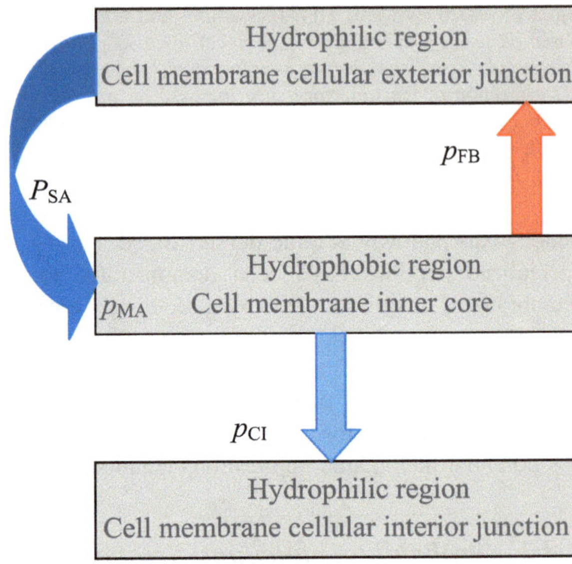


**Fig. 3.15** Cell internalization of the peptides across cell membrane. This model diagram can be compared with a water tank getting filled up with an inlet big pipe (with a probability $P_{SA}$ of the agents to be in) while leaking through two outlets

on the surface of the cell. This analysis can be extended to calculating the cell surface adsorption probability $P_{SA}$ for cell surface adsorbates, e.g., NPs, peptides, drugs, that freely move on the surface of the cell. For doing, so we need to just calculate the expression of 'free energy' of interaction what can then be used in Boltzmann's formula for calculating probability.

As an example, let us consider the free energy of interaction of an adsorbate; e.g., NP with a defect or interaction site (IS) in the cell surface lipid membrane background environment is $\Delta G_{NP\text{-}IS}$. The cell surface adsorption probability $P_{SA}$ would therefore follow this equation:

$$P_{SA} \approx \mathrm{Exp}\{-\Delta G_{NP-IS}/k_B T\} \tag{3.7}$$

Here, $k_B$ is the Boltzmann constant and $T$ is the absolute temperature of the interaction site. $\Delta G_{NP\text{-}IS}$ depends on biophysical properties of the interacting NP and IS, and the environment.

The membrane adsorption probability is the probability of the agent to get trapped inside membrane's hydrophobic core region. This trapping probability is either permanent for some agents (we have assumed here $p_{MA}$) or transient for others; let us consider this transient membrane trapping probability as $p_{MA,tr}$. The main difference between these two can be calculated monitoring the agents residing time inside membrane. In case of former, the residing time is infinity while for latter it is a finite time. When a peptide forms an ion channel across the cell membrane, the channel lifetime may indicate the status of the probability. Either way, the probability depends on the agent–membrane interaction energetics. If the free energy of interaction of a hydrophobic membrane (HM) residing NP with the membrane constituents is assumed to be $\Delta G_{NP\text{-}HM}$, we can apply Boltzmann statistics to calculate the membrane adsorption probability $p_{MA}$ as:

$$P_{\mathrm{MA}} \approx \mathrm{Exp}\{-\Delta G_{\mathrm{NP-HM}}/k_{\mathrm{B}}T\} \tag{3.8}$$

Calculation of the transient membrane trapping probability $p_{\mathrm{MA,tr}}$ follows the identical protocol as that of $P_{\mathrm{MA}}$. The only difference would be in the free energy of binding $\Delta G_{\text{NP-HM}}$ for the two cases what determines the difference in the agents' stability inside the membrane hydrophobic core.

A detailed analysis on how to calculate this type of free energy has been presented in Ashrafuzzaman and Tuszynski (2012a, b) for the case of peptide accumulation in the planar lipid membrane. Thus, calculation of all the necessary probability functions as explained here is possible by knowing biophysical properties of the NP and the NP adsorbing cell membrane.

### 3.4.3 Model Membrane Diffusion of Inserted Peptides

Cell membrane constituents generally move laterally in the 2D plane of the membrane. This lateral diffusion of membrane-bound agents is a prerequisite for many biological functions. The diffusivity which in physical form measured as mobility or inverse friction of such membrane-bound species can be influenced by many biological factors (Guo and Gai 2010) including a few biophysical properties of the agent itself.

The lateral diffusion of the cell membrane lipids and the membrane-adsorbed biomolecules, such as proteins and peptides, is found to control the timescale within which key molecular events take place (Liebman and Pugh 1979; Pastan and Willingham 1981; Hackenbrock 1981; Koppel 1982; Gross and Webb 1986; Peters 1988) in biological systems. Membrane-bound antimicrobial peptides (AMPs) need to diffuse laterally to find each other and form the functional oligomeric structures. Here, the diffusion rate of the membrane-bound peptides may limit the rate of oligomer formation. Similarly, the lateral diffusion rate of membrane-bound $\alpha$-helices plays an important role in determining the folding time of MPs, while the two-stage model of MP folding (Popot and Engelman 2000) suggests that secondary structure formation precedes the formation of tertiary structures. The two-stage model (compare between stage I and stage II), as presented in Fig. 3.16, explains how diffusion drives side-to-side helix and interactions cause association. This interaction finally results in a especially functional protein representation in a membrane. Here to mention that the interaction driven energies have recently been calculated by us for the case of association of peptides e.g., gramicidin A (gA) or alamethicin monomers to form $\beta$-helix or toroidal pore in a model membrane system considering the charge property based peptide-peptide and peptide-lipid interactions (Ashrafuzzaman and Tuszynski 2012a, b).

Schematic sketches were put forward by Borman and Engelman in 1992 (Borman and Engelman 1992) to explain transmembrane signal transduction with the help of MP structural changes that draws displacement, rotation, push and pull mechanisms. Here, two models, namely single-helix model and oligomerization model, have been considered for transmembrane diffusion of information. The single-helix model

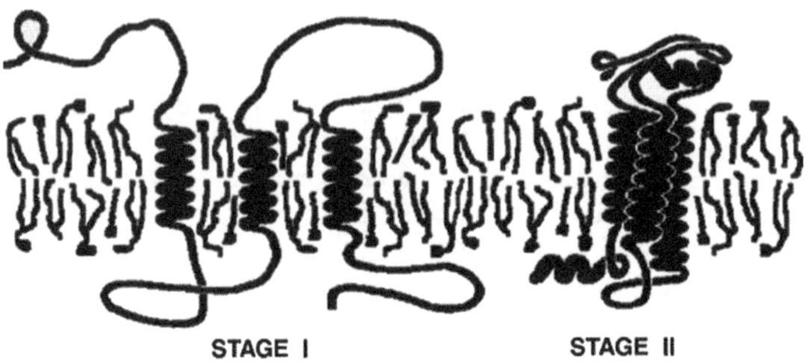


**Fig. 3.16** Two-stage model. Individual helices of a polytopic protein are postulated to be stable separately as domains in a lipid bilayer. Their stability as domains is a consequence of the hydrophobic effect and main-chain hydrogen bonding. Other interactions then drive side-to-side helix association, resulting in a functional protein. Specific folding energy is provided mainly by packing of the preformed helices with each other, by loop structures, and by interactions with prosthetic groups. Additionally, ion pairs and hydrogen bonds between helices are sometimes found, and general contributions are made by interactions with the lipid environment. See Popot and Engelman (1990, 2000)

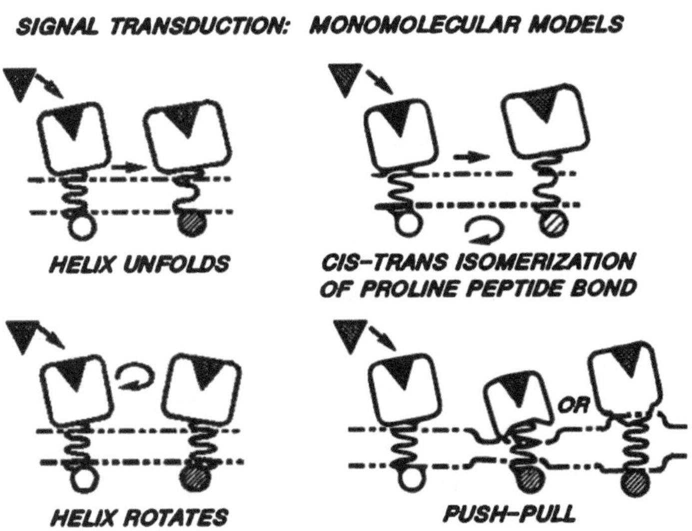


**Fig. 3.17** Signal transduction: single-helix model. Each diagram depicts the conformation of the receptor before and after the ligand interactions. Four mechanisms are as follows: Helix unfolds, proline peptide bonds isomerize, helix rotates, and then the helix is pushed or pulled

suggests four mechanisms of transmitting information through the bilayer via a single-helical domain (see the explanation in the caption of Fig. 3.17), whereas the latter model suggests that signaling through the transmembrane regions may offer the diffusion through oligomerization as explained in the caption of Fig. 3.18.

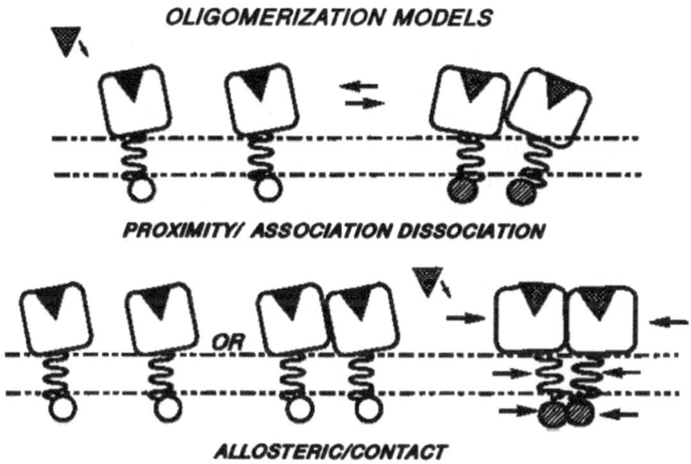


**Fig. 3.18** Oligomerization models. The upper schematic diagram suggests an association/dissociation equilibrium that usually favors oligomerization of proximal receptors in the presence of ligands (top, right). The bottom diagram suggests for possibility of receptors to exist as monomers or oligomers and may oligomerize and/or alter their conformations by an allosteric mechanism

#### 3.4.3.1 Monitoring Diffusion of Membrane-Bound Peptides: General Perspective

The mobility of a membrane-bound species is experimentally found to be sensitive to its conformation and location (Guo and Gai 2010). The diffusion measurement can provide useful information regarding the conformational distribution of membrane-bound peptides. The mobility of a membrane-bound species is also found to depend on its concentration. An analysis of direct measurements on dynamic parameters related to diffusion properties and surroundings that influence the diffusivity/mobility may provide important understanding to the problem. We shall address this issue here.

A few years ago, Guo and Gai (2010) attempted to provide a better understanding of how the mobility of a membrane-bound biological molecule gets dictated by the conformation and location of the agent. They studied the diffusion properties of a pH low insertion peptide (pHLIP) in model membranes using fluorescence correlation spectroscopy. Two distinguishable results were reported as follows:

(i) When the pHLIP peptide was located on the membrane surface, its lateral diffusion was characterized by a distribution of diffusion times, the characteristic of which depends on the peptide–lipid ratio.
(ii) As the pHLIP adopted a well-defined transmembrane $\alpha$-helical conformation, the peptide still was observed to exhibit heterogeneous diffusion, and the distribution of diffusion times was found to be independent of the peptide–lipid ratio.

Fluorescence intensity fluctuations arising from the fluorescent molecules diffusing into and out of a small confocal volume are measured in fluorescence correlation spectroscopy (FCS). It helps measure the diffusion time and hence the diffusion constant of the agents in diffusion (Magde et al. 1974; Haustein and Schwille 2007). In addition, observing single molecules and probing the diffusion heterogeneity are performed using FCS (Burns et al. 2005; Zhang and Granick 2005; Guo et al. 2007). To address these, Guo and Gai performed FCS to measure pHLIP diffusion using two lipid bilayer systems, namely supported lipid bilayer constructed of phospholipid 1-palmitoyl-2-oleoyl phosphatidylcholine (POPC) and giant unilamellar vesicles (GUVs). For detailed preparation technique, read from Guo and Gai (2010).

#### 3.4.3.2 Diffusion of pHLIP Bound to a Supported Lipid Bilayer

A 2D-constructed membrane is generally utilized as supported lipid bilayer for diffusion measurements. Figure 3.19a presents FCS measurement data, with an acquisition time of 40 s, and indicates that the pHLIP diffusion is very heterogeneous. Figure 3.19b indicates that increasing the peptide concentration appears to decrease the heterogeneity of the diffusion time ($\tau_D$) and also leads to a decrease in the average peptide diffusion time.

FCS measurements using a fluorescent-tracer-labeled lipid (i.e., Texas Red DHPE) without pHLIP show that the diffusion of the lipid is heterogeneous. This result is in agreement with that of Burns et al. (2005). It also suggests that the

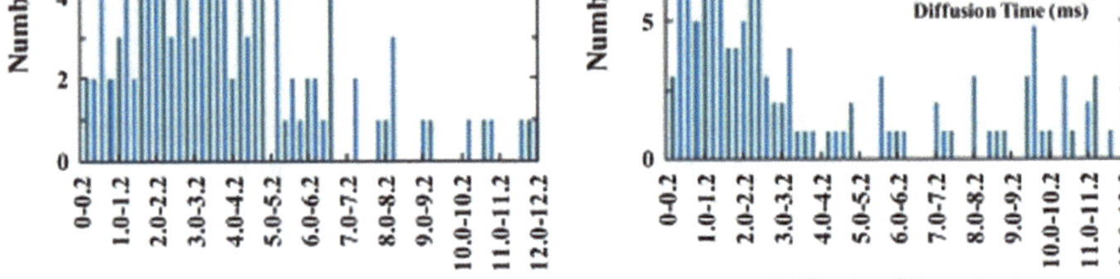


Fig. 3.19 Diffusion time ($\tau_D$) distributions of TMR-labeled pHLIP bound to a supported lipid bilayer of POPC at pH 8 and different total peptide concentrations: **a** 1 nM and **b** 1 μM. (*Inset*) Diffusion time ($\tau_D$) distribution of Texas Red 1,2-dihexadecanoyl-*sn*-glycero-3-phosphoethanolamine, triethylammonium salt (Texas Red DHPE) (0.002 mol%) in supported lipid bilayer of POPC at pH 8. See Guo and Gai (2010)

heterogeneous pHLIP diffusion observed in the current case could arise from interactions between peptide and the substrate or between lipid and the substrate. Nevertheless, control experiments showed that the characteristic diffusion times of pHLIP in buffer, obtained from multiple repeated measurements, are only distributed within a very narrow time range. The pHLIP diffusion constant is determined to be $(1.3 \pm 0.1) \times 10^{-6}$ cm$^2$/s in aqueous solution, which is consistent with those obtained on other peptides of similar size (Tjernberg et al. 1999; Sengupta et al. 2003).

#### 3.4.3.3 Diffusion of pHLIP Bound to Giant Unilamellar Vesicles

The giant unilamellar vesicles (GUVs) used in this study have diameters within 5–50 μm. This dimension remains static on the FCS experimental timescale. Figure 3.20a–c shows that for POPC membranes the pHLIP diffusion times at pH 8 have a distribution whose position and width depend on the concentration of peptide or the ration of peptide–lipid. The distribution becomes bimodal, and the average diffusion time becomes shorter at relatively high concentration. A similar concentration dependence is also reported for POPC/POPG membranes at pH 8 but with overall less pronounced change. In contrast, at pH 4 the $\tau_D$-distribution of pHLIP shows little or no dependence on concentration of the peptide; see Fig. 3.20d–f. The pH dependence comparison is also better presented in Fig. 3.21. These results altogether indicate that the location and conformation of a membrane-bound peptide (in this case pHLIP) are important determinants of the diffusibility/mobility of the peptide.

The above analysis of the results of the diffusion of pHLIP in two model membranes, namely supported lipid bilayers and giant unilamellar vesicles, clearly suggests for a heterogeneous peptide diffusion. The extent of the heterogeneity depends on pH and peptide concentration. These are two important determinants of the peptide diffusion coefficients. These results and conclusions found here are obvious considering various biophysical properties the membrane-bound peptides pose to have that dictate the dynamic peptide interactions with membrane constituents. The interaction mechanisms are found readily explained in our earlier published book 'Membrane Biophysics' (Ashrafuzzaman and Tuszynski 2012a).

### *3.4.4 Therapeutic Applications of CPP for Enhanced Cell Uptake of Agents*

In therapeutic applications of stem cells, the delivery of proteins is considered the safest. Proteins can regulate gene expression transiently without considerable alterations in genomics. Although it is quite difficult to accurately measure efficiency or quantity of the intracellular protein uptake, a comparison study of CPP-conjugated protein delivery system was recently performed using seven

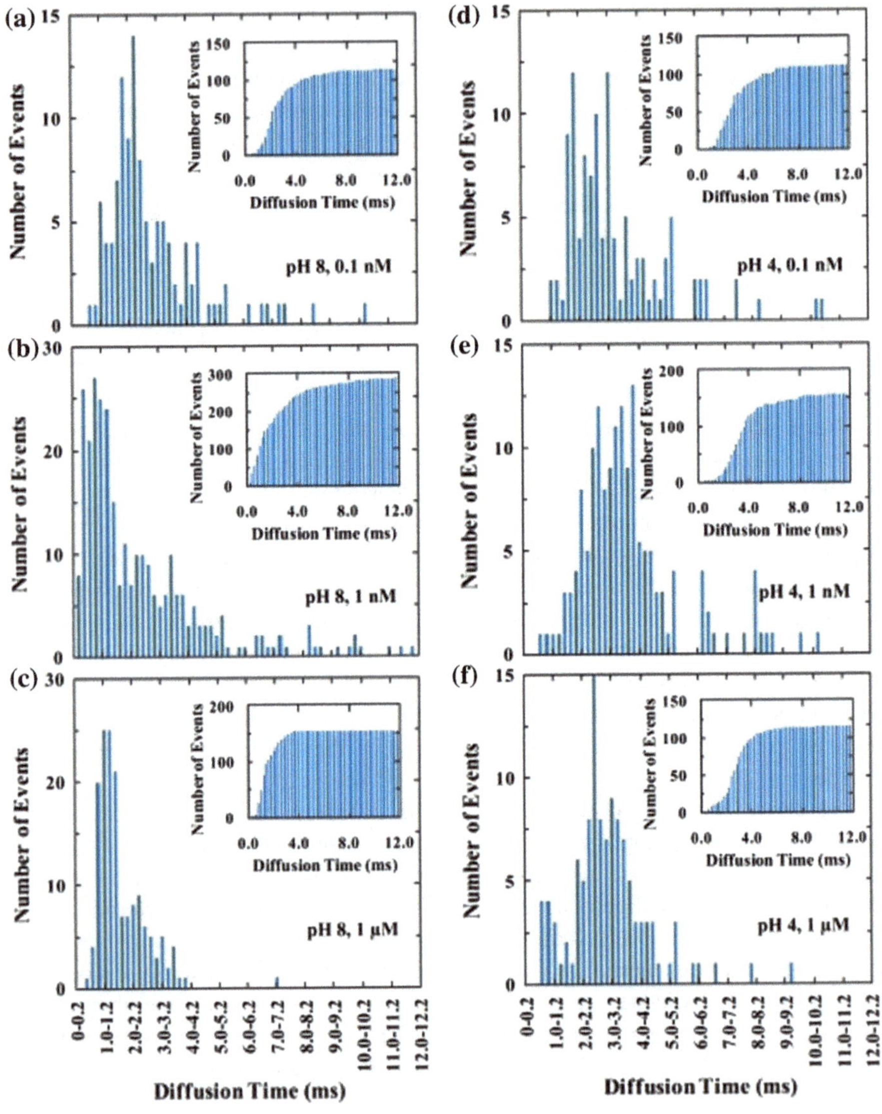


**Fig. 3.20** Diffusion time ($\tau_D$) distributions of TMR-labeled pHLIP bound to the lipid bilayer of GUVs measured at different total peptide concentrations and pH values, as indicated. (*Insets*) Corresponding cumulative distribution of the diffusion times (Guo and Gai 2010)

arginines and streptolysin O (SLO)-mediated system (Jo et al. 2014). This study was chosen to explain how a CPP might involve itself in creating a better cell uptake mechanism for important biological agents' cell internalization in comparison with other mediators, e.g., an exotoxin streptolysin.

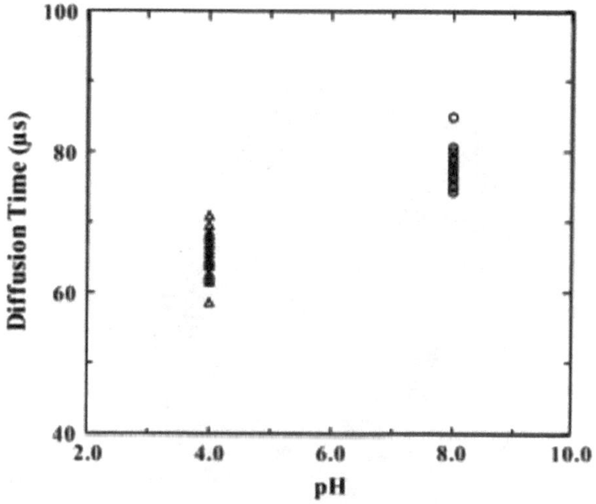


**Fig. 3.21** Diffusion times of TMR-labeled pHLIP in water at pH 4 and 8, respectively (Guo and Gai 2010)

To compare CPP- and SLO-mediated protein delivery systems, the investigators used GFP and ESRRB proteins, which are known to regulate pluripotency-related genes, for delivery into human bone marrow stromal cells (hBMSCs) and human testicular stromal cells (hTSCs). CPP-conjugated protein delivery was reported more efficient, lower cytotoxicity, and higher biological activity than SLO-mediated protein delivery system. These results thus suggest that delivery of CPP-conjugated proteins is an efficient tool for introducing biologically active proteins into cells. Enhanced and efficient cellular adsorption of agents with the help of CPP may have important implications in clinical cell-based therapy.

As said here, Jo et al. performed a few experiments to examine the intracellular protein uptake of the two delivery systems, CPP-conjugated and SLO-mediated (Jo et al. 2014). Suspended hTSCs and hBMSCs were treated with the following:

(i) 10 μg of CPP-GFP for CPP-conjugated protein delivery system, or
(ii) 10 μg of GFP for 1 hr after SLO-mediated pore formation for 50 min for SLO-mediated protein delivery system.

The efficiency of protein uptake was then compared with two treatments. Regarding uptake percentage of cells, the study reports to have observed a high efficiency of protein delivery using both of two different protein delivery systems, which is nearly 99%. The investigators then performed amount quantification of total intracellular protein uptake by confocal laser microscopy analysis. Results are presented in Fig. 3.22.

As shown in Fig. 3.22a, CPP-conjugated GFP was observed to deliver efficiently into hTSCs and hBMSCs. On the other hand, the SLO-mediated GFP was also observed to deliver but with a lower delivery efficiency compared with CPP-conjugated GFP. ImageJ software (NIH) was then used to measure the intensity of the intracellular GFP signal by performing the semiquantitative analysis of GFP uptake. The reported intensity of the CPP-GFP signal was found over

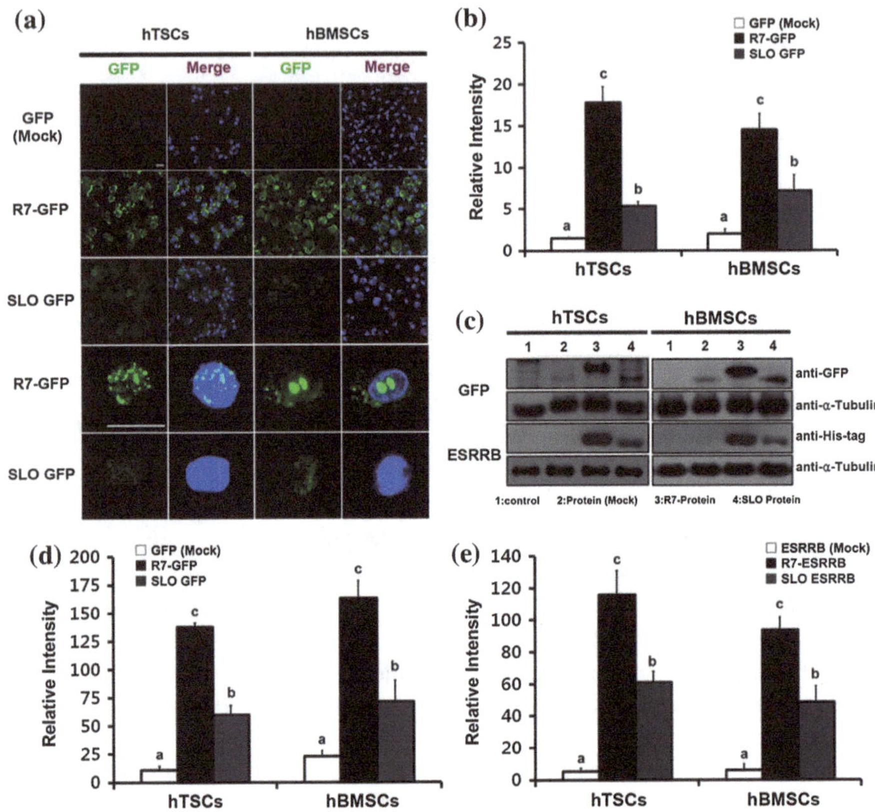


**Fig. 3.22** **a** Transduction of GFP and R7-GFP was detected by confocal microscopy. GFP or R7-GFP were visualized in green. Nuclei were counterstained with DAPI, and the images were merged (the top 3 rows show 400 × magnification, and the bottom 2 rows show 1000 × magnification plus 3 × zoom). Scale bars represent 20 μm. **b** Relative intensity of GFP. **c** Quantification of delivery of protein (GFP and ESRRB) and CPP-conjugated protein (R7-GFP and R7-ESRRB) was confirmed by Western blot. Lane 1, nontreated control; Lane 2, mock protein control (GFP and ESRRB); Lane 3, CPP-conjugated protein (R7-GFP and R7-ESRRB); Lane 4, SLO-mediated protein (GFP and ESRRB). All samples were normalized to α-tubulin. Full-length blot images are available in the online link for supplementary materials (see Jo et al. 2014). Relative intensities are shown for both GFP (**d**) and ESRRB (**e**). Data are presented as means ± SEM of three replicates. a, b, c Different superscripts represent significant differences ($p < 0.05$). For details, see Jo et al. (2014)

twofold that of the SLO-mediated GFP signal in hTSCs and hBMSCs (see Fig. 3.22b).

Western blot analysis was performed to investigate the difference in quantity of the two delivery methods. These results were similar to those of the confocal microscopy image data (see Fig. 3.22c–e). To compare the delivery efficiency of a large protein, 50 kDa ESRRB was transduced into hTSCs and hBMSCs. Although ESRRB weighs twofold more than GFP, the delivery efficiency was not decreased when compared with that of GFP delivery. Additionally, the cellular uptake of the CPP-ESRRB protein was more efficient than that of the SLO-mediated ESRRB protein (see Fig. 3.22a–e).

Jo et al. (2014) performed in vitro cytotoxicity assay to evaluate the cytotoxicity of the two protein delivery systems using two different assays as follows:

**First**, a cell viability assay was performed. Live cells were detected with calcein-AM (green signal), and dead cells were detected with ethidium homodimer-1 (red signal) (see Fig. 3.23a). The viability of the CPP-conjugated protein delivery system was 90.0% ± 1.26 in hTSCs and 85.9% ± 1.10 in hBMSCs, compared to the control. However, the viability of the SLO-mediated protein delivery system was 84.0% ± 0.70 in hTSCs and 76.4% ± 0.85 in hBMSCs. Comparison indicates that the pore-forming toxin significantly reduces cellular viability (see Fig. 3.23b).

**Second**, the cell apoptosis was investigated by the TUNEL assay. Although the CPP-conjugated protein delivery system induced apoptosis in only a few cells, the SLO-mediated protein delivery system induced apoptosis in over 4 and 10% of cells (see Fig. 3.23c).

Jo et al. then performed a comparison of the biological activity of ESRRB-delivered cells using two different protein delivery systems (Jo et al. 2014). To compare the biological activity of CPP-ESRRB and SLO-mediated ESRRB delivery, hTSCs and hBMSCs were treated, and the cells were collected 24 h after delivery. The biological activity of ESRRB was measured as the alteration of the expression levels of pluripotency-related genes using qRT-PCR. The expression levels of OCT4, SOX2, and NANOG showed a significant increase in the CPP-ESRRB-delivered cells compared with the cells treated by SLO-mediated ESRRB delivery (see Fig. 3.24a, b). To measure biological activity of ESRRB in other way, proliferation rate of control, CPP-conjugated and SLO-mediated ESRRB-delivered hTSCs and hBMSCs were calculated. The groups of CPP-ESRRB-delivered hTSCs and hBMSCs showed a significantly increased proliferation rate (see Fig. 3.24c). Differentiation potential was then examined to detect biological activity of delivered ESRRB. All three groups of hTSCs and hBMSCs were induced in vitro differentiation into three mesodermal lineage cells: adipogenic, chondrogenic, and osteogenic cells. In adipogenic differentiation, we confirmed elevated expression level of C/EBPα and PPARγ which are adipogenic-specific markers in both CPP-conjugated and SLO-mediated ESRRB-delivered cells compared to control cells. Importantly, CPP-ESRRB-delivered hTSCs and hBMSCs showed significantly higher expression levels in both markers than SLO-mediated ESRRB-delivered cells (see Fig. 3.24d). In case of chondrogenic differentiation, COMP and SOX9 were used as chondrogenesis-specific markers, and CPP-ESRRB-delivered hTSCs showed much higher expression level in COMP compared to control and SLO-mediated ESRRB-delivered cells. Besides, hBMSCs showed significantly increased efficiency in SOX9 expression level in both CPP- and SLO-mediated ESRRB-delivered cells, but significantly higher expression was detected in CPP-ESRRB-delivered cells than the SLO-mediated ESRRB-delivered cells (see Fig. 3.24e). However, no

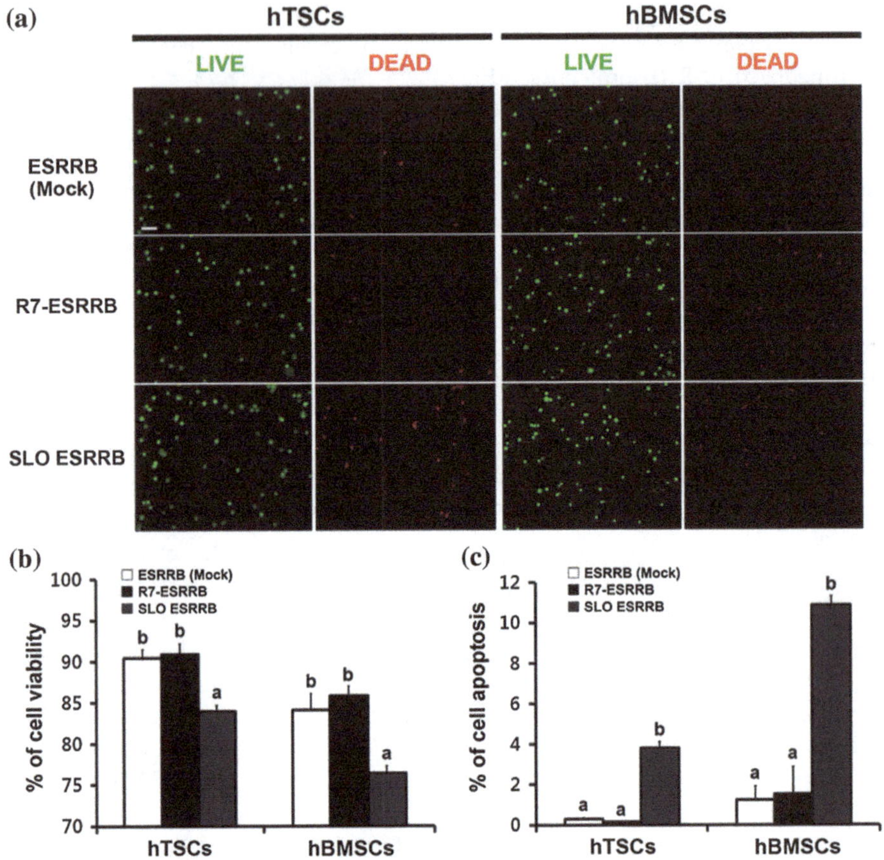


**Fig. 3.23** **a** Cell viability assay; live cells (green) and dead cells (red) were detected by fluorescence microscopy. **b** Live cells and dead cells were counted, and the ratio of live to dead cells was calculated. **c** Apoptosis was examined by the TUNEL assay. Scale bars represent 100 μm. Data are presented as means ± SEM of six (**b**) and three (**c**) replicates. a, b Different superscripts represent significant differences ($p < 0.05$). For details, see Jo et al. (2014)

significant difference was found in osteogenesis differentiation in all groups (Fig. 3.24f). Collectively, delivery of the large CPP-ESRRB protein was shown to be efficient and results in measurable biological activity levels.

### *3.4.5 CPP Selection*

CPPs vary greatly in size, amino acid sequence, and charge but they share the common feature that they have the ability to rapidly translocate the plasma membrane and enable delivery to the cytoplasm or nucleus (Lindgren et al. 2000).

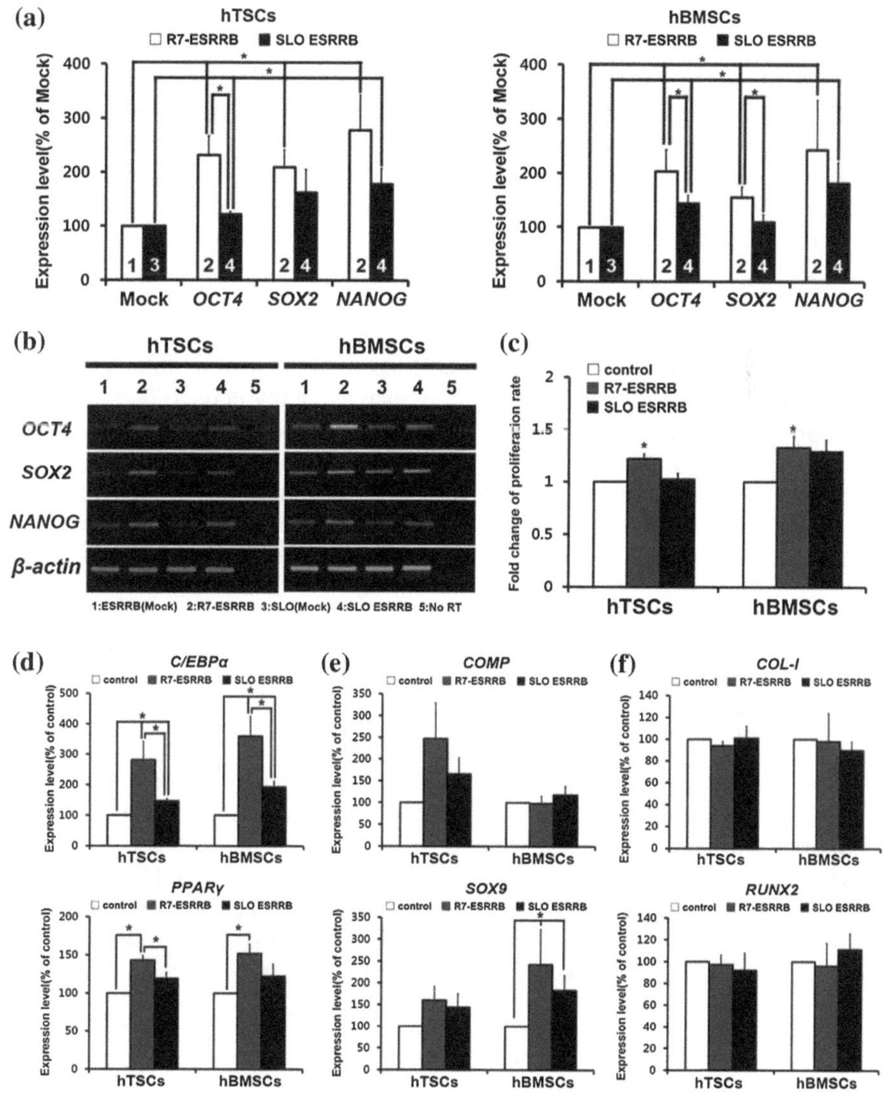


**Fig. 3.24** **a** Quantitative real-time RT-PCR and **b** RT-PCR analysis of pluripotency-related genes. Expression levels of OCT4, SOX2, and NANOG in hTSCs and hBMSCs were analyzed 24 h after R7-ESRRB or SLO-mediated ESRRB delivery. Each gene was normalized to β-actin as a housekeeping control (1: ESRRB(mock), 2: R7-ESRRB, 3: SLO(mock), 4: SLO ESRRB). **c** Altered proliferation rate of ESRRB-delivered hTSCs and hBMSCs was calculated as fold change of population doubling number. In vitro, differentiation analysis of the ESRRB-delivered cells was performed by quantitative real-time RT-PCR of **d** adipogenesis-specific markers C/EBPα and PPARγ, **e** chondrogenesis-specific markers COMP and SOX9, and **f** osteogenesis-specific markers COL-I and RUNX2. Data are presented as means ± SEM of four (**a**), six (**b**), three (**c**), and four (**d–f**) replicates. *$p < 0.05$. For details, see Jo et al. (2014)

The idea to use CPPs as delivery vectors originates from so-called membrane shuttling proteins such as the *Drosophila* homeobox protein antennapedia, the HIV-1 transcriptional factor TAT, and the capsid protein VP22 from HSV-1. Three decades ago, TAT was found to rapidly translocate over the cellular membrane, into the cytoplasm (Green and Loewenstein 1988). The same properties were later shown for a *Drosophila* homeobox protein. A short, 16 amino acid (aa) peptide derived from the third loop of the antennapedia protein was found responsible for the cellular translocation of the whole protein (Derossi et al. 1994). This pioneering work initiated the whole field to use peptides as efficient delivery vectors for bioactive compounds what we now know as cell-penetrating peptides.

Since the CPPs use as delivery vectors, a myriad of peptides has been reported to have the cell-penetrating properties. These peptides originate from different classes, either naturally occurring peptide sequences such as virally derived (TAT, VP22), from transcription factors (pAntp), chimeric peptides (transportan, MGP) or synthetic (polyarginines, Pep-1).

In 2013, Bechara and Sagan presented an overview on CPPs' 20 years (Bechara and Sagan 2013). According to the origin, CPPs are now classified into three distinguishable main classes:

(i) Peptides derived from proteins;
(ii) Chimeric peptides that are formed by the fusion of two natural sequences;
(iii) Synthetic CPPs which are rationally designed sequences usually based on structure–activity studies.

For a list on CPP selection, see Table 3.5.

Other attempts to classify CPPs, in spite of their diversity, were based on the physicochemical characteristics of the sequences, e.g., their amphipathicity (Ziegler 2008) or their hydrophobicity (Madani et al. 2011). A review summarizes the different classifications and the physicochemical properties of the so-far described CPPs (Milletti 2012). Even though CPPs have a great sequence variety, it is possible to identify three major classes: cationic, amphipathic, and hydrophobic. A broad overview of the current CPPs landscape is presented in Fig. 3.25 (see Milletti 2012).

**Table 3.5** Origin and sequences of some of the studied CPPs

| Peptide | Origin | Sequence | Reference |
|---|---|---|---|
| *Protein-derived* | | | |
| Penetratin | Antennapedia (43–58) | RQIKIWFQNRRMKWKK | Derossi et al. (1994) |
| Tat peptide | Tat(48–60) | GRKKRRQRRRPPQ | Vives et al. (1997) |
| pVEC | Cadherin (615–632) | LLIILRRRIRKQAHAHSK | Elmquist et al. (2001) |
| *Chimeric* | | | |
| Transportan | Galanin/ mastoparan | GWTLNSAGYLLGKINLKALAALAKKIL | Pooga et al. (1998) |

(continued)

**Table 3.5** (continued)

| Peptide | Origin | Sequence | Reference |
|---|---|---|---|
| MPG | HIV-gp41/ SV40 T-antigen | GALFLGFLGAAGSTMGAWSQPKKKRKV | Morris et al. (1997) |
| Pep-1 | HIV-reverse transcriptase/ SV40 T-antigen | KETWWETWWTEWSQPKKKRKV | Morris et al. (2001) |
| *Synthetic* | | | |
| Polyarginines | Based on Tat peptide | $(R)_n$; $6 < n < 12$ | Mitchell et al. (2000), Futaki et al. (2001) |
| MAP | de novo | KLALKLALKALKAALKLA | Oehlke et al. (1998) |
| $R_6W_3$ | Based on penetratin | RRWWRRWRR | Delaroche et al. (2007) |

The data presented in Fig. 3.25 contain more than 100 diverse CPPs (presented in Milletti (2012)) corresponding to the CPPs derived from heparan-, RNA- and DNA-binding proteins, signal peptides, AMPs, viral proteins, natural proteins, and peptide libraries. Most CPPs in the set have a net positive charge (83%); anionic CPPs do not form a class of their own, and they are assigned to different classes on a case-by-case basis; amphipathic CPPs, which comprise both cationic and anionic peptides, form the largest class (44%). Only 15% of the peptides are classified as hydrophobic (Milletti 2012).

### 3.4.6 *Cell Internalization Mechanisms of CPPs: a Model Demonstration*

Attempts have been made in order to elucidate the true mechanisms of peptide-mediated uptake. But the results are divergent. It seems apparent that different peptides utilize different cell uptake pathways (Thorén et al. 2003).

CPP translocation mechanisms suggested in early studies that the internalization of these peptides was not inhibited by depletion of the cellular adenosine triphosphate (ATP) pool, low temperature (+4 °C), or inhibitors of endocytosis (Langel 2002). Chemical modifications of the peptide sequences, such as the synthesis of retro, enantio, or retroenantio analogs, appeared not to affect the internalization properties (Derossi et al. 1996). Translocation was thought to result from direct transfer through the cell membrane lipid bilayer. Figure 3.26 demonstrates the CPP translocation through membrane (Bechara and Sagan 2013).

The inverted micelle formation was the proposed mechanism for these uptakes. Here, cationic residues interact with negatively charged plasma membrane followed

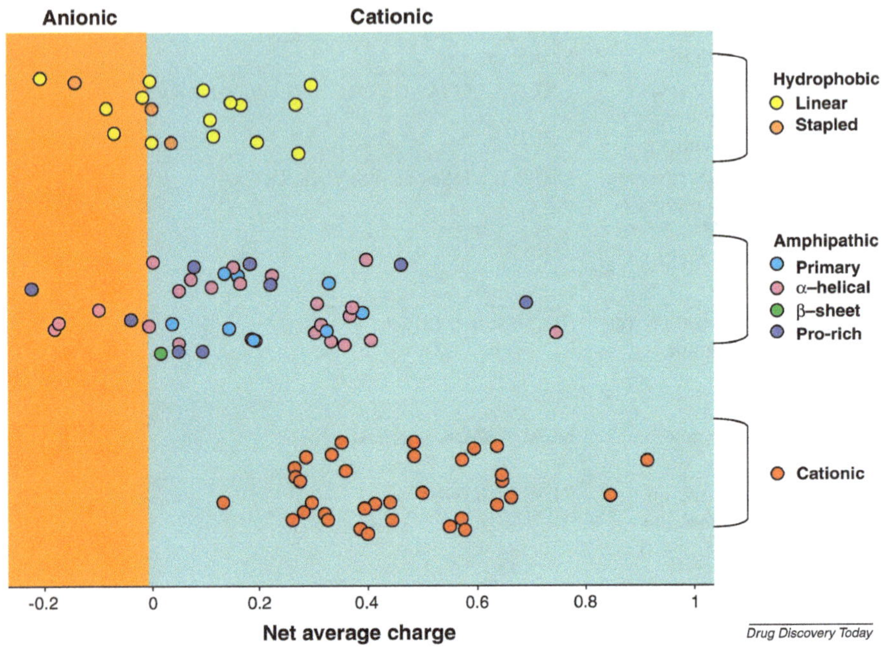


**Fig. 3.25** Distribution of CPPs by their net average charge and class. Anionic CPPs can be classified as hydrophobic or amphipathic CPPs. By contrast, many cationic CPPs are highly charged peptides, without any amphipathic arrangement or hydrophobic character. Copied from Milletti (2012)

by invagination of tryptophans into the membrane, inducing inverted micelle formation (Derossi et al. 1996). The mechanism of inverted micelle formation seems to explain some aspects of CPP translocation and still appear to hold for some peptides used in cellular delivery of bioactive molecules (see Fig. 3.26).

Direct translocation involves plasma membrane destabilization, in an energy- and temperature-independent manner. Evoked at first as the mechanism of membrane internalization of CPPs and then refuted as an artifact of fixation, direct translocation was later confirmed using fluorescence in living cells, quantification of the uptake at 4 °C and in the absence of endocytosis (e.g., ATP depletion), and using various biophysical approach in model systems (Jiao et al. 2009; Saalik et al. 2011; Thoren et al. 2004; Binder and Lindblom 2003; Elmquist and Langel 2003). Noteworthy, while endocytosis in inhibited at 4 °C, direct translocation is also decreased for membrane dynamics and fluidity is affected at such low temperature (Zaro and Shen 2003, 2005). Thus, assessing direct translocation at low temperatures in living cells leads to an underestimation of this latter. Various hypotheses have been reported to explain direct translocation of CPPs across the lipid bilayer.

CPP translocation in other studies is considered mostly an energy-dependent process (Vives et al. 2003). Mechanisms may involve extracellular heparan sulfate

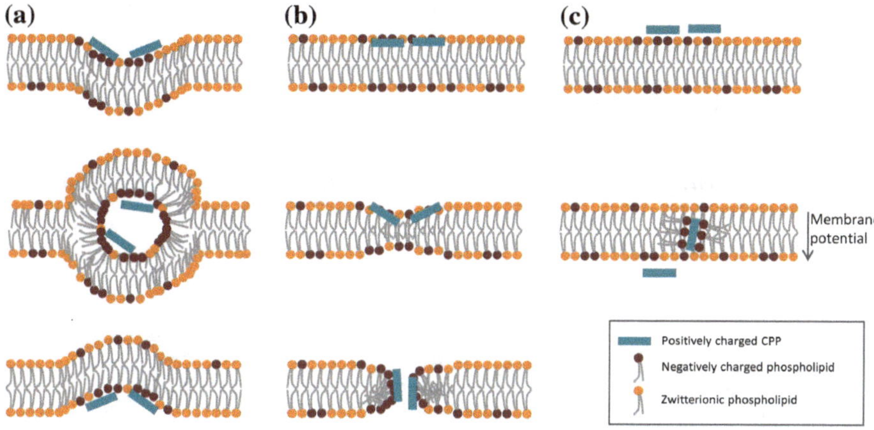


**Fig. 3.26** Examples of the proposed mechanisms for direct translocation. **a** Inverted micelle formation. **b** Pore formation. **c** Adaptive translocation

(Console et al. 2003) and different types of endocytosis (Richard et al. 2003; Fittipaldi et al. 2003) such as macropinocytosis, clathrin-dependent endocytosis, caveoale-dependent endocytosis or clathrin endocytosis, and caveoale-independent endocytosis. These processes do not necessarily contradict each other rather demonstrating that peptide-mediated membrane translocation is mediated by several different pathways simultaneously. Different peptides may also utilize diverse uptake mechanisms depending on their cargo and biophysical properties (Fischer et al. 2002). The endosomal pathways contribute to the major route of uptake of CPPs, although several reports show an uptake independent of endocytosis (Thorén et al. 2003; Henriques et al. 2005; Deshayes et al. 2004). The cell uptake mechanism could also be altered by different cargoes if the peptides form a stable complex with its cargo and if the cargo is covalently bound to a CPP or how the cargo is attached (Silhol et al. 2002). For schematic representation, see Fig. 3.27.

## 3.5 Bacterial Cell Uptake of Antibiotics: Antibiotic Influx Across Bacterial Membrane

Examining the bacterial outer membrane (OM) permeability requires attention to address the antibiotic uptake by cells (Nikaido 2003). Pathway of influx of antibiotics was inspected rigorously in a review by Nikaido (2003). Agents that contain a strong acid group, a quaternary ammonium group, or multiple charged groups are expected to have difficulty in crossing the bilayer regions of the OM. This is especially so because the lipopolysaccharides (LPS)-containing bilayer of the OM shows permeability to lipophilic probes that is about two orders of magnitude lower than that of the conventional phospholipid bilayer membranes (Plesiat

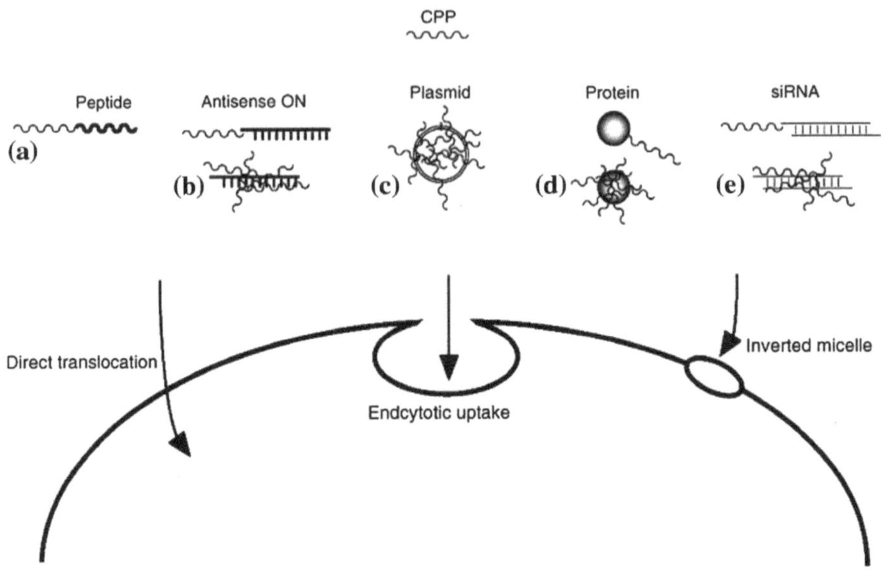


**Fig. 3.27** Suggested uptake mechanisms for CPPs and examples of delivered cargoes. (A) CPP and peptide in single amino acid chain. (B) ON either in complex or covalently linked. (C) Plasmid in complex by electrostatic interaction. (D) Protein either as fusion protein or in complex with CPP. (E) siRNA, covalently linked or as complex

and Nikaido 1992). Increased susceptibility in the mutant or in the PMBN-treated bacterium indicates strongly that the major diffusion pathway is through the OM bilayer. In Fig. 3.28, pretty old sketch on the schematic representation of bacterial membrane is shown (Chopra and Ball 1982); for a simpler schematic diagram; see Galdiero et al. (2012). There are three principal layers, namely the OM, the peptidoglycan cell wall, and the inner membrane 'IM'. The two membrane layers delimit two aqueous compartments, the cytoplasm and the periplasm.

Bacterial OMs, like other biological membranes, contain channel-forming proteins for membrane permeabilization. One such nonspecific diffusion channel-forming protein, porin (see Fig. 3.29), was discovered in 1976 (Nakae 1976). Porins were found in every species of gram-negative bacteria investigated and even in a group of 'gram-positive' bacteria, the *Corynebacterium-Nocardia-Mycobacterium* complex, which produces a lipid-rich, bilayer-like 'cell wall.' Bacterial OM-residing porins are found with the β-barrel structures.

Taking the theoretical considerations together with the available experimental data, it is likely that small agents (*β*-lactams, tetracycline, chloramphenicol, and fluoroquinolones) use mainly the porin channels for penetration, at least in the *Enterobacteriaceae*, with their high-permeability porins. In contrast, large, lipophilic agents, such as macrolides, rifamycins, novobiocin, and fusidic acid, find difficulties in diffusing through the porin channels, and therefore even the slow diffusion across the lipid bilayer becomes significant.

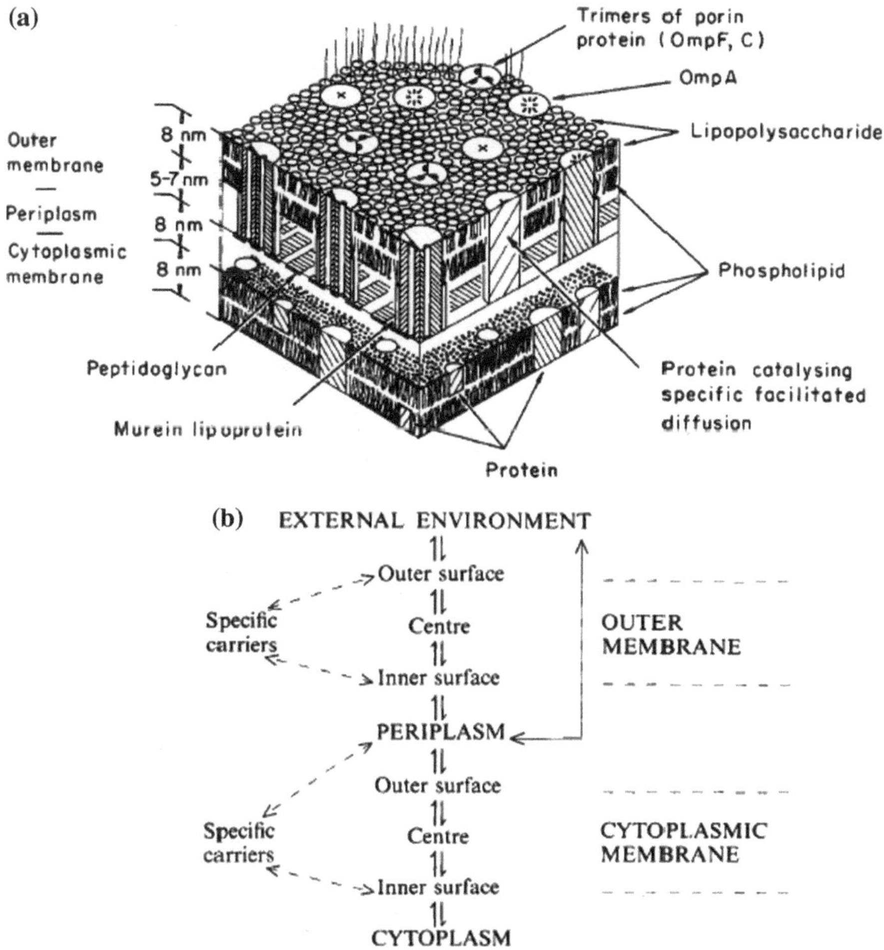


**Fig. 3.28** Schematic models depicting envelope structure (**a**) in light of model structure. **b** The antibiotic entry routes in gram-negative bacteria. Copied from Chopra and Ball (1982)

Porins exert a profound influence on the intrinsic level of antibiotic resistance in gram-negative bacteria. Thus, *Pseudomonas aeruginosa* and other members of the fluorescent pseudomonads are known to show intrinsic resistance to a wide range of antibiotics, in comparison with the members of the *Enterobacteriaceae*. This is mainly due to the use of a slow porin, rather than a classical trimeric porin, as the major channel. OM's low permeability becomes even more effective in preventing the influx of antibiotics, thanks to the synergistic effect of the multidrug efflux complex, which extrudes drugs directly into the medium (Nikaido 1994, 1996). Mechanisms that determine the OM barrier properties through the modification of porin and other effective channel expression and/or function may have a huge impact on the sensitivity of gram-negative bacteria to antibiotics.

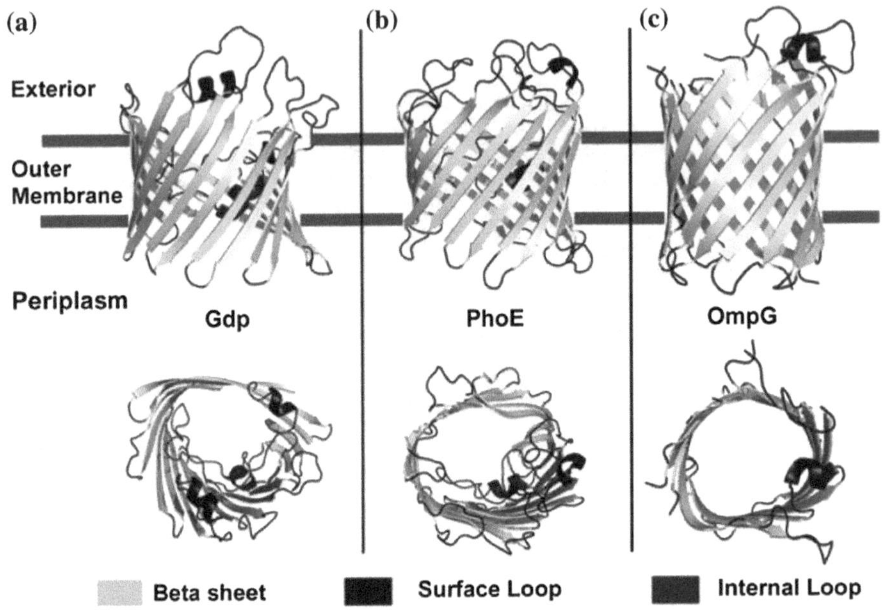


**Fig. 3.29** 3D structures of three different porins: Gdp from *Rhodobacter capsulatus*; PhoE and OmpG from *Escherichia coli*. Surface and internal loops are shown in dark gray. The extracellular space is located at the top of the figure and the periplasmic space at the bottom (Galdiero et al. 2012)

**Spectrometric Insights into the Bacterial Cell Uptake of Antibiotics**
Bacterial multidrug resistance has always been a significant health issue. Understanding the membrane permeation of antibiotics in clinically relevant bacterial pathogens is crucial. Last year, Cinquin et al. (2015) published a paper to demonstrate the microspectroscopic insights into the subject. Getting a medically relevant required antibiotic concentration inside a bacterium is a pivotal step for most antibacterials. Spectrometric methodology has been developed for drug detection inside bacteria, and recent studies are dedicated to bacterial cell imaging. Cinquin et al. (2015) demonstrated a method to quantify the timescale of antibiotic accumulation in living bacterial cells. Tunable ultraviolet excitation provided by DISCO beamline (Synchrotron Soleil) combined with microscopy allows spectroscopic analysis of the antibiotic signal in individual bacterial cells. Here, real-time quantification of drug has been performed.

Methods of quantifying the kinetics of antibiotic accumulation at the single bacterium level and the role of membrane-associated mechanisms of resistance on this process have been demonstrated in the techniques (we avoid presenting the detail that is published in Jamme et al. 2010, 2013). The techniques combine the spatial resolution of microscopy with the high spectral definition required to distinguish drug-related signal from autofluorescence due to the UV excitation. Deep ultraviolet (DUV) photons, provided by synchrotron radiation, bring excellent tunability and sufficient flux to follow label-free antibiotics inside cells.

To obtain reliable measurements on each individual bacterium, segmentation of the signal needed to be carried out to provide an appropriate 'outline' of the cells. In the case studies of Cinquin et al. using three antibacterial agents, the bacterium covered a relatively small area; therefore, the amount of information per pixel was sensitive to the area selected for analysis (Cinquin et al. 2015). Using a 100 X objective, the bacterium was only represented by a few hundred pixels. Manually drawing a contour around the cell led to bias that under- or overestimated the area of the bacterium (see in Fig. 3.30). To avoid adding noise from fragments, unfocused, or aggregated bacteria, measurements were performed only in bacteria with an area between 70 and 300 pixels.

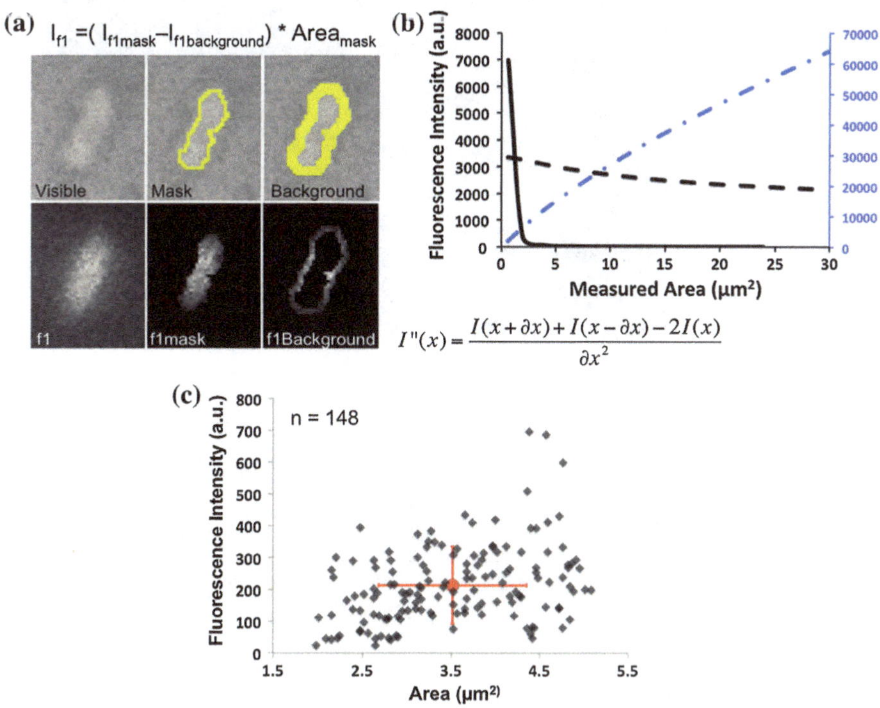


**Fig. 3.30** Segmentation and sorting strategies. **a** DUV microscope images. Top panels (from left to right), a bacterium under «visible» transmitted light, the automatic computed mask, and the background mask. Lower panels, the emission intensity of bacterium under 275 nm excitation (30 s acquisition) in the filter 1 (327–353 nm) that selects the autofluorescence. To measure the fluorescence of the bacterium, its total intensity of fluorescence is subtracted by the average intensity selected by the background mask reported for the whole bacteria and divided by the bandwidth of the used filter. **b** Total intensity of fluorescence depending of area centered on a bacterium (blue dotted line), average intensity of fluorescence (black dotted curve) and second derivative of the total intensity of fluorescence calculated using the formula written under the graph (black curve). **c** Fluorescence intensity from autofluorescence channel for each bacterium against size. For detailed explanation, see Cinquin et al. (2015)

Image acquisitions of untreated bacteria were performed to measure the extent of autofluorescence collected by filter (see Fig. 3.30). A second set of images were acquired with bacteria incubated with drugs at different time points in the presence or absence of the efflux blocker carbonyl cyanide m-chlorophenylhydrazone (CCCP). The cross talk correlation effect (Trón et al. 1984) was used to finalize the fluorescence intensity (see Fig. 3.31). Correcting for cross talk was fundamental since the emissions were close to each other in space with significant overlap between the autofluorescence emission and the antibiotic emission.

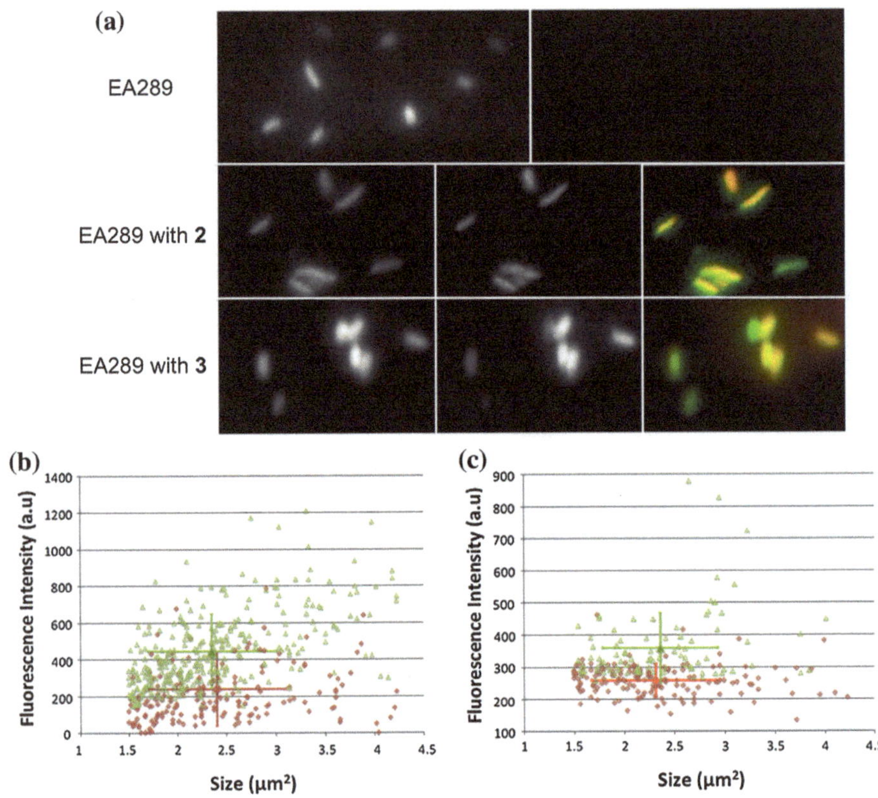


**Fig. 3.31** Fluorescence intensity using cross talk correlation (see Cinquin et al. 2015). **a** Fluorescence images of EA289 bacteria excited with a wavelength of 275 nm, untreated (top panel), and incubated with an antibacterial agent (middle panel) or with another antibacterial agent (lower panel): emission from filter 1 (327–353 nm; left panel); emission from filter 2 (370–410 nm for compound **2** and 420–480 nm for compound **3**; middle panel) corrected from cross talk effect; merge of the two precedent images: filter 1 in green and filter 2 in red (right panel). **b** Fluorescence corrected from cross talk effect in function of its size from each bacterium treated with compound **2** (red diamonds) or with compound **2**+ CCCP (green triangles) and **c** with compound **3** (red diamonds) or with compound **3**+ CCCP (green triangles). The red dots with bars and green dots with bars are the average with standard deviations for each population. For details, see Cinquin et al. (2015)

Translocation through the bacterial membrane (influx) and the activity of efflux pumps (efflux) may need to be jointly addressed to fully understand the required intracellular concentration of antibiotic close to the target. It is important to visualize the accumulation rate of an antibacterial agent, to determine its concentration inside the bacterial cell. Assays suggest that the maximum (steady state) of accumulation is reached at 20 min in EA298 (see data in Cinquin et al. 2015). Besides, more pinpointed developments are yet to be made to address the sub-cellular localization of the antibiotic molecules inside various bacterial cell compartments including membrane, cytoplasm.

## References

G. A. Duncan, D. H. Fairbrother, and M. A. Bevan. Diffusing colloidal probes of cell surfaces. Soft Matter, 2016,12, 4731–4738.

A. E. Nel, L. Mädler, D. Velegol, T. Xia, E. M. V. Hoek, P. Somasundaran, F. Klaessig, V. Castranova, and M. Thompson. Understanding biophysicochemical interactions at the nano–bio interface. Nature Materials 8, 543–557 (2009).

E. C. Wang and A. Z. Wang. Nanoparticles and their applications in cell and molecular biology. Integr Biol (Camb). 2014 Jan; 6(1): 9–26.

E. C. Cho, Q. Zhang, and Y. Xia. The effect of sedimentation and diffusion on cellular uptake of gold nanoparticles. Nat. Nanotechnol. 2011 Apr 24; 6(6): 385–391.

B. Halamoda-Kenzaoui, M. Ceridono, P. Colpo, A. Valsesia, P. Urbán, I. Ojea-Jiménez, S. Gioria, D. Gilliland, F. Rossi, A. K. Ovaskainen. 2015. Dispersion Behaviour of Silica Nanoparticles in Biological Media and Its Influence on Cellular Uptake. PLoS ONE 10(10): e0141593.

D A Berk, A Clark, Jr, and R M Hochmuth. Analysis of lateral diffusion from a spherical cell surface to a tubular projection. Biophys J. 1992 Jan; 61(1): 1–8.

S. Zhang, H. Gao, and G. Bao. Physical Principles of Nanoparticle Cellular Endocytosis. ACS Nano, 2015, 9 (9), pp 8655–8671.

Allen TM, Cullis PR. Drug delivery systems: entering the mainstream. Science, 2004, 303: 1818–1822.

Connor EE, Mwamuka J, Gole A, Murphy CJ, Wyatt MD. Gold nanoparticles are taken up by human cells but do not cause acute cytotoxicity. Small, 2005, 1: 325–327.

Gao H, Shi W, Freund LB. Mechanics of receptor-mediated endocytosis. Proc Natl Acad Sci USA, 2005, 102: 9469–9474.

Ghosh P, Han G, De M, Kim CK, Rotello VM. Gold nanoparticles in delivery applications. Adv Drug Delivery Rev, 2008, 60: 1307–1315.

Meyers MA, Mishra A, Benson DJ. Mechanical properties of nanocrystalline materials. Prog Mater Sci, 2006, 51: 427–556.

Shubayev VI, Pisanic II TR, Jin S. Magnetic nanoparticles for theragnostics. Adv Drug Delivery Rev, 2009, 61: 467–477.

Geng Y, Dalhaimer P, Cai S, Tsai R, Tewari M, Minko T, Discher DE. Shape effects of filaments versus spherical particles in flow and drug delivery. Nat Nanotechnol, 2007, 2: 249–255.

Paciotti GF, Myer L, Weinreich D, Goia D, Pavel N, McLaughlin RE, Tamarkin L. Colloidal gold: a novel nanoparticle vector for tumor directed drug delivery. Drug Delivery, 2004, 11: 169–183.

Ai J, Biazar E, Jafarpour M, Montazeri M, Majdi A, Aminifard S, Zafari M, Akbari HR, Rad HG. Nanotoxicology and nanoparticle safety in biomedical designs. Int J Nanomedicine, 2011, 6: 1117–1127.

Curtis J, Greenberg M, Kester J, Phillips S, Krieger G. Nano-technology and nanotoxicology. Toxicol Rev, 2006, 25: 245–260.

Yue T, Zhang X. Signal transduction across cellular membranes can be mediated by coupling of the clustering of anchored proteins in both leaflets. Phys Rev E, 2012, 85: 011917.

Ashrafuzzaman, M., 2015a. Diffusion across cell phase states. Biomedical Sci. Today. 1:e4.

Ashrafuzzaman, M., 2015b. Phenomenology and energetics of diffusion across cell phase states. Saudi J. of Biol. Sci., 22: 666–673.

Lapotko DO, Lukianova-Hleb EY, Oraevsky AA. Clusterization of nanoparticles during their interaction with living cells. Nanomedicine, 2007, 2: 241–253.

Ginzburg VV, Balijepalli S. Modeling the thermodynamics of the interaction of nanoparticles with cell membranes. Nano Lett, 2007, 7: 3716–3722.

Yang K, Ma YQ. Computer simulation of the translocation of nanoparticles with different shapes across a lipid bilayer. Nat Nanotechnol, 2010, 5: 579–583.

Lu F, Wu SH, Hung Y, Mou CY. Size effect on cell uptake in well-suspended, uniform mesoporous silica nanoparticles. Small, 2009, 5: 1408–1413.

Davda J, Labhasetwar V. Characterization of nanoparticle uptake by endothelial cells. Int J Pharm, 2002, 233: 51–59.

Qhobosheane M, Santra S, Zhang P, Tan W. Biochemically functionalized silica nanoparticles. Analyst, 2001, 126: 1274–1278.

Nativo P, Prior IA, Brust M. Uptake and intracellular fate of surface-modified gold nanoparticles. ACS Nano, 2008, 2: 1639–1644.

Pan Y, Neuss S, Leifert A, Fischler M, Wen F, Simon U, Schmid G, Brandau W, Jahnen-Dechent W. Size-dependent cytotoxicity of gold nanoparticles. Small, 2007, 3: 1941–1949.

Li Y, Yuan H, von dem Bussche A, Creighton M, Hurt RH, Kane AB, Gao H. Graphene microsheets enter cells through spontaneous membrane penetration at edge asperities and corner sites. Proc Natl Acad Sci USA, 2013, 110: 12295–12300.

J.-P. Chapel and J.-F. Ersatile. electrostatic assembly of nanoparticles and polyelectrolytes: Coating, clustering and layer – by-layer processes, Berret, Curr. Opin. Colloid Interface Sci., 2012,17, 268.

Gindy ME, Prud'homme RK. Multifunctional nanoparticles for imaging, delivery and targeting in cancer therapy. Expert Opin Drug Deliv. 2009;6(8):865–878.

Ruoslahti E, Bhatia SN, Sailor MJ. Targeting of drugs and nanoparticles to tumors. J Cell Biol. 2010;188(6):759–768.

Sarabjeet Singh Suri, Hicham Fenniri, and Baljit Singh, Nanotechnology-based drug delivery systems. Journal of Occupational Medicine and Toxicology, 2007, 2:16.

P. Couvreur, G. Barratt, E. Fattal, P. Legrand, C. Vauthier, Nanocapsule technology: a review, Crit. Rev. Ther. Drug Carrier Syst., 19 (2002), pp. 99–134.

Rajesh Singh, James W. Lillard Jr., Nanoparticle-based targeted drug delivery. Exp Mol Pathol. 2009 June; 86(3):215–223.

A. Satoh, Introduction to Practice of Molecular Simulation: Molecular Dynamics, Monte Carlo, Brownian Dynamics, Lattice Boltzmann and Dissipative Particle Dynamics, Elsevier, Amsterdam, 2011.

Alder, B. J.; T. E. Wainwright (1959). "Studies in Molecular Dynamics. I. General Method". J. Chem. Phys. 31 (2): 459.

M. Nedyalkova, S. Madurga, S. Pisov, I. Pastor, E. Vilaseca and F. Mas, J. Chem. Phys., 2012, 137.

Stoll, S. (2014). Computer Simulations of Soft Nanoparticles and Their Interactions with DNA-Like Polyelectrolytes.

Ashrafuzzaman, M., and J. Tuszynski. 2012a. Membrane Biophysics, Springer, Heidelberg, Germany.

Ashrafuzzaman, M., and J. A. Tuszynski. 2012b. Regulation of channel function due to coupling with a lipid bilayer, J. Comput. Theor. Nanosci. 9: 564–570.

Ashrafuzzaman, M. C.-Y. Tseng, and J.A. Tuszynski. 2014. Regulation of channel function due to physical energetic coupling with a lipid bilayer. Biochemical and Biophysical Research Communications. 445:463–468.

D A Mannock, R N Lewis, R N McElhaney, M Akiyama, H Yamada, D C Turner, and S M Gruner. Effect of the chirality of the glycerol backbone on the bilayer and nonbilayer phase transitions in the diastereomers of di-dodecyl-beta-D-glucopyranosyl glycerol. Biophys J. 1992 Nov; 63(5): 1355–1368.

D. A. Mannock, R. N. McElhaney, P. E. Harper, and S. M. Gruner. Differential scanning calorimetry and X-ray diffraction studies of the thermotropic phase behavior of the diastereomeric di-tetradecyl-beta-D-galactosyl glycerols and their mixture. Biophys J. 1994 Mar; 66(3 Pt 1): 734–740.

Ashrafuzzaman M., Beck H. (2004a) in Vortex dynamics in two-dimensional Josephson junction arrays, (University of Neuchatel, http://doc.rero.ch/record/2894?ln=fr), ch 5, p 85.

M Ashrafuzzaman, H Beck. (2004b). Vortex dynamics in dilute two-dimensional Josephson junction arrays. Journal of magnetism and magnetic materials 272, 284–285.

Nadia Ruthardt, Don C Lamb and Christoph Bräuchle. Single-particle Tracking as a Quantitative Microscopy-based Approach to Unravel Cell Entry Mechanisms of Viruses and Pharmaceutical Nanoparticles. Molecular Therapy (2011) 19 7, 1199–1211.

Andrew I. Shevchuk, Phil Hobson, Max J. Lab, David Klenerman, Nina Krauzewicz, Yuri E. Korchev. Imaging Single Virus Particles on the Surface of Cell Membranes by High-Resolution Scanning Surface Confocal Microscopy. Biophysical Journal, Volume 94, Issue 10, 15 May 2008, Pages 4089–4094.

Gerard D. Byrne, Driton Vllasaliu, Franco H. Falcone, Michael G. Somekh, and Snjezana Stolnik. Live Imaging of Cellular Internalization of Single Colloidal Particle by Combined Label-Free and Fluorescence Total Internal Reflection Microscopy. Mol. Pharmaceutics 2015, 12, 3862–3870.

Nuri Oh and Ji-Ho Park. Endocytosis and exocytosis of nanoparticles in mammalian cells. Int J Nanomedicine. 2014; 9(Suppl 1): 51–63.

Pelkmans L, Kartenbeck J, Helenius A. Caveolar endocytosis of simian virus 40 reveals a new two-step vesicular-transport pathway to the ER. Nat Cell Biol. 2001; 3:473–483.

Lakadamyali M, Rust MJ, Babcock HP, and Zhuang X. Visualizing infection of individual influenza viruses. Proc Natl Acad Sci USA. 2003;100:9280–9285.

Rust MJ, Lakadamyali M, Zhang F, and Zhuang X. Assembly of endocytic machinery around individual influenza viruses during viral entry. Nat Struct Mol Biol. 2004;11:567–573.

Ewers H, Smith AE, Sbalzarini IF, Lilie H, Koumoutsakos P, and Helenius A. Single-particle tracking of murine polyoma virus-like particles on live cells and artificial membranes. Proc Natl Acad Sci USA. 2005;102:15110–15115. [PMC free article] [PubMed].

Payne CK. Imaging gene delivery with fluorescence microscopy. Nanomedicine (Lond) 2007;2:847–860. [PubMed].

Brandenburg B, Lee LY, Lakadamyali M, Rust MJ, Zhuang X, and Hogle JM. Imaging poliovirus entry in live cells. PLoS Biol. 2007;5:e183. [PMC free article] [PubMed].

Watson P, Jones AT, and Stephens DJ. Intracellular trafficking pathways and drug delivery: fluorescence imaging of living and fixed cells. Adv Drug Deliv Rev. 2005;57:43–61.

Rémy-Kristensen A, Clamme JP, Vuilleumier C, Kuhry JG, and Mély Y. Role of endocytosis in the transfection of L929 fibroblasts by polyethylenimine/DNA complexes. Biochim Biophys Acta. 2001;1514:21–32.

Kopatz I, Remy JS, and Behr JP. A model for non-viral gene delivery: through syndecan adhesion molecules and powered by actin. J Gene Med. 2004;6:769–776. [PubMed].

Rejman J, Bragonzi A, and Conese M. Role of clathrin- and caveolae-mediated endocytosis in gene transfer mediated by lipo- and polyplexes. Mol Ther. 2005;12:468–474. [PubMed].

Rinne J, Albarran B, Jylhävä J, Ihalainen TO, Kankaanpää P, Hytönen VP. et al. (2007 Internalization of novel non-viral vector TAT-streptavidin into human cells BMC Biotechnol 71.

Lundin P, Johansson H, Guterstam P, Holm T, Hansen M, Langel U. et al. (2008) Distinct uptake routes of cell-penetrating peptide conjugates Bioconjug Chem 192535–2542.2542 [PubMed].

Mäe M, Andaloussi SE, Lehto T, and Langel U. Chemically modified cell-penetrating peptides for the delivery of nucleic acids. Expert Opin Drug Deliv. 2009;6:1195–1205. [PubMed].

Helenius A, Kartenbeck J, Simons K, and Fries E. On the entry of Semliki forest virus into BHK-21 cells. J Cell Biol. 1980;84:404–420.

Marsh M, and Bron R. SFV infection in CHO cells: cell-type specific restrictions to productive virus entry at the cell surface. J Cell Sci. 1997;110 (Pt 1):95–103.

Döhner K, and Sodeik B. The role of the cytoskeleton during viral infection. Curr Top Microbiol Immunol. 2005;285:67–108.

von Gersdorff K, Sanders NN, Vandenbroucke R, De Smedt SC, Wagner E, and Ogris M. The internalization route resulting in successful gene expression depends on both cell line and polyethylenimine polyplex type. Mol Ther. 2006;14:745–753.

Marsh M, and Helenius A. Virus entry: open sesame. Cell. 2006;124:729–740.

Godinez WJ, Lampe M, Wörz S, Müller B, Eils R, and Rohr K. Deterministic and probabilistic approaches for tracking virus particles in time-lapse fluorescence microscopy image sequences. Med Image Anal. 2009;13:325–342.

Sbalzarini IF, and Koumoutsakos P. Feature point tracking and trajectory analysis for video imaging in cell biology. J Struct Biol. 2005;151:182–195.

Saxton MJ, and Jacobson K. Single-particle tracking: applications to membrane dynamics. Annu Rev Biophys Biomol Struct. 1997;26:373–399.

Karla de Bruin, Nadia Ruthardt, Katharina von Gersdorff, Ralf Bausinger, Ernst Wagner, Manfred Ogris, and Christoph Bräuchle. Cellular Dynamics of EGF Receptor–Targeted Synthetic Viruses. Molecular Therapy (2007) 15 7, 1297–1305.

Bausinger R, von Gersdorff K, Braeckmans K, Ogris M, Wagner E, Bräuchle C. et al. (2006 The transport of nanosized gene carriers unraveled by live-cell imaging Angew Chem Int Ed Engl 451568–1572.1572.

Sauer AM, de Bruin KG, Ruthardt N, Mykhaylyk O, Plank C, andBräuchle C. Dynamics of magnetic lipoplexes studied by single particle tracking in living cells. J Control Release. 2009;137:136–145.

Urs F. Greber. Virus and Host Mechanics Support Membrane Penetration and Cell Entry. J. Virol. April 2016 vol. 90 no. 8 3802–3805.

Mercer J, Schelhaas M, Helenius A. 2010. Virus entry by endocytosis. Annu Rev Biochem 79:803–833. https://doi.org/10.1146/annurev-biochem-060208-104626. CrossRefMedlineGoogle Scholar.

Suomalainen M, Greber UF. 2013. Uncoating of non-enveloped viruses. Curr Opin Virol 3:27–33. https://doi.org/10.1016/j.coviro.2012.12.004. CrossRefMedlineGoogle Scholar.

Yamauchi Y, Greber UF. 15 February 2016, posting date. Principles of virus uncoating: cues and the snooker ball. Traffic https://doi.org/10.1111/tra.12387.

Bao G, Suresh S. 2003. Cell and molecular mechanics of biological materials. Nat Mater 2:715–725. https://doi.org/10.1038/nmat1001.

Stefania Luisoni, Maarit Suomalainen, Karin Boucke, Lukas B. Tanner, Markus R. Wenk, Xue Li Guan, Michał Grzybek, Ünal Coskun, Urs F. Greber. Co-option of Membrane Wounding Enables Virus Penetration into Cells. Cell Host & Microbe, Volume 18, Issue 1, 8 July 2015, Pages 75–85.

M. Luo. Influenza Virus Entry. In Viral Molecular Machines, Advances in Experimental Medicine and Biology 726 (Editors: Michael G. Rossmann and Venigalla B. Rao). Springer Science +Business Media (www.springer.com), 2012. Page 201–223.

Bottcher C, Ludwig K, Herrmann A, van Heel M, Stark H (1999) Structure of influenza haemagglutinin at neutral and at fusogenic pH by electron cryo-microscopy. FEBS Lett 463:255–259 Link: http://www.sumanasinc.com/webcontent/animations/content/influenza.html.

Peter Järver, Ülo Langel. (2006) Cell-penetrating peptides—A brief introduction. Biochimica et Biophysica Acta (BBA) – Biomembranes. Volume 1758, Issue 3, March 2006, Pages 260–263.

M. Lindgren, M. Hallbrink, A. Prochiantz, Ü. Langel. Cell-penetrating peptides. Trends Pharmacol. Sci., 21 (2000), pp. 99–103.

Lin Guo and Feng Gai. Heterogeneous Diffusion of a Membrane-Bound pHLIP Peptide. Biophys J. 2010 Jun 16; 98(12): 2914–2922.

Liebman P.A., Pugh E.N. The control of phosphodiesterase in rod disk membranes: kinetics, possible mechanisms, and significance for vision. Vision Res. 1979;19:375–380. [PubMed].

Pastan I.H., Willingham M.C. Journey to the center of the cell: role of the receptosome. Science. 1981;214:504–509. [PubMed].

Hackenbrock C.R. Lateral diffusion and electron transfer in the mitochondrial inner membrane. Trends Biochem. Sci. 1981;6:151–154.

Koppel D.E. Measurement of membrane protein lateral mobility. In: Hesketh T.R., Kornberg H.L., Metcalfe J.C., Northcote D.H., Pogson C.I., Tipton K.F., editors. Techniques in the Life Sciences, B4/II. Lipid and Membrane Biochemistry. Elsevier Biomedical Press; County Clare, Ireland: 1982.

Gross D., Webb W.W. Molecular counting of low-density lipoprotein particles as individuals and small clusters on cell surfaces. Biophys. J. 1986;49:901–911. [PMC free article] [PubMed].

Peters R. Lateral mobility of proteins and lipids in the red cell membrane and the activation of adenylate cyclase by β-adrenergic receptors. FEBS Lett. 1988;234:1–7. [PubMed].

Popot J.-L., Engelman D.M. Helical membrane protein folding, stability, and evolution. Annu. Rev. Biochem. 2000;69:881–922.

Popot J-L, Engelman DM. Membrane protein folding and oligomerization: the two-stage model. 1990. Biochemistry 29:4031–37.

B. J. Bormann and D. M. Engelman. INTRAMEMBRANE HELIX-HELIX ASSOCIATION IN OLIGOMERIZATION AND TRANSMEMBRANE SIGNALING. Annu. Rev. Biophys. Biomol. Slruel. 1992.21:223–42.

Magde D., Elson E.L., Webb W.W. Fluorescence correlation spectroscopy. II. An experimental realization. Biopolymers. 1974;13:29–61. [PubMed].

Haustein E., Schwille P. Fluorescence correlation spectroscopy: novel variations of an established technique. Annu. Rev. Biophys. Biomol. Struct. 2007;36:151–169. [PubMed].

Burns A.R., Frankel D.J., Buranda T. Local mobility in lipid domains of supported bilayers characterized by atomic force microscopy and fluorescence correlation spectroscopy. Biophys. J. 2005;89:1081–1093. [PMC free article] [PubMed].

Zhang L., Granick S. Slaved diffusion in phospholipid bilayers. Proc. Natl. Acad. Sci. USA. 2005;102:9118–9121. [PMC free article] [PubMed].

Guo L., Chowdhury P., Gai F. Heterogeneous and anomalous diffusion inside lipid tubules. J. Phys. Chem. B. 2007;111:14244–14249.

Tjernberg L.O., Pramanik A., Rigler R. Amyloid β-peptide polymerization studied using fluorescence correlation spectroscopy. Chem. Biol. 1999;6:53–62. [PubMed].

Sengupta P., Garai K., Maiti S. The amyloid β peptide (Aβ1–40) is thermodynamically soluble at physiological concentrations. Biochemistry. 2003;42:10506–10513.

Junghyun Jo, Soomin Hong, Won Yun Choi & Dong Ryul Lee. Cell-penetrating peptide (CPP)-conjugated proteins is an efficient tool for manipulation of human mesenchymal stromal cells. Scientific Reports 4, Article number: 4378 (2014).

M. Green, P.M. Loewenstein. Autonomous functional domains of chemically synthesized human immunodeficiency virus tat trans-activator protein. Cell, 55 (1988), pp. 1179–1188.

D. Derossi, A.H. Joliot, G. Chassaing, A. Prochiantz. The third helix of the Antennapedia homeodomain translocates through biological membranes. J. Biol. Chem., 269 (1994), pp. 10444–10450.

Chérine Bechara, Sandrine Sagan. Cell-penetrating peptides: 20 years later, where do we stand? FEBS, Volume 587, Issue 12, 19 June 2013, Pages 1693–1702.

E. Vives, P. Brodin, B. Lebleu. A truncated HIV-1 Tat protein basic domain rapidly translocates through the plasma membrane and accumulates in the cell nucleus. J. Biol. Chem., 272 (1997), pp. 16010–16017.

A. Elmquist, M. Lindgren, T. Bartfai, U. Langel. VE-cadherin-derived cell-penetrating peptide, pVEC, with carrier functions. Exp. Cell Res., 269 (2001), pp. 237–244.
M. Pooga, M. Hallbrink, M. Zorko, U. Langel. Cell penetration by transportan. FASEB J., 12 (1998), pp. 67–77.
M.C. Morris, P. Vidal, L. Chaloin, F. Heitz, G. Divita. A new peptide vector for efficient delivery of oligonucleotides into mammalian cells. Nucleic Acids Res., 25 (1997), pp. 2730–2736.
M.C. Morris, J. Depollier, J. Mery, F. Heitz, G. Divita. A peptide carrier for the delivery of biologically active proteins into mammalian cells. Nat. Biotechnol., 19 (2001), pp. 1173–1176.
D.J. Mitchell, D.T. Kim, L. Steinman, C.G. Fathman, J.B. Rothbard. Polyarginine enters cells more efficiently than other polycationic homopolymers. J. Pept. Res., 56 (2000), pp. 318–325.
S. Futaki, T. Suzuki, W. Ohashi, T. Yagami, S. Tanaka, K. Ueda, Y. Sugiura. Arginine-rich peptides. An abundant source of membrane-permeable peptides having potential as carriers for intracellular protein delivery. J. Biol. Chem., 276 (2001), pp. 5836–5840.
J. Oehlke, A. Scheller, B. Wiesner, E. Krause, M. Beyermann, E. Klauschenz, M. Melzig, M. Bienert. Cellular uptake of an alpha-helical amphipathic model peptide with the potential to deliver polar compounds into the cell interior non-endocytically. Biochim. Biophys. Acta, 1414 (1998), pp. 127–139.
D. Delaroche, et al.. Tracking a new cell-penetrating (W/R) nonapeptide, through an enzyme-stable mass spectrometry reporter tag. Anal. Chem., 79 (2007), pp. 1932–1938.
A. Ziegler. Thermodynamic studies and binding mechanisms of cell-penetrating peptides with lipids and glycosaminoglycans. Adv. Drug Deliv. Rev., 60 (2008), pp. 580–597.
F. Madani, S. Lindberg, U. Langel, S. Futaki, A. Graslund. Mechanisms of cellular uptake of cell-penetrating peptides. J. Biophys., 2011 (2011), p. 414729.
F. Milletti. Cell-penetrating peptides: classes, origin, and current landscape. Drug Discov. Today, 17 (2012), pp. 850–860.
P.E. Thorén, D. Persson, P. Isakson, M. Goksor, A. Onfelt, B. Norden. Uptake of analogs of penetratin, Tat(48–60) and oligoarginine in live cells. Biochem. Biophys. Res. Commun., 307 (2003), pp. 100–107.
Ü. Langel. Cell-Penetrating Peptides: Processes and Applications. CRC Press, Boca Raton, FL (2002).
D. Derossi, S. Calvet, A. Trembleau, A. Brunissen, G. Chassaing, A. Prochiantz. Cell internalization of the third helix of the Antennapedia homeodomain is receptor-independent. J. Biol. Chem., 271 (1996), pp. 18188–18193.
C.Y. Jiao, D. Delaroche, F. Burlina, I.D. Alves, G. Chassaing, S. Sagan. Translocation and endocytosis for cell-penetrating peptide internalization. J. Biol. Chem., 284 (2009), pp. 33957–33965.
P. Saalik, A. Niinep, J. Pae, M. Hansen, D. Lubenets, U. Langel, M. Pooga. Penetration without cells: membrane translocation of cell-penetrating peptides in the model giant plasma membrane vesicles. J. Control Release, 153 (2011), pp. 117–125.
P.E. Thoren, D. Persson, E.K. Esbjorner, M. Goksor, P. Lincoln, B. Norden. Membrane binding and translocation of cell-penetrating peptides. Biochemistry, 43 (2004), pp. 3471–3489.
H. Binder, G. Lindblom. Charge-dependent translocation of the Trojan peptide penetratin across lipid membranes. Biophys. J., 85 (2003), pp. 982–995.
A. Elmquist, U. Langel. In vitro uptake and stability study of pVEC and its all-D analog Biol. Chem., 384 (2003), pp. 387–393.
J.L. Zaro, W.C. Shen. Quantitative comparison of membrane transduction and endocytosis of oligopeptides. Biochem. Biophys. Res. Commun., 307 (2003), pp. 241–247.
J.L. Zaro, W.C. Shen. Evidence that membrane transduction of oligoarginine does not require vesicle formation. Exp. Cell Res., 307 (2005), pp. 164–173.
E. Vives, J.P. Richard, C. Rispal, B. Lebleu. TAT peptide internalization: seeking the mechanism of entry. Curr. Protein Pept. Sci., 4 (2003), pp. 125–132.

S. Console, C. Marty, C. Garcia-Echeverria, R. Schwendener, K. Ballmer-Hofer. Antennapedia and HIV transactivator of transcription (TAT) "protein transduction domains" promote endocytosis of high molecular weight cargo upon binding to cell surface glycosaminoglycans. J. Biol. Chem., 278 (2003), pp. 35109–35114.

J.P. Richard, K. Melikov, E. Vives, C. Ramos, B. Verbeure, M.J. Gait, L.V. Chernomordik, B. Lebleu. Cell-penetrating peptides. A reevaluation of the mechanism of cellular uptake. J. Biol. Chem., 278 (2003), pp. 585–590.

A. Fittipaldi, A. Ferrari, M. Zoppe, C. Arcangeli, V. Pellegrini, F. Beltram, M. Giacca. Cell membrane lipid rafts mediate caveolar endocytosis of HIV-1 Tat fusion proteins. J. Biol. Chem., 278 (2003), pp. 34141–34149.

R. Fischer, T. Waizenegger, K. Kohler, R. Brock. A quantitative validation of fluorophore-labelled cell-permeable peptide conjugates: fluorophore and cargo dependence of import. Biochim. Biophys. Acta, 1564 (2002), pp. 365–374.

S.T. Henriques, J. Costa, M.A. Castanho. Translocation of beta-galactosidase mediated by the cell-penetrating peptide pep-1 into lipid vesicles and human HeLa cells is driven by membrane electrostatic potential. Biochemistry, 44 (2005), pp. 10189–10198.

S. Deshayes, A. Heitz, M.C. Morris, P. Charnet, G. Divita, F. Heitz. Insight into the mechanism of internalization of the cell-penetrating carrier peptide Pep-1 through conformational analysis. Biochemistry, 43 (2004), pp. 1449–1457.

M. Silhol, M. Tyagi, M. Giacca, B. Lebleu, E. Vives. Different mechanisms for cellular internalization of the HIV-1 Tat-derived cell penetrating peptide and recombinant proteins fused to Tat. Eur. J. Biochem., 269 (2002), pp. 494–501.

Hiroshi Nikaido. Molecular Basis of Bacterial Outer Membrane Permeability Revisited. Microbiol. Mol. Biol. Rev. December 2003 vol. 67 no. 4 593–656.

Plesiat, P., and H. Nikaido. 1992. Outer membranes of gram-negative bacteria are permeable to steroid probes. Mol. Microbiol. 6:1323–1333.

IAN CHOPRA and PETER BALL. Transport of Antibiotics into Bacteria. Adv Microb Physiol. 1982;23:183–240.

Stefania Galdiero, Annarita Falanga, Marco Cantisani, Rossella Tarallo, Maria Elena Della Pepa, Virginia D'Oriano, and Massimiliano Galdiero. Curr Protein Pept Sci. 2012 Dec; 13(8): 843–854.

Nakae, T. 1976. Outer membrane of Salmonella. Isolation of protein complex that produces transmembrane channels. J. Biol. Chem. 251:2176–2178.

Nikaido, H. 1994. Prevention of drug access to bacterial targets: permeability barriers and active efflux. Science 264:382–388.

Nikaido, H. 1996. Multidrug efflux pumps of gram-negative bacteria. J. Bacteriol. 178:5853–5859.

Bertrand Cinquin, Laure Maigre, Elizabeth Pinet, Jacqueline Chevalier, Robert A. Stavenger, Scott Mills, Matthieu Réfrégiers & Jean-Marie Pagès. Microspectrometric insights on the uptake of antibiotics at the single bacterial cell level. Scientific Reports 5, Article number: 17968 (2015).

Jamme F, Villette S, Giuliani A, Rouam V, Wien F, Lagarde B, Réfrégiers M. Synchrotron UV fluorescence microscopy uncovers new probes in cells and tissues. Microsc. Microanal. 16, 507–14 (2010).

Jamme F, Kascakova S, Villette S, Allouche F, Pallu S, Rouam V, Réfrégiers M. Deep UV autofluorescence microscopy for cell biology and tissue histology. Biol. Cell 105, 277–88 (2013).

L Trón, J Szöllósi, S Damjanovich, S H Helliwell, D J Arndt-Jovin, and T M Jovin. Flow cytometric measurement of fluorescence resonance energy transfer on cell surface. Quantitative Evaluation of the Transfer Efficiency on a Cell-by-Cell Basis. Biophys. J. 45, 939–46 (1984).

# Chapter 4
# Nanoscale Processes Giving Rise to Ion Pores

Agents such as drugs, peptides, biomolecules that are classified as cell active agents (CAAs) get distributed on the cell surface while they are administered to reach out to cellular targets. Study on cell surface morphology may help us address the distribution of CAAs. Cell surface separates hydrophobic cell membrane core from the cell surrounding water environment. The cell membrane's outer surface is hydrophilic. In an attempt made by any type of agents to enter into the cell, at the first point of entry, the agents have to get transported naturally or via carriers across the hydrophilic–hydrophobic boundary. The dynamics related to the transport of this agent follows various laws of physics, such as general diffusion, random distribution, controlled directional flow, and above all cell surface interaction-based adsorption. Statistical mechanics concepts play crucial roles behind understanding the physics of cell surface adsorption of drugs. Although it is a biological process to understand the phenomenon, one needs to deal with physics theories and necessary experimental techniques.

There are various CAAs which do not only penetrate into cellular environment but also act upon certain sections in specific cellular compartments, such as general cell membrane, specific cell membrane proteins and lipids, general mitochondria, specific mitochondrial membrane, membrane proteins and lipids, cytosolic proteins, nuclear membrane. Among them, a few are found to create specific structures, e.g., ion channels or pores across hydrophobic membrane boundary. Two types of studies may help understand the underlying physics behind the action of agents in cell surface regions. They are either direct observation-based studies or indirect. The direct assays include imaging of the agents on cell surface to locate their mostly temporal presence. During imaging phase, their instantaneous coexistence with the cell surface structure may be detected. The other indirect method of detection involves techniques that help understand the effects of agents due to their often partial and instantaneous presence in various regions. Both require clear understanding of the statistical nature of the distribution of agents on cell surface. Phenomenology and energetics are two aspects we need to specifically address. This chapter is dedicated to doing so using a few mostly physics-based techniques.



M. Ashrafuzzaman, *Nanoscale Biophysics of the Cell*,
https://doi.org/10.1007/978-3-319-77465-7_4

## 4.1 Cell Surface Absorption of Biomolecules. A Case Study Using Chemotherapy Drugs

Recently, we produced a few sets of important data on cell surface binding of drugs taking chemotherapy drug (CD) colchicine as example (see Ashrafuzzaman et al. 2016, submitted). We shall avoid providing details regarding the cell surface inspection techniques as they are well-established ones and can be found in any focused textbooks addressing on these techniques. Here we provide a few important phase diagrams to help us demonstrate mainly two things, namely (i) the general structure of cell surface and (ii) the structure of surface constructed drug clusters as a result of cell surface adsorption of drug agents. Knowledge of the drug cluster structures will help us put forward a few novel hypotheses regarding mechanisms behind creation of drug clusters on cell surface and subsequent initiation of drug-specific ion pores/channels inside cell surface/cell membrane.

CDs were recently reported to induce ion pores inside lipid bilayer membranes by us (Ashrafuzzaman et al. 2011, 2012) as like as peptide-induced channels (Andersen 1983; Huang 1986; Latorre and Alvarez 1981; Ludtke et al. 1996; Matsuzaki et al. 1996). We wish to address here the cell surface binding of these membrane-active agents (MAAs). Atomic force microscopy (AFM) is a cantilever imaging technique that can address the general surface profile of cells. Surface-adsorbed nanoparticles (NPs) are monitored utilizing AFM techniques by measuring the shifts in localized surface heights (Binnig et al. 1986; Schneider et al. 1997; Bhushan 2004; Kuznetsova et al. 2007; Franz and Puech 2008). Here we have used two different AFMs, namely TappingMode AFM (TM-AFM) for cell surface morphology study and MultiMode 8 AFM (MM-AFM) for adhesion force measurements. Details on both techniques along with other cell surface inspection techniques are presented in Ashrafuzzaman et al. (2016).

AFM images show cell surface geometry and help measure the surface bumps or landscapes at a low nanometer (nm) dimension. Besides addressing surface geometry, AFM technique is also capable of measuring the adhesion force (which can be derived into adhesion energy) related to the cell surface distribution of NPs. We therefore considered this AFM technique to address the statistical mechanical nature of the possible cell surface adsorption of CDs by measuring primarily the cell surface roughness and adhesion forces between CDs and cell surface as a result of surface presence of drugs.

Living cells generally react to biological alterations such as cell spreading and shape alterations of membranes and nuclei (Caille et al. 2002; Laurent et al. 2005). AFM is a versatile tool (Binnig et al. 1986) in biological sciences (Binnig et al. 1986; Schneider et al. 1997; Bhushan 2004; Kuznetsova et al. 2007; Franz and Puech 2008; Caille et al. 2002; Laurent et al. 2005; Preiner et al. 2015) which is capable of resolving molecular details of cell surface under ambient condition (Langer et al. 2000; Madl et al. 2006). It helps address the nanoscale organization and dynamics of cell membranes and cell walls, measuring cell mechanics and cell adhesion (Heinisch et al. 2012). General cell action and specific cell-based target site delivery of agents perhaps require to fulfill two primary aspects. Firstly, the agents must experience considerable absorbance on the cell surface region while they are delivered in the hydrophilic extracellular liquid environment. Secondly, the agents must ensure reasonable statistical presence in and beyond hydrophobic cell membrane core regions (Ashrafuzzaman et al. 2006, 2008) through of course membrane interaction mechanisms (Ashrafuzzaman and Tuszynski 2012a, b; Ashrafuzzaman et al. 2012). Phenomenological address and understanding of the molecular mechanisms behind agents' cell membrane action, e.g., the membrane permeabilization, are crucial aspects. Ion channels as monitored, usually from electrophysiology recording of ion currents across lipid membranes doped with various types of naturally occurring and synthetic membrane proteins (MPs) or antimicrobial peptides (AMPs), may be utilized to investigate whether a protein–lipid complex creates an ion flowing pore/channel across a membrane (e.g., see Andersen 1983; Huang 1986; Latorre and Alvarez 1981; Ludtke et al. 1996; Matsuzaki et al. 1996; Ashrafuzzaman et al. 2006, 2008; Ashrafuzzaman and Tuszynski 2012a, b). Using the identical electrophysiology technique, we recently investigated CDs to inspect if they induce events inside membrane analogous to those due to AMP-induced ion channels (Ashrafuzzaman et al. 2011, 2012). CDs colchicine and taxol were chosen as example agents due to their premier use in drug discovery (e.g., see Schiff et al. 1979; Callen 1985; Seidemann et al. 1987; Fisherman et al. 1992; Holmes et al. 1994). Molecular mechanism of drug effects on membranes was elucidated by performing in silico molecular dynamic (MD) simulations of various CD–lipid complexes (see Ashrafuzzaman et al. 2012). A thorough theoretical analysis on general CAA diffusion across cell membrane and other inter- and intrastructural cell boundaries has recently been made (Ashrafuzzaman 2015a, b). Here theoretical parameters related to quantitative distribution with statistical nature of CAAs have been demonstrated via measuring their mobility in cell media. All these various techniques help us understand the mechanisms of general actions and the statistical nature of distribution of CAAs and MAAs, e.g., CDs on or across cell surface regions, indirectly. A direct measurement is therefore due. AFM is chosen especially for this purpose.

Colchicine, a potential CD candidate, has a long history of use in immune system diseases (see Callen 1985; Seidemann et al. 1987). Colchicine won the approval of USA Food and Drug Administration (FDA) as a drug for treating acute gout and Familial Mediterranean fever in 2009. It is reported to inhibit leukocyte–endothelial cell adhesion (Rosenman et al. 1991) and T cells activation (Mekory et al. 1989) through binding to tubulin dimers, and thus prevents polymerization into microtubules (Schiff et al. 1979). Colchicine is also found to bind with nuclear periphery and to disorder the nuclear membrane phospholipid bilayers (Borisy and Taylor 1967; Agutter and Suckling 1982).

Another CD taxol has been used as agent for ovarian, lung, and breast cancer chemotherapy (Balasubramanian and Straubinger 1994; Matsumoto and Sakai 1979; Sonee et al. 1998). Due to its molecular action involving stabilization of microtubules (Schiff et al. 1979), it is especially found suitable for combination therapy (Fisherman et al. 1992). Taxols incorporated into liposomes penetrate into the acyl chain domain which causes alterations of the physical properties, e.g., phospholipid phase transitions, lipid order parameters, fluidity, of membranes (Balasubramanian and Straubinger 1994).

Our recent investigations suggest that CDs colchicine and taxol induce ion pores across cell membranes. We predict specifically that the ‘toroidal type’ lipid-lined ion pores are induced by CDs (Ashrafuzzaman et al. 2011, 2012). Due to CDs’ known effects on membrane’s electrical and physical properties (Ashrafuzzaman et al. 2011, 2012; Agutter and Suckling 1982; Balasubramanian and Straubinger 1994; Matsumoto and Sakai 1979; Sonee et al. 1998; Shiba et al. 1988; Mons et al. 2000), it may be important from cytotoxicity point of view to address their general cell surface binding. But the mechanisms that play the roles behind cell surface binding and subsequent effects across cell membrane are not yet clear. Although ion pores are observed, the pre-pore conditions while CDs interact with cell surface have not been demonstrated earlier until we have recently performed the AFM imaging assays (Ashrafuzzaman et al. 2016).

AFM imaging is performed in this new approach to address the cell surface binding of CDs using colchicine as an example agent. In the cell assay studies, we wished to achieve a thorough statistical mechanical understanding on cell surface distribution of the agents and energetics behind (Ashrafuzzaman et al. 2016). In doing so, we aimed to achieve three fundamental objectives, namely as follows:

(i) firstly, observe and measure the statistical significance of the cell surface binding of CDs,
(ii) secondly, understand the nature of the surface distribution (randomly or creating clusters) of drug molecules,
(iii) finally, measure the surface energetics of drugs.

This remarkable statistical mechanical address of drug binding on cell surface may produce a clearer picture regarding how MAAs act upon cell surface and produce membrane actions known as general drug effects or often as cytotoxicity of drugs that target various cellular structures.

### *4.1.1 Cell Culture and Colchicine Treatment of Cells*

Although this section will be explained quite in details in Ashrafuzzaman et al. (2016), I wish to brief it here for the readership of this book. We have our cell culture facility at Biochemistry Department, King Saud University. We maintained the human colorectal adenocarcinoma cells, LoVo, in Dulbecco’s Modified Eagle’s Medium (DMEM; Gibco), supplemented with 10% fetal bovine serum (FBS; Gibco) and 50 units/mL penicillin-50 μg/mL streptomycin (Gibco). At 75–80% confluence, cells were treated without (control) or with colchicine (Sigma-Aldrich) at concentration 1 or 100 μM for 24 h. Cells were then trypsinized, washed in phosphate-buffered saline (PBS; Gibco), and counted using handheld automated cell counter, Scepter (Millipore) fitted with 60 μm sensor to a final concentration of $5 \times 10^5$ cells/mL in PBS for AFM investigations.

### 4.1.2 AFM Mapping of the Cell Surface

A sharp tip is mounted to the end of a flexible cantilever. The tip is brought in contact with the surface, and images are achieved by scanning the surface of the sample. While scanning the force between the tip and the surface leads to bending the cantilever. Consequently, deflection is detected by a laser beam reflected at the end of the cantilever onto a sensitive detector. A three-dimensional (3D) topographic surface image is achieved by plotting the deflection as a function of the *X-Y* position. In a common mode of operation, a feedback loop is used to maintain a constant deflection, while the topographic information is obtained from the cantilever vertical displacement (Bhushan 2004; Franz and Puech 2008; Kuznetsova et al. 2007). Here we have used two different AFMs, namely TappingMode AFM (TM-AFM) for cell surface morphology study and MultiMode 8 AFM (MM-AFM) for adhesion force measurements.

Quantitative adhesion force images were recorded on a MM-AFM, Bruker (Ashrafuzzaman et al. 2016). The Antimony (n)-doped Si tip with spring constant of 200 N/m was used under PeakForce Quantitative Nanomechanical Mapping (QNM) mode. PeakForce QNM is a special kind of imaging mode of AFM based on PeakForce Tapping technology. In this technique, each tip–sample interaction is considered to extract quantitative nanomechanical properties such as elastic modulus, adhesion, deformation, and dissipation. This technique allows mapping the above-mentioned properties quantitatively and at high resolution at every pixel and at the same rate of the normal topography map. Each adhesion force image consists of $512 \times 512$ pixels, and each pixel indicates the quantitative adhesion force extracted from the force versus separation plot in a single nanoindentation.

### 4.1.3 AFM Image Inspections. CDs Form Clusters on Cell Surface

Figure 4.1 (top panel) presents AFM topography images demonstrating the profiles such as height, amplitude error, and phase of a cancer cell on $SiO_2$ substrate. In the background of top panel figures, a few visible small blobs could be reminiscent of solvent molecules used to make the solution of cells.

Our aim was to check whether the colchicine molecules bind to the cancer cell membrane. The colchicine-treated cells were deposited onto the $SiO_2$ surface, and AFM images were recorded. Figure 4.1 (bottom panel) presents the AFM topography images of a single colchicine-treated cell. Here the shape of the cell is distorted, and this could be due to colchicine induction to the cell. It is seen from the bottom panel phase image (see Fig. 4.1c) that the cell is sitting on the liquid. In contrast to the top panel, the bottom panel image of a single cell (Fig. 4.1) shows a few blobs of different sizes on the cell membrane. This is clearly visible in 3D cell images (see Fig. 4.2).

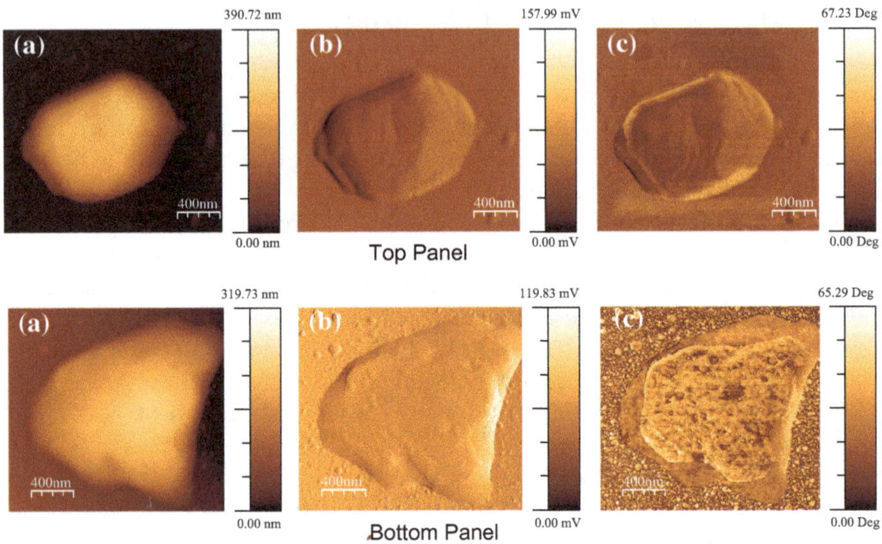


**Fig. 4.1** A single cell TM-AFM image on $SiO_2$ substrate with and without colchicine binding: top panel (bare cell) **a** AFM height image. The height of the cell is presented in color scale 0 (dark fields)–390.72 nm (light fields). **b** AFM amplitude error image. The amplitude of the cell is presented in color scale 0 (dark fields)–157.99 mV (light fields). **c** AFM phase image of the cell. The phase of the cell is presented in color scale 0 (dark fields)–67.23° (light fields); bottom panel (1 μM colchicine-treated cell) **a** AFM height image. The height of the cell is presented in color scale 0 (dark fields)–319.73 nm (light fields). **b** AFM amplitude error image. The amplitude of the cell is presented in color scale 0 (dark fields)–119.83 mV (light fields). **c** AFM phase image of the cell. The phase of the cell is presented in color scale 0 (dark fields)–65.29 Deg (light fields). To appear in Ashrafuzzaman et al. (2016)

In order to measure the height of different blobs on the colchicine-treated cells, a high-resolution AFM height image was recorded (see Fig. 4.3). Compare the particle height with colchicine molecule dimension as shown in Fig. 4.4. As mentioned in Fig. 4.5, the height profiles are recorded at the centers of the corresponding clusters shown in Fig. 4.3a.

We further wished to observe the particle and cluster distribution by plotting the cluster frequency in a histogram plot (see Fig. 4.3) (Ashrafuzzaman et al. 2016). We can calculate the total number of clusters of colchicine molecules ($N_{\text{cluster}}$) on cell surface using the following formula:

$$N_{\text{cluster}} = \sum_{i=0}^{n} f_i \tag{4.1}$$

The total number of cell surface-adsorbed colchicine molecules ($N_{\text{col}}$) can be deducted using the following formula:

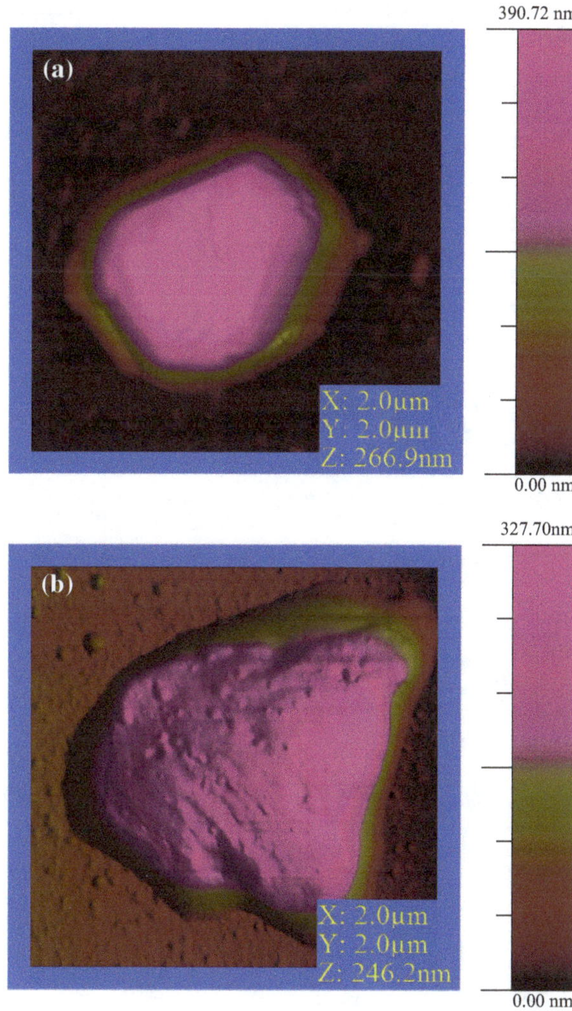


**Fig. 4.2** 3D TM-AFM height image of a single cell in Fig. 4.1a (top and bottom panels) with and without colchicine binding on $SiO_2$ substrate. **a** A single cell without colchicine binding. The height of the cell is presented in color scale 0 (dark fields)–390.72 nm (light fields). **b** A 1 µM colchicine-treated cell. The height of the cell is presented in color scale 0 (dark fields)–327.70 nm (light fields). It is clearly seen that the cell is sitting on the liquid. This will appear in Ashrafuzzaman et al. (2016)

$$N_{col} = \sum_{i=0}^{n} f_i h_i / a_{col} \tag{4.2}$$

By assuming colchicine molecules as hard spheres, we can calculate the total colchicine accumulated along the line of a cluster height $h_i$ to be $h_i/a_{col}$ where $i$ = 1, 2, 3, etc.

$N_{cluster}$ and $N_{col}$ are plotted in Fig. 4.6. The mechanism of cell surface adsorption of colchicines shows strong tendency of saturation between 1 and 100 µM (*X*-axis in log scale, Fig. 4.6).

Figure 4.3a indicates that the size and number of colchicine clusters increase with increasing colchicine concentrations. The particle heights $h_i$s (*X*-axis values on

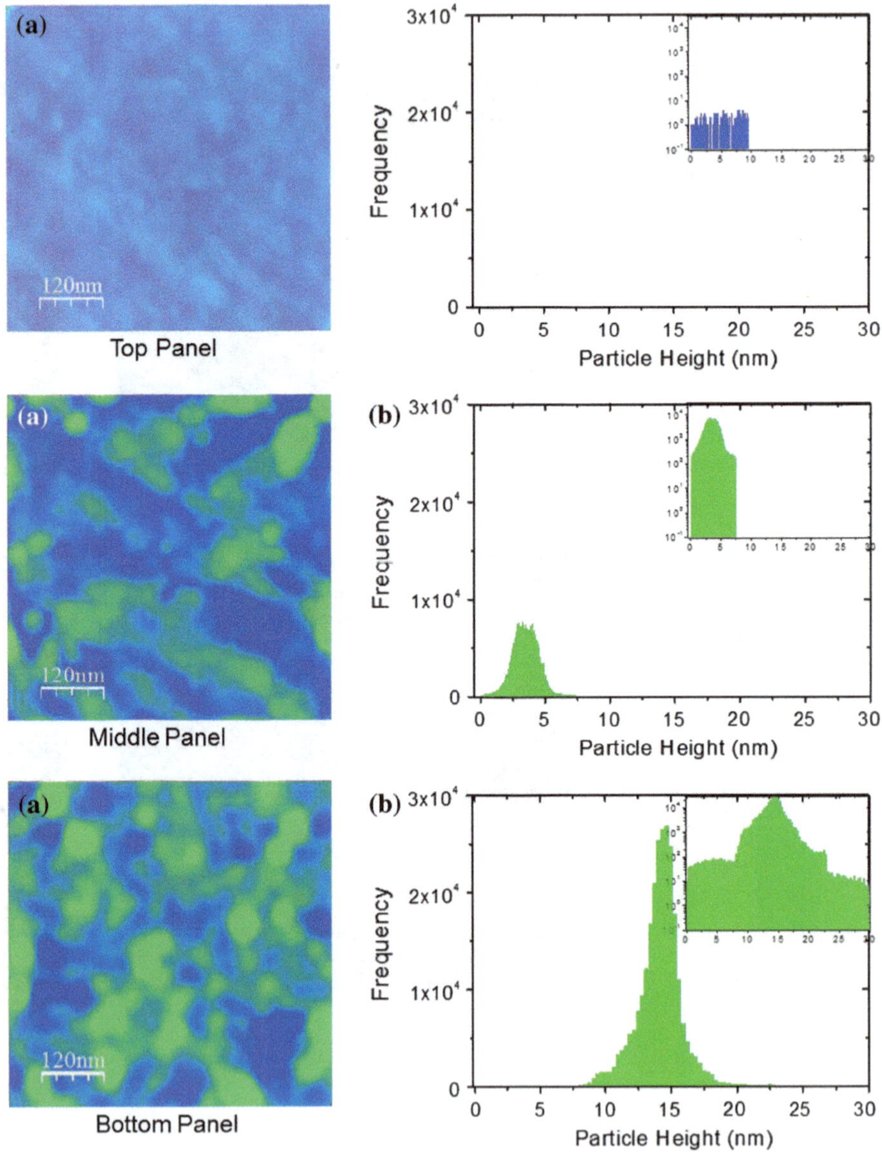


**Fig. 4.3** Top panel **a** high-resolution TM-AFM image of a control cell membrane. **b** Histograms are based on counted clusters on the section of cell surface shown in figure (a). Each histogram represents the counted number of clusters (called frequency, let us assume it as $f_i$, plotted along *Y*-axis. Here $i = 1, 2, 3$, etc.) against a height profile (known as particle height here in Fig. 4.3b, let us assume it as $h_i$, plotted in nm scale along *X*-axis) shown in Fig. 4.3b. Middle and bottom panels are data for cells treated with 1 and 100 μM colchicine, respectively. Histograms were plotted using program Origin 9.1 after detecting the numbers using program WSxM. The origin of the detected few events (visible in log plots shown in insets of Fig. 4.3b) for control cell surface is nothing but noise. This will appear in Ashrafuzzaman et al. (2016)

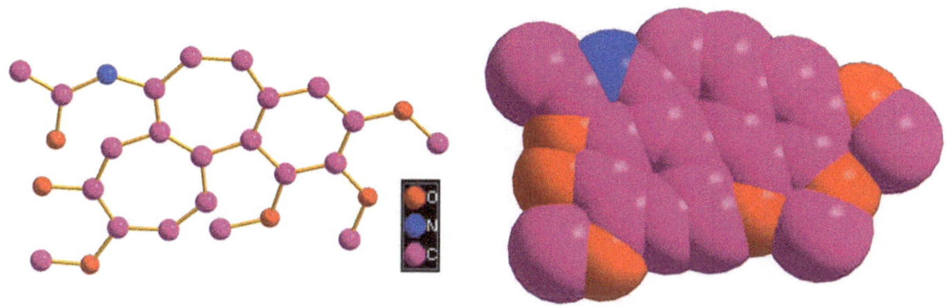


**Fig. 4.4** Diameter ($d_{col}$) of a single colchicine molecule is approximately $d_{col} \approx 1$ nm. Minimum 1 nm and maximum 1.18 nm depending on the orientation have been measured using crystallography software 'Diamond'. Left: a ball–stick representation of a colchicine molecule. Right: a space-filling representation of the same molecule. Hydrogen is omitted for clarity

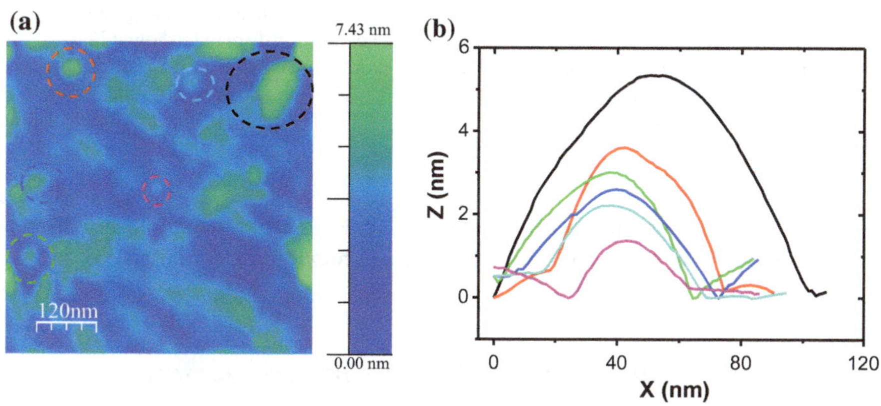


**Fig. 4.5** **a** A high-resolution AFM height image of a 1 µM colchicine-treated cell wall bound to $SiO_2$ substrate surface. The height of the cell wall is presented in color scale 0 (dark fields)–7.43 nm (light fields). **b** A number of height profiles recorded through the center of the circles marked in Fig. 4.5a. Here we marked the blobs (or clusters of colchicine molecules) of different sizes by circles of various colors. A number of height profiles were recorded through the centers of the circles in Fig. 4.5a and presented in Fig. 4.5b, which ranges approximately between 1.1–5.5 nm. Therefore, it can be concluded that the colchicine molecules bind to the cell membrane either in single or cluster form. This will appear in Ashrafuzzaman et al. (2016)

which histograms stand, see Fig. 4.3b) also increase quickly with increasing colchicine concentrations which are reflected in the histogram positions in the wide range in *X*-axis values. This represents higher values of $h_i$s due to clustering relative to single colchicine dimension which is less than 1.1 nm (see Fig. 4.4). The colchicine molecules are therefore assumed to be detected in mostly accumulated clusters with different dimensions when the cells are treated with higher colchicine molarity. Figure 4.6 shows the quantitative trend of the drug adsorption on the cell surface. As said earlier, the mechanism of cell surface adsorption of colchicine

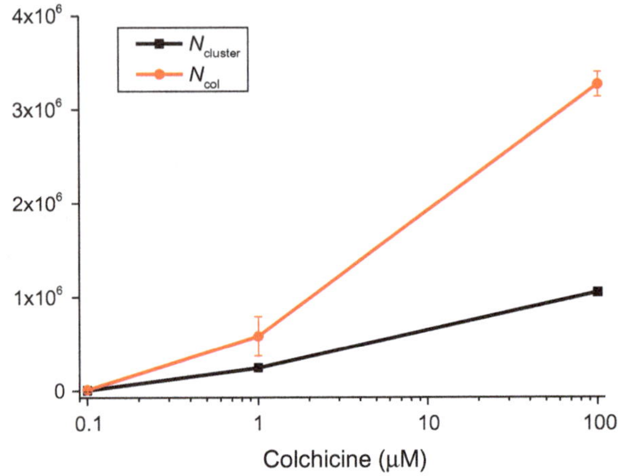


**Fig. 4.6** Cell surface-adsorbed colchicine molecules have been measured and plotted in this figure. Lowest *X*-axis point 0.1 should be read 0, and we needed this display to show it in log scale. The number each histogram represents is the counted number of clusters against a height profile (known as particle height in Fig. 4.3b and shown in Fig. 4.5). The height profiles are recorded at the centers of the corresponding clusters shown in Fig. 4.3a. This will appear in Ashrafuzzaman et al. (2016)

molecules shows strong tendency of drug molecule adsorption saturation which is reflected in Fig. 4.6, between 1 and 100 μM colchicine concentrations.

### 4.1.4 *Adhesion of Colchicine Clusters on Cell Surface*

In order to understand the interaction between colchicine drug and cell membrane, quantitative adhesion force was measured at every pixel of the topography map by using a PeakForce Quantitative Nanomechanical Mapping (QNM) mode of Bruker MM-AFM. The samples were completely dried, consequently no liquid was present on $SiO_2$ surface, and colchicines were observed in clusters on the cell surface. High-resolution MM-AFM images of a single colchicine-treated cell are presented in Fig. 4.7.

Quantitative adhesion force ($F_{ads}$) between probe and surfaces ($SiO_2$, drug, and cell membrane) is been extracted from Fig. 4.7c. We used a NanoScope Analysis software 1.5. This adhesion force is then converted to adhesion energy ($\Gamma_{ads}$) by using the following equation (Grierson et al. 2005):

$$\Gamma_{ads} = \frac{F_{ads}}{2\pi R} \tag{4.3}$$

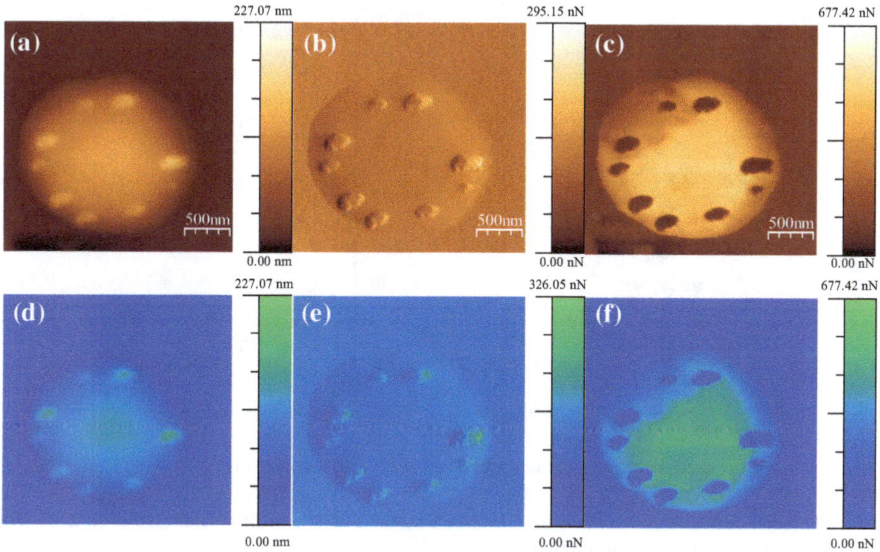


**Fig. 4.7** MM-AFM images of a colchicine bound cell on $SiO_2$ substrate. **a** Height image. The height of the cell is presented in color scale 0 (dark fields)–227.07 nm (light fields). **b** Force error image. The force of the cell is presented in color scale 0 (dark fields)–295.15 nN (light fields). **c** Adhesion image of the cell. The adhesion of the cell is presented in color scale 0 (dark fields)–677.42 nN (light fields). The corresponding 3D figures are presented in (**d**), (**e**), (**f**), respectively. This will appear in Ashrafuzzaman et al. (2016)

Here $R$ is the radius of the AFM tip which is approximately 8 nm. The adhesion energy $\left(\Gamma_{\mathrm{drug-cell}}\right)$ between drug and cell membrane has been calculated by using the following equation (Koenig et al. 2011):

$$\Gamma_{\mathrm{drug-cell}} = 2\left(\gamma_{\mathrm{drug}} \times \gamma_{\mathrm{cell}}\right)^{1/2} \tag{4.4}$$

Here $\gamma$ is the surface energy. Using the same equation, we calculated the surface energy of AFM tip $\left(\gamma_{\mathrm{tip}} = 2.52\ \mathrm{J.m^{-2}}\right)$, where surface energy of $SiO_2$ surface was used, $\gamma_{\mathrm{SiO2}} = 0.15\ \mathrm{J.m^{-2}}$ (Koenig et al. 2011). The adhesive energy between colchicine clusters and cell membranes $\left(\Gamma_{\mathrm{drug-cell}}\right)$ has been calculated from quantitative adhesive force measurements by AFM for five different cells (35 clusters over sizes 97–590 nm). $\Gamma_{\mathrm{drug-cell}}$ is plotted against cluster size in Fig. 4.8. $\Gamma_{\mathrm{drug-cell}}$ is observed to decrease with increase of cluster size. We also calculated $\Gamma_{\mathrm{drug-cell}}$ per single colchicine drug and find that the adhesion energy per colchicine decreases with cluster size increase before gradually saturating between drug clusters with 225–375 nm size (Fig. 4.8).

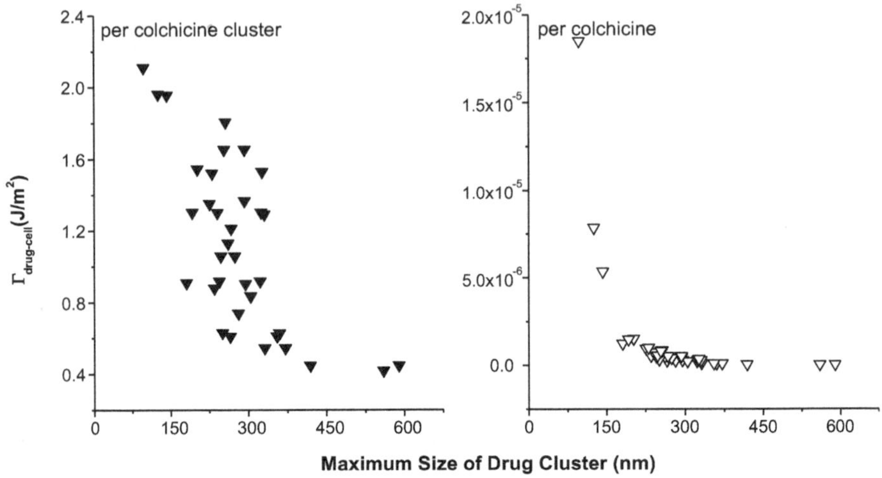


**Fig. 4.8** Adhesion energy between drug and cell surface versus drug cluster size plot. This will appear in Ashrafuzzaman et al. (2016)

## 4.2 Drug Clustering on Cell Surface Follows General Chemistry and Physics Formalisms

Drug clustering that has been phenomenologically demonstrated here follows general rules of physics and chemistry. Strong statistical mechanics formalisms are found active. We shall address on a few such rules that have been recently detected in mainly our own research findings.

### *4.2.1 Drug Clusters Appear with Statistical Geometric Randomness*

AFM data and phase diagrams demonstrate the cell surface binding of CDs. Strong statistical nature in surface adsorption of drugs has been observed. Drugs are found independently, as well as in clusters with various cluster sizes in their cell surface-bound states. The cluster number and size increase with the cell incubating drug molarity. The cluster heights represent the number of colchicine molecules accumulated therein. This suggests for creation of drug complexes on cell surface as a result of binding. The clustering appears with quite randomness especially from the geometric perspectives. The size distribution and their height profiles appear with randomness. With the increase of the cell incubating drug molarity, the range of geometric dimensions, e.g., cluster size/diameter, cluster height, covers wide range though we find more and more bigger clusters to appear. The adhesion energy per colchicine was though found to decrease with cluster size increase before gradually saturating between drug clusters with 225–375 nm size (Fig. 4.8). This suggests that although drug clustering is phenomenologically a random process as it appears in our findings the inherent mechanisms are likely to follow some fundamental scientific rules that are universal in nature. The calculated adhesion forces provide hint on the physical nature of the clusters but pinpointed calculations of energies of the clusters considering the physical properties of the drug molecules and cell surface provide further details on molecular mechanisms that are active in the process of clustering of drugs. Here in Sect. 4.2.2, we shall address on how a few simple molecular dynamic (MD) simulations helped us to detect the predicted fundamental rules that might be active to create drug–lipid energetic bindings (Ashrafuzzaman et al. 2012, 2014).

### *4.2.2 Charge Property-Based Interactions Make Inherent Rules Inside Drug Clusters*

Our recent studies suggest that CDs interact with lipids due to mainly charge properties (Ashrafuzzaman et al. 2012). In MD simulations of drug–lipid complexes, van der Waals (vdW) and electrostatic (ES) interactions are primarily detected, providing vdW and ES energies, respectively (see Fig. 4.9; details are presented later in current chapter) (Ashrafuzzaman et al. 2012). Theoretical screened Coulomb and vdW interactions have been developed to address the mechanisms of ion pore/channel formation due to MAAs peptides, CDs, etc., inside lipid bilayer membranes (Ashrafuzzaman et al. 2012, 2014; Ashrafuzzaman and Tuszynski 2012a, b). These charge-based interaction mechanisms and the related energies depend largely on the participating agents' mutual organization. The theory and the MD simulation-based information together suggest that a primarily charge property-based interaction portfolio may be behind the creation of a cluster among the participating agents and that the cluster formation perhaps leads to creation of the MAA-induced ion pores. We shall analyze this in detail later in this chapter.

#### 4.2.2.1 *In Silico* MD Simulation Modeling of CD–Lipid Interactions. Methods

We used MD methodology in order to model in silico CD–lipid interactions. Based on the Monte Carlo concept, we considered five different relative locations and orientations randomly generated in each CD–lipid complex, created by a CD and a lipid molecule, as initial structures for MD simulations to increase sampling size for better statistical analysis. For each location- and orientation-specific complex, a 6 ns explicit water MD simulation at a temperature of 300 K in an aqueous solution at pH 7 was performed. We applied the software package Amber 11 (Case et al. 2010). The Amber force field ff03 was especially used. The explicit water TIP3P model was used to simulate solvent effects. The force field parameters for CDs and

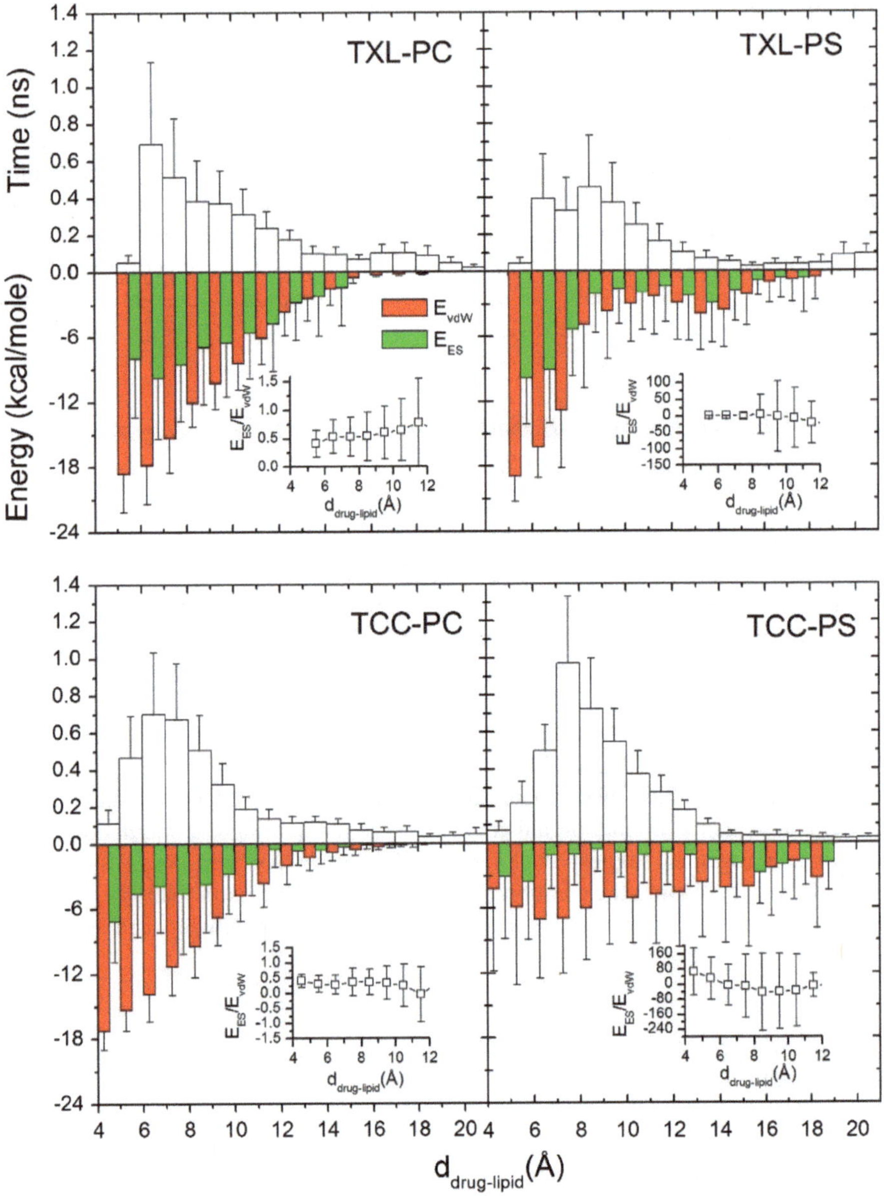


lipids (PC and PS) were generated using an Amber module antechamber (Wang et al. 2004, 2006). Note that the force field parameters for CDs are similar to the ones generated for colchicine (Huzil et al. 2010) and taxol (Freedman et al. 2009), respectively. Both studies in Freedman et al. (2009), Huzil et al. (2010) have shown these parametrizations lead to simulations to be consistent with experiments. Therefore, we can expect to observe similar results based on these parametrizations

◀**Fig. 4.9** Energies are detected from MD simulation on drug–lipid complexes. Simulations have been performed on two of the major cell membrane constituent lipids phosphatidylcholine (PC) and phosphatidylserine (PS) to address their interaction with cell membrane-adsorbed CDs: taxol (TXL) and thiocolchicoside (TCC). In all four histogram plots (upper panel) of time versus $d_{\text{drug-lipid}}$, the time durations when drug/lipid stay together (height) within a distance (width) during 6 ns simulations are presented. Lower panels show the histograms of nonbonded van der Waals (vdW) energy ($E_{vdW}$) and electrostatic (ES) interactions energy ($E_{ES}$). To avoid color conflict, $E_{vdW}$ and $E_{ES}$ are shown to occupy half-half widths though each half represents the whole width of the corresponding histogram (for details, see Ashrafuzzaman et al. 2012)

in current studies. Twenty complexes were energy minimized using the steepest descent method for the first ten cycles and then followed by a conjugate gradient for another 1000 cycles. We then applied Langevin dynamics during the process of heating up the system for 200 picosecond (ps) with the energy-minimized complex. Drug and lipid molecules were restrained using a harmonic potential with a force constant k = 100 N/m. Afterward, we introduced pressure regulation to equilibrate water molecules around the complex and to reach an equilibrium density for another 200 ps in addition to temperature regulation. The MD production run was continued for 6 ns. Given five various initial structures for each lipid–drug pair, a total 30 ns simulation result was analyzed to gain insights into the direct interactions of the corresponding pair. Note that the phospholipid was gently restrained with a harmonic potential with a force constant k = 10 N/m, applied only to the phosphate during the production runs. The purpose of this restraint is to mimic a single phospholipid being 'restrained' in membrane, while both headgroup and two tails of such lipid still possess certain degrees of freedom.

#### 4.2.2.2 *In Silico* MD Simulation Modeling of CD–Lipid Interactions. Understanding the Statistics and Detection of Energies

Figure 4.10 shows the MD results on drug–lipid distances using five CD–lipid complexes as initial structures as shown in the inset. It plots the separation distance of centers of mass of CD and lipid molecules, $d_{\text{drug-lipid}}$, against simulation time t (in ns order). Two popular CD molecules have been used in the MD simulations: taxol (TXL) and thiocolchicoside (TCC). Similarly, two important cell membrane lipids that are also popularly used in various model membrane studies are used: PC and PS.

Note that $d_{\text{drug-lipid}}$ was used as the simplest property to quantify the effects of CD–lipid interactions. It shows that $d_{\text{drug-lipid}}$ fluctuates around 10 Å, similarly to the initial setting in two to three simulations in TXL-PC, TCC-PC and TXL-PS, TCC-PS complexes. Drugs and lipids were observed to be gradually separated in most of the simulations.

Figure 4.11 plots snapshots of CD–lipid complex at the beginning (left) and at sixth ns (right) of the simulations. Note that it only presents the case that a CD likely to bind with the single lipid. Although TXL in both blue and red cases shown in Fig. 4.10 starts with different orientations, the simulations indicate TXL likely to bind to similar location than near phosphate group domain shown in Fig. 4.11a.

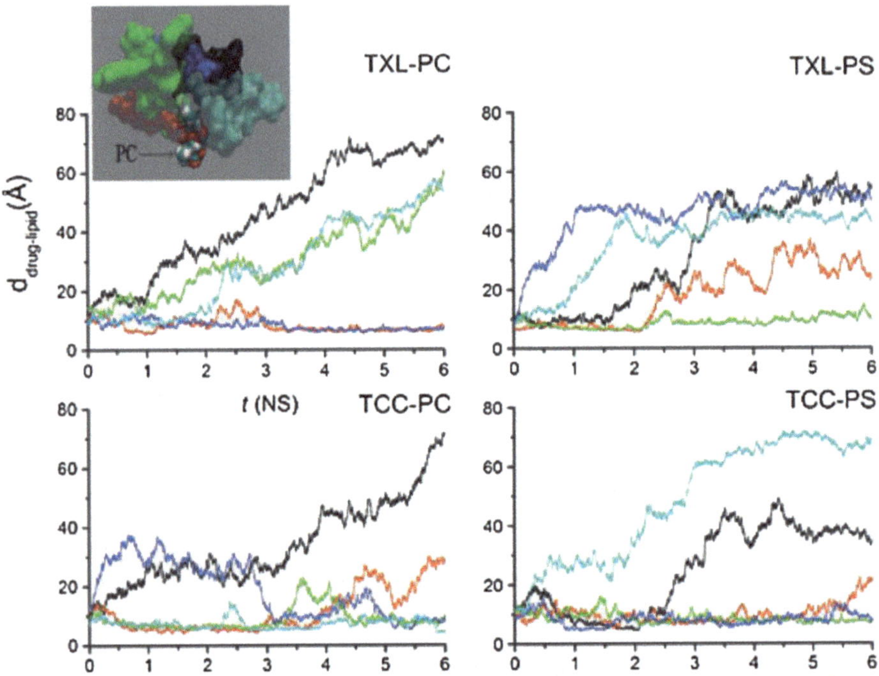


**Fig. 4.10** MD simulation results representing the change of the CD–lipid center of mass distance $d_{drug\text{-}lipid}$ with time. Five curves with different colors represent five independent initial CD–lipid complexes. The inset (only TXL-PC is shown) shows the cartoon representations of initial structures of five complexes with TXL following the color of the corresponding curve. This figure is copied from Ashrafuzzaman et al. (2012)

We also observe TCC (cyan case in Fig. 4.10) tends to bind to similar location (Fig. 4.11b). Yet Fig. 4.11a, b shows TXL and TCC (green case in Fig. 4.10) likely insert into the cavity between two short tails of PS, respectively.

The solvent-accessible (SA) area of the complex in the MD simulations was used to investigate whether the hydrophobic effects contribute to CD–lipid binding. Figure 4.12 shows SA areas in all four cases against $d_{drug\text{-}lipid}$. When both drug and lipid molecules are completely separated, we can expect them to be entirely exposed to solvent; i.e., the corresponding SA areas are at a maximum. Fig. 4.12 shows that the SA areas in all four cases are unchanged between the start and the investigated 20 Å length. This suggests that within this range used here the drug–lipid complexes stay at an equilibrium solvation condition.

Histograms of $d_{drug\text{-}lipid}$ from all five 6 ns simulations and the corresponding energy contributions from two nonbonded interactions, van der Waals ($E_{vdW}$) and electrostatic forces ($E_{ES}$) versus $d_{drug\text{-}lipid}$, are shown in Fig. 4.9. The histogram of $d_{drug\text{-}lipid}$ shows that both TXL and TCC spent more than 2 ns within 6 Å < $d_{drug\text{-}lipid}$ < 10 Å and away from lipids most of the time (see the upper panels in Fig. 4.12). It suggests the possibility for drugs to briefly bind with lipids.

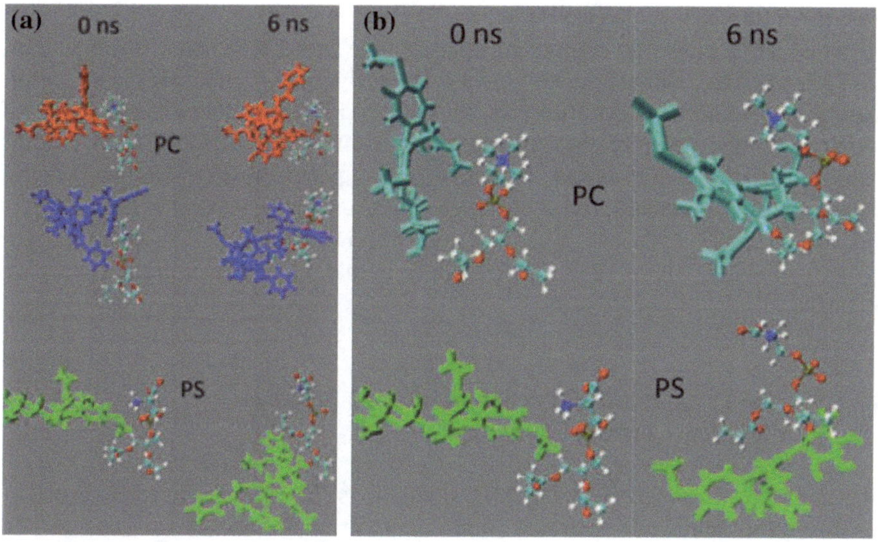


**Fig. 4.11** Snapshots of MD simulations. CD–lipid complex at the beginning (left) and sixth ns (right) of the simulation. The colorings of CDs are corresponded to the color definitions used in Fig. 4.10. **a** It shows structures of TXL-PC (upper and mid panel) and TXL-PS (bottom); **b** it shows structures of TCC-PC (top) and TCC-PS (bottom). This figure is copied from Ashrafuzzaman et al. (2012)

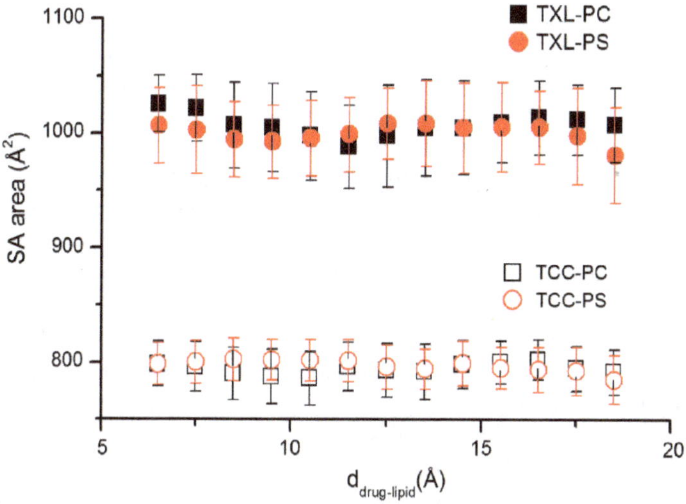


**Fig. 4.12** Solvent-accessible (SA) areas for four complexes are plotted against the CD–lipid center of mass distance $d_{\text{drug-lipid}}$. It shows SA areas are independent of $d_{\text{drug-lipid}}$ for four complexes

Figure 4.9 (top panel) indicates that TXL likely favors the interaction with PC over PS while TCC shows no significant lipid-specific preference. Both $E_{ES}$ and $E_{vdW}$ for TXL and TCC interacting with PC are inversely proportional to $d_{drug\text{-}lipid}$, while there are no such trends in either TXL-PS or TCC-PS cases. Below $d_{drug\text{-}lipid} < 6$ Å which is on the order of the lipid headgroup dimension, the CD–lipid binding stability drastically decreased (see the upper panel in Fig. 4.9) and below $d_{drug\text{-}lipid} < 4$ Å no stability was observed because there was no structure found at this low distance value (see Fig. 4.10). It is shown that both $E_{ES}$ and $E_{vdW}$ are strongly effective within 12 Å ($E_{vdW}$ slightly dominant as shown in the inset plots of Fig. 4.9).

#### 4.2.2.3 The Drug–Lipid Physical Interactions Stabilize the Drug Clusters Energetically

The main purpose of performing MD simulations that are presented here was to discover the physical energies that play primary role(s) to ensure the time-dependent physical coexistence between CDs and lipids in membranes. MD simulations were performed to investigate the stability of the pairwise coexistence (due to physical interactions) of the ion-channel-forming agents and lipids.

MD results suggest that the CD–lipid complex fluctuates within a separation over a period of time. These results suggest that both TXL and TCC likely bind with PC and PS given appropriate initial conditions. Through our in-depth analysis, we found evidence suggesting that the hydrophobic effect is unlikely to contribute to the distance-dependent CD–lipid binding. The analysis of energy contributions from two nonbonded interactions, $E_{vdW}$ and $E_{ES}$, versus $d_{drug\text{-}lipid}$ revealed crucial insights into the cause of the observed stability of the CD–lipid complexes. Both $E_{vdW}$ and $E_{ES}$ appear to be the main contributors to the energetic CD–lipid binding, and vdW interactions contribute slightly more than ES interactions as the drug and lipid approach closer. Binding stability generally is found to decrease quickly with increasing $d_{drug\text{-}lipid}$ within 6 Å < $d_{drug\text{-}lipid}$ < 16 Å at which considerable structures were observed (see Fig. 4.10). Both vdW and ES interactions contribute comparably with both energies decreasing with increasing $d_{drug\text{-}lipid}$. Beyond a 16 Å separation, the interactions become negligible. The CD–lipid stability is still observed due to the effects of a combination of possible sustained presence of long-range $E_{ES}$ (though $E_{vdW} \approx 0$, data not shown) and other energies possibly induced by the surrounding environment. Large standard deviations (see Fig. 4.9) are suggestive of the conformational space of the CD–lipid complexes not being completely explored in MD simulations. Nonetheless, this incompleteness does not preclude our proposed interpretation. Importantly, the CD–lipid interactions resemble the protein–lipid vdW and ES interactions found in MD simulations of a β-helical gramicidin A (gA) channel in phospholipid membranes (Woolf and Roux 1994). This suggests that the CD–lipid interactions resemble these mechanisms, namely the vdW and ES interactions as found in the channel-forming AMP–lipid interactions.

Both CDs are found to permeabilize lipid bilayer membranes by creating ion pores, a mechanism that is usually exerted on lipid membranes by various AMPs. In Chap. 6, we shall address on ion channels in brief. Readers are also encouraged to learn more on ion channels from our earlier published book 'Membrane Biophysics' (Ashrafuzzaman and Tuszynski 2012b). The MD simulation-based discovery of common types of primary physical interactions (vdW and ES) between pore-inducing agents and lipids for both CDs and AMPs (Ashrafuzzaman et al. 2012, 2014) also suggests that these two different classes of biologically active molecules may act on membranes using common molecular actions. Therefore, CDs are likely to generate certain AMP type mechanisms.

A mechanistic understanding of the drug–phospholipid interactions can be gained from the novel picture of interaction energetics between MAAs and a lipid bilayer membrane (Ashrafuzzaman and Tuszynski 2011, 2012a, b, c; Ashrafuzzaman 2011). A screened Coulomb interaction (SCI) model has been used to represent the interaction of MAAs with lipids in the bilayer, due to the presence of the distributions of the localized charges on both MAAs and phospholipids of all types. When any MAA and phospholipid approach each other to form a MAA–phospholipid complex (e.g., the case of CD-PS/PC interaction in the case of CD–lipid clusters on cell surface), the localized charges present in the complex experience each other's electric fields. In this electrostatic interaction scenario, the charges interact with each other not only directly (via Coulomb interactions) but also indirectly through other charges in the vicinity. Hence, the interaction between any two localized charges becomes screened, so the interaction takes the traditional form of the SCI (Ashrafuzzaman and Beck 2004). Furthermore, the interaction energy may change due to many other parameters, such as variations in the membrane's electrical conditions, the presence of hydrocarbons, lipid density. Most importantly, the total energy of the complex appears to release some energy what we call free energy to let the cluster be stable.

In general, the binding energy between CD and phospholipid in a membrane is due to the sum of the traditional Lennard-Jones and screened Coulomb potentials (Ashrafuzzaman and Tuszynski 2011, 2012a, b, c; Ashrafuzzaman 2011; Ashrafuzzaman and Beck 2004). A change of drug–phospholipid binding stability which appears through the distance-dependent probability (see Fig. 4.9 in our MD simulation) is mainly due to the change of drug–phospholipid coupling energy which varies mainly according to the effects of the hydrophobic coupling to be filled with localized single, double, etc., charges, respectively, that are present within the interaction field in a drug–phospholipid complex. In the case of the interaction of any CD with a phospholipid bilayer, the SCI extends beyond the nearest-neighbor phospholipids to other phospholipids residing in the vicinity, exactly following the protocol presented in references (Ashrafuzzaman and Tuszynski 2011, 2012a, b, c; Ashrafuzzaman 2011; Ashrafuzzaman and Beck 2004).

To better understand the generalized drug–lipid (PS or PC) interactions, more MD studies are needed to focus on CDs and phospholipid membrane rather than a single phospholipid. But as mentioned earlier, the total energy of the CD–lipid cluster should be somewhat additive of the individual CD–lipid pairwise interactions energies. This deeper understanding of the interactions will provide clear insights into the membrane effects of CDs.

We have demonstrated here the formation of drug molecule clusters on a cell surface (Ashrafuzzaman et al. 2016). Cell surface general binding and cluster-forming phenomena may not directly prove that these molecular processes are sole causes of ion pore formation inside cell membrane. But the nature of such drug adsorption mechanisms on cell surface may suggest that due to this kind of drug adsorption and accumulation, the localized mechanical and electrical properties are influenced which may lead to causing the energetics and corresponding conformations to be regulated (He et al. 1995). This is clearly evident in the pattern of energies and stability of drug–lipid complexes shown in Fig. 4.9 (Ashrafuzzaman et al. 2012). The reasonable minimum energy conformation may statistically appear due to phenomenological drug binding and clustering on cell surface regions, a hint on which is found in the trend of adhesive energy calculated here (see Fig. 4.8).

In broader conclusion, CD molecules have been found to bind statistically on cell surface. Both random drug binding and drug clustering have been detected. CDs are reported to create ion pores in planar lipid bilayer membrane due to charge property-based drug–lipid interactions. Cell surface general adsorption of drugs and their permanent-type binding may suggest for quite strong interactions with certain cell surface agents. The adhesive energy of drug molecules is found to be quite strong. Its decreasing trend with increasing cluster size hints for the better binding condition in bigger clusters satisfying a fundamental physics concept which equivalently suggests that bigger clusters should represent for deeper potential wells, as a result higher chance for a drug molecule of getting trapped inside. The vdW and Coulomb or screened Coulomb (in special system, e.g., ion channels in membrane Ashrafuzzaman and Tuszynski 2012a; Ashrafuzzaman et al. 2012) interactions in drug–lipid complexes may appear as prime reasons behind strong binding of CDs on cell surface. General drug binding with lipid(s) and clustering of drugs on cell surface certainly follow some specific energetics. The drug–lipid interaction energetics may be the molecular mechanisms, but the broader energetic picture can be a bit complicated. The free energy per drug in a drug cluster is perhaps different (expected to be smaller) than that in case of a single drug, a hint is already equivalently observed here in adhesive energy measurements. Appropriate theoretical analysis is underway. Favorable energetics in an adsorbed drug cluster may lead to creation of an ion channel. Thus, this chapter has addressed the pre-channel condition of cell surface-active agents as they approach to cell membrane and start creating a complex with membrane constituents like lipids (Ashrafuzzaman et al. 2016).

## 4.3 Hypotheses on Cell Surface Drug Clustering

Cell membrane-permeable ion pores allow materials and information to pass through them following some fundamental physics. Some of the physics-related parameters are predictable or even detectable or measurable using valid computational and experimental techniques. Chapter 6 will address on the ion channel phenomena and related observable parameters based on physics principles. But the mechanisms behind ion channel formation are yet unknown or to the best of our knowledge quite little or poorly known. The phenomenological AFM inspection and explanations based on energetics on cell surface clustering of ion-channel-forming drugs provide some clues (Ashrafuzzaman et al. 2016) but not rules that might be considered as dictating aspects of ion channel formation. We therefore attempt to put forward a few hypotheses for further investigation purposes. Behind creation of ion pores, general CAAs or specific MAAs such as AMPs, CDs are required to distribute on cell surface following some general biophysical rules. Our hypotheses may help design scientific investigations toward discovering necessary hidden rules that paly behind creating drug clusters, forming ion channels or other events.

Careful inspection of earlier presented AFM images (Ashrafuzzaman et al. 2016) suggests that the CDs are distributed on cell surface following a statistical nature and a condensed matter pattern. Section 4.1 was dedicated to explaining the nature of drug clusters on cell surface. The drug clusters have various sizes, and they are randomly distributed all over the cell surface. Here we consider the hypotheses:

**Hypothesis I, Drug Clustering** *Drugs cluster on cell surface following a cell surface adsorption of the drug molecules.* This adsorption happens due to localized driving forces caused by interactions between drugs and cell surface membrane ingredients (mostly lipids). The cluster is therefore a lipid-mediated accumulation of drug molecules.

**Hypothesis II, Initiation of Channel Creation** *Creation of drug clusters initiates the process of ion channel formation.* Ion channels are of various types following different structural formalisms (discussed in Chap. 6) constructed inside a lipid bilayer membrane depending on the properties of the channel-forming agents and membrane ingredients like lipids, available membrane proteins in the vicinity and other ingredients, membrane dielectric conditions, membrane mechanical properties.

**Hypothesis III, Ion Channel Size Estimation** *The post-drug cluster created ion channel cross-sectional changes proportionally as the drug cluster size changes. The larger is the cluster size, the bigger the ion channel pore cross section and vice versa.* The ion channels are generally formed with different cross-sectional areas.

## 4.4 Theoretical Background on Drug Clustering and Related Modeling

We have put forward a few hypotheses here considering our inspected information on drug clusters. In light of them, we can assume some energy states of drug clusters. Let us assume $E_1$, $E_2$, $E_3$, …, $E_n$ are equilibrium energy of corresponding states 1, 2, 3, …, $n$, or generally we can assume $E_i$ is the energy of state $i$. Here energy is defined as the total energy of the cluster constructed as a result of the drug–lipid interactions. The energy can be considered as the sum of interaction energies like electrostatic, van der Waals in a hydrophobic membrane background in the interface of cellular exterior and cell membrane surface. Our MD simulations detect some of these energies for the case of CDs (see Sects. 4.1–4.2) (Ashrafuzzaman et al. 2012).

Let us construct the virtual sketch of drug clusters as shown in Fig. 4.13. These sketches have been drawn in light of AFM-detected drug clusters on cell surface.

### *4.4.1 Drug Cluster Trapping of Drug Molecules*

We shall apply the statistical mechanics concept on the distribution of the drug molecules on cell surface in a thermodynamic state/condition. The Boltzmann distribution is a probability distribution that gives the probability that a system will be in a certain state as a function of that state-specific energy and the temperature of the system. The Boltzmann probability or trapping probability for a drug molecule on cell surface to diffuse into a cluster $i$ ($i$ is 1, 2, 3, 4, etc.) is defined as

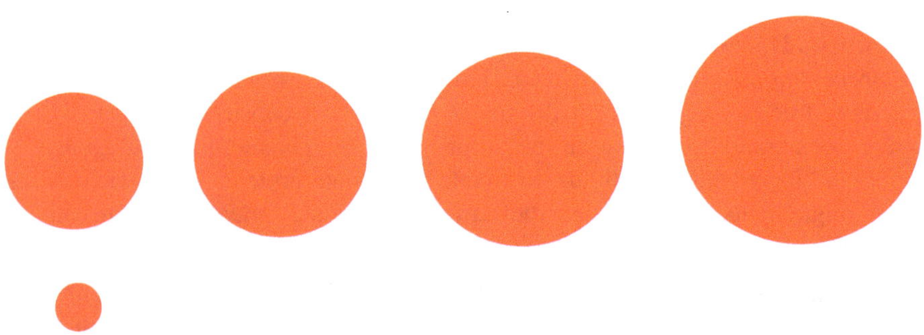

**Fig. 4.13** Two-dimensional (2D) representation of the drug clusters (top): Left to right (small to large) are three-dimensional (3D) spherical (looks like circle in 2D view) drug clusters 1, 2, 3, and 4. AFM data analysis presented in this chapter suggests that as the cluster size increases the height of the cluster may not increase proportionally, so consideration of 2D structure may be a quite reasonable model. Smallest sphere at the bottom (left) is the size of a drug molecule (we will often refer it just as a 'particle' here). Let us assume that $n_1$, $n_2$, $n_3$, and $n_4$ number of drug molecules construct the drug clusters 1, 2, 3, and 4, respectively. Here, $n_1 < n_2 < n_3 < n_4$. Let us assume that the energy per drug molecule is $\epsilon_1$, $\epsilon_2$, $\epsilon_3$, $\epsilon_4$ in cluster 1, 2, 3, 4, respectively. That makes the free energy per drug molecule in cluster 1, 2, 3, 4 is $\Delta\epsilon_1$, $\Delta\epsilon_2$, $\Delta\epsilon_3$, $\Delta\epsilon_4$, respectively, relative to the cell surface background where the energy per drug molecule is $\epsilon_0$. Here, $\Delta\epsilon_i = \epsilon_i - \epsilon_0$ with $i$ = 1, 2, 3, 4, etc. Here 'free energy' per drug molecule refers to the energy that is released by the molecule to get trapped inside the drug cluster or the energy that is gained by the molecule to get released from the cluster

$$p_i \sim \mathrm{Exp}\{-\Delta\epsilon_i/k_B T\} \tag{4.5}$$

Here $k_B$ is the conventional Boltzmann constant and $T$ is the absolute temperature. The four probability functions $p_1$, $p_2$, $p_3$, and $p_4$ corresponding to four clusters in Fig. 4.13 are

$$p_1 \sim \mathrm{Exp}\{-\Delta\epsilon_1/k_B T\} \tag{4.5.1}$$

$$p_2 \sim \mathrm{Exp}\{-\Delta\epsilon_2/k_B T\} \tag{4.5.2}$$

$$p_3 \sim \mathrm{Exp}\{-\Delta\epsilon_3/k_B T\} \tag{4.5.3}$$

$$p_4 \sim \mathrm{Exp}\{-\Delta\epsilon_4/k_B T\} \tag{4.5.4}$$

In case when the drug clusters exist independently on cell surface, we can sketch their equivalent energy conformations as shown in Fig. 4.14.

In case of a single particle on cell surface falling into a cluster, the particle can be considered to transit from a potential well with energy state $E_0$ to any potential well i with energy state $E_i$, where $i$ = 1, 2, 3, …, n (Fig. 4.15). As explained earlier, this transition lets the particle to transfer from an energy state where binding energy per

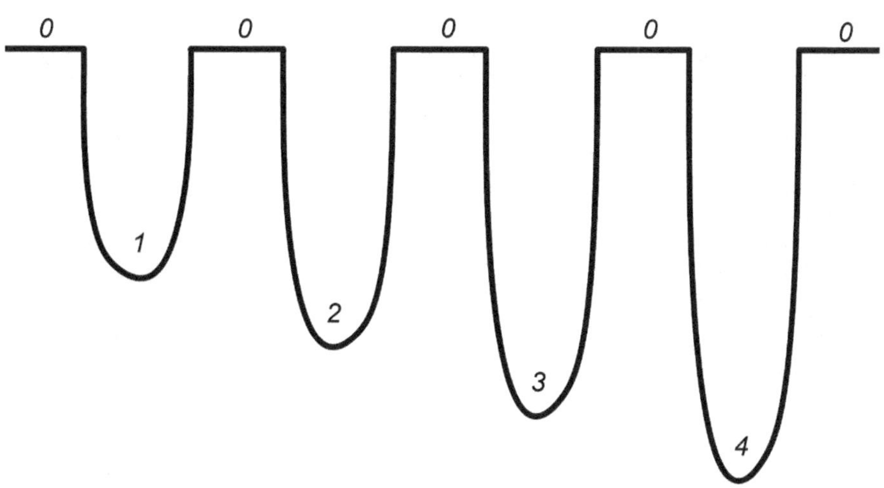


**Fig. 4.14** Discrete drug clusters 1, 2, 3, 4 exist with different potential wells relative to equilibrium cell surface background with 0 equilibrium potential. The potential well depth increases with the number of particles in the cluster. The smallest potential well (not sketched here) with smallest well depth equivalent to an energy state $E_0$ exists with single particle in the well/cluster

particle is $\epsilon_0$ to another energy state where binding energy per particle is $\epsilon_i$ through compromise of binding energy equal to $\Delta\epsilon_i$. Here $\Delta\epsilon_i = \epsilon_0 - \epsilon_i$.

Two important questions we wish to address are as follows:

i. How can we calculate the energy per particle in a potential well as shown in Fig. 4.15?
ii. Does each particle contribute equal amount of energy into any potential well irrespective of the cluster size corresponding to the potential well?

The above two questions are important to understand the nature of the energetics of the drug cluster. The drug accumulation into or release from the drug cluster causes the cluster to increase or decrease in size. The drug trapping probability (see Eq. 4.5: Eqs. 4.5.1–4.5.4) directly depends on the drug molecule energetics inside drug clusters. As the drug clusters appear with various size distribution, we need to understand the drug trapping probability into any drug cluster considering the per drug potential energy inside a drug cluster with specific size as this is basically the measure of drug cluster size-specific free energy of a drug that is to be considered during transition. Before answering the above-mentioned two questions we would therefore require to present a theoretical calculation on energy of a single particle in a potential well. This energy is $E_0$ when the well represents an energy state constructed by a particle alone in the well but may hold another value while the energy

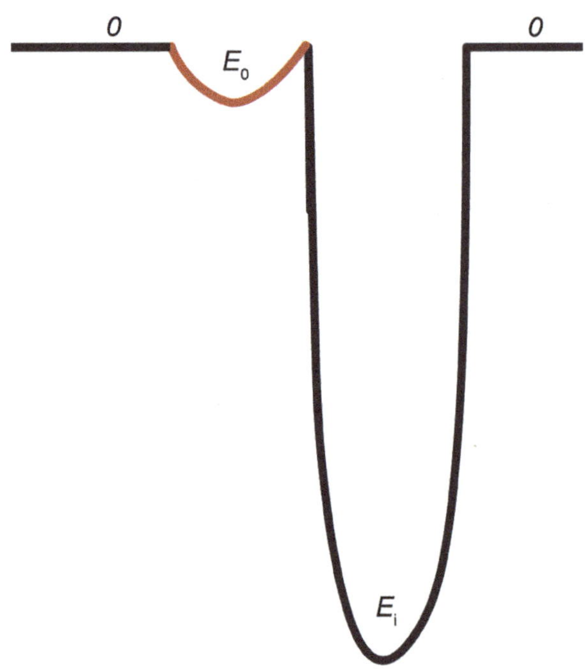


**Fig. 4.15** Single drug falling into a drug cluster with energy state $E_i$. This is equivalent to a drug falling from an energy state $E_0$ (distinguished with red color bottom) into another energy state $E_i$. The vice versa is also possible while drug gets released from a cluster. As before, 0 is nothing but the cell surface background energy without the presence of any NP

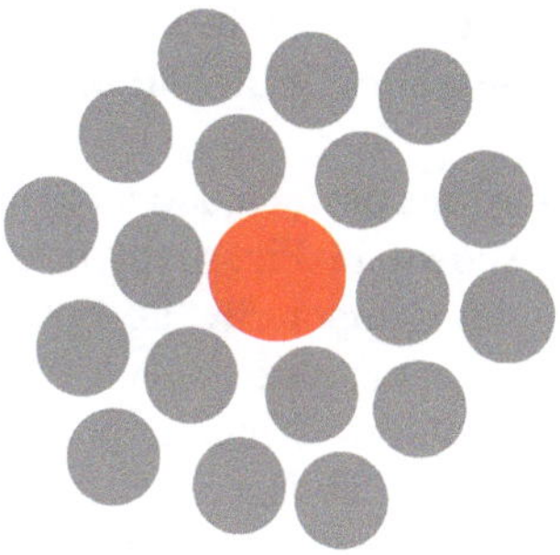

**Fig. 4.16** Here we have sketched a possible model diagram of the accumulation of a drug particle inside two-dimensional (2D) lipid monolayer for demonstration of interaction pattern purposes only. It does not show any specific phase structure of lipid membrane that is explained in details in our previously published book 'Membrane Biophysics' (Ashrafuzzaman and Tuszynski 2012b). Gray color spheres represent lipid headgroups, and red color sphere is the drug particle. This is nothing but the smallest form of the earlier mentioned drug clusters which is nothing but just a scenario where single drug particle interacts with lipids (mainly) on cell surface

state is constructed by more than 1 particle. An independently moving single particle on cell surface is usually bound to lipid membrane via interactions with lipids on membrane monolayer. The particle is under interactions with the lipids on approximately a two-dimensional plain [similar to the analogy presented in Ashrafuzzaman and Tuszynski 2012a, b). This scenario can be sketched as in Fig. 4.16.

The challenge is to develop a theoretical form for the binding energy of a drug particle $\epsilon_0$ in cluster created between the drug particle and lipids on a two-dimensional plane. We have recently developed a theory on interacting peptides with lipids in a lipid monolayer (see Ashrafuzzaman and Tuszynski 2012a, b). This theoretical platform can be generalized to address this theoretical issue of drug particle interaction with lipids on a cell surface.

### *4.4.2 Binding Energy of Drug Molecules Inside Drug Clusters*

We have assumed earlier that the energy per drug molecule (drug molecule's binding energy) in a drug cluster (assigned as *i*th cluster) is $\epsilon_i$, where $i = 1, 2, 3, \ldots$, etc. We have also assumed that a single drug molecule surrounded by lipids mimicking the free movement of drugs/NPs on cell surface experiences a cell surface binding energy $\epsilon_0$. We therefore need to calculate theoretically the forms of both $\epsilon_0$ and $\epsilon_i$. As mentioned earlier, we can do so by revising our previous calculations on energetics of ion channels accumulated inside a lipid membrane (Ashrafuzzaman and Tuszynski 2012a, b).

#### 4.4.2.1 Independently Mobile Drug Molecule's Binding Energy on Cell Surface. Theoretical Calculation of $\epsilon_0$

In this case, a drug particle may be considered to adjust its position on a cell surface following the conformation shown in Fig. 4.16. A drug particle is expected to coexist with surrounding lipids temporarily as both drug and lipid molecules continuously diffuse to new locations due to their physical dynamics. As in cell membrane lipid arrangement follows standard phase behavior (for details, see Ashrafuzzaman and Tuszynski 2012b), accommodation of drugs within a phase may be possible through some perturbations in the lipid organization phases. Here a drug particle is considered to interact with its nearest-neighbor lipids using direct Coulomb interaction mechanism. The drug particle is assumed to interact with distant lipids using SCI mechanism. The farther the lipid, the higher-order screening is required to get the lipid interacted with the drug particle (for details, see Ashrafuzzaman and Tuszynski 2012a, b). To better understand this drug–lipid interactions, we may revise the random coexistence of drug and lipids in Fig. 4.16 and present the model in a virtual 'shell model' form as shown in Fig. 4.17. In this latter symmetric arrangement, we have assumed that the drug locates itself at the center of a circle and lipids arrange at various radial distances $r_{LL}$, $2r_{LL}$, $3r_{LL}$, etc.

Shell Model to Address Drug–Lipid Screened Coulomb Interactions

Figure 4.17 shows the possible arrangement of drug with lipids on a symmetric backbone what we brand here (for the sake of presentation) as the 'shell model' to explain the drug–lipid interactions easily. Here we are presenting 2D sketch of the shell model presentation of the drug–lipid arrangement on cell surface as generally understandable that drugs move locally on a 2D plane. Importantly, a single drug molecule is more likely to find its match in the surface of the cell membrane, probably in the hydrophilic–hydrophobic boundary. The nearest-neighbor curve holds lipids that are interacted with the central drug molecule using direct Coulomb interaction (DCI) what we call as 'zeroth-order SCI interaction line'. Subsequent curves hold lipids that are interacted with the central drug molecule using SCI what we call as 'first-order SCI interaction line', 'second-order SCI interaction line', 'third-order SCI interaction line', etc. (see Fig. 4.17).

In SCI methodology, we assume each particle, drug or lipid, to appear with some charges on them. The drug–lipid complex is therefore nothing but equivalently can be considered as a complex of charges. We can apply the strategy that we developed recently to calculate the energy of such a system (Ashrafuzzaman and Tuszynski 2012a, b). We shall brief it here.

In a many-charge system, each charge interacts with another one following SCI formula, recently developed by us (see in Ashrafuzzaman and Tuszynski 2012) in Fourier space:

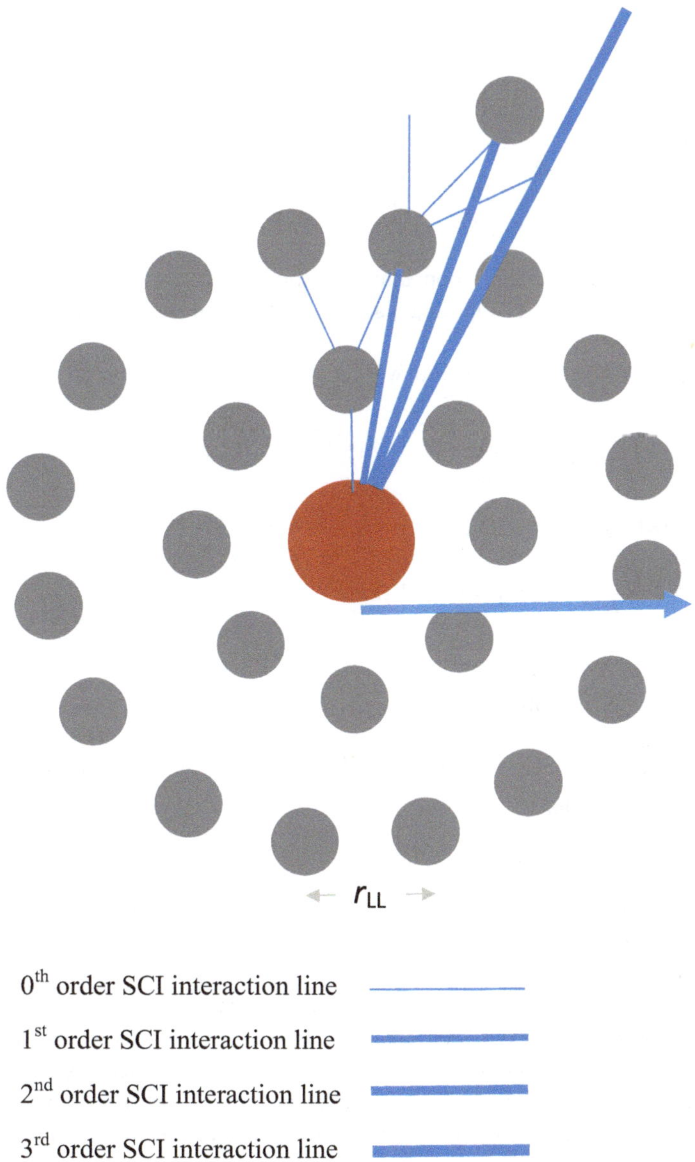


**Fig. 4.17** Different screening orders in SCI are shown (modeled here with interaction lines having various thicknesses) in the interaction of a drug particle with surrounding lipids on cell surface as shown in earlier Fig. 4.16. $r_{LL}$ is average interlipid distance. The outward line with an arrow represents that there is possibility for the drug particle to extend its interactions with lipids that are as far as infinite distant. This is valid as intercharge interactions (explained in Coulomb interaction formula) extend to infinite distance

$$Vsc(k) = \frac{V(k)}{1 + \left(\frac{1}{2\pi k_B T} n\right) V(k)} \tag{4.6}$$

Let us assume,

$$f(T, n) = \frac{1}{2\pi k_B T} n \tag{4.7}$$

Here $k_B$ is the Boltzmann's constant, $T$ is the absolute temperature, and $n$ is the number density of the participating particles, namely drug and lipids in this case (Fig. 4.17).

Using Taylor's series expansion we get:

$$\begin{aligned} Vsc(k) &= V(k) + V(k)f(T,n)V(k) + V(k)\frac{(f(T,n)V(k))^2}{2!} + V(k)\frac{(f(T,n)V(k))^3}{3!} + \cdots \\ &= V(k) + f(T,n)V(k)V(k) + \frac{(f(T,n))^2}{2} V(k)V(k)V(k) \\ &+ \frac{(f(T,n))^3}{6} V(k)V(k)V(k)V(k) + \cdots \end{aligned} \tag{4.8}$$

We can conclude the following:

$$\begin{aligned} Vsc(k) &= \text{0th order SCI} + \text{1st order SCI} + \text{2nd order SCI} + \text{3rd order SCI} \\ &+ \text{4th order SCI} + \cdots \end{aligned} \tag{4.9}$$

Numerical Computation on Energetics to Address Drug–Lipid Screened Coulomb Interactions

The drug particle interacts with neighboring lipids following SCI mechanisms whose theoretical forms have been deducted as presented in Eqs. (4.8) and (4.9). To calculate the first-order, second-order, third-order, etc., SCI energies represented by the interaction lines in Fig. 4.17, we need to perform a Fourier transformation on the respective forms deducted here. The form of Fourier transform follows from the following equation:

$$V_{sc}(r) = \int d^3 k e^{(ik.r)} V_{sc}(k) \tag{4.10}$$

To solve this complicated integration, one may perform numerical computations (NCs). Mathematica 10 or earlier versions can be used to make a general program and perform the NCs (for details, see Ashrafuzzaman and Tuszynski 2012a, b).

Two things are especially required to consider, namely, charge and distance of the interacting particles, drug and lipids here. As an assumption, we may consider the lowest possible distance between a drug and lipid to be equal to the average interlipid distance, $r_{LL} \sim 0.8$ nm (one may read Mannock et al. 1992, 1994 for a few dimension-related basic parameters). Regarding charges on the lipids, we know some of the lipids are zwitterionic, others bear net charges—either positive or negative (e.g., see Afanasenkau and Offenhäusser 2012; Abe and Shayman 2009). All these three types of lipids, e.g., PC, PS, and DOTAP, bear distribution of charges within their structures (see Fig. 4.18).

When a lipid approaches another one and reaches within a short distance, the charge distribution of one lipid experiences electrical polarization caused by the presence of the charge(s) of the other one and vice versa. Here short distance can be referred to a distance within which the charges interact with or exert influence on each other. As a result, we may consider two scenarios:

(i) Overall distribution of charges by combining all polarized charges of the participating lipids where all charges fall within each other's electric fields, or,
(ii) Consider each lipid (zwitterionic or charged) to have a net nonzero charge (either positive or negative), and these two lipids fall within each other's net nonzero charge created electric fields.

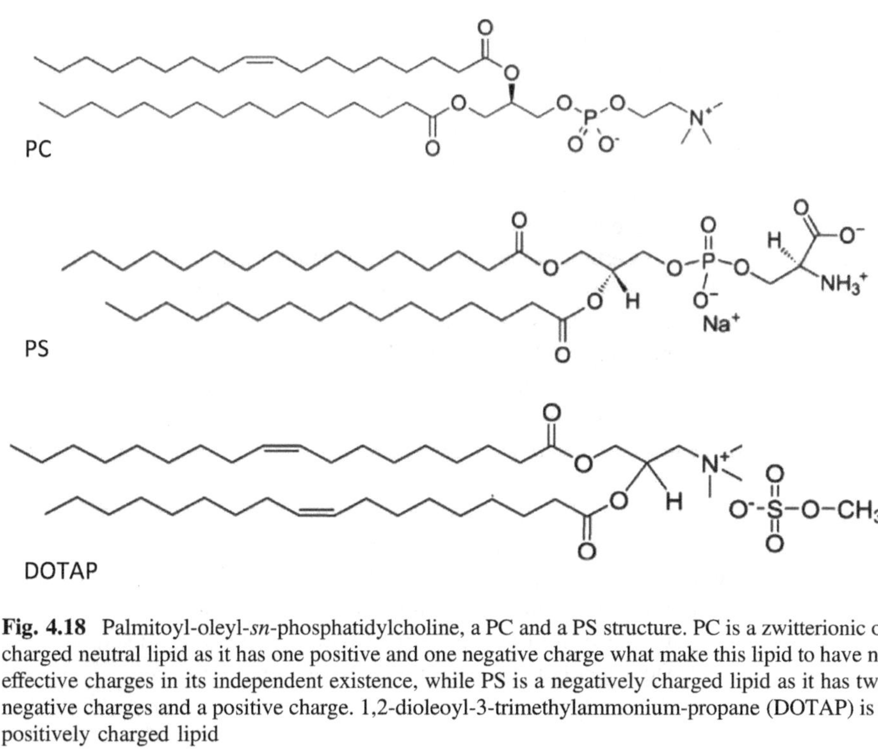


**Fig. 4.18** Palmitoyl-oleyl-*sn*-phosphatidylcholine, a PC and a PS structure. PC is a zwitterionic or charged neutral lipid as it has one positive and one negative charge what make this lipid to have no effective charges in its independent existence, while PS is a negatively charged lipid as it has two negative charges and a positive charge. 1,2-dioleoyl-3-trimethylammonium-propane (DOTAP) is a positively charged lipid

The above-mentioned modeling on charge distribution may be considered valid in the case schematized in Fig. 4.17 where lipids participate to create a drug–lipid cluster or structure. Here participating lipids and drugs are considered to create a structure due to perhaps the clustering of charges among the participating lipid and drug particles. Inside the equivalent charge, cluster charges interact with each other causing a many-charge interaction phenomenon. In such scenario, we have developed our SCI model calculations (Ashrafuzzaman and Tuszynski 2012a, b). Equation (4.10) can be used for calculating the energies corresponding to various SCI orders as presented in Eqs. (4.8) and (4.9).

We performed the NCs for peptide–lipid binding energy calculation to address the channel energetics in membrane (Ashrafuzzaman and Tuszynski 2012a, b). We used two different kinds of peptides gramicidin A (gA) and alamethicin (Alm) which are known to induce ion channels, gA and Alm channels, with different structures in lipid bilayer membrane. Here we also considered a peptide with charge and geometry accumulating inside a lipid monolayer, so the end result is as follows: A peptide gets surrounded by a layer of lipids (see Ashrafuzzaman and Tuszynski 2012a, b). This is quite equivalent to the analogy model sketched earlier for a general drug creating a cluster with lipids in a lipid membrane monolayer; see Fig. 4.17. We can consider the peptide–lipid binding energetics conceptually equivalent to general drug–lipid binding energetics.

Figures 4.19 and 4.20 demonstrate the energetics (in arbitrary units) of a drug (equivalently, a gA molecule in form of a gA channel) in lipid bilayers with different lipid screening orders and lipid dimensions using our theoretical calculation. In Figs. 4.19 and 4.20, $G_I$ and $G_{II}$ ($G_I > G_{II}$) represent energy levels at conformational states I and II, equivalent to two conformational states gA monomers exist in free form (no channel formation: $M_I$) and gA dimer form (channels formed: $D_{II}$), respectively. Formation of a channel with any level of stability requires an energetic transition (reduction) $\Delta G_{I,II}$ ($=G_I–G_{II}$). Although the energy minima appears at different values of reaction coordinates, we have chosen the transition $G_I \leftrightarrow G_{II}$ at a certain value of reaction coordinate (as shown in Fig. 4.20) to address how the corresponding back-and-forth conformational changes between gA monomers and dimers ($M_I \leftrightarrow D_{II}$) may become regulated due to $\Delta G_{I,II}$ which depends mainly on the bilayer physical properties for a certain channel type. The binding energy between two gA monomers (data not shown) alone in a gA channel is many orders of magnitude smaller than the binding energy of the gA channel with the bilayer at the channel bilayer interface. $\Delta G_{I,II}$ (see Figs. 4.19 and 4.20) therefore represents the amount of energy gA monomers need to compensate to form a stable gA channel which arises mainly from the hydrophobic binding between the gA channel and the bilayer at the two bilayer channel interfaces. The smaller the value of $\Delta G_{I,II}$, the higher the stability of gA channels. We observe that the value of $\Delta G_{I,II}$ for the second-order lipid screening is orders of magnitude higher than that for the first-order lipid screening (higher orders of lipid screening in our theoretical approach account for higher values of free length $d_0$-$l$ which is a measure of mismatch between lipid membrane thickness $d_0$ and accumulated channel length $l$). Knowing the effective values of charges (in units of coulomb) on

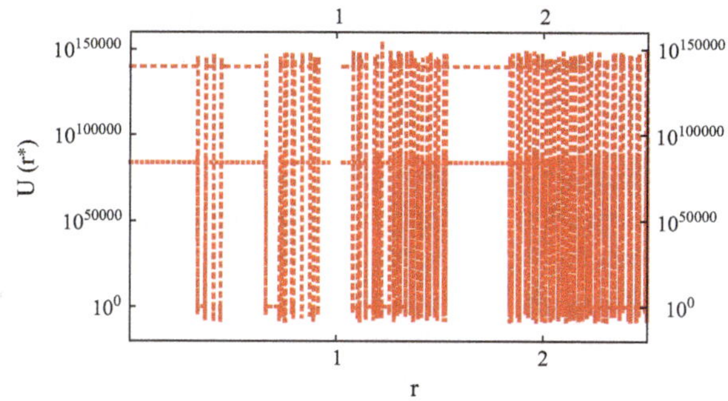


**Fig. 4.19** Energy versus reaction coordinate (using Eq. (4.10) here and hereafter) plot for gA channels in lipid bilayer energetics at different order of screening (single- and double-dashed curves are for first- and second-order screening, respectively). Both curves undergo transitions between two states I and II that are represented by higher and lower energies $G_I$ and $G_{II}$, respectively. Only real parts of the energies have been considered. For simplicity, the mentioned U(r*), the additional drug–lipid potential energy called Lennard-Jones potential (Ashrafuzzaman and Tuszynski 2012a, b) has been subtracted to present only the screened Coulomb potential plot here and other energy plots in all subsequent figures. $q_L/q_{gA} = 0.005$, $(1/\varepsilon_0)q_Lq_{gA} \approx 1$ has been chosen (here and in Fig. 4.20) for simplicity, $r_{LL} = 7.74597$ Å. In the plot, the energy at r = 0 Å has been excluded to avoid the associated singularity. Numerical integration here (and hereafter) has been performed using Mathematica 9 within $(-k_{max}2\pi/r_{LL}, k_{max}2\pi/r_{LL})$, where $k_{max} = 100$ and the step size for integration dr = 0.001 have been taken as judicious choices

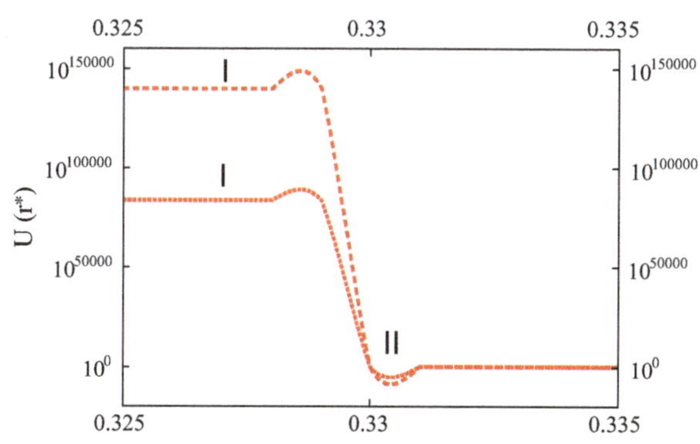


**Fig. 4.20** Energy versus reaction coordinate plot for gA channels in lipid bilayer energetics at different orders of screening (single- and double-dashed curves are for the first- and second-order screening, respectively). I and II represent levels with free energies $G_I$ and $G_{II}$, respectively, where gA monomers exist as free (no channel formed) and gA dimmer (gA channel formed). $q_L/q_{gA} = 0.005$, $r_{LL} = 7.74597$ Å. *Ad hoc* assumptions on $q_{gA}$ ~ electron charge and other relevant parameters give an estimate of $G_I$ and $G_{II}$ to be $10^{-1}$ and $10^{-8}$ in first-order and $10^5$ and $10^{-4}$ in second-order lipid screening in unit of kJ/mole which seriously depends on $q_L$ as $d_0$ increases. The energy orders for $G_I$ and $G_{II}$ as mentioned here are also valid approximations for the corresponding free energy levels presented elsewhere (Ashrafuzzaman and Tuszynski 2012a, b)

a gramicidin monomer, $q_{gA}$, and that of a lipid's headgroup region, $q_L$, one can readily calculate and show in real energy unit, e.g., joule (J), using our SCI theory, that $G_I$ has values which are drastically reduced, and hence the value of $\Delta G_{I,II}$ collapses as the value of $d_o$-$l$ approaches 0 Å. For example, making an ad hoc assumption that $q_{gA}$ and $q_L$ should be on the order of a few electron charges, we find $\Delta G_{I,II}$ to be on the order of kJ/mole for the first-order lipid screening which closely corresponds to the hints on the phenomenological bilayer deformation energy in another study (Ashrafuzzaman et al. 2006). But the same ad hoc assumption ensures that $\Delta G_{I,II}$ increases to the order of $10^5$ kJ/mole for the second-order lipid screening. This drastic increase in bilayer deformation energy requirements for stable channel formation with increasing the bilayer thickness channel length mismatch causes gA channel formation to be extremely hard at a higher order of lipid screening. Beyond a certain level of hydrophobic bilayer channel mismatch, the deformation energy reaches values which are outside a biological binding energy scale which suggests that at this high energy level the β-helical gA channels must experience exponential growth in its instability and finally may undergo a structural transition which in fact is experimentally observed.

As said earlier, the gA here represents a drug particle, equivalently as presented in Fig. 4.17. The numerical calculation-based finding of required higher energies corresponding to higher SCI order actually explains how a drug molecule may extend its interaction to surrounding lipids. Usually, the closest ones experience higher binding. This is why a freely moving drug molecule on a cell membrane surface apparently may be not totally free but under the influence of some kind of short-range interactions with surrounding lipids. The interaction bourn energetics follows some specific rules as well. We shall address them here based on our understanding of two different kinds of ion-channel-forming molecules gA and Alm.

### Drug–Lipid Interaction Energies Follow Specific Power Laws

Figures 4.21 and 4.22 show how lipid charge relative to the charge of the channel-forming peptides causes changes in the values of both $G_I$ and $G_{II}$ (see Fig. 4.21) and consequently $\Delta G_{I,II}$ (see Fig. 4.22). We observe about three-times higher values of both $G_I$ and $G_{II}$ for both the first and second order of lipid screening for an Alm channel with three monomers in the 'barrel-stave' pore than that for a gA channel as shown in Fig. 4.21. Quantitatively similar (threefold increases) higher values of $\Delta G_{I,II}$ (see Fig. 4.22) are observed for both lipid screening orders for an Alm channel than those for a gA channel. The three-times higher values of $G_I$, $G_{II}$, and $\Delta G_{I,II}$ in an Alm channel with three monomers are obvious; because in this specific Alm channel conformation there are six channel bilayer interaction sites while a gA channel always has only two interaction sites with the bilayer (see Ashrafuzzaman and Tuszynski 2012a, b). This suggests also that the lipid interactions with drug molecules follow drug molecule specificity meaning depends on the drug molecule physical properties. It seems that the

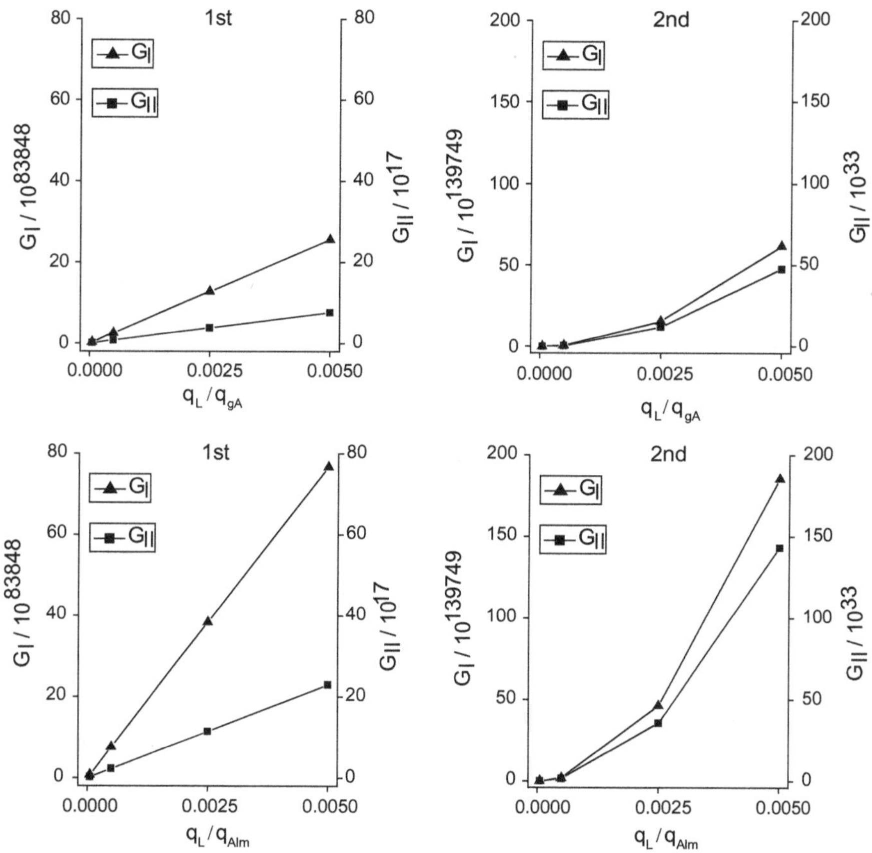


**Fig. 4.21** Energy ($G_I$ and $G_{II}$) versus $q_L/q_{gA}$ for gA channel (upper panel) or $q_L/q_{Alm}$ for Alm channel (lower panel) in lipid bilayer energetics at the first (1st) and second (2nd) order of lipid screening. $r_{LL}$ = 7.74597 Å

interaction energy between monomers in both gA and Alm channels becomes irrelevant in comparison with the binding energy between the channel and the bilayer. We also observe that $G_I$, $G_{II}$, and $\Delta G_{I,II}$ increase with the increase of lipid–peptide charge ratio following specific rules:

$$G_I, G_I, \Delta G_{I,II} \propto (q_L/q_M)^s \tag{4.11}$$

where $s$ = 1, 2, etc., for the first-order, second-order, etc., lipid screening, respectively, for both gA and Alm channels. Here, $q_M$ stands for gA ($q_{gA}$) or Alm ($q_{Alm}$) monomer charges.

In Fig. 4.23, we observe a modest and linearly proportional effect of lipid dimension $r_{LL}$ on $\Delta G_{I,II}$ for both the first- and second-order lipid screening with three-times higher effects for Alm channels than for gA channels. Figure 4.24 shows that for both gA and Alm channels the reaction coordinate at which we have

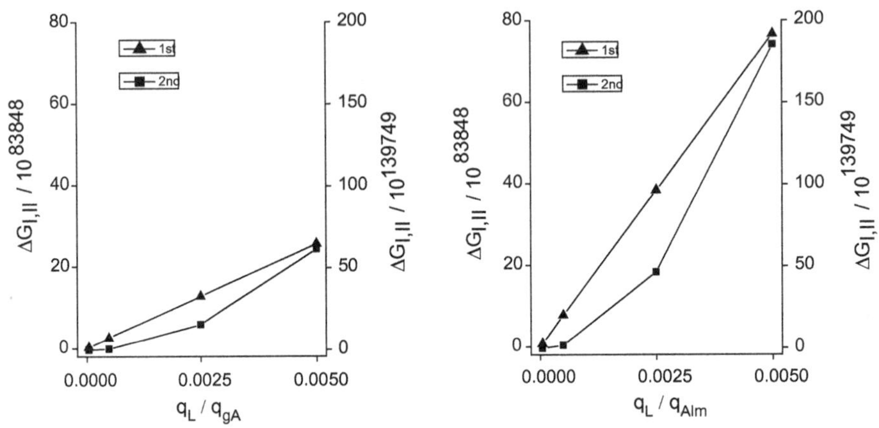


**Fig. 4.22** $\Delta G_{I,II}$ versus $q_L/q_{gA}$ (left panel) or $q_L/q_{Alm}$ (right panel) plot in lipid bilayer energetic at the first and second order of lipid screening. Here, $r_{LL}$ = 7.74597 Å

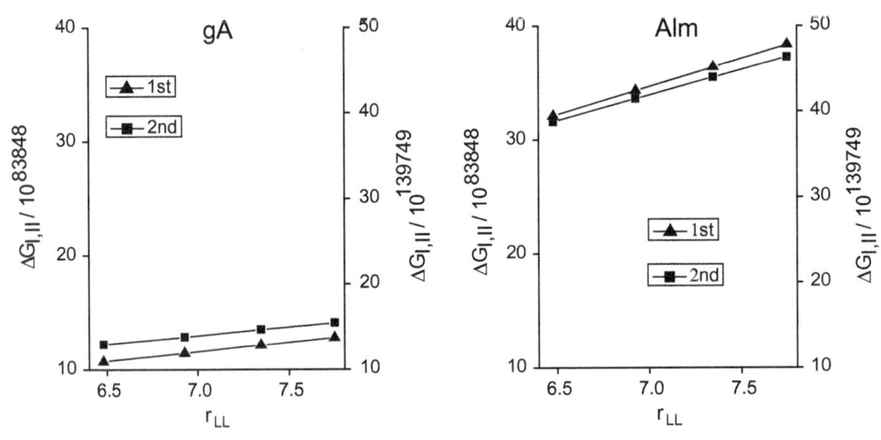


**Fig. 4.23** $\Delta G_{I,II}$ versus $r_{LL}$ plot for gA channel (left panel) and Alm channel (right panel) in lipid bilayer energetics at the first and second order of lipid screening. Here, $q_L/q_M$ = 0.0025

shown the calculation of $\Delta G_{I,II}$ increases (shifts toward higher values) in linear proportionality to $r_{LL}$ for both the first- and second-order lipid screening.

The dramatic theoretical result is illustrated in Fig. 4.25 where we observe that $\Delta G_{I,II}$ exponentially increases with the increase of $d_0$-$l$ (here we have rephrased the order of screening by $d_0$-$l$):

$$\Delta G_{I,II} \propto e^{d_0 - l} \tag{4.12}$$

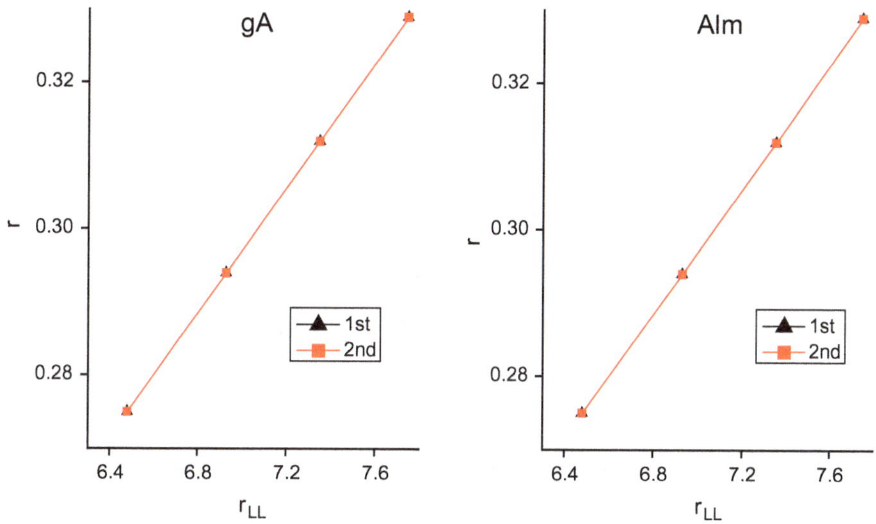


**Fig. 4.24** Reaction coordinate at which we have considered the $\Delta G_{I,II}$ versus $r_{LL}$ plot for gA channel (left panel) and Alm channel (right panel) in lipid bilayer energetics at the first and second order of lipid screening. Here, $q_L/q_M = 0.0025$

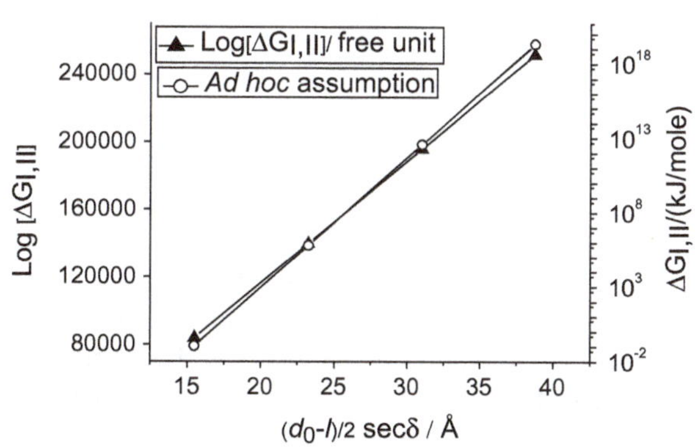


**Fig. 4.25** Log$\left[\Delta G_{I,II}\right]$ versus $d_0$-$l$ plot. Here ($d_0$-$l$) secδ is the distance covered by the lipid headgroups in the deformed regions of the bilayer at the bilayer gA channel interaction sites. δ (for mathematical simplicity can be considered to appear with a constant value within 0–90° in a particular lipid bilayer membrane for all participating lipids in all orders of screening at the gA channel bilayer interface) the angle at which lipids in the deformed portion of the bilayer couples with the extension of gA channel length. $\Delta G_{I,II}$ increases exponentially with $d_0$-$l$. $\Delta G_{I,II}$ at lower values of $d_0$-$l$ (e.g., $d_0$-$l \approx 0$) or at other higher values of $d_0$-$l$ can be extrapolated from the plot. For a certain type of lipid with fixed lipid charge $\Delta G_{I,II} \propto e^{d_0-l}$, consequently the dissociation force imposed by the bilayer on gA channel ($F_{dis}$) increases exponentially with $d_0$-$l$, i.e., $F_{dis} = -\left(-\partial/\partial(d_0-l)\Delta G_{I,II}(d_0-l)\right) \propto e^{(d_0-l)}$. As a result, gA channel lifetime $\tau$ decreases exponentially with the increase of $d_0$-$l$. *Ad hoc* assumptions on $q_{gA} \sim$ electron charge and other relevant parameters give an estimate of $\Delta G_{I,II}$ / (kJ/mole) (O) which seriously depends on $q_L$ as $d_0$ increases. Experimental gA channel stability predicts results perhaps falling within second-order screening ($d_0$-$l$ < 40 Å) (Ashrafuzzaman and Tuszynski 2012a, b)

Consequently, the dissociation force

$$F_{\text{dis}} = -\left(-\partial/\partial(d_0 - l)\Delta G_{\text{I,II}}(d_0 - l)\right)$$

also follows an exponential relation:

$$F_{\text{dis}} \propto e^{(d_0 - l)} \tag{4.13}$$

which is very different from the dissociation force calculated based on the 'bilayer elastic model' (see Wallace et al. 1981; Katsaras et al. 1992; Durkin et al. 1993; Andersen et al. 1998; Miloshevsky and Jordan 2004) where $F_{\text{dis}}$ on gA channel has been reported to change linearly with the change of $d_0$-$l$ (Ashrafuzzaman et al. 2006). Due to sufficient experimental validation and analysis on an appropriate theoretical platform, we believe our SCI model is valid over bilayer elastic model to explain the drug–lipid interactions inside cell membrane.

The above ion channel–lipid membrane interaction energetics suggests that a drug when it coexists with lipids inside a lipid layer the energy it experiences can be theoretically addressed using SCI model calculations. The various order screened energy line can be theoretically deducted by expanding the equivalent exponential term of the binding energy as deducted here in case of calculations performed for channel membrane binding. Our SCI model calculations therefore help calculate the binding energy of drugs inside lipid membrane. Here the calculated $\Delta G_{\text{I,II}}$ is a measure of our previously mentioned binding energy $\boldsymbol{\epsilon_0}$. The value of $\boldsymbol{\epsilon_0}$ varies for various drug particle types.

## Field of Interaction of a Drug with Surrounding Lipids Is Finite!

A drug in a cluster (see Fig. 4.17) interacts with neighboring lipids following SCI methodology. The direct interaction happens between drug molecule and lipids that stay adjacent to each other. Beyond the nearest-neighbor distance, the interaction extends. But according to SCI, the interaction becomes weaker. A thorough theoretical and computational analysis has been presented earlier. We shall schematize this scenario here using, e.g., gA interactions with lipids on membrane (for details, see also the Ashrafuzzaman and Tuszynski 2012).

The gA coupling extends to surrounding lipids up to a certain radial length. The condition is better presented in the schematic 2D diagrams in Figs. 4.26 and 4.27. In Fig. 4.26, the gA dimer (gA channel) is schematized to adjust inside lipid bilayer membrane. In Fig. 4.27, the length up to which gA dimer coupling effect extends is modeled. The length is shown up to the downward-pointing arrow beyond where the membrane readjusts to its normal condition which is free from the influence of the drug particle (peptide here). Within this radial length (Fig. 4.27), a number of lipids usually exist all around over 360° on a plane that are under the influence of a single drug. These lipids under the influence of a single drug and the drug molecule itself together determine the total energy of the potential well represented by a single drug particle on

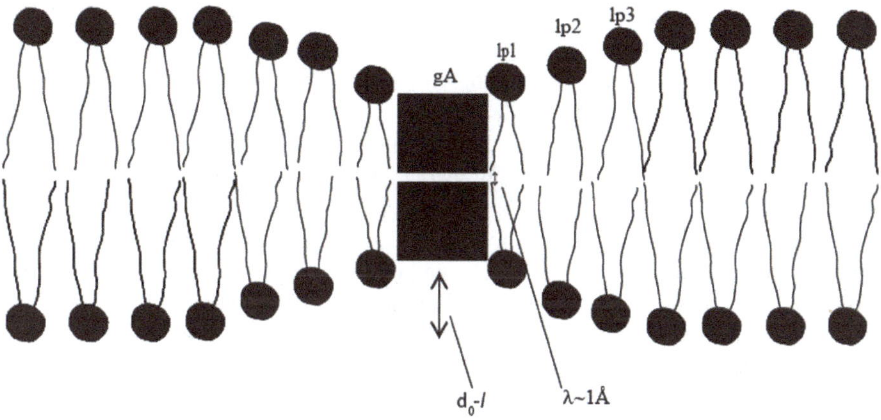


**Fig. 4.26** Gramicidin A monomer (gA) and lipids on each monolayer make a chain (with chain reaction) with continuous bending until the equilibrium membrane thickness reaches. A gramicidin A (gA) monomer interacts with lp1 (direct Coulomb interaction which is the zeroth-order term in the expansion of the SCI), with lp2 (first-order SCI), with lp3 (second-order SCI), etc. The interactions extend to all directions on each lipid monolayer surface

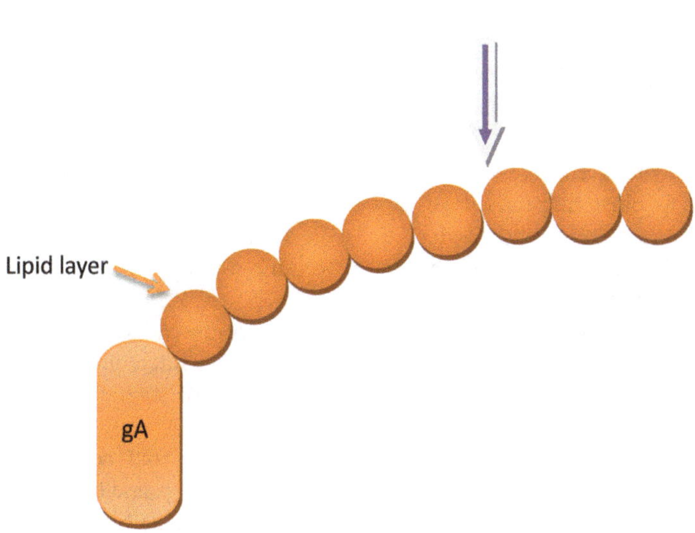


**Fig. 4.27** Gramicidin A (gA) monomer in a channel is assumed to find a lipid (just the headgroup is schematically mentioned) on the perturbed region of the bilayer next to it with bear Coulomb interaction but the next neighboring lipid with first-order SCI and so on. The gA monomer perhaps no longer can extend its interaction beyond the lipid on the right side of the downward-pointing arrow where the bilayer regains the form of its unperturbed thickness. Here we have shown only the lipid headgroup regions because most probably the lipid headgroup regions are responsible for the effective localized charges in lipids

a cell surface as sketched earlier in Fig. 4.15. A single drug is sketched here to fall into a drug cluster with energy state $E_i$ from an energy state $E_0$. This energy $E_0$ is the total energy of the cluster created by the single drug and lipids within distance up to the downward arrow (see Fig. 4.27) in all directions on an approximately two-dimensional membrane plane. Calculation of the total binding energy may follow the shell model arrangement of the drug–lipids (see Fig. 4.17) and related numerical analysis that are presented earlier.

Figure 4.27 shows radial distance of the lower pointing arrow from the gA molecule, let us consider it $\xi_{radial}$, and it can be assumed as a correlation length what determines how many lipids may fall within the interaction field created by the drug molecule around it. As per Fig. 4.17 (shell model) where the lipids are assumed to align on a circular symmetric shell in approximate 2D field (membrane surface in case of Fig. 4.27), $\xi_{radial}$ follows an approximate relation:

$$\xi_{radial} = n\, r_{LL} \tag{4.14}$$

Here for simplicity $n$ may be assumed as an arbitrary integer that determines the SCI orders to calculate the energy following Eqs. (4.8)–(4.10) and using Mathematica 10 programming for numerical simulations. $\xi_{radial}$ of course depends on various other things, e.g., the type of lipid phase, temperature.

#### 4.4.2.2 Clustered Drug Molecules' Binding Energy on Cell Surface. Theoretical Calculation of $\epsilon_i$

In a drug cluster (see Fig. 4.28), both drugs and lipids participate to construct the clusters. This clustering may have various different geometric conditions. Relative concentrations of drug and lipid particles in a cluster may determine qualitative differences in values of parameters that determine a cluster's physical conditions. Some analyses are presented here.

Figure 4.28a represents the random drug adsorption by cell surface. Here the drug association with lipid monolayer happens through interaction of independent drugs with surrounding lipids. The binding energy $\epsilon_0$ of such a drug molecule is derived earlier.

Figure 4.28b represents a condition in which substantial adsorption of drug particles by cell surface in lipid monolayer has caused drugs and lipids initiating the creation of a 'drug–lipid cluster' where both drugs and lipids locate randomly. The calculation of drug particle binding energy $\epsilon_i$ in a drug–lipid cluster is a bit complicated that require consideration of three different distances units, namely lipid–lipid distance, $r_{LL}$, drug–lipid distance, $r_{DL}$, and drug–drug distance, $r_{DD}$, respectively (see Fig. 4.29). Figure 4.29 is sketched to show different screening orders in SCI in the interaction of a drug particle with surrounding lipids or drugs on cell surface for mixed drug cluster as shown in earlier Fig. 4.28b.

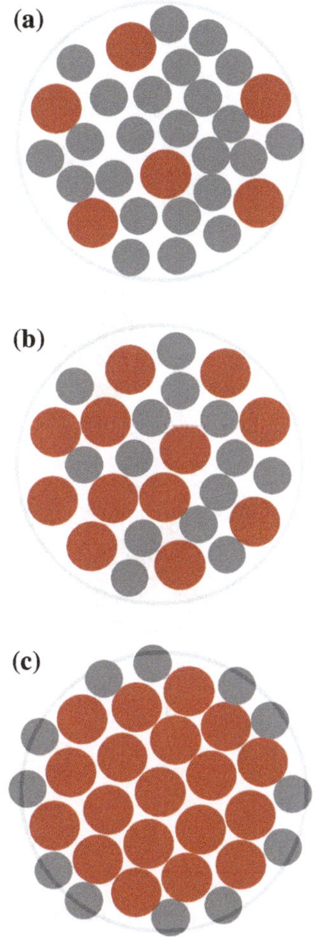


**Fig. 4.28** Drugs penetrate into cell surface membrane that may initiate creation of clusters. **a** Drugs randomly penetrate into lipid monolayer on cell surface. No considerable interdrug particles communication is yet established. **b** Substantial adsorption of drug particles by cell surface in lipid monolayer causes drugs and lipids initiating the creation of a so called (I wish to name it) drug–lipid cluster where both drugs and lipids locate randomly. **c** Drugs stay together and make a cluster which is surrounded by lipids. I wish to call it a 'drug–drug cluster'. Red sphere represents drug particle, and brown sphere represents lipid molecule. The circular boundary may be considered for a circle with radius $\xi_{\text{radial}}$ that determines the cluster size. The model is a 2D representation though the drug–lipid and drug–drug clusters actually appear in AFM phase imaging as three-dimensional as presented earlier in Figs. 4.2, 4.3, 4.5, and 4.7

Figure 4.28c represents a condition in which drugs stay together and make a cluster which is surrounded by lipids what I call a 'drug–drug cluster'. The calculation of drug particle binding energy $\epsilon_i$ in a drug–drug cluster requires consideration of only one distance unit drug–drug distance, $r_{DD}$, (see Fig. 4.30) instead of three different distance units $r_{LL}$, $r_{DL}$, and $r_{DD}$, required for the drug–lipid cluster (see Fig. 4.29). Figure 4.30 is sketched to show different screening orders in SCI in the interaction of a drug particle with surrounding drugs on cell surface for a drug–drug cluster as shown in earlier Fig. 4.28c.

For the cases in Figs. 4.29 and 4.30, the NCs follow exactly identical protocols as used in Sect. 4.4.2.1 to calculate drug binding energy $\epsilon_i$ except for the use of different charge distributions, distance units, etc. Only interested researchers may repeat the computation for their focused research projects following the protocols presented here and in references that are quoted.

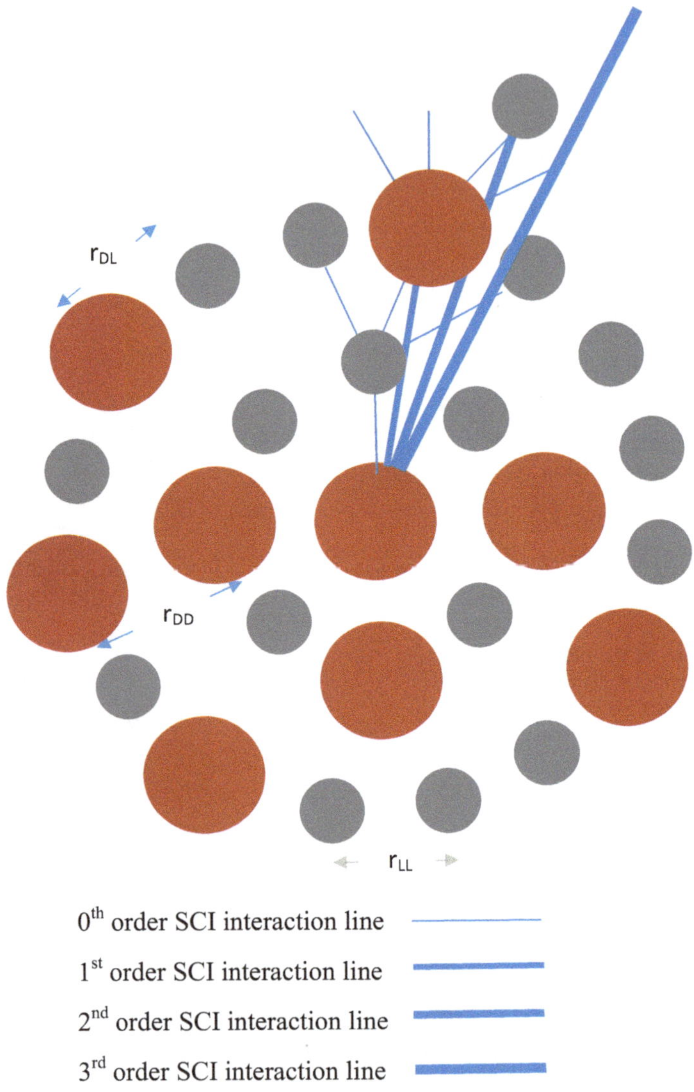


**Fig. 4.29** Different screening orders in SCI are shown (modeled here with interaction lines having various thicknesses) in the interaction of a drug particle with surrounding lipids or drugs on cell surface for mixed drug cluster as shown in earlier Fig. 4.28b. As before, we have sketched an equivalent shell model presentation in a 2D view. Here average interparticle spaces can be three different distances, namely lipid–lipid distance, $r_{LL}$, drug–lipid distance, $r_{DL}$, and drug–drug distance, $r_{DD}$, respectively

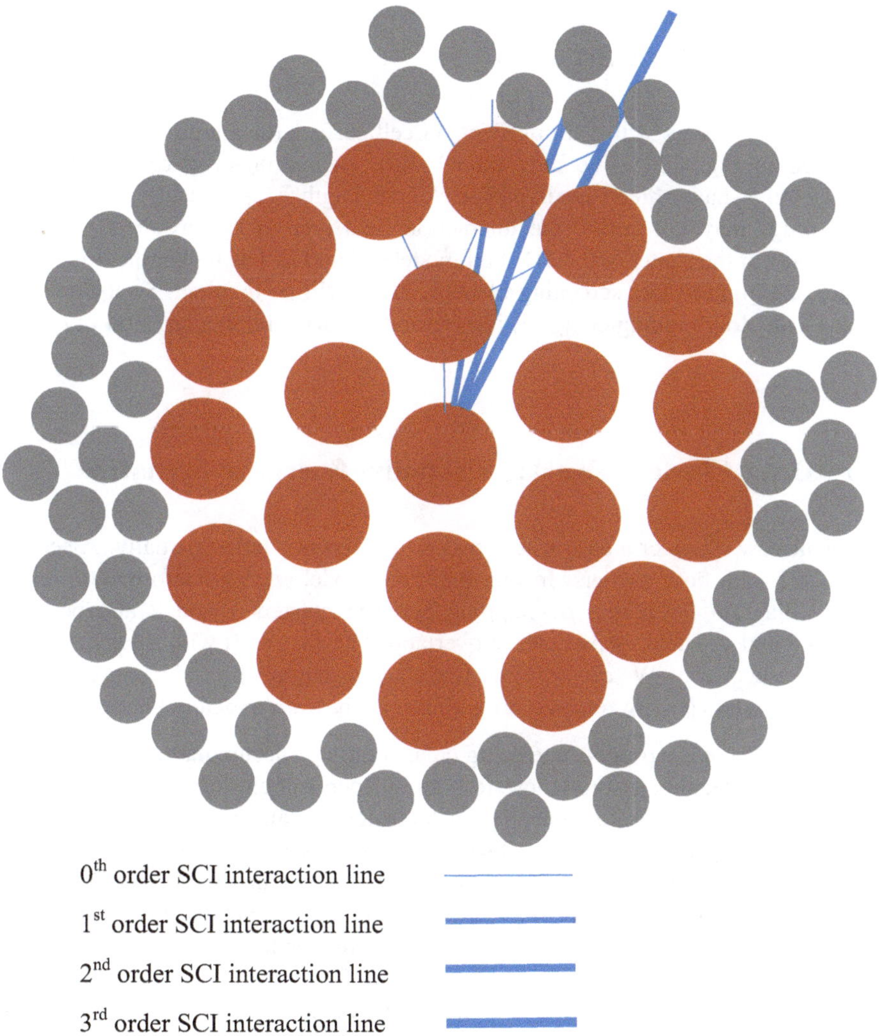


**Fig. 4.30** Different screening orders in SCI are shown (modeled here with interaction lines having various thicknesses) in the interaction of a drug particle with surrounding drugs for unitary drug cluster on cell surface as shown in earlier Fig. 4.28c. As before, to explain the interaction with various screening order we have sketched an equivalent shell model presentation for the central unitary drug cluster in a 2D view. Here the central unitary drug cluster is modeled to be surrounded by the hydrophobic lipid environment

## 4.5 Drug–Drug Clusters. Size Distribution and Energetic Scenario

Drugs create clusters with various sizes on cell surface (detailed in models I have presented in Fig. 4.31). We may model three such drug–drug clusters here in Fig. 4.31 that have three values of correlation length $\xi_{radial}$ that determines how many drugs may fall within the interaction field created by any drug molecule (considered at the center in the 2D shell model) around it. The bigger the drug–drug cluster, the higher-order screening of interactions will be required. So the average value of the drug binding energy $\epsilon_i$ is expected to vary with the variation of cluster sizes.

## 4.6 Drug Clusters Overlap Partially. Energetic Scenario

We shall now consider a case where two drug clusters overlap partially. A section may share with both clusters. In Fig. 4.32, we model energy conformation representing diagram for partial overlapping between clusters 1 and 2 and between clusters 3 and 4. The drug particles at overlapped regions exist with energy states 0' and 0", respectively. We can consider two cases as reasons behind creation of such an overlapped state between two drug clusters. They are explained here.

Case (i). There is a pseudostate being created through accumulation of a few drug molecules, just too close to both independent clusters, so particles from the pseudostate have hoping tendency to fall into either of the independent drug clusters. Pseudostate means a drug cluster where drug molecules are just close to each other but not with enough mutual bonding so that they can be considered as a complete drug cluster. This is perhaps the brief state that either initiates to form a cluster or disintegrates into individual drug particles.

Case (ii). Two independent drug clusters, due to their dynamic nature/movement, approach to each other and overlap partially by sharing sections of their clusters.

For former case (i), the values of *0'* or *0"* represent little lower than the absolute *0* dedicated for cell surface potential energy. Here *0'* or *0"* is the total binding energy of the pseudostate creating accumulated drugs. For the latter case (ii), the values of *0'* or *0"* represent values that are (close to or) between $E_1$ and $E_2$ or between $E_3$ and $E_4$, respectively. Theoretically, both cases (i) and (ii) are possible to exist, but in reality, the case (ii) is quite near impossible to be created. As two drug clusters are formed, they take themselves to their own independent energy states with completely different physics inside isolating them from the rest of the cell surface. Here, we also need to consider that the cluster formation is a classical mechanical and not a quantum mechanical phenomenon. So drug particles in an independent state can

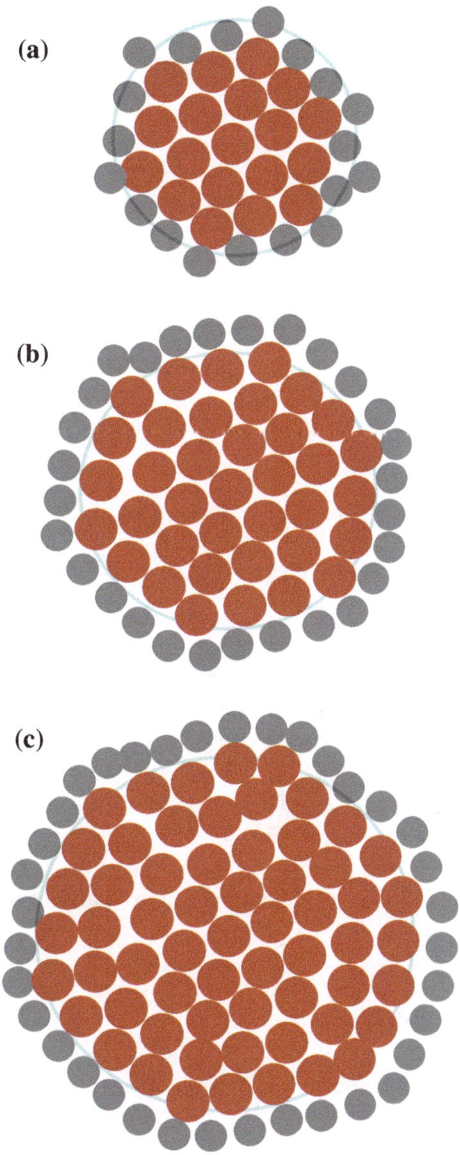


**Fig. 4.31** Drug–drug clusters with three different sizes, small to big, are A, B, C, respectively. All three clusters are surrounded by continuum of lipids layers. I just show one layer here

only escape to the cell surface and certainly are unable to tunnel into any nearby drug clusters. If at the edge of a cluster a few drug molecules suddenly accumulate through weak bonding and next to this accumulated pseudostate there exists the edge of another independent drug cluster, there is a possibility to raise an overlapping as explained in case (i). The lifetime of such case (i) should be quite short or case (i) exists very briefly before either converting (the overlapped clusters) into either two independent clusters; e.g., overlapped clusters shown in Fig. 4.32 get

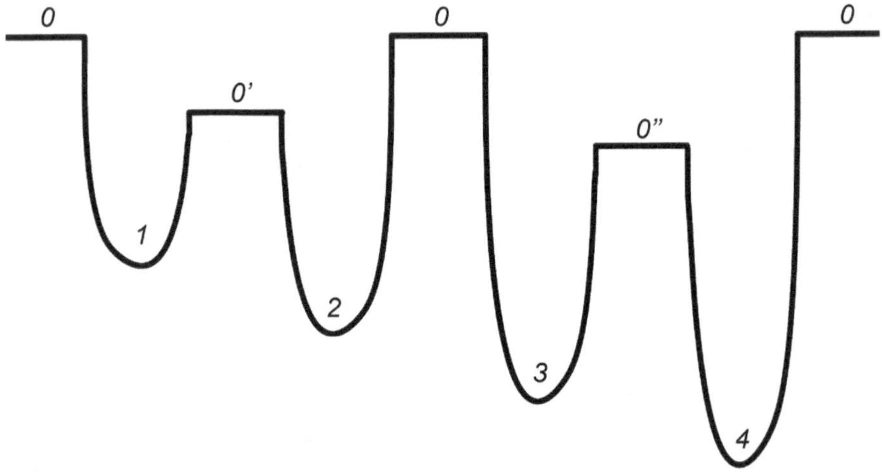


**Fig. 4.32** This is a specific case where a few drug clusters overlap. Here clusters 1 and 2, and clusters 3 and 4 partially overlap. The overlap section between cluster 1 and 2 exists with energy state 0', and between 3 and 4 exists with energy state 0"

converted into independent drug clusters 1' and 2' and 3' and 4' (not shown here), respectively. Here, 1', 2', 3', and 4' exist with higher number of drug molecules than that of 1, 2, 3, and 4 (shown in Fig. 4.32) and higher amount of total energy than $E_1$ and $E_2$, $E_3$ and $E_4$, respectively. The other possibility is that the overlapped clusters 1 and 2, and 3 and 4 accumulate more tightly and convert into just two new bigger clusters. Due to quite high level of energetic independence, the latter possibility is quite slim leaving the former possibility a more reality. These explanations are just made based on mostly philosophical thoughts than any realistic experimental validation which is absent to date.

## References

Md. Ashrafuzzaman, Z. Khan, M. Alanazi, M.S. Alam. 2016. Cell surface binding and lipid interactions behind chemotherapy drug induced ion pore formation in membranes. Submitted.

Ashrafuzzaman, M., M. Duszyk, and J. Tuszynski. 2011. Chemotherapy drug molecules thiocochicoside and taxolpermeabilize lipid bilayer membranes by forming ion channels. J. Phys.: Conf. Series. 329:1–16.

Ashrafuzzaman, M., C.-Y. Tseng, M. Duszyk, J. Tuszynski. 2012. Chemotherapy drugs form ion pores in membranes due to physical interactions with lipids. Chem. Biol. Drug Des. 80: 992–1002.

Andersen, O. S., 1983. Ion movement through gramicidin A channels. Studies on the diffusion-controlled association step. Biophys. J. 41: 147–65.

Huang, H. W., 1986. Deformation free energy of bilayer membrane and its effect on gramicidin channel lifetime. Biophys. J. 50:1061–1071.

Latorre, M., and O. Alvarez. 1981. Voltage-dependent channels in planar lipid bilayer membranes. Physiol. Rev. 6:77–150.

Ludtke, S. J., K. He, W. T. Heller, T. A. Harroun, L. Yang, and H. W. Huang. 1996. Membrane pores induced by magainin. Biochemistry. 35: 13723–28.

Matsuzaki, K., O. Murase, H. Tokuda, N. Fujii, and K. Miyajima. 1996. An antimicrobial peptide, magainin 2, induced rapid flip-flop of phospholipids coupled with pore formation and peptide translocation. Biochemistry. 35: 11361–68.

Binnig, G., C. F. Quate, and C Gerber. 1986. Atomic force microscope. Physical Review Letters 56: 930–933.

Schneider, S. W., K. C. Sritharan, J. P. Geibel, H. Oberleithner, and B. P. Jena. 1997. Surface dynamics in living acinar cells imaged by atomic force microscopy: Identification of plasma membrane structures involved in exocytosis. Proc. Natl. Acad. Sci. USA. 94: 316–321.

Bhushan, B (Ed). 2004. Springer Handbook of Nanotechnology. ISBN 978-3-642-02525-9.

Kuznetsova, T. G., M. N. Starodubtseva, N. I. Yegorenkov, S. A. Chizhik, and R. I. Zhdanov. 2007. Atomic force microscopy probing of cell elasticity. Micron. 38: 824–833.

Franz, C. M., and P. H. Puech. 2008. Atomic force microscopy: a versatile tool for studying cell morphology, adhesion and mechanics. Cellular and Molecular Bioengineering.1:289–300.

Caille, N., O. Thoumine, Y. Tardy, and J. J. Meister. 2002. Contribution of the nucleus to the mechanical properties of endothelial cells. Journal of Biomechanics. 35:177–187.

Laurent, V. M., S. Kasas, A. Yersin, T. E. Schaffer, S. Catsicas, G. Dietler, A. B. Verkhovsky, and J. J. Meister. 2005. Gradient of rigidity in the lamellipodia of migrating cells revealed by atomic force microscopy. Biophysical Journal. 89: 667–67.

Preiner, J., A. Horner, A. Karner, N. Ollinger, C. Siligan, P. Pohl, and P. Hinterdorfer. 2015. High-speed AFM images of thermal motion provide stiffness map of interfacial membrane protein moieties. Nano Lett. 15, 759–763.

Langer, M. G., A. Koitschev, H. Haas, U. Rexhausen, J.K.H. Hörber, and J. P. Ruppersberg. 2000. Mechanical stimulation of individual stereocilia of living cochlear hair cells by atomic force microscopy. Ultramicroscopy. 82: 269–278.

Madl, J., S. Rhode, H. Stangl, H. Stockinger, P. Hinterdorfer, G. J. Schütz, and G. Kada. 2006. A combined optical and atomic force microscope for live cell investigations. Ultramicroscopy. 106: 645–651.

Heinisch, J. J., P. N. Lipke, A. Beaussart, S. E. K. Chatel, V. Dupres, D. Alsteens, and Y. F. Dufrêne. 2012. Atomic force microscopy–looking at mechanosensors on the cell surface. Journal of Cell Science. 125: 4189–4195.

Ashrafuzzaman, M., O. S. Andersen, and R. N. McElhaney. 2008. The antimicrobial peptide gramicidin S permeabilizes phospholipid bilayer membranes without forming discrete ion channels. Biochim. Biophys. Acta.1778: 2814–22.

Ashrafuzzaman, M., and J. A. Tuszynski. 2012a. Regulation of channel function due to coupling with a lipid bilayer, J. Comput. Theor. Nanosci. 9: 564–570.

Ashrafuzzaman, M., and J. Tuszynski. 2012b. Membrane Biophysics, Springer, Heidelberg, Germany.

Schiff, P. B., J. Fant, and S. B. Horwitz. 1979. Promotion of Microtubule Assembly in vitro by Taxol. Nature. 277: 665–666.

Callen, J. P., 1985. Colchicine is effective in controlling chronic cutaneous vasculitis in lupus erythematosus. J. Am. Acad. Dermatol.13: 193–200.

Seidemann, P., B. Fjellner, and A.Johannesson. 1987. Psoriatic arthritis treated with oral colchicine. J. Rheumatol. 14: 777–79.

Fisherman, J., M. McCabe, and M. Hillig. 1992. Phase I study of taxol and doxorubucin (Dox) with G-CSF in previously untreated metastatic breast cancer. Proc. Am. SOC. Clin. Oncol. 1175A.

Holmes, F. A., A.P. Kudelka, J. J. Kavanagh, M. H. Huber, J. A. Ajani, and V. Valero,in: G.I. Georg, T. T. Chen, I. Ojima, and D. M. Vyas (Eds.). 1994. Taxane Anticancer Agents: Basic Science and Current Status, Vol. ACS Symposium Series 583, American Chemical Society, Washington, DC:31–57.

Ashrafuzzaman, M., 2015a. Diffusion across cell phase states. Biomedical Sci. Today. 1:e4.

Ashrafuzzaman, M., 2015b. Phenomenology and energetics of diffusion across cell phase states. Saudi J. of Biol. Sci., 22: 666–673.

Rosenman, S. J., A. A. Ganji, W. M. Gallatin. 1991. Contact dependent redistribution of cell surface adhesion and activation molecules reorganization. FASEB J. 5: 1603.

Mekory, Y. A., D. Baram, A. Goldberg, and A. Klajman. 1989. Inhibition of delayed hypersensitivity in mice by colchicines: Mechanism of inhibition of contact sensibility in vivo. Cell. Immunol. 120: 330–40.

Borisy, G. O., and E. W. Taylor. 1967. The mechanism of action of colchicine: Colchicine binding to seaurchin eggs and the mitotic apparatus. J. Cell. Biol. 34: 533–48.

Agutter, P. S., and K. E. Suckling. 1982. Effect of colchicine on mammalian liver nuclear envelope and on nucleo-cytoplasmic RNA transport. Biochim. Biophys. Acta. 698:223–229.

Balasubramanian, S. V., and R. M. Straubinger. 1994. Taxol-lipid interactions: taxol-dependent effects on the physical properties of model membranes. Biochemistry. 33: 8941–8947.

Matsumoto, G., and H. Sakai. 1979. Microtubules inside the plasma membrane of squid giant axons and their possible physiological function. J. Membrane Biol.50: 1–14.

Sonee, M., E. Barron, F. A. Yarber, and S. F. Hamm-Alvarez. 1998. Taxol inhibits endosomal-lysosomal membrane trafficking at two distinct steps in CV-1 cells. Am. J. Physiol. Cell Physiol.44:1630–39.

Shiba, M., E. Watanabe, S. Sasakawa, Y. Ikeda. 1988. Effects of taxol and colchicines on platelet membrane properties. Thromb Res.52: 313–23.

Mons, S., F. Veretout, M. Carlier, I. Erk, J. Lepault, E. Trudel, C. Salesse, P. Ducray, C. Mioskowski, and L. Lebeau. 2000. The interaction between lipid derivatives of colchicines and tubulin: Consequences of the interaction of the alkaloid with lipid membranes. Biochim. Biophys. Acta.1468:381–95.

D. S. Grierson, E. E. Flater, and R. W. Carpick. 2005. Accounting for the JKR-DMT transition in adhesion and friction measurements with atomic force microscopy. J. Adhesion Sci. Technol. 19: 291–311.

S. P. Koenig, N. G. Boddeti, M. L. Dunn, and J. C. Bunch. 2011. Ultrastrong adhesion of graphene membranes. Nature Nanotechnology. 6: 543–546.

Ashrafuzzaman, M. C.-Y. Tseng, and J.A. Tuszynski. 2014. Regulation of channel function due to physical energetic coupling with a lipid bilayer. Biochemical and Biophysical Research Communications. 445:463–468.

Case DA, Darden TA, Cheatham, TE III, Simmerling CL, Wang J et al. (2010) AMBER 11. University of California, San Francisco, USA.

Wang J, Wolf RM; Caldwell JW, Kollman PA, Case DA. Development and testing of a general AMBER force field. J. of Comp. Chem., 2004; 25: 1157–1174.

Wang J, Wang W, Kollman PA, Case DA (2006). Automatic atom type and bond type perception in molecular mechanical calculations. J. of Mol. Graphics and Modelling; 25, 247260.

Huzil JT, Mane J, Tuszynski JA. Computer assisted design of second generation colchicine derivatives, Interdisciplinary Sciences- Computational Life Sciences., 2010;2:169–174.

Freedman H, Huzil JT, Luchko T, Luduena RF, Tuszynski JA. Identification and Characterization of an Intermediate Taxol Binding Site Within Microtubule Nanopores and a Mechanism for Tubulin Isotype Binding Selectivity, J. of Chem. Info. and Modeling., 2009:49;424–436.

Woolf TB, Roux B. (1994) Molecular dynamics simulation of the gramicidin channel in a phospholipid bilayer. Proc. Natl. Acad. Sci., USA.;91: 11631–35.

Ashrafuzzaman M, Tuszynski J (2011) Ion pore formation in lipid membranes due to complex interactions between lipids and channel formping peptides or biomolecules. In HB of Nanosci., Eng. & Tech.; Goddard, Brenner, Lyshevki and Iafrate, Eds.; Taylor and Francis (CRC press), New York, USA.

Ashrafuzzaman M (2011) Antimicrobial peptides modulate bilayer barrier properties using a variety of mechanisms of actions, Science against microbial pathogens: communicating current research and technological advances (Microbiology Book Series, No.3, Ed: Antonio Méndez-Vilas), Formatex Res. Cen., Spain; Vol 2, 938–950.

Ashrafuzzaman M, Tuszynski J. 2012c. Ion pore formation in lipid bilayers and related energetic considerations, Curr. Med. Chem.;19:1619–34.

Ashrafuzzaman M., Beck H. (2004) in Vortex dynamics in two-dimensional Josephson junction arrays, (University of Neuchatel, http://doc.rero.ch/record/2894?ln=fr), ch 5, p 85.

He K, Ludtke SJ, Huang HW, Worcester DL. (1995) Antimicrobial peptide pores in membranes detected by neutron in-plane scattering. Biochemistry;34: 15614–18.

D A Mannock, R N Lewis, R N McElhaney, M Akiyama, H Yamada, D C Turner, and S M Gruner. Effect of the chirality of the glycerol backbone on the bilayer and nonbilayer phase transitions in the diastereomers of di-dodecyl-beta-D-glucopyranosyl glycerol. Biophys J. 1992 Nov; 63(5): 1355–1368.

D. A. Mannock, R. N. McElhaney, P. E. Harper, and S. M. Gruner. Differential scanning calorimetry and X-ray diffraction studies of the thermotropic phase behavior of the diastereomeric di-tetradecyl-beta-D-galactosyl glycerols and their mixture. Biophys J. 1994 Mar; 66(3 Pt 1): 734–740.

Dzmitry Afanasenkau and Andreas Offenhäusser. Positively Charged Supported Lipid Bilayers as a Biomimetic Platform for Neuronal Cell Culture. Langmuir, 2012, 28 (37), 13387–13394.

Akira Abe and James A. Shayman. The role of negatively charged lipids in lysosomal phospholipase A2 function. J Lipid Res. 2009 Oct; 50(10): 2027–2035.

Md Ashrafuzzaman, M A Lampson, D V Greathouse, R E Koeppe II, and O S Andersen. Manipulating lipid bilayer material properties using biologically active amphipathic molecules. Journal of Physics: Condensed Matter 18: S1235–S1255 (2006).

B. A. Wallace, W. R. Veatch and E. R. Blout, Conformation of gramicidin A in phospholipid vesicles: circular dichroism studies of effects of ion binding, chemical modification, and lipid structure. Biochemistry 20 (1981) 5754–5760.

J. Katsaras, R. S. Prosser, R. H. Stinson and J. H. Davis, Constant helical pitch of the gramicidin channel in phospholipid bilayers, Biophys. J. 61 (1992) 827–830.

J. T. Durkin, L. L. Providence, R. E. Koeppe II and O. S. Andersen, Energetics of heterodimer formation among gramicidin analogues with an NH2-terminal addition or deletion consequences of missing a residue at the join in the channel. J. Mol. Biol. 231, (1993) 1102–1121.

O. S. Andersen, C. Nielsen, A. M. Maer, J. A. Lundbæk, M. Goulian and R. E. Koeppe II, Gramicidin channels: molecular force transducers in lipid bilayers, Biol. Skr. Dan. Vid. Selsk. 49 (1998) 75–82.

G. V. Miloshevsky and P. C. Jordan, Gating gramicidin channels in lipid bilayers: reaction coordinates and the mechanism of dissociation, Biophys. J. 86 (2004) 92–104.

# Chapter 5
# Molecular Machines of the Cell

Cell is a chemical machine, first proposed by Loeb in 1906. Wilson's opinion suggested that 'the specificity of each kind of cell depends essentially on what we call its organization, i.e., upon the construction of the cell machine' (Wilson 1925). Details on the understanding of the cell by Loeb and Wilson have been presented in Taylor (2001). Cellular constituents make varieties of most important molecular machines (MMs). It is now quite clear that the most complex MMs are proteins within the cells. Therefore, it is like existence of independent MMs within a big machine 'cell.' Many of the essential cell-based processes, including transcription, translation, folding of proteins, degradation of proteins, are all naturally carried out by various MMs.

Molecular machines may fall in two general categories.

(i) *Natural molecular machines*. Molecules that naturally exist inside the cell and participate in constructing 'natural molecular machines.'
(ii) *Artificial molecular machines*. The other type of machines is referred as 'artificial molecular machines' constructed by the external agents independently inside a biological organ, section or structure or in combination with a naturally existed biostructure inside or outside the cell.

Besides big biomolecules, there are various small molecules, such as different peptides participating in constructing MMs. Both types of natural and artificial MMs follow the general rules of structures, and structure-specific functions that are applicable in case of any general machine that we know of. These machines have mechanical, electrical, and chemical properties, so they do follow formalisms that are raised due to the mechano–electrical–chemical properties.

This chapter will address in quite details on various types of MMs that are studied using biophysical, biochemical, medical, engineering techniques for understanding their roles in biological systems, and possible use in addressing various cell-based diseases, and applications in drug design/discovery. To obtain a quite complete picture, we shall address on a few example general MMs like ribosome and myosin as well as address on various ion channels that either



M. Ashrafuzzaman, *Nanoscale Biophysics of the Cell*,
https://doi.org/10.1007/978-3-319-77465-7_5

naturally exist or artificially constructed in mainly the areas of cell-based plasma membranes and mitochondrial membranes.

## 5.1 General Understanding of the Molecular Machines

For general understanding of the structure and function of MMs, we shall use specific example cases. In-depth analysis of a few biomolecule structures and functions will help understand how these systems may be compared to general machines. Here we use ribosome to address the general features of a MM.

### *5.1.1 Ribosome. A Complex Molecular Machine Inside Living Cell*

As an example, we shall address here how the ribosome, an essential MM, works inside the cell. Ribosome is often considered as a supramolecular factory that synthesizes proteins in all cells. It assembles molecules like polymeric proteins whose sequence is controlled by the sequence of messenger RNA (mRNA) molecules. Translocation of transfer RNA (tRNA) and mRNA through the ribosome is one of the most cell-based dynamic events during protein synthesis. The translocation is catalyzed by elongation factor G (EF-G) and driven by Guanosine-5′-triphosphate (GTP) hydrolysis. Major unresolved questions are: 'how the movement is induced and what the moving parts of the ribosome are.' That means to understand properly the molecular mechanism of the movement is to be understood. Various studies addressed these questions to get into the insights (e.g., see Dunkle and Cate 2010; Shoji et al. 2009; Frank et al. 2010; Rodnina et al. 2011).

Time-resolved cryoelectron microscopy reveals trajectories of tRNA movement through ribosome. Due to thermal fluctuations, the ribosome spontaneously samples a large number of conformational states. The spontaneous movement of tRNAs through the ribosome is loosely coupled to the motions within the ribosome. EF-G stabilizes conformational states prone to translocation and promotes a conformational rearrangement of the ribosome (unlocking) that accelerates the rate-limiting step of translocation: The movement of the tRNA anticodons on the small ribosomal subunit. EF-G acts as a Brownian ratchet providing directional bias for movement at the cost of GTP hydrolysis. Figure 5.1 addresses how tRNA moves through the ribosome (for details, see Rodnina et al. 2011).

The two tRNA molecules bound to the A and P sites of the ribosome in the pre-translocation state (PRE) move to the P and E sites, respectively, and the movement is promoted by EF-G (Fig. 5.1). Spontaneous dissociation of the deacylated tRNA from the E site results in a post-translocation complex with a single tRNA in the P site (POST). The domains of EF-G (1–5) are indicated. In the absence

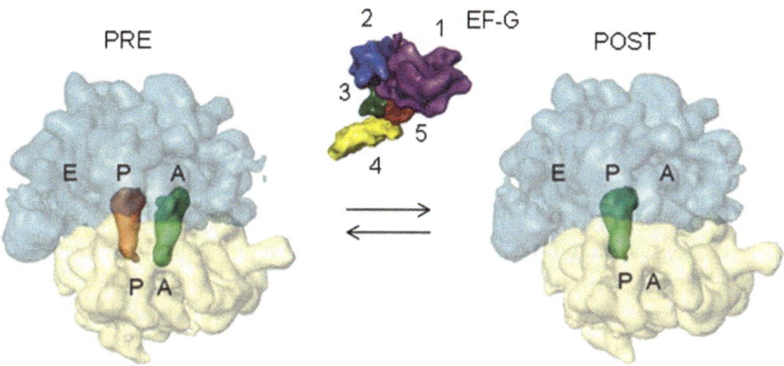


**Fig. 5.1** Movement of tRNA through the ribosome With due permission, copied from (cf. Rodnina et al. 2011)

of EF-G, the reaction can proceed in forward or backward directions, albeit slowly, following the thermodynamic gradient.

### *5.1.2 tRNA Moves Spontaneously Through the Ribosome*

This spontaneous translocation can proceed in both forward and backward directions, implying that its basic features are inherent to the ribosome–tRNA complex itself (see Shoji et al. 2006; Fredrick and Noller 2003; Gavrilova et al. 1976; Konevega et al. 2007). Movement of tRNAs through the ribosome is directed solely by thermal fluctuations and the thermodynamic gradient. Compared to the reaction promoted by EF-G, this spontaneous thermal fluctuation-driven movement is pretty slow. In the transition from one canonical ribosome-binding site to the other, the tRNAs acquire intermediate configurations that are known as the so-called hybrid states.

Single-molecule fluorescence resonance energy transfer (FRET) studies indicate that the ribosome spontaneously fluctuates between classic or nonrotated and hybrid or rotated states (Cornish et al. 2008; Blanchard et al. 2004). The time-resolved cryo-electron microscopic (cryo-EM) analysis of spontaneous tRNA retro-translocation (Fischer et al. 2010) reveals trajectories of the tRNAs and accompanying conformational fluctuations of the ribosome. Eight most distinct states from the 50 cryo-EM reconstructions of tRNAs at different stages of retro-translocation are presented in Fig. 5.2. Here the numbers above the arrows represent equilibrium constants ($K_{eq}$), i.e., the ratio between states $n$ and $n + 1$ for the rapid transitions between the various post- and pre-states. The transition between the ensembles of post- and pre-states (indicated by square brackets), which entails the movement of the tRNA anticodon stem–loops on the 30S ribosomal subunit, is slow in comparison.

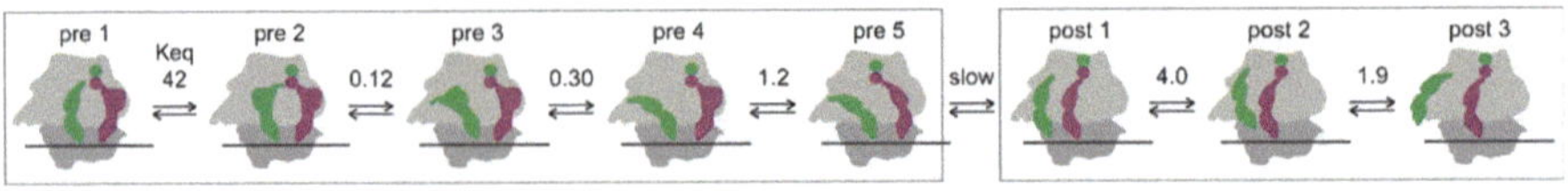


**Fig. 5.2** Trajectories of tRNA movement are schematized here (cf. Rodnina et al. 2011)

The brief explanation on ribosomal MM provides some example information about how biological machines work. A book was recently published on Molecular Machines in Biology (Frank 2011). This volume focuses on important MMs whose architecture is known and whose functional principles have been established by tools of biophysical imaging, such as X-ray crystallography and cryoelectron microscopy, and fluorescence probing, such as single-molecule FRET. Interested and focused readers may consider reading this book to enhance their knowledge on various cell-based MMs, their structures, functions, and possible roles in cellular conditions and specific diseases. In this chapter, our scope is limited to go to the details of this subject.

## 5.2 Physical Condition of Molecular Machines

Most of the MMs are mechanical entities though locally created natural electrical engines often appear to feed the machines with transiently produced electrical energies and thus help them function molecularly. Of course, the thermal condition in which the machine resides is a generally important determinant of the MM's functionality. For example, motor protein myosin is responsible for muscle contraction, kinesin moves cell-based cargo along microtubules away from the nucleus, dynein is engaged in producing axonemal beating of motile cilia and flagella. These naturally grown biological machines are on one hand way more complex in structure than the human made any kind of machines to date. On the other hand, these machines are often found to work naturally requiring no external interferences other than the locally created complex network among various inputs, outputs, and core stages structural sections and established bioinformatics based communications thereof.

The motor proteins bind to a polarized cytoskeletal filament and use the energy derived from repeated cycles of ATP hydrolysis to move steadily along it. A nice book 'Molecular Biology of the Cell' was written more than a decade ago in which analysis on molecular motors as MMs was made quite in details (see Alberts et al. 2002).

Different motor proteins coexist in every eukaryotic cell. They usually differ in three ways:

(i) the type of filament they bind to (either actin or microtubules),
(ii) the direction in which they move along the filament, and
(iii) the 'cargo' they carry.

Many motor proteins are known to carry big cell constituents/organelles which are even membrane-enclosed—such as mitochondria, Golgi stacks, or secretory vesicles—to their target locations in the cell. Other motor proteins cause cytoskeletal filaments to slide against each other. This process generates force that drives phenomena as muscle contraction, ciliary beating, and cell division.

Kinesin superfamily proteins, important molecular motors, directionally transport various cargos, including membranous organelles, protein complexes, and mRNAs. The mechanisms by which different kinesins recognize and bind to specific cargos, as well as how kinesins unload cargo and determine the direction of transport, have been identified (Hirokawa et al. 2009). Molecular genetic experiments uncover important and unexpected roles for kinesins in the regulation of such physiological processes as higher brain function, tumor suppression, and developmental patterning (see Hirokawa et al. 2009).

## 5.3 Cargo Recognition and Cargo-Mediated Regulation of Molecular Machine Myosin

Biological system exists mainly with various dynamic compounds inside various levels including in cell. Motion is one important hallmark of dynamic character of a body. Organized motions are observed to be hallmarks of living organisms. Such motions range from collective cell movements during development stages and muscle contractions at the macroscopic scale all the way down to cellular cargo including various biomolecules and organelles that are engaged in transportation and mechanoforce sensing at more microscopic scales. Energy required for such molecular-level biological motions is almost invariably provided by cellular chemical fuels in the form of nucleotide triphosphate. Biological systems have independently evolved through designing a group of nanoscale engines as molecular motors, to convert cellular chemical fuels into mechanical energy.

Various molecular motors such as cytoskeleton motors (myosin, kinesin, and dynein), nucleic acid-based motors, cellular membrane-based rotary motors are on continuous dynamic motions due to localized energetic interactions, energetic conversions, energy transfer, or exchange, etc., among within the various parts of such nanomachines, as well as with the surrounding biological materials. We shall address some aspects of myosins, as an example demonstration of a cytoskeleton motor.

Mechanochemical parameters are known to define the motor activity of each myosin. Some of these are general machine-related parameters like duty ratio, processivity, power stroke/stepping, velocity, and directionality. These parameters have been defined quite in details for myosin, see O'Connell et al. (2007).

- Duty ratio is defined as the proportion of the ATPase cycle that the motor domain remains strongly bound to actin.
- Processivity is a mechanochemical property closely associated with duty ratio and refers to the number of catalytic cycles a motor can perform before diffusing

away from its track. For highly processive motors, hundreds of rounds of ATP binding and hydrolysis can occur before release from actin while nonprocessive motors will bind actin once per ATPase cycle and release.

- Muscle myosin II in various nucleotide-bound conformations support the hypothesis that this region undergoes a rotational movement, thereby acting as a lever arm that exerts a power stroke, during the weak to strong (ADP·Pi → ADP) binding transition.
- The velocity at which a myosin moves along an actin filament is defined by three parameters, the displacement per ATP hydrolyzed (step size *d*), the duty ratio (*f*), and the overall cycling rate (actin-activated ATPase). The velocity *V* stands as O'Connell et al. (2007),

$$V = \frac{d \cdot k_{\mathrm{ATPase}}}{f} \tag{5.1}$$

Independent measurements of these three parameters can be used to calculate a velocity that is frequently in good agreement with experimentally determined values, although there are exceptions.

- Actin filaments have an inherent polarity with a fast growing plus (barbed) end and a slower growing minus (pointed) end. Filaments within the cytoplasm are also organized in a polarized fashion, with plus ends typically oriented toward the plasma membrane and minus ends toward the interior region. The direction of myosin movement on actin, therefore, has profound implications for function. As an example, a myosin involved in trafficking vesicles to the plasma membrane during secretion would necessitate plus end-directed movement. Directionality is therefore considered an important factor.

Table 5.1 summarizes many of the known values for the above-described parameters (see O'Connell et al. 2007) where a survey has been made and results presented based on various studies quoted therein. Due to limited scopes here we shall avoid providing detailed explanations on all of these studies. Interested readers are encouraged to go through these references (see in Table 5.1).

$$V(\rho) = V_{\max}\left[1 - (1 - f)^{\rho \cdot \bar{A}}\right] \tag{5.2}$$

Here

$V$ is the average filament velocity as a function of density $\rho$,
$V_{\max}$ is the maximum gliding velocity,
$f$ is the duty ratio, and
$\bar{A}$ is the area of interaction (defined as the product of twice the filament reach and mean filament length)

**Table 5.1** Mechanochemical properties of selected myosins (cf. O'Connell et al. 2007)

| Class | | Duty ratio[a] | Landing rate | Rate-limiting step | Velocity[b] (μm/s) | Directionality |
|---|---|---|---|---|---|---|
| I | Myo1a | 0.05 (Jontes et al. 1997) | | Pi release (Ostap et al. 1996) | 0.05 (Wolenski et al. 1993), 0.1 (Collins et al. 1990) | Plus (Mooseker et al. 1989) |
| | Myo1e | | | Pi release (El Mezgueldi et al. 2002) | | |
| | Myo1b | | | | 0.137 (Lin et al. 2005) | |
| II | Skeletal muscle | 0.038 (Harris et al. 1993), 0.05 (Uyeda et al. 1990) | | Pi release | 6.6 (Harris et al. 1993), 6.9 (Uyeda et al. 1990) | Plus |
| | Smooth muscle | 0.04 (Harris et al. 1993) | | Pi release | 0.58 (Harris et al. 1993), 0.546 (Warshaw et al. 1990) | Plus |
| | Nonmuscle myosin IIA | 0.05[c] (Kovacs et al. 2003) | | Pi release (Kovacs et al. 2003) | 0.3 (Wang et al. 2000) | Plus |
| | Nonmuscle myosin IIB | 0.82[c] (Rosenfeld et al. 2003), 0.4[c] (Wang et al. 2003) | | ADP release (Rosenfeld et al. 2003), Pi release (Wang et al. 2003) | 0.092 (Pato et al. 1996) | Plus |
| III | Myo3 | | | | 0.11 (Komaba et al. 2003) | Plus (Komaba et al. 2003) |
| V | Myo5a | 0.7[c] (De La Cruz et al. 1999) | 1 (Mehta et al. 1999), 1.4 (Krementsova et al. 2006) | ADP release (De La Cruz et al. 1999) | 0.311 (Mehta et al. 1999) | Plus (Cheney et al. 1993) and (Wolenski et al. 1993) |
| | Myo5b | 0.79[c] (Watanabe et al. 2006) | 1 (Watanabe et al. 2006) | ADP release (Watanabe et al. 2006) | 0.22 (Watanabe et al. 2006) | Plus (Watanabe et al. 2006) |
| | S.c. Myo2p | 0.2 (Reck-Peterson et al. 2001) | 5 (Reck-Peterson et al. 2001) | | 4.5 (Reck-Peterson et al. 2001) | |

(continued)

**Table 5.1** (continued)

| Class | | Duty ratio[a] | Landing rate | Rate-limiting step | Velocity[b] (μm/s) | Directionality |
|---|---|---|---|---|---|---|
| | D.m. MyoV | 0.1[c] (Toth et al. 2005) | | Pi release (Toth et al. 2005) | 0.46 (Toth et al. 2005) | |
| VI | Myo6 (monomer) | 0.8[c] (De La Cruz et al. 2001) | | ADP release (De La Cruz et al. 2001) and (Morris et al. 2003) | 0.058 (Wells et al. 1999), 0.131 (Morris et al. 2003) | Minus (Wells et al. 1999) |
| | Myo6 (dimer) | | | ADP release (Morris et al. 2003) | 0.4 (Yoshimura et al. 2001), 0.307 (Morris et al. 2003) | |
| VII | Myo7a | 0.9[c] (Watanabe et al. 2006) | | ADP release (Watanabe et al. 2006) | 0.16 (Inoue and Ikebe 2003), 0.19 (Udovichenko et al. 2002) | Plus (Inoue and Ikebe 2003) |
| | Myo7b | 0.8 (Henn et al. 2005; Yang et al. 2005) | | ADP release (Henn et al. 2005; Yang et al. 2005) | | |
| IX | Myo9b | 1 (Post et al. 2002) | 1 (Post et al. 2002; Inoue et al. 2002) | ATP hydrolysis (Nalavadi et al. 2005; Kambara and Ikebe 2005) | 0.015 (Post et al. 1998), 0.038 (Post et al. 2002), 0.08 (Inoue et al. 2002), 1.1 (Nishikawa et al. 2006) | Minus (Inoue et al. 2002), Plus (O'Connell and Mooseker 2003) |
| X | Myo10 | 0.16 (Kovacs et al. 2005), 0.6, (Homma and Ikebe 2005) | | | 0.3 (Homma et al. 2001) | Plus (Homma et al. 2001) |
| XI | Myosin XI | | 1 (Tominaga et al. 2003) | | 4.6 (Tominaga et al. 2003) | Plus (Tominaga et al. 2003) |
| XIV | TgMyoA | | | ATP hydrolysis or Pi release, (Herm-Gotz et al. 2002) | 5.2 (Herm-Gotz et al. 2002) | Plus (Herm-Gotz et al. 2002) |

[a]Determined using the following equation (O'Connell et al. 2007)
[b]Kron and Spudich assay (see Kron and Spudich 1986) at maximum motor density
[c]Duty ratio of single-headed construct

### 5.3.1 Molecular Machine Myosin Dimerization

Several mechanisms can help bring myosin monomers (e.g., Myo6 motors) in close proximity to facilitate dimerization (see Fig. 5.3). For example, a number of proteins interact with the tail domain of Myo6. Binding of two Myo6 heavy chains to one of these could induce a functional dimer (Fig. 5.3b). Alternatively, interactions with vesicular cargo could cluster enough Myo6 molecules to result in coiled-coil formation into a true dimer (Fig. 5.3c) or a functional dimer of two motors in close proximity (Fig. 5.3d).

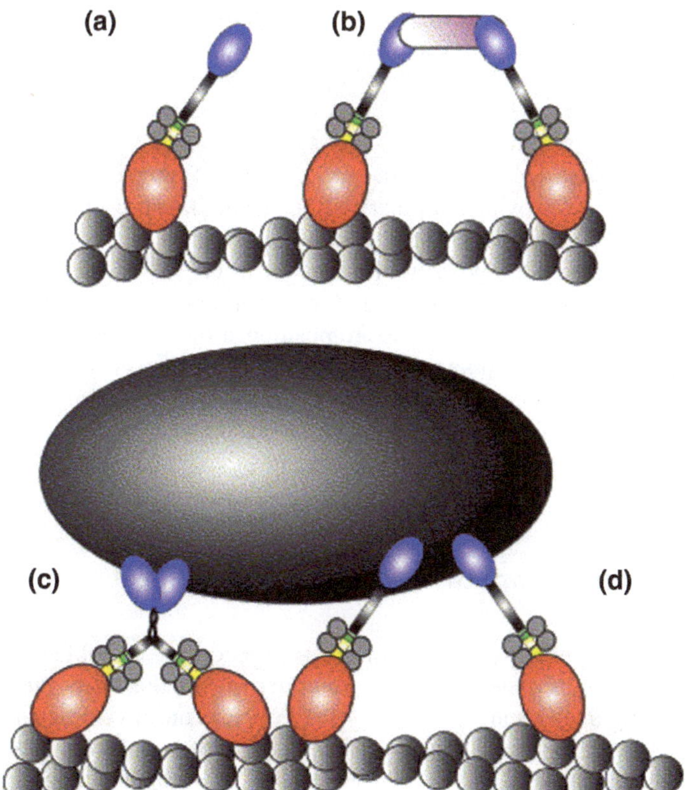


**Fig. 5.3** Possible states and mechanisms of Myo6 oligomerization in vivo. **a** Myo6 might function exclusively as a monomer or **b** binding of two Myo6 motors to the same (or complexed) tail interacting protein(s) could induce the formation of a functional dimer. **c** Vesicular cargo could act as a recruitment platform for Myo6 with dimerization occurring once enough motors are concentrated **c**, or two monomers in close proximity could work together as a functional dimer **d** (cf. O'Connell et al. 2007)

### 5.3.2 Molecular Machine Myosin Movement in Specific Direction

Time-lapse imaging of the polarity-tagged filaments by surface-bound motors reveals the directionality of the motor. A myosin whose direction of movement is known should be used to verify the accuracy of the polarity tag since annealing events in solution and subsequent breakage events during in vitro motility can affect the apparent location of the tag.

Myosins are mostly plus end-directed in nature (see Table 5.1). Single-headed Myo6 expressed in baculovirus was the first myosin described to move toward the minus end of actin, consistent with its proposed roles in trafficking vesicles toward the interior during clathrin-mediated endocytosis and anchoring the membrane between stereocilia of the inner ear. Tail-less BV-expressed Myo9b is also reported to move to the minus end of actin. However, native Myo9b moves to the plus end of actin and it is not clear why this discrepancy exists. We wish not to present details due to limited scope in this book.

Figure 5.4 presents a case study. In Fig. 5.4a, cells from the alga *Nitella* are cut open and pinned down with the polarized actin bundles in each half exposed to solution. The movement of myosin-coated beads along the cables is compared to a control motor to assess the polarity of the actin. In Fig. 5.4b, nucleation from the plus ends of actin bundles from *Limulus* acrosomal processes creates a 'broom' of known polarity. Myosin-coated beads are positioned on the broom with an optical trap. In Fig. 5.4c, in the gliding filament assay, actin filaments are fluorescently tagged on one end by polymerization off of the minus end of seeds whose plus ends are occluded by gelsolin (gray sphere). These are imaged moving on a coverslip coated with myosin, and the trailing end indicates the directionality of the myosin.

### 5.3.3 Unconventional Myosin Structure and Function

Lu et al. (2014) recently demonstrated the action of a subfamily of actin filament-based motors called unconventional myosins. Other than muscle myosin II, the remaining myosins are collectively referred to as unconventional myosins. In general, myosins can use ATP to fuel two types of mechanomotions, namely dynamic tethering actin filaments with various cellular compartments or structures and actin filament-based intracellular transport. This group has been successful to address a few features that may be considered important to know. Figure 5.5 demonstrates the summary of the understandings.

Some myosins (e.g., myosin-VI) can function as both cellular transporters and mechanical tethers. To function as a processive transporter, myosins need to form dimers or multimers. To be a mechanical tether, a monomeric myosin is sufficient. It has been shown for myosin-VI that its cellular cargo proteins can play critical roles in determining the motor properties. Dab2, an adaptor protein linking

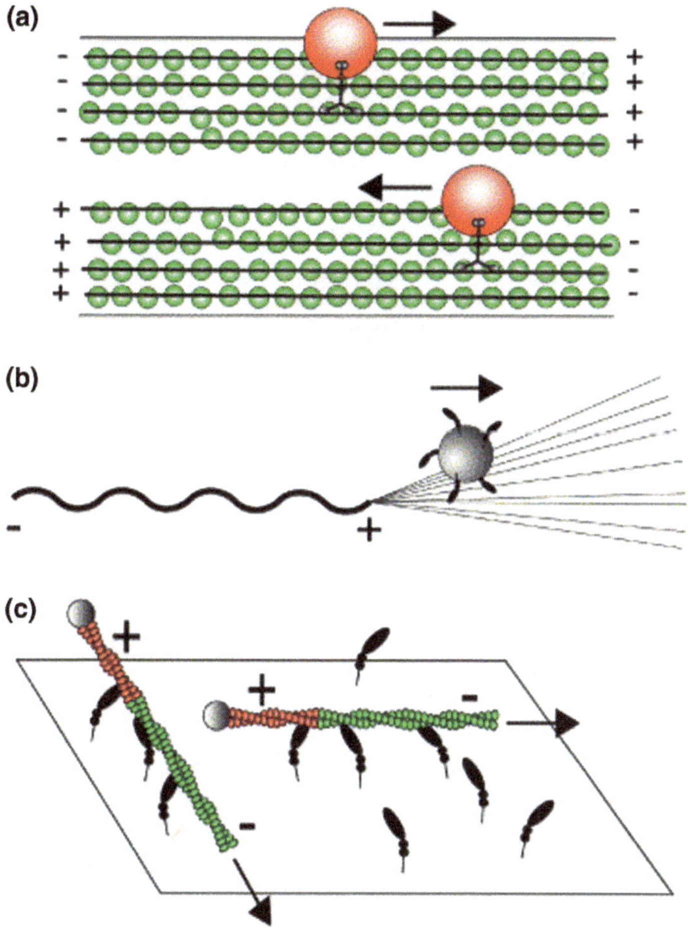


**Fig. 5.4** Assays to determine myosin directionality (cf. O'Connell et al. 2007)

endocytic vesicles with actin filament-bound myosin-VI, can induce the motor to form a transport-competent dimer. Such a cargo-mediated dimerization mechanism has also been observed in other myosins including myosin-V and myosin-VIIa.

The tail domains of myosins are very diverse both in lengths and their protein domain compositions and thus enable motors to engage a broad range of different cellular cargoes. Remarkably, the cargo-binding tail of one myosin alone often can bind to multiple distinct target proteins. A series of atomic structures of myosin-V/cargo complexes (solved recently) reveals that the globular cargo-binding tail of the motor contains a number of nonoverlapping target recognition sites for binding to its cargoes including melanophilin, vesicle adaptors RILPL2, and vesicle-bound GTPase Rab11. The structures of the MyTH4–FERM tandems from myosin-VIIa and myosin-X in complex with their respective targets reveal that MyTH4 and FERM domains extensively interact with each other forming structural and

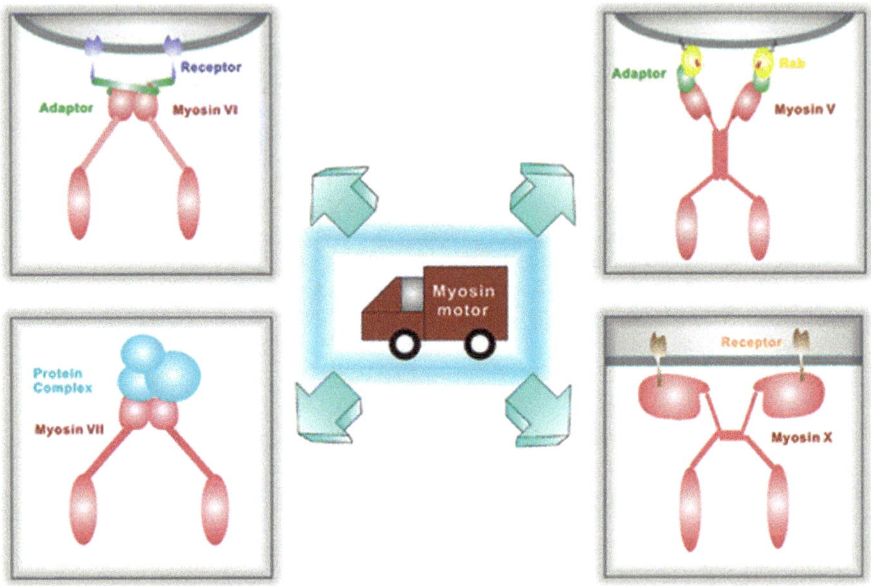


**Fig. 5.5** Role of various myosins is demonstrated in these model presentations (cf. from Lu et al. 2014)

functional supramodules in both motors and demonstrate that the structurally similar MyTH4–FERM tandems of the two motors display totally different target binding modes. These structural studies also shed light on why numerous mutations found in these myosins can cause devastating human diseases such as deafness and blindness, intellectual disabilities, immune disorders, and diabetes.

### *5.3.4 Structural Basis of Cargo Recognition by Myosin and Related Function*

Myosin-X is an important unconventional myosin that is critical for cargo transportation to filopodia tips. Myosin-X is also utilized in spindle assembly by interacting with microtubules. A couple of years ago, Hirano and colleagues (2011) presented results on a series of structural and biochemical studies of the myosin-X tail domain cassette, consisting of myosin tail homology 4 (MyTH4) and 4.1 and ezrin/radixin/moesin (FERM) domains in complex with its specific cargo, a netrin receptor deleted in colorectal cancer (DCC). The MyTH4 domain was folded into a helical Vps27p/Hrs/STAM (VHS)-like structure and was associated with the FERM domain. It was reported that an unexpected binding mode of the DCC peptide to the subdomain C groove of the FERM domain was observed, which was distinct from previously reported β–β associations found in radixin–adhesion molecule complexes. Direct interactions between the MyTH4–FERM cassette and tubulin C-terminal acidic tails were also revealed, and a positively charged patch of the MyTH4 domain was identified, which was involved in tubulin binding. DCC binding was found to interfere with integrin binding. The molecular basis by which myosin-X facilitates alternative dual binding to cargos, and microtubules was elucidated in this study (Hirano et al. 2011).

Hirano et al. (2011) found domains and motifs in myosin-X and DCC and their physical properties. In Fig. 5.6a, Myosin-X consists of an N-terminal motor domain, three IQ motifs, a coiled-coil (CC) region, three PH domains, and MyTH4 and FERM domains. The tandem myosin tail homology 4 (MyTH4) and 4.1 and ezrin/radixin/moesin (FERM) domain cassette has central role in cargo recognition. The tandem comprising the MyTH4 and FERM domains frequently appears in related myosins (myosin-IV, VII, X, XII, XIV, and XV) and is referred to as the MyTH4–FERM cassette. Interestingly, this cassette is found in one of three ancestral myosins in the earliest eukaryotes, and myosin-X is conserved in vertebrates (Richards and Cavalier-Smith 2005). Three PH domains contain one split PH domain (PHn and PHc) and sequentially appear as PHn–PH1–PHc–PH2. In Fig. 5.6b, the DCC cytoplasmic tail possesses three conserved regions, P1, P2, and P3 (orange ovals), which might facilitate protein–protein interactions. The transmembrane helix (TH) is represented by a gray rectangle. In Fig. 5.6c, isothermal titration calorimetry (ITC) profiles and thermodynamic data for binding experiments at 20 °C are presented. Injection of 6 μl of 733 μM DCC peptide into 50 μM MyTH4–FERM cassette solution was made. The protein exhibited a $K_D$ value of 0.53 ± 0.06 μM, $\Delta H$ of −4.98 ± 0.04 kcal/mol, and $\Delta S$ of 11.7 cal/mol. Raw data for 30 sequential injections (the upper panel) and the plot of the heat evolved (kcal) per mole of DCC peptide added, corrected for the heat of the peptide dilution, against the molar ratio of the peptide to the cassette. The data (filled squares) were fitted using the software 'one set of sites,' and the solid line represents the best fit. In biochemical studies, ITC showed a relatively strong affinity with a dissociation constant $K_D$ value of 0.53 μM. In Fig. 5.6d, hydrodynamic property of the complex determined by analytical ultracentrifugation (AUC) measurements is addressed. The distribution of apparent molecular mass obtained from sedimentation velocity analysis of the complex shows a mono-modal peak with apparent molecular mass of 62.7 ± 8.6 kDa. The purified myosin-X MyTH4–FERM cassette exists as a monomer in solution as suggested by AUC analysis, which showed a mono-modal peak corresponding to a monomer (62.7 kDa) close to the calculated (63.3 kDa). MyTH4–FERM cassette and the DCC P3 peptide form a stable complex, which could be purified using conventional gel filtration. Gel filtration and AUC showed a mono-modal peak corresponding to a 1:1 complex (data not shown). Contrary to the cargo-loaded dimerization of myosin-VI (Yu et al. 2009), no cargo-mediated dimerization was observed for myosin-X through the MyTH4–FERM cassette.

Besides mediating the biological function of selective cargo transportation on actin cables, myosin-X directly interacts with microtubules and has a key role to play in spindle assembly during meiosis to ensure delivery of replicated chromosomes to daughter cells following cell division (Weber et al. 2004; Woolner et al. 2008). This

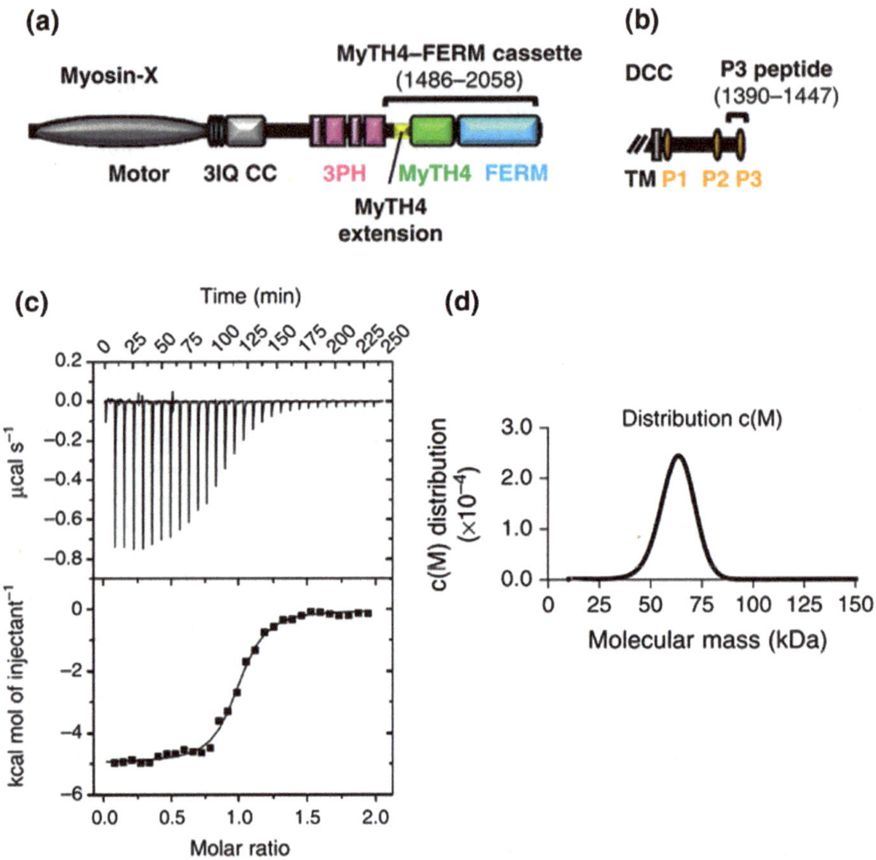


**Fig. 5.6** Diagrams of domains and motifs found in myosin-X and DCC and their physical properties. **a** an N-terminal motor domain, three IQ motifs, a coiled-coil (CC) region, three PH domains and MyTH4 and FERM domains. **b** The DCC cytoplasmic tail possesses three conserved regions, P1, P2, and P3 (orange ovals). The TH is represented by a gray rectangle. **c** ITC profiles and thermodynamic data for binding experiments at 20 °C. **d** Hydrodynamic property of the complex determined by AUC measurements (cf. Hirano et al. 2011)

surprising myosin-X function is mediated by a direct interaction between microtubules and the MyTH4–FERM cassette. However, the manner by which myosin-X recognizes microtubules has remained unclear. Myosin-X has been found to have a role in integrin-dependent spindle orientation (Toyoshima and Nishida 2007).

A finding related to the DCC αP3 helix that was docked into the groove between the β5C strand and the α1C helix of subdomain C of the myosin-X FERM domain (see Hirano et al. 2011). In sharp contrast to this α-helix groove docking (see Fig. 5.7a), binding partners of ERM proteins have been shown to bind this groove of FERM domains by forming an additional antiparallel β–β association with the β5C strand (Hamada et al. 2003; Takai et al. 2007, 2008; Terawaki et al. 2007;

Mori et al. 2008) (see Fig. 5.7b). Another adhesion molecule CD44 and a membrane-associated protease neutral endopeptidase 24.11 (NEP) preserve the Motif-1$\beta$ conserved sequence Ile-Asn (IN) and form a short $\beta$ strand followed by a reverse turn (Terawaki et al. 2007; Mori et al. 2008). These binding motifs differ from the DCC/neogenin motif that forms the P3 $\alpha$-helix. Compared with radixin subdomain C, myosin-X possesses a shorter $\beta$5C strand (four residues vs. six residues of radixin) and shorter $\alpha$1C helix (five turns vs. six turns of radixin), and these are loosely associated to form a larger groove for $\alpha$-helix accommodation, given the absence of bulky nonpolar residues at the groove surface (see Fig. 5.7f).

ERM proteins bind the $Na^+/H^+$ exchanger regulatory factor-1, 2 (NHERF-1 and -2), which are key cytoplasmic proteins involved in the anchoring of ion channels and receptors to the actin cytoskeleton through binding to ERM proteins. These adaptor proteins preserve FERM-binding Motif-2 (MDWxxxxx(L/I)Fxx(L/F)) and form an $\alpha$-helix that docks into the groove formed by the $\beta$-sandwich loops of subdomain C (Terawaki et al. 2006), which displays no similarity to the DCC/neogenin P3 motif (see Fig. 5.7c). The myosin-X FERM domain lacks the PtdIns $(4,5)P_2$-binding site found in the radixin FERM domain as previously described (see Fig. 5.7d).

It was one of the most exciting findings in cell biology of cytoskeletons that myosin-X has the ability to function as a motorized link between actin filaments and microtubules in spindle assembly during meiosis (Weber et al. 2004; Toyoshima and Nishida 2007; Woolner et al. 2008). Only a single report of *Xenopus* myosin-X has shown the direct binding of myosin-X to microtubules (Weber et al. 2004). The study of Hirano et al. (2011) confirmed that result in a mammalian myosin-X but also revealed that at least the major part of the novel myosin-X–microtubule interaction is mediated by the MyTH4 domain and the C-terminal acidic tails of tubulins. Interestingly, these acidic tails of $\alpha\beta$-tubulins are also collectively referred to as E-hooks, which interact with the kinesin K-loop for processive motor movement (Okada and Hirokawa 2000; Hirokawa et al. 2009). It is well known that $\alpha$-tubulin acidic tails possess a terminal tyrosine residue at the C-terminus, which is recognized by the CAP-Gly domains of microtubule-binding proteins such as CLIP-170 (Honnappa et al. 2006; Mishima et al. 2007). The MyTH4–FERM cassette has no homologous region to the CAP-Gly domain and binds both $\alpha$- and $\beta$-tubulin acidic tails.

The study of Hirano et al. (2011) suggests that the interactions of the MyTH4–FERM domain with microtubules and DCC are mutually exclusive, in spite of the fact that the binding sites are not overlapping. This is particularly important as myosin-X is now reported to have several different binding partners, implying that under different cellular conditions, a different subset of proteins may be associated with myosin-X. It is noteworthy that integrin binding also interferes microtubule binding. It is speculated that the primary motor function carrying cargos and the linker function linking an actin filament and a microtubule are two alternative functions of myosin-X.

We could continue addressing on various similar biologically active MMs, but our limited scope forces to prioritize and address a few other machines ‘ion

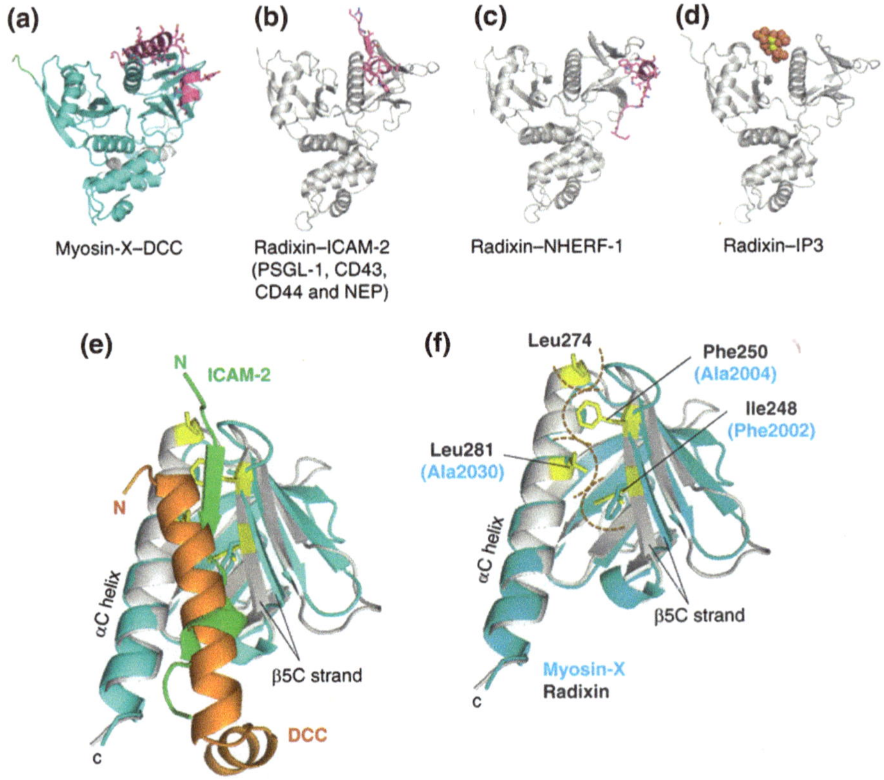


**Fig. 5.7** Peptide recognition by FERM domains. Comparison of complex structures of the myosin-X and radixin FERM domains bound to their binding partners. **a** The FERM domain of the MyTH4–FERM cassette bound to the DCC P3 peptide of the current structure. **b** The radixin–ICAM-2 complex (1J19) exhibits the prototypic binding mode of adhesion molecule recognition by β–β association. A similar binding mode (Motif-1α recognition) is found in both the radixin–PSGL-1 (2EMT) and radixin–CD43 (2EMS) complex structures. The radixin–CD44 (2ZPY) and radixin–NEP (2YVC) complexes also display this binding mode with the modified recognition motif Motif-1β. **c** The radixin–NHERF complex (2D10) displays the second class of binding mode with the NHERF peptide docked into the β-sandwich loops. **d** The radixin–IP3 complex (1GC6) reveals the PtdIns(4,5)$P_2$-binding site between subdomains A and C. **e** Superimposition of DCC P3 (orange) and ICAM-2 (green) peptides bound to the myosin-X (cyan) and radixin (gray) FERM domains, respectively. The ICAM-2 peptide forms a short β strand followed by a 1-turn $3_{10}$ helix. **f** Superimposition of the myosin-X (cyan) and radixin (gray) FERM domains bound to DCC P3 and ICAM-2 peptides, respectively. The bound peptides are omitted for clarity. Nonpolar residues of the radixin FERM domain are highlighted as stick models (yellow) (cf. Hirano et al. 2011)

channels' later in this chapter. Continued research has been pursued to not only address MMs' structure and function aspects but also their genetic origins, mutations in pinpointed structure regions, and related disease condition, etc. Search is continued to find answers to two important aspects about the biological machines inside cells:

(i) How are biological machines constructed naturally and how do they function?
(ii) What happens to these machines' structure and function due to the effects of rise of diseases in biological system, especially in cell?

When the answers to these two questions are found for any specific biological machine, we may then ask another question:

(iii) How can we repair any possible structural or functional perturbation in a biological machine?
We often solve this question (iii) through searching agents and right physiological conditions for repairing such biological disorders.

Knowing the details of the biological machines is one of the top priorities of current biomedical sciences. A few answers to the mentioned questions will be discovered in our subsequent chapters. Before that we wish to address on a few natural and artificial MMs existed or constructed in plasma membranes and mitochondrial membranes in cell. Our priority on ion channel is due to that ion channels are prime structures which are found engaged in regulating important cellular transport mechanisms for various agents that often define the life of the biological system. The wellness of cells, so is perhaps of the biological life, or turning into disease state largely relies on the condition of ion channels in membranes.

## 5.4 Ion Channels Induced by Natural Agents in Cell Membrane. Natural Membrane-Based Molecular Machines

Ion channels are fascinating molecular micromachines responsible for electrical signaling in biology. Beginning of the concept of ion channels dates back to as many as two centuries. During mid-nineteenth century (in 1850), Helmholtz was first to successfully determine the velocity of the electric signal on a nerve cell. Even earlier, during the first half of the nineteenth century, various investigations were pointing slowly toward some kind of an underlying chemical process that was taking place in nerve and muscle tissues as opposed to simple conduction or any kind of supernatural 'vitalism'; i.e., forces unique only to living things.

In 1759, Michel Adanson was the first to understand the way an electric eel stores electricity within itself. Afterward, battery inventor Volta noticed that there existed similarities in organization between the fish's electric organs and the batteries he built. Galvani's publication of results using electrical current in 1791 to make a dead frog's legs twitch appeared as an encouraging point for many to try and explain how electrical current helps nerve cells transmit messages to muscles.

In 1817, the Swedish chemist Berzelius suggested that the electric eel's current was nothing but due to an organic chemical process. In 1827, Italian physicist Leopoldo Nobili appeared as the first ever scientist to have been successful to use an instrument galvanometer that he refined and measured the current in a frog. Here to mention that 'current' in biological system refers to the 'ionic current.' This is the current (defined as the flow of charges) due to the flow of various ions such as sodium, potassium.

The above explanations created the background of searching for currents in biological systems especially in cellular sectors. During couple of decades in the late twentieth century, many ground-breaking discoveries were made. Now we have got a tremendous amount of focused study-based information on various specific ion channels. We shall directly go to explain various features of a few important ones in this chapter.

### *5.4.1 Molecular Micromachine. A Voltage-Gated Ion Channel Closing-Opening*

We find a few transmembrane proteins forming ion channels which are activated by the changes of electrical membrane potential. These ion channels are generally known as voltage-gated ion channels (VGICs).

The proteins' functioning as VGICs show three remarkable properties that enable nerve cells to conduct an electric impulse (Lodish et al. 2000):

(i) Voltage gating: opening in response to changes in the membrane potential,
(ii) Channel closing and inactivation, and
(iii) Ion specificity: like other ion channels, exquisite specificity for those ions that will permeate and those that will not.

Lodish et al. provided a description on the molecular analysis of the VGIC proteins, starting with techniques for measuring their functional properties (for details see Lodish et al. 2000). One universal property that emerges from molecular cloning is that all VGICs, be they $Na^+$, $K^+$, or $Ca^{2+}$ channels, are related in structure and in function. To obtain a general idea about how a VGIC works, we may observe Fig. 5.8.

### *5.4.2 Dynamics of a Voltage-Gated Ion Channel. Movement of Voltage Sensor Domains*

VGICs are able to regulate ion flux mechanism on a millisecond (ms) timescale. In sensing the membrane voltage change, each ion channel uses four voltage sensor domains (VSDs) connected to a central ion-conducting pore domain. The fourth transmembrane segment (S4) of each VSD carries several positively charged amino acid residues responsible for VSD gating (Borjesson and Elinder 2008). At least three elementary charges per VSD must traverse outwards through the membrane electric field to open a channel that corresponds to a considerable

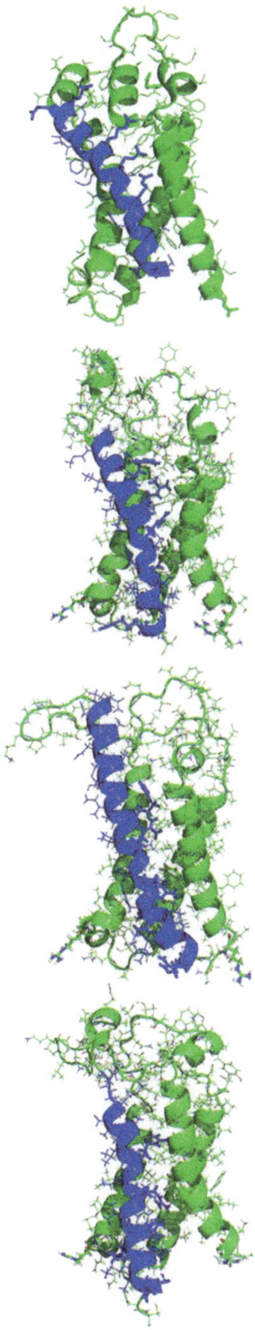

**Fig. 5.8** Deactivation process for a voltage sensor from a VGIC. From top to bottom, the sensor starts in a fully activated state corresponding to a depolarized membrane. As hyperpolarization is applied, the voltage sensor domain moves through at least two intermediate states in which the charged arginine side chains in the S4 helix (blue) move one position down for each (leaving the next arginine in the charge transfer center) before it reaches a fully relaxed down state. In the down state, a linker causes the voltage sensor domain to push inward on the pore domain (not shown), which in turn will close the pore (cf. Lindahl 2012)

displacement of the S4 helix. To demonstrate this, we copy from Henriona et al. (2012) Fig. 5.9 (see specifically Fig. 5.9a) (Schoppa et al. 1992; Zagotta et al. 1994; Keynes and Elinder 1998).

S4's positive charges make salt bridges with negative countercharges on their move through the VSD (Keyness and Elinder 1998; Long et al. 2007; Papazian et al. 1995; DeCaen et al. 2008, 2009). There is a proposal that the VSD might undergo a conformational alteration following the opening, when the channel relaxes to an inactivated, that is closed, state (Villalba-Galea et al. 2008). Besides conferring voltage dependence to ion channels, VSDs also regulate enzymes (Murata et al. 2005) and act as voltage-gated proton channels (Sasaki et al. 2006; Ramsey et al. 2006). Experimental data suggest that each voltage sensor (in particular the S4 helix) must adopt at least one intermediate state between the distinguishable conducting and the resting states (Zagotta et al. 1994; Keynes et al. 1998; Schoppa and Sigworth 1998).

In regard to exploring ion channel kinetics, structural constraints in a working ion channel can be explored by introducing pairs of cysteines that, if close enough,

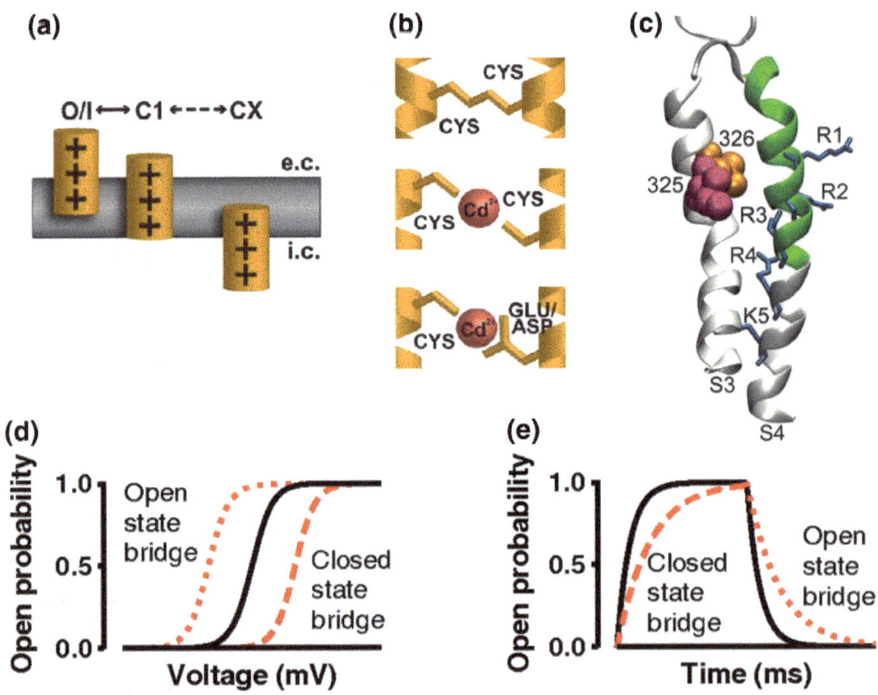


**Fig. 5.9** Probing VSD with $Cd^{2+}$ bridges. **a** Multiple molecular states of the S4 voltage sensor. O, open; I, inactivated; C, closed states. **b** Three types of interactions. Covalent disulfide bonds (*Upper*), $Cd^{2+}$ bridges coordinating two cysteines (*Middle*), and $Cd^{2+}$ bridges coordinating one cysteine and two or several glutamates/aspartates (*Lower*). **c** Residues 325 (magenta) and 326 (orange) are depicted as space filling. Residues 355–369 (green) are investigated with respect to interactions with 325 and 326. Blue sticks represent the gating charges. (**d** and **e**), possible effects of $Cd^{2+}$ bridges (red) on voltage and time dependence of the open probability (cf. Henriona et al. 2012)

can be linked by disulfide bonds (see Fig. 5.9b, upper). This may alter the ion channel kinetics (DeCaen et al. 2008, 2009; Larsson and Elinder 2000; Broomand and Elinder 2008). In a 2012 study, Henriona and co-investigators used weaker $Cd^{2+}$ bridges that are more likely to catch the channel in native conformations (Webster et al. 2004; Campos et al. 2007; Li et al. 2011; Lin et al. 2011) (sulfur distances <6.5 Å (Rulisek and Vondrasek 1998) (see Fig. 5.9b, middle) (Henriona et al. 2012)). Their lower strength means the interactions can break and reform, which makes it possible to maintain the channel in a working state. $Cd^{2+}$ bridges are found to normally form between cysteines and histidines as well as between negatively charged glutamates and aspartates (Rulisek and Vondrasek 1998). The $Cd^{2+}$ bridges between one cysteine and multiple glutamates or aspartates were also explored (see Fig. 5.9b, lower).

Henriona and co-investigators used the voltage-gated Shaker K channel expressed in oocytes from the frog *Xenopus laevis* to search for possible interactions (Henriona et al. 2012). Residues 325 and 326 in S3 are close to the gating charges R1 (=R362), R2 (=R365), and R3 (=R368) in S4 in the open state (Broomand and Elinder 2008). To probe for movements within the VSD, exploration was made for interactions between 325 and 326 and a long stretch of residues in S4 (355–369; green segment in (Fig. 5.9c) using $Cd^{2+}$ bridges. An interaction in the open state is expected to shift the voltage dependence in a negative direction and to slow down closing (see Fig. 5.9d, e). Out of 49 mutations, 20 confirm interactions were reported (Henriona et al. 2012).

This investigation by Henriona and co-investigators followed three steps where they (Henriona et al. 2012):

(i) started by identifying several strong Cys-$Cd^{2+}$-Cys bridges,
(ii) used these interactions to build molecular models, and
(iii) used several weaker Cys-$Cd^{2+}$-Glu/Asp interactions to assess predictions generated by the models.

These constraints and derived models provided detailed information covering a complete cycle of conformations for a voltage sensor. Figure 5.10 shows molecular models of VSD states and the gating process.

The studies of Henriona and co-investigators and proposed models have been found to cover a complete sequence of transformations from the deepest closed state to an open state. This set suggests that the VSD activation involves S4 moving at least 12 Å relative to the rest of the VSD, transferring three to four charges across the membrane, from the O to the C3 state, and that it could be possible to reach an even more closed C4 state in some cases with larger structural changes (although this would require a 17 Å motion). The motion can primarily be defined as a sliding helix, and for O through C3, there is a constant-length $3_{10}$-helix region moving along the sequence of S4 to maintain its location in the hydrophobic region around F290 of the VSD. The Cole-Moore effect where significant hyperpolarization makes it harder to subsequently activate the channel (Cole and Moore 1960), C3 can be considered the default resting state, supported by recent simulations of ωomega;-currents (Khalili-Araghi et al. 2012), while C4 might only be reached

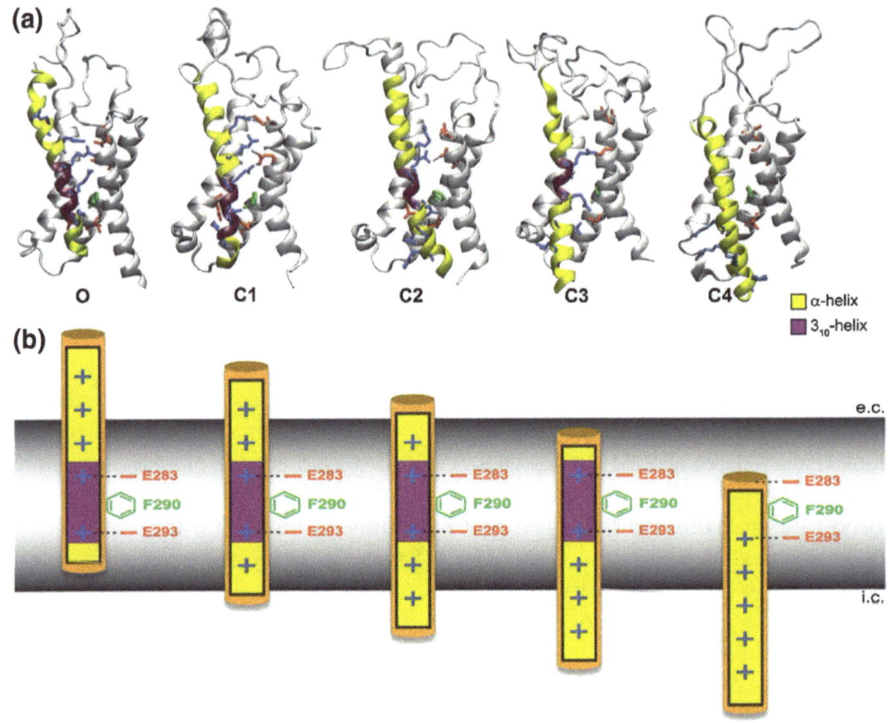


**Fig. 5.10** Molecular models of VSD states and the gating process. **a** Metal-ion constraints and molecular simulation relaxation predict each C1 through C3, and, under some conditions, C4, state to correspond to one additional arginine side chain in S4 (blue sticks) translating across the hydrophobic zone lock formed by F290 (green sticks), forming salt bridges to negatively charged residues in S1–S3 (red sticks; E247 in S1 and E283 in S2 above F290, and E293 in S2 and 316D in S3 below F290. The region of S4 close to F290 adopts $3_{10}$-helix (purple) in all models; however, C4 with the rest of S4 in α-helix (yellow). This suggests a gating model where a virtually constant-length $3_{10}$-helix region slides along the sequence of S4 without any net free energy cost that avoids overall rotation of the entire S4 helix during gating—only the outer ends rotate. **b** Cartoon highlighting some of the features from A. O, open; I, inactivated; C, closed states (cf. Henriona et al. 2012)

after long hyperpolarization. Thus, the model of a more natural motion from O through C3, while C3 to C4 requires a larger change, could help reconcile many of the different constraints observed in experiments.

### *5.4.3 Conformational Free Energy Associated with the VSD Activation*

Delemotte and co-investigators recently made an effort to address the conformational free energy associated with the VSD activation (see Delemotte et al. 2015).

This study shows that the atomistic description of VSD activation obtained by molecular dynamics simulations and free-energy calculations is consistent with the phenomenological models adopted to account for the macroscopic observables measured by electrophysiology. Among possible schemes for computing the free energy, the authors have chosen here the well-tempered metadynamics, a method that enhances sampling by adding a repulsive history-dependent potential along a set of a few reaction coordinates (RCs) (for details, see Barducci et al. 2008).

The investigators have been successful to bridge the gap between top-down and bottom-up approaches and reconcile the macroscopic and microscopic views of VSD activation dynamics by using two distinct levels of coarse graining. They first used well-tempered metadynamics to bias the exploration of the configurational space along aptly defined collective variables (CVs) describing the activation process; they then used the ensemble of configurations explored in the metadynamics run to calculate the free-energy profile along the experimentally relevant degree of freedom, the gating charge $Q$ (see Fig. 5.11).

The complexity of this simulation task requires an extensive use of the high-performance computing facilities; in particular, for the specific transition, $\sim 4$ μs of molecular dynamic (MD) trajectory were collected. However, thanks to massively parallel simulations, here the free-energy surface was calculated using 42 semi-independent walkers, completion of the calculations can be obtained in few days on large supercomputers. Figure 5.12 presents the free-energy landscape of the ε/δ-transition: Models describing the transition from putative resting state ε and first intermediate δ are thus investigated using two CVs.

We wish not to detail the simulation study and related analytical expressions for the conformational free energy associated with the VSD activation here due to limited scope of this book. The readers are encouraged to go through the original article of Delemotte et al. (2015) at their leisure and academic interests.

### *5.4.4 General Remarks About Naturally Occurring Molecular Machines-Channels*

In earlier section, we have addressed the general properties of natural ion channel VGIC and its VSD activation mechanisms. There are a lot and various types of natural ion channels in cell. Their coordinated actions and generalized functions make the health of cell or abnormality what emerges often as disease state.

Physiological conditions and transports of cell are maintained and regulated by channels like chloride channels, potassium channels, sodium channels, calcium channels, proton channels, various nonselective cation channels. The channel-gating mechanisms are highly dependent on localized channel hosting cellular compartments' electrical properties, such as free charges, electrical polarity, resting potential.

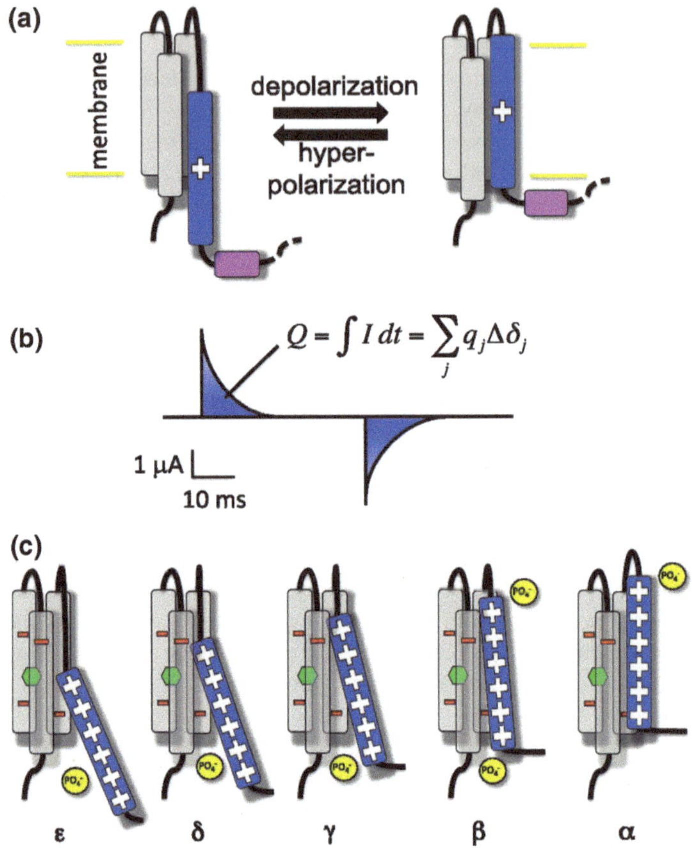


**Fig. 5.11** Activation of voltage sensor domains. **a** VSDs are formed by four transmembrane helices. Upon changes in the membrane potential, the positively charged S4 (blue) moves across the membrane relative to a static S1–S3 bundle (gray), transmitting the electrical signal to a linker peptide (purple). **b** This movement is reported by the measurement of transient currents called gating currents. The time integral of these, the gating charge $Q$, can be expressed as the sum of the contributions of the charges of the system. **c** Cartoon depiction of the stepwise activation of the Kv1.2 VSD. From the most resting ($\varepsilon$) to the most activated conformation ($\alpha$), S4 proceeds in a ratchet-like upward motion in which its positively charged residues jump from a negative binding site to the next. The negative charges of S1–S3 are depicted in red and the ones of the lipid headgroups in yellow. The hydrophobic gasket at the center of the VSD is represented by a green hexagon (cf. Delemotte et al. 2015)

Mitochondria are a host of many natural MMs that act as channels. Mitochondrial outer membrane permeabilization (MOMP) is a complex multistep process (Ganesan et al. 2012). Bax channel can induce both transient permeabilization, detected by protein release, and more substantial long-lasting permeabilization, measured by the rate of oxidation of added cytochrome c (Ganesan et al. 2012). Other naturally occurring mitochondria-based natural MMs are, e.g., channels produced by Bcl 2, ceramide.

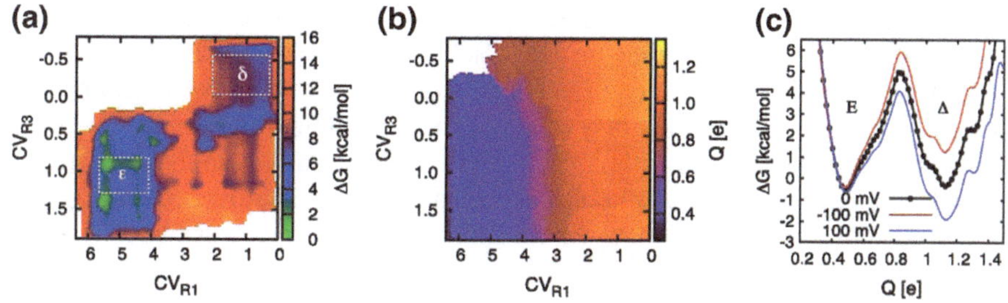


**Fig. 5.12** Free-energy landscape of the $\varepsilon/\delta$-transition. **a** Free-energy map in 2D $CV_{R1}/CV_{R3}$ space. Regions of low free energy are depicted in cold colors (green to blue) and regions of high free energy in hot colors (orange to red). **b** Mapping of the average value of the gating charge $Q$ as a function of $CV_{R1}$ and $CV_{R3}$. **c** Reweighted free-energy profile along $Q$. The free-energy profiles at −100 and +100 mV were obtained by adding a linear component to the profile obtained under 0 mV using $\Delta G(Q,V) = \Delta G(Q,V=0) - QV$. Starting with the geometrical definition of the two putative activation intermediates $\varepsilon$ and $\delta$, we have obtained two larger ensembles of structures corresponding to stable thermodynamic states of the experimental (voltage-clamped) ensemble E and Δ. All free energies are reported in kcal/mol. For details, see Delemotte et al. (2015)

Our main aim in this chapter is to provide a few example cases. So we avoid to list all channel-based MMs that are found to date. Of course, readers will get more on the channels in another Chap. 8 where we have addressed disease states.

## 5.5 Ion Channels Induced by Cell Membrane Active External Agents. Artificial Construction of Molecular Machines in Cell Membranes

The primary role of a cell membrane is to create barrier against materials transferring between cellular exterior and interior regions. But the presence of certain natural or artificial agents (especially during the time of treatment) like membrane proteins (MPs), antimicrobial peptides, drugs occasionally induce transient or stable transport properties into cell membranes. Artificial agents like antibiotics, drugs are found to destine at membranes and cause formation of random disorders or semi-stable pores. These membrane disruption mechanisms are often found highly dynamic, time dependent, and specific to the agents inducing them. The events also fall in different classes due to the diversities in their structures and mechanisms. We shall discuss the details of a few of various classes of such membrane-based active events what we also call MMs of the membrane that are induced by artificial agents.

Bilayer-spanning proteins primarily catalyze the selective transfer of materials and information across biological membranes. In this process, MPs undergo conformational changes, e.g., the opening/closing transitions in ion channels (see Unwin and Ennis 1984; Perozo et al. 1999, 2002), the shift in substrate binding site accessibility in conformational carriers, and ATP-driven pumps (see Toyoshima and Mizutani 2004). The protein conformational changes involve the protein/bilayer interface, where the protein is coupled to the bilayer through hydrophobic interactions. They then perturb the bilayer immediately adjacent to the protein (see Israelachvili 1977; Sackmann 1984; Gruner 1991; Andersen et al. 1992; Brown 1994). That is the protein conformational changes involve not only rearrangements within the protein, but more importantly also interactions with the environment, particularly with the host bilayer. Some of these phenomena have been schematically addressed in Fig. 5.13 in light of the investigations on the gating mechanisms observed in mechanosensitive channels in model membranes (see Perozo et al. 2002). Here the mechanosensitive channels are predicted to act as membrane-embedded mechanoelectrical switches. The switches induce opening of large water-filled pores that couple hydrophobically with lipid bilayers. This pore bilayer coupling or binding forces bilayer to deform near the pore opening to adjust the mismatch between pore length and the bilayer's resting thickness. The elastic properties of bilayer (Helfrich 1973) perhaps ensure that due to a possible continuous bending in the lipid monolayers near both longitudinal edges of the pore, the bilayer does not disintegrate. A structural change in the channel-forming agents is also a prediction made in the model picture (Fig. 5.13). We have made this prediction due to mainly the observed structural rearrangements in the mechanosensitive channels during the back-and-forth transitions between channel's closed and open states (Perozo et al. 2002). This structural changes may be due to rotation, bending, translation, etc. Behind the structural changes within both pore and lipid monolayers near the pore, there exists a driving force. This driving force causes the coupling between pores/channels and bilayer through creating structural changes in both lipid layers and the channels. We have recently discovered the origin of this driving force to be the coupling energy originating from the interactions due to primarily the localized electrical properties of channel-forming agents (MPs, AMPs, etc.) and lipids on the bilayer (Ashrafuzzaman and Tuszynski 2012a, b, c). In our earlier book 'Membrane Biophysics' (see in Ashrafuzzaman and Tuszynski 2012b), Chap. 4 was dedicated to addressing the whole picture of energetics related to the channel formation or in general functions of MPs in membranes.

The transport properties of a lipid bilayer membrane depend on the membrane constituents. But the hydrophobic MP-lipid coupling which raises certain lipid-protein, protein-lipid–protein, or lipid–protein–lipid complexes truly controls most of the processes occurring across membranes. AMPs are active in changing the biophysical properties of cell membranes. They often interact with lipids and create AMP-lipid complexes which lead to the creation of protein-lined or lipid-lined classified/well-structured ion channels/pores, less structured ion flowing pores, localized disorders or defects, etc. It is like creating an artificial MM that works with different degree of functionality. In formation of such special structures in a membrane involving AMPs and lipids, various mechanisms are observed which are mostly specific to AMPs. The complexes often appear with various distinguishable structures: A few of them are, for example, linear β-helix created by, e.g., gramicidin A (gA), barrel stave pore created by, e.g., alamethicin, toroidal pore created by, e.g., magainin, melittin, lipidic channels created by, e.g., ceramides (example of a non-AMP-induced channel), defects created by, e.g., gramicidin S. In this chapter, we shall address about the structures of those mentioned

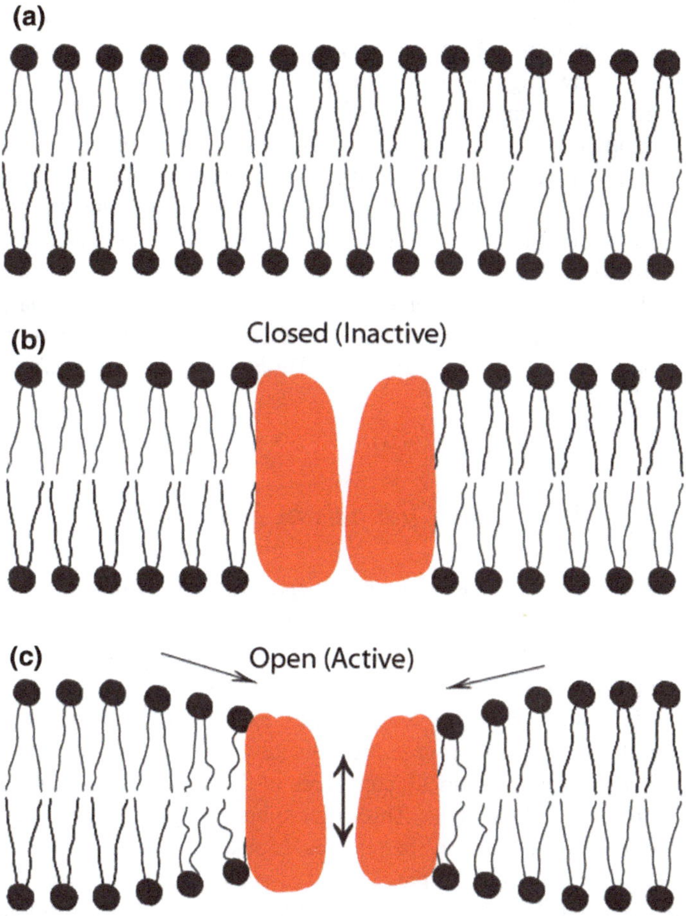


**Fig. 5.13** Lipid bilayer membrane exists with an average constant thickness unless any random MP or physically ordered channel induces any kind of bilayer deformations. In diagram A, a plain lipid bilayer membrane has been schematized. Lipid bilayer behaves like a barrier against ion flow across the membrane. In B, a lipid bilayer membrane with a possible inactive ion channel (in red color) within it has been schematized. In this case, despite the presence of ion channel, lipid bilayer still behaves as a barrier against ion flow across the membrane. The channel in this case is apparently considered to be inactive or nontransporting agent. In C, a lipid bilayer membrane with a possible active ion channel has been schematized. In this case, unlike the cases represented by A and B, the bilayer conducts ions across the membrane. Decrease or increase in bilayer thickness near the channel interface as the channel undergoes the conformational transition from closed to open states or vice versa is an important hypothesis in this cartoon representation. A structural change (especially size) in the channel-forming agents is also a prediction. These schematic diagrams have been drawn in light of the gating mechanisms observed in mechanosensitive channels in model membranes (Perozo et al. 2002)

membrane-disrupting events and the primary mechanisms which dictate the formation and function of them (see also Ashrafuzzaman and Tuszynski 2012a). Based on our existing knowledge on various model studies, we shall create a complete platform to address the antimicrobial effects of a group of AMPs. As mentioned earlier, this chapter will be constructed considering the structural aspects of the membrane transport properties changing events induced mainly by AMPs and sometimes by even certain classes of lipids. All these membrane events not only follow certain structural complexities due to just the biophysical coexistence between channel-forming agents and lipid membrane, like for example mentioned in Fig. 5.13, but that the phenomena follow complex energetics. We shall address the structural aspects here in brief, and the energetic aspects will be referred in Chap. 5 of our previous book 'Membrane Biophysics' (see Ashrafuzzaman and Tuszynski 2012b).

### *5.5.1 Protein-Lined Ion Channels in Membranes*

In protein-lined channels, it is generally considered that channel-forming peptides align along the channels. Ions flow through the longitudinal axis of the channel's interior region between cellular exterior and interior regions, and they most likely experience interactions with channel-forming peptides. Only at the entry and exit levels of the channels, ions are expected to experience interactions with lipids (especially with lipids' headgroups). A general cartoon illustration is shown in Fig. 5.13. Usually linear, cylindrical, etc., types of structures in channels are found in the class of protein-lined channels. This class refers to highly ordered peptide structures in association with lipids in membranes. Lipids play important role in the regulatory phase of the channels but the creation of such channels primarily depends on the properties (chemical, electrical, mechanical including geometry and material, etc.) of both the channel-forming peptides and lipids on the membrane. Membrane-based hydrocarbons have a considerable regulatory role too. The number and type of the amino acid sequences and other constituents in the peptides, geometrical sizes (length, cross section, etc.) of the peptides, and charge properties of the participating atoms, and the effective final charge of the peptides in the hydrophobic membrane environment, etc., play the crucial role in the construction of the protein-lined channels. Best examples are found in two ion channels, namely gA and alamethicin channels. Therefore, this section will be dedicated to addressing the issues related to protein-lined channels using gA and alamethicin channels as examples.

Recently, a MM 'biosensor' was explored out of gA ion channel by Cornell and co-investigators (see Moradi-Monfared et al. 2012). These investigators have performed the operation and construction of a high-resolution gA ion channel biosensor. This biosensor achieves high sensitivity detection by modulating the flow of ions in response to target molecules.

In artificial membrane, the inserted biological ion channels have been used to sequence single strands of DNA (Branton et al. 2008), to open the pore of a nonselective efflux channel (Kocer et al. 2005; Sukharev and Anishkin 2004; Louhivuori et al. 2010) and to effect the light-driven production of ATP by $F_0F_1$-ATPase (Steinberh-Yfrach et al. 1998). Feringa and co-investigators reported the switching of the 'mechanosensitive channel of large conductance' (MscL) of *Escherichia coli*, which can be opened in response to the introduction of charged entities at a certain location in the protein channel (Kocer et al. 2005). Initially, a light-cleavable group was used to release acetate anions, and thus open the channel, with a reversible version, utilizing a spiropyran switch. Ion channel proteins modified postsynthetically to be light switchable have been inserted into living cells, and successfully opened and closed (Banghart et al. 2004, 2009; Volgraf et al. 2006; Stein et al. 2013; Schoenberger et al. 2014, 2014; Kramer et al. 2005; Mourot et al. 2012). Unnatural, light-switchable amino acids have also been incorporated into ion channel proteins (Szymanski et al. 2013; Lougheed et al. 2004; Banghart et al. 2006). This discussion (see also Erbas-Cakmak et al. 2015) shows that the ion channels artificial or natural have huge machinery roles in cellular transport mechanisms and many important functions.

#### 5.5.1.1 Protein-Lined Gramicidin A Channels in Lipid Membranes

A gA channel is a simple dimer of two right-handed, $\beta^{6.3}$-helical subunits or gA monomers (Arseniev et al. 1986; Ketchem et al. 1997; Townsley et al. 2001). This channel is formed by reversible, trans-bilayer association of the subunits (O'Connell et al. 1990): $M_{\text{left}} + M_{\text{right}} \underset{k_{-1}}{\overset{k_1}{\rightleftarrows}} D$; here $M$ and $D$ represent gA monomer and dimer, respectively. The subscripts (left and right) denote monomers residing in each bilayer leaflet. Here, $k_1$ and $k_{-1}$ are the two rate constants determining the channel appearance rate ($f_g = k_1 \cdot [M]^2$; $[M]$ being the monomer concentration) and channel lifetime ($\tau = 1/k_{-1}$). A model demonstration has been presented in Fig. 5.14. Peptides residing inside membrane occasionally approach to each other and depending on the bilayer environment associate with each other and make a dimer with a very short lifetime which spans over a considerable range in millisecond (ms) order. The range of channel lifetimes depends on the strength of the hydrophobic coupling between channel and the bilayer. The readers may learn the details of the various properties to be considered to understand the ordered channel versus nonchannel phases in our earlier publication (see Ashrafuzzaman and Tuszynski 2012b, Chap. 5).

#### 5.5.1.2 Protein-Lined Alamethicin Channels in Lipid Membranes

Barrel stave pores are formed by alamethicin channels (Boheim 1974; He et al. 1995). In this type of a pore, the peptides align in a way such that they make a cylinder where the peptides stay on the surface of the cylinder. Many conductance states depending on the number of monomers involved in forming the cylindrical

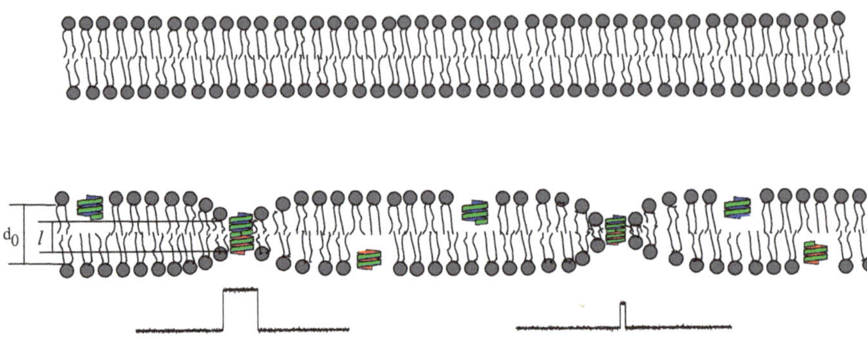


**Fig. 5.14** Bilayer deforms at the bilayer gA channel coupling area which incurs an energetic cost (see Chap. 5, Ashrafuzzaman and Tuszynski 2012b). This schematic diagram showing gA channels in lipid bilayers follows identical morphology as presented in Fig. 5.14. The upper panel shows a lipid bilayer without any integral MP. Lower panel shows a bilayer with integral gA monomers and dimers of different lengths. When gA channels are formed inside a lipid bilayer under a transmembrane potential difference between two sides, the bilayer conducts a current pulse with a specific average pulse width (proportional to the gA channel lifetime) and height (proportional to the gA channel conductance) depending on the gA channel type (the number of amino acids in the structures of gA monomers). Two types of gA monomers are schematically structured here to produce two gA channels of different lengths ($l$). $d_0$ is the unperturbed thickness of the bilayer (Ashrafuzzaman et al. 2006; Ashrafuzzaman and Tuszynski 2012a)

channel is possible. A model diagram is shown and explained in Fig. 5.15. Here the channel is assumed to be formed due to inter monomer binding. The channel also experiences hydrophobic coupling with lipid monolayers in its both longitudinal terminals. Unlike gA channels where we observe only one-ordered gA dimer state, there is no unique cylindrical alamethicin channel state as evidence. Alamethicin channels with various cylindrical states are possible. Perhaps a transition between different cylindrical channel states is also a reality. These are modeled in Fig. 5.15 (Ashrafuzzaman and Tuszynski 2012a) in light of previous models (see Boheim 1974; He et al. 1995). A phenomenological explanation on how the conformational transitions between different energy states in a barrel stave pore occur (Fig. 5.15) has recently been addressed (Ashrafuzzaman and Tuszynski 2012a). The readers may learn more about the stability, energetics, and regulation of the channel conformational states in Chap. 5 of book 'Membrane Biophysics' (see Ashrafuzzaman and Tuszynski 2012b).

## 5.5.2 *Lipid-Lined Ion Channels in Membranes*

In lipid-lined channels, it is generally considered that channel-forming peptides interact with lipid membrane and the participating agents create the alignment of lipids along the channels. Ions flow through the opening connecting the cellular exterior and interior regions and perhaps through the lipid regions avoiding the

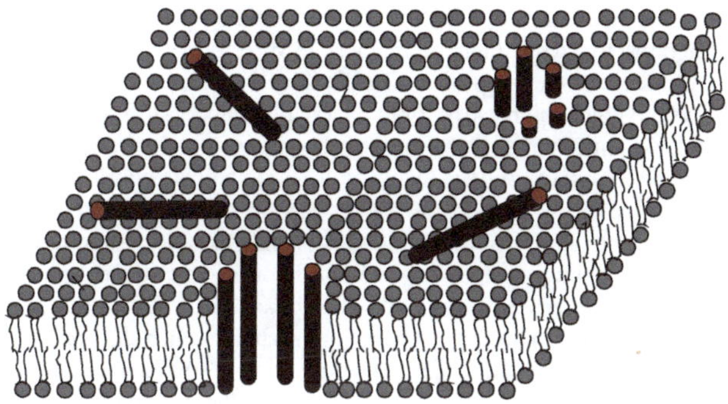

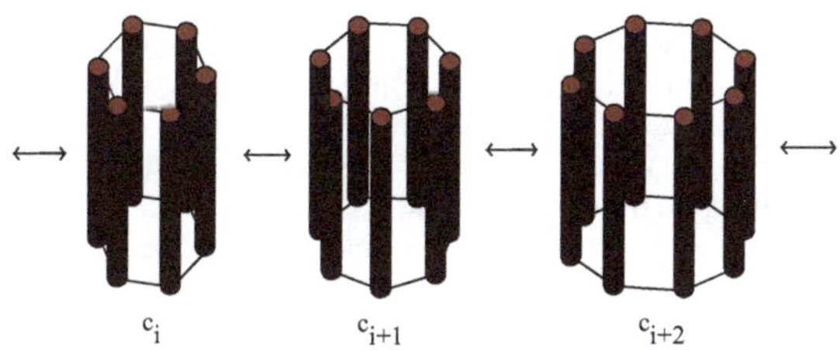
$c_i$
$c_{i+1}$
$c_{i+2}$

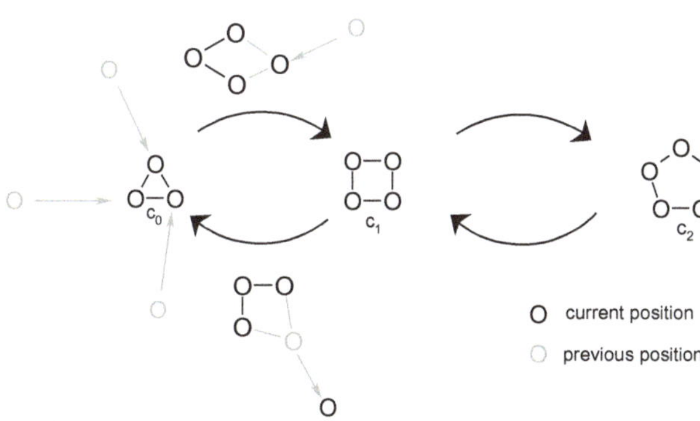
$c_0$
$c_1$
$c_2$
current position
previous position

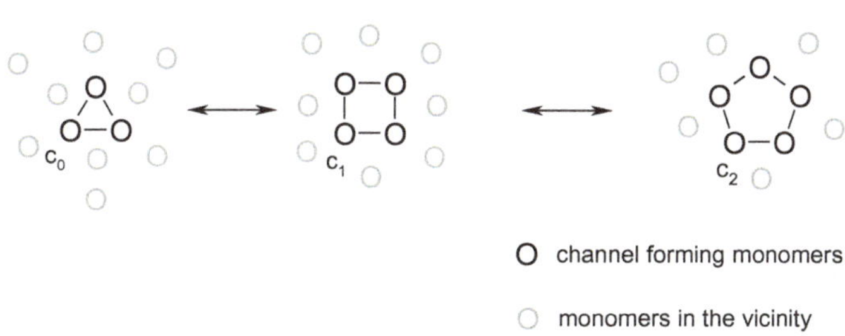
$c_0$
$c_1$
$c_2$
channel forming monomers
monomers in the vicinity

◂**Fig. 5.15** Barrel stave model for alamethicin channel formation inside lipid bilayers. Cylindrical rods are schematic diagrams for alamethicin monomers in three-dimensional view (in two-dimensional view along the cylinder axis they appear as circles). Top panel shows three-dimensional (3D) view, middle panel shows the transition between different conduction pores of alamethicin channels, and the two lower panels (two-dimensional (2D) view of the channels only from membrane surface where alamethicin monomers are seen only along their longitudinal direction so they appear as circles) represent the two possible mechanisms on how inter-channel conduction-level transformation happens. In the bottom panel, we consider that the monomers already exist in a structured form of alamethicin channel where the pore radius changes by reorganization of the channel-forming monomers. The structural reorganization perhaps enforces addition/subtraction of monomers into/from the channels without considerable requirement of monomers' translation. Other 2D view illustrates a possible model of alamethicin channel formation and transformation between different conduction levels where the pore radius increases by addition of monomer(s) from the surrounding space where monomers randomly move into the channel. The reduction in the pore radius occurs through release of the monomers from the cylindrical surface of the channels. Both of the models in 2D views are valid explanations of the upper 3D structures of alamethicin channels. In-depth structural studies are much needed to establish either of them. Taking three monomers in the 0th conductance level is an arbitrary choice, but our reverse calculation using experimental values of cylindrical alamethicin pore conductances and the theoretical values of the cross-sectional areas of different alamethicin pores hint that three monomers perhaps form the 0th conductance level. Faded circles and bonds in 2D views are shown to distinguish their inactivity in the channel's conduction mechanism or their former positions

peptides involved in creating channels. The lipid alignment causes an opening which may look like a long cylindrical lipid aligned channel where the membrane thickness may not dramatically change. The other possibility is that the membrane thickness slowly vanishes at the channel opening which can equivalently be considered as a broken membrane condition. In many investigations, this broken membrane structure has been predicted as a model for the lipid-lined channels. Figure 5.16 explains schematic diagram of this condition. These structures are found to be induced by both AMPs, e.g, magainin (Matsuzaki et al. 1996; Ludtke et al. 1996), melittin (Yang et al. 2001), colicin (Sobko et al. 2006) or by other nonantimicrobial agents, e.g., the recently discovered pores induced by chemotherapy drugs thiocolchicoside and taxol molecules (Ashrafuzzaman et al. 2011, 2012). The magainin-, melittin-, and colicin-induced toroidal pores can also be considered as protein-lined channels because here the peptides are always associated with the lipid headgroups result in the lipid monolayers bending continuously inward, so that the pore is lined by both peptides and lipid headgroups. We have recently observed the lipid-lined ion channels being induced by chemotherapy drugs in model membrane studies (see Fig. 5.16) (Ashrafuzzaman et al. 2011; Ashrafuzzaman et al. 2012).

### 5.5.3 *Lipidic Channels in Membranes*

We shall describe an ion channel which is formed by lipids. Although lipids primarily exist across the lipid layers in a membrane, ceramides behave differently. Ceramides form channels called ceramide channels due to special organization of

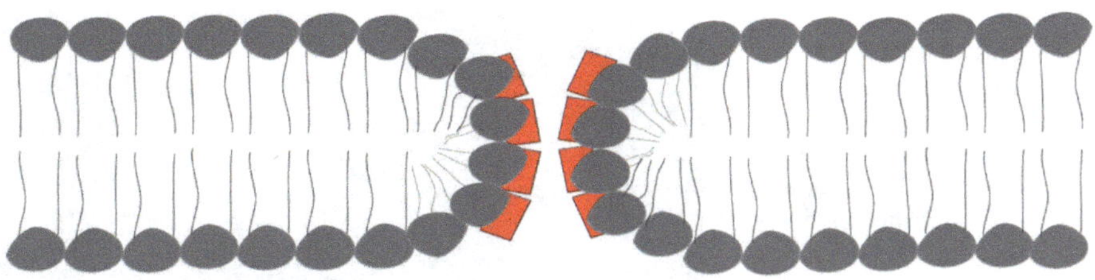

**Fig. 5.16** Chemotherapy drugs TCC and TXL-induced proposed toroidal-type ion pores in a lipid bilayer membrane possibly associated with a spontaneous change of the pore cross section. This novel model is explained in recent publications (Ashrafuzzaman and Tuszynski 2012b; Ashrafuzzaman et al. 2012)

ceramides in phospholipid membranes. Ceramide channel is also an example of a lipidic channel. A ceramide is a lipid molecule composed of the amino acid sphingosine and a fatty acid. Ceramides exist in great concentrations in plasma membrane and act as signaling molecules for a number of cellular functions. Ceramides may also have a role in certain pathological states, including cancer, obesity, diabetes, inflammation. Understanding of ceramide organization in membrane is therefore an important field which has enormous medical interests. A model structure of ceramide channel is presented in Fig. 5.17. The ceramide channels look more like alamethicin's barrel stave pore but have no resemblance with the structures of other channels like toroidal or β-helical types described earlier in this chapter. That means the lipid membrane adjusts with the complex of ceramide molecules at the channel membrane interface and creates no vanishing of the membrane thickness like what is proposed in the case of toroidal pores. Detailed understanding about the mechanism of the lipidic channels still to be discovered, but some insights are already found in research articles published by many groups, especially a few mentionable here (Siskind et al. 2003; Anishkin et al. 2006; Samanta et al. 2011).

(A) The channel is slightly tilted to illustrate the columns that span the membrane, each consisting of six ceramide monomers. The pore is lined by hydroxyl groups. The hydrocarbon tails are oriented parallel to the plane of the membrane. The columns are arranged in an antiparallel fashion so that the carbonyl oxygen of the amide linkage (red) is only visible in every other column. The pore diameter of this 48-column channel is 10 nm.

(B) A model of a segment of a smaller ceramide channel showing how it might interface with the phospholipid membrane. Note the slightly hourglass shape of the pore and the distorted phospholipids (lighter colors) that cover the hydrocarbon chains of the ceramides at the end of the channel. The structure of this interface is an illustration of the results reported from molecular dynamic simulations by the same group (Anishkin et al. 2006). This figure with description has been taken with publisher's permission from Samanta et al. (2011).

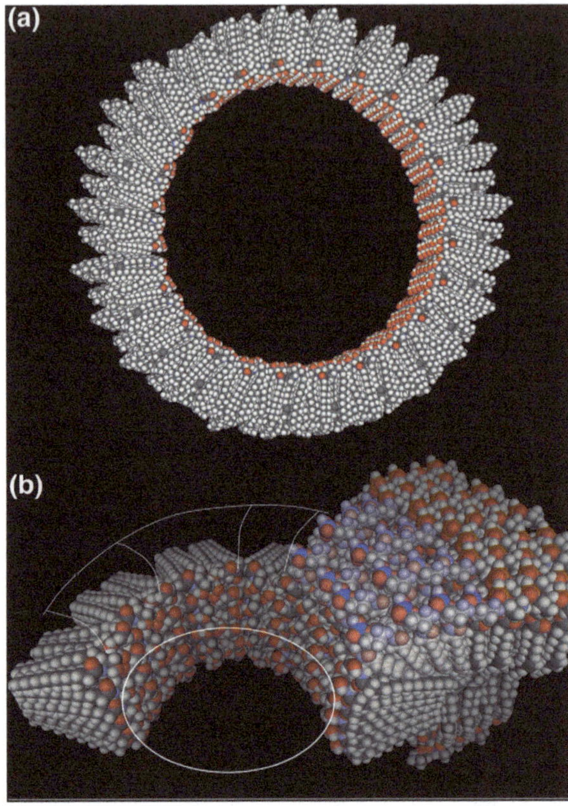


**Fig. 5.17** These model diagrams on the $C_{16}$-ceramides in membranes (ceramide channel) were provided by my collaborator Professor Marco Colombini which was published earlier (Samanta et al. 2011). Besides this article, earlier papers of the group, e.g., Siskind et al. (2003) can also be read to understand details about the ceramide channel phenomena in lipid membranes

## 5.6 Molecular Machines of the Mitochondria

Mitochondria are organelles of eukaryotic cells and depend for their biogenesis on a set of MMs for mainly transport of various proteins. Most of these transported proteins are destined to cytosol. The MMs of the mitochondria drive essential biological processes. How these MMs evolved is a fundamental question to date. Recently, Clements and co-investigators identified in α-proteobacteria the component parts of a mitochondrial protein transport machine in mitochondria derived from an α-proteobacterial endosymbiont (Clements et al. 2009). In bacteria, the machine components are found in the inner membrane, topologically equivalent to the mitochondrial proteins. Although the bacterial proteins have been reported to function in simple assemblies, relatively little mutation would be required to convert them to function as a protein transport machine.

The mitochondria in cells are evident to have evolved from intracellular bacteria (Muller and Martin 1999; Emelyanov 2003; Gray et al. 1999). The natural conversion of these intracellular bacteria into mitochondria required the evolution of protein transport machines (Dolezal et al. 2006). Clements et al. proposed that simple 'core' machines were established in the first eukaryotes by drawing on preexisting bacterial proteins that had previously provided distinct functions (Clements et al. 2009). Subsequently, over time, in a stepwise process in keeping with Darwinian evolution, additional modules would have been added to the core machines to enhance their function. This proposition is supported by three findings:

(i) protein components found in bacteria are related in sequence to the components of mitochondrial protein transport machines, but
(ii) these bacterial proteins are not found as part of protein transport machines, and
(iii) some apparently 'primitive' organisms found today have protein transport machines that function with only one or few component parts.

### 5.6.1 Protein Transport in Mitochondrial Molecular Machines. General Features

The process of protein transport into mitochondria requires the action of four membrane-embedded MMs, namely translocase of the outer membrane (TOM), translocase of the inner membrane (TIM) TIM22, TIM23, and sorting and assembly machinery (SAM) complexes (Emelyanov 2003; Gray et al. 1999; Dolezal et al. 2006; Murcha et al. 2007), each composed of up to eight distinct protein subunits. However, bacteria do not import proteins across outer and inner membranes, and the TOM and TIM23 complexes that provide this protein import function to mitochondria do not have counterparts in bacteria. The TIM23 complex is specifically responsible for protein transport across the mitochondrial inner membrane (MIM). This protein transport machine has been studied rigorously in yeast, in plants, and in humans and is found to be composed of a conserved set of protein subunits that associate together to form the MM (Murcha et al. 2007; Pfanner and Chacinska 2002; Neupert and Herrmann 2007). The following three of the subunits are both found in enough representative groups to suggest they are present in all eukaryotic organisms and are essential for cell viability in yeast:

(i) Tim23, a simple transmembrane protein that forms the channel through which protein substrates pass into the mitochondrial matrix;
(ii) Tim44, found on the inner face of the mitochondrial membrane where it interacts with both Tim23 and Hsp70, thereby docking the protein import motor to the Tim23 channel; and
(iii) Tim14/presequence translocase-associated motor (PAM), Pam18 subunit that interacts with several proteins in the TIM23 complex and directly stimulates the ATPase activity of Hsp70, thereby activating the motor to drive protein transport (Emelyanov 2003; van der Laan et al. 2006; Bolender et al. 2008; Mokranjac and Neupert 2008).

Clements et al. showed that α-proteobacteria have a protein of the Tim44 family that functions in membrane quality control and a Tim14/Pam18 protein that functions in a distinct process (Clements et al. 2009). Together with the LivH amino acid transporter, these component parts would have provided 'pre-adaptation' to bacteria ahead of a need for protein import. The proteins, TimA and TimB, function in distinct protein complexes in bacteria, yet evolved to serve as modules of a protein transport machine in mitochondria (see Fig. 5.18) (Clements et al. 2009).

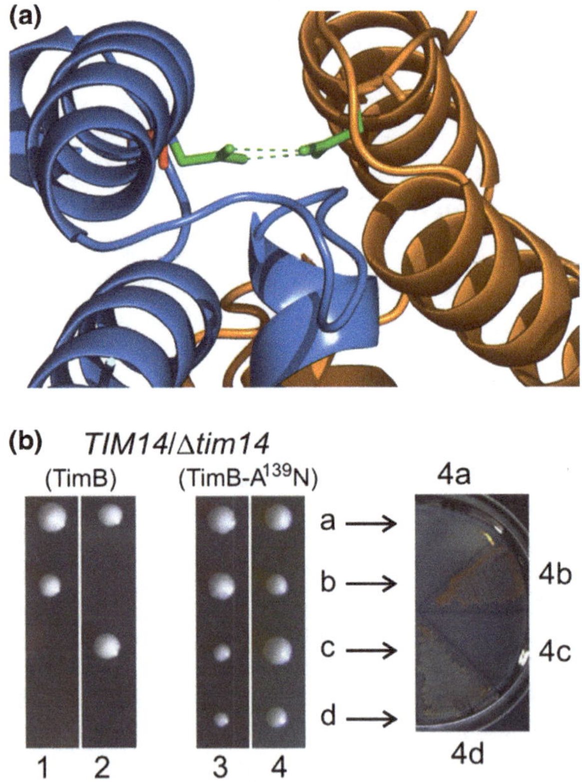


**Fig. 5.18** TimB can be converted to function in mitochondrial protein import. **a** Model of *Pb*TimB-A$^{139}$N (blue) binding to Tim16 (orange) based on the published Tim14-Tim16 interaction (2GUZ). In Tim14, the asparagine residue (green) forms a pair of hydrogen bonds with an asparagine in Tim16. In *Cc*TimB and *Pb*TimB, the native residue at this position is an alanine (red). **b** TimB from *C. crescentus* and *P. bermudensis* were engineered for expression in yeast by adding an N-terminal mitochondrial targeting sequence and transmembrane domain. A point mutation was introduced at residue 139 of *Pb*TimB (TimB-A$^{139}$N) to replace the alanine with an asparagine residue. After transformation with plasmids carrying the engineered TimB constructs, Tim14/Δ*tim14* yeast cells were sporulated. Tetrads were dissected: two spores of tetrads 1 and 2 formed viable colonies; four spores of tetrads 3 and 4 formed viable colonies. Arrows indicate subculturing the cells from tetrad 4 to measure growth phenotype, verifying that the two smaller colonies are Δ*tim14* cells kept viable by *Pb*TimB(A$^{139}$N)

Like the TIM23 complex in the inner membrane, the TOM complex in the outer mitochondrial membrane is composed of multiple components, as seen in Fig. 5.18. Three essential and ubiquitous subunits are found here: the Tom40 channel, and Tom22, and Tom7 subunits (Dolezal et al. 2006). In yeast and other fungi, additional Tom5 and Tom6 subunits are also found as part of the core TOM complex, and additional import receptors are sometimes present to maximally enhance protein transport (Neupert and Herrmann 2007), so is the higher efficiency of the machine's functionality ensured.

### *5.6.2 Protein Transport in Mitochondria. Common Principles and Mechanisms*

Mitochondrial protein translocases are found not to be independently operating units, but in fact closely cooperate with each other. Dudek et al. recently reviewed the latest studies that indicate how the pathways for mitochondrial protein biogenesis are embedded into a functional network of various other physiological processes, such as energy metabolism, signal transduction, and maintenance of mitochondrial morphology (Dudek et al. 2013). The highlights of the analysis are as follows:

(i) Mitochondrial protein translocases operate as highly dynamic MMs.
(ii) Multiple import machineries mediate the sorting and assembly of nuclear-encoded mitochondrial precursor proteins.
(iii) Close cooperation between translocation machineries of inner and outer membrane is crucial for mitochondrial protein sorting.

Figure 5.19 shows how different targeting signals direct nuclear-encoded precursor proteins on specific transport routes to their final localization within mitochondria. Most of the mechanisms described here were initially discovered in *S. cerevisiae*, but the vast majority of the protein machineries involved were later found to be highly conserved in higher eukaryotes (see Fig. 5.19) (Dudek et al. 2013). Except for a few α-helical outer MPs, virtually all precursors initially enter mitochondria by passing a common entry gate formed by the translocase of the outer membrane (TOM complex). Upon translocation across the outer membrane, different transport routes exist that are specifically required for transport of a subset of precursor proteins to their final destination.

Figure 5.20 addresses on multiple mechanisms that exist to integrate proteins with different transmembrane topologies into the OM. Guided by the β-signal precursor, proteins are delivered from the small Tim chaperones to the SAM complex (TOB complex) in the outer mitochondrial membrane for folding and insertion into the lipid bilayer (Wiedemann et al. 2003; Kutik et ak. 2008; Paschen et al. 2003; Gentle et al. 2004). For the biogenesis of the Tom40 precursor, the SAM complex is found to associate with the Mdm10 protein (Meisinger et al. 2004, 2007).

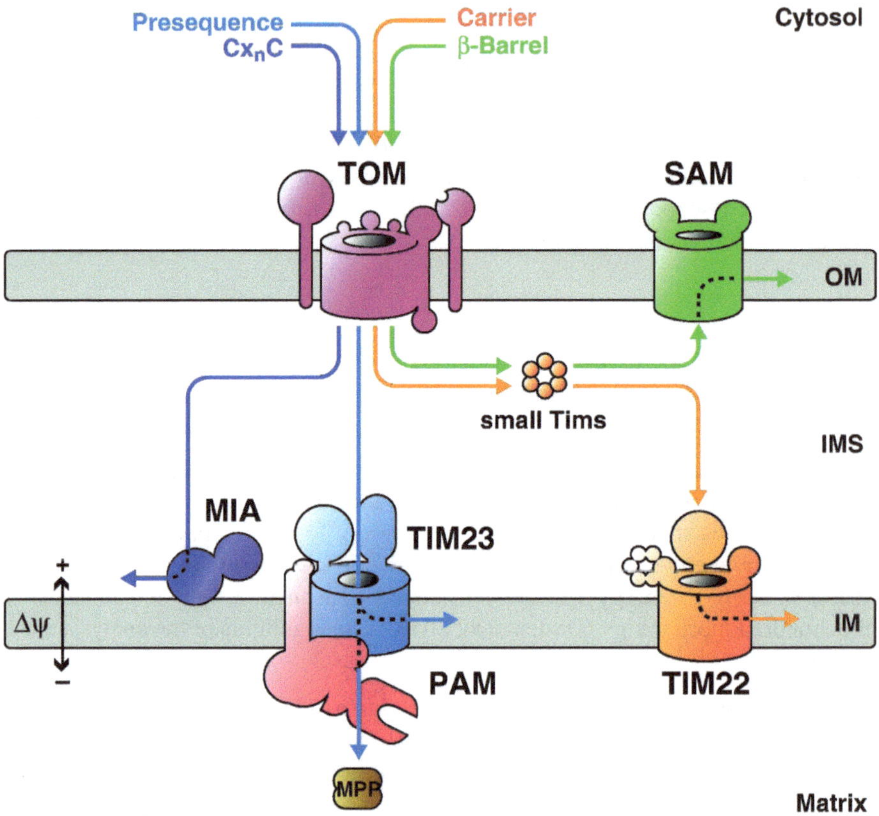


**Fig. 5.19** Different targeting signals direct nuclear-encoded precursor proteins on specific transport routes to their final localization within mitochondria. After translocation of precursors through the general translocase of the outer membrane (TOM complex), distinct downstream import pathway diverges in the intermembrane space (IMS): Biogenesis of β-barrel proteins of the outer membrane (OM) requires the small Tim chaperones of the IMS and the sorting and assembly machinery (SAM). Proteins of the IMS that contain cysteine-rich signals ($Cx_nC$) are imported via the mitochondrial intermembrane space import and assembly (MIA) pathway. Carrier proteins of the inner membrane (IM) are transported with the help of the small Tims and the translocase of the inner membrane 22 (TIM22 complex). Presequence-containing proteins are inserted into the inner membrane or imported into the matrix by the translocase of the inner membrane 23 (TIM23 complex; presequence translocase). Matrix translocation requires the activity of the presequence translocase-associated import motor (PAM). Presequences are proteolytically removed by the mitochondrial processing peptidase (MPP) upon import. $\Delta\psi$, membrane potential across the inner mitochondrial membrane Copied from Dudek et al. (2013) with due copyright permission

Figure 5.21 describes the modular rearrangements and coupling to different partner protein complexes. Three membrane-integral subunits are found to form the catalytic core of the TIM23 complex (TIM23CORE): Tim23, Tim17, and Tim50.

Tim23 forms a $\Delta\psi$-dependent protein-conducting pore across the mitochondrial inner membrane (van der Laan et al. 2007; Truscott et al. 2001; Meinecke et al. 2006;

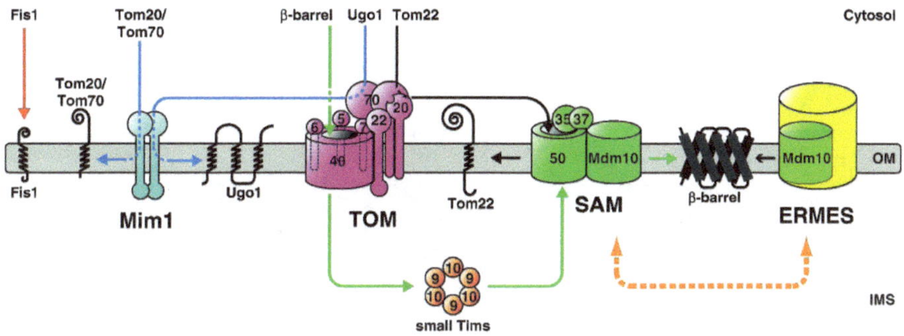


**Fig. 5.20** Multiple mechanisms exist to integrate proteins with different transmembrane topologies into the outer mitochondrial membrane (OM). A few α-helical proteins of the outer membrane, like Fis1, seem to insert without the help of proteinaceous translocases. OM proteins with an N-terminal α-helical membrane anchor, like Tom20 or Tom70, and multi-spanning α-helical OM proteins, like Ugo1, depend on Mim1 for membrane integration. Tom22 is recognized by the TOM complex receptors and subsequently inserted into the OM via the SAM complex. The SAM complex also mediates the membrane integration of β-barrel OM proteins, which are handed over from the TOM to the SAM complex by the small Tim chaperones Tim9–Tim10. Mdm10 has a dual function in the SAM complex and in the endoplasmic reticulum (ER)–mitochondria encounter structure (ERMES). Further, ERMES proteins have been implicated in the biogenesis of β-barrel proteins as well. IMS, intermembrane space Copied from Dudek et al. (2013) with due copyright permission

Alder et al. 2008). The N-terminal IMS region of Tim23 (Tim23IMS) is part of the preprotein receptor domain of the TIM23 complex (Komiya et al. 1998; Bauer et al. 1996). Tim17 helps in stabilization and regulation of the channel formed by Tim23 and the differential sorting of preproteins (Chacinska et al. 2005, 2010; Meier et al. 2005; Martinez-Caballero et al. 2007). Tim50 perhaps helps maintain the Tim23 channel across the inner membrane in a closed state to prevent ion leakage and dissipation of $\Delta\psi$ in the absence of a preprotein (Meinecke et al. 2006). We would leave the details for the readers who may consider reading the review article by Dudek et al. for better understanding of this subject (Dudek et al. 2013).

## 5.7 Molecular Machines 'Pumps' in the Membrane

The movement of sodium and potassium ions across the cell membrane is an active transport process involving the hydrolysis of ATP to provide the necessary energy. This process involves an enzyme referred to as $Na^+/K^+$-ATPase. The process is responsible for maintaining the excess amount of $Na^+$ outside the cell and the excess amount of $K^+$ ions on the inside. The sodium–potassium pump is an important contributor to action potentials produced by nerve cells. This kind of MMs is called P-type ion pumps because the ATP interactions phosphorylate the transport protein and cause a change in its conformation.

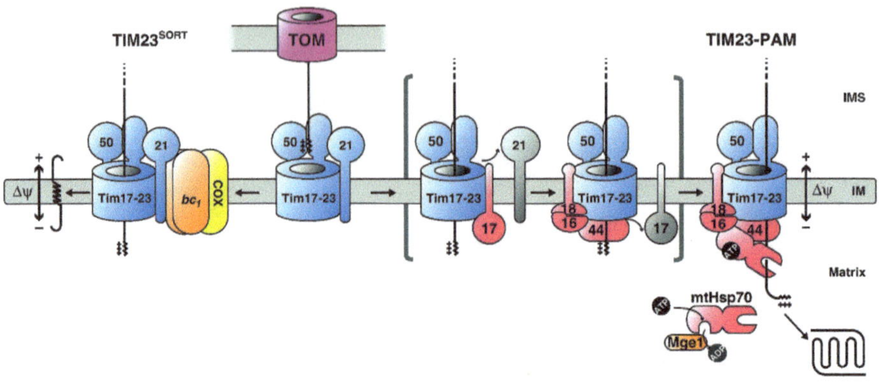


**Fig. 5.21** Modular rearrangements and coupling to different partner protein complexes trigger functional switches in the presequence translocase (TIM23 complex). The membrane potential ($\Delta\psi$) across the inner mitochondrial membrane (IM) drives the transfer of positively charged presequences from the TOM complex into the protein-conducting pore of the TIM23 complex in a process that involves Tim21. For membrane integration of preproteins, the TIM23–Tim21 complex (TIM23$^{SORT}$) associates with respiratory chain supercomplexes composed of cytochrome $bc_1$ complexes and cytochrome *c* oxidase (COX). Matrix translocation requires the recruitment and activation of the presequence translocase-associated import motor (PAM). In the working model illustrated here, Pam17 displaces Tim21 from the TIM23 core complex and together with Tim44 triggers the binding of the Pam16–Pam18 module. Pam17 itself is released again during later assembly steps of PAM. In the fully active PAM machinery, mtHsp70 cooperates with Pam16–Pam18, Tim44, and Mge1 to mediate ATP-driven import of preproteins into the mitochondrial matrix. TIM23 complex: Tim17–Tim23 (Tim17–23), Tim50 (50), Tim21 (21); PAM: Tim44 (44), Pam16 (16), Pam17 (17), Pam18 (18); IMS, intermembrane space Copied from Dudek et al. (2013) with due copyright permission

Besides acting upon transport of ions like sodiums, potassiums, etc., pumps are also found to act upon drug transport phenomena. In a recent review, Sun et al. summarized the pharmacological exploitation of efflux pump inhibitors as a promising antidrug resistance intervention (Sun et al. 2014). Efflux pumps are found in almost all bacterial species. Genes encoding this class of proteins can be located on chromosomes or plasmids (Poole 2007; Piddock 2006). Bacterial efflux pumps are classified into five families:

(i) the resistance-nodulation-division (RND) family,
(ii) the major facilitator superfamily (MFS),
(iii) the adenosine triphosphate (ATP)-binding cassette (ABC) superfamily,
(iv) the small multidrug resistance (SMR) family (a member of the much larger drug/metabolite transporter (DMT) superfamily), and
(v) the multidrug and toxic compound extrusion (MATE) family (Poole 2007; Piddock 2006; Putman et al. 2000).

Except for the RND superfamily, found only in gram-negative bacteria, efflux systems of the other four families: MFS, ABC, SMR, and MATE are widely distributed in both gram-positive and gram-negative bacteria (Handzlik et al. 2013).

Depending on the specific classes they belong to, efflux pumps are either single-component transporters or multiple-component systems. They contain not only an inner membrane transporter, but also an outer membrane channel and a periplasmic adaptor protein, such as the RND-type efflux pumps (Li and Nikaido 2004). RND family-based pumps have been found to be associated extensively with clinically significant antibiotic resistance, such as AcrB in *E. coli* and *Salmonella typhimurium* and MexB in *Pseudomonas aeruginosa.* In gram-positive bacteria, the clinically significant efflux pumps are members of the MFS family, for example NorA in *Staphylococcus aureus* and PmrA in *Streptococcus pneumoniae* (Piddock 2006). Here as a general demonstration, we present in Fig. 5.22 the structure of AcrB and the functional rotation mechanism of the drug extrusion by the RND efflux system AcrAB-TolC. Only one of the three protomers bound with the substrate minocycline or doxorubicin (see Fig. 5.22a). The substrate binding pocket in the protomer is located in the porter domain, which is consisted of four subdomains, PN1, PN2, PC1, and PC2. In the protomer, the pore-forming helix of PN1 is inclined and blocks the potential exit from the substrate-binding pocket (see in Fig. 5.22a, red). The access protomer exists as the loose (L) conformation, because it possesses a vacant binding site, waiting for the binding of the second substrate. Based on these information, the three-step functional rotation mechanism of drug

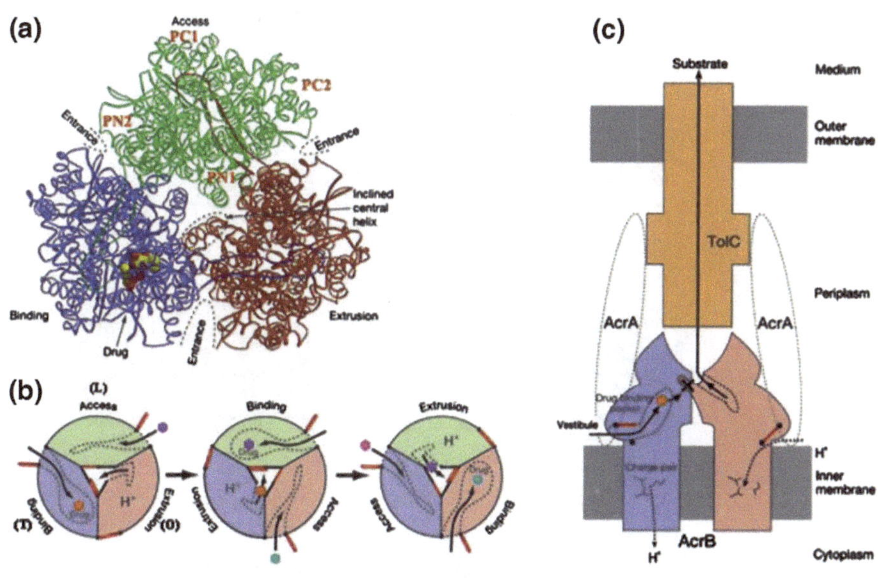


**Fig. 5.22** Structure of AcrB and the functional rotation mechanism of the drug extrusion by the RND efflux system AcrAB-TolC. The three protomers are colored as green, blue, and red, respectively. **a** Top view of a ribbon representation with a drug (yellow and red) in the binding protomer. **b** Top view from the distal side of the cell. Drugs are illustrated as hexagons. Red lines represent the entrance and exit sites of each protomer. The drug-binding pocket and translocation pathway are in dotted lines. **c** Side view of the AcrAB-TolC efflux system parallel to the membrane plane. Drugs are illustrated as hexagons. Red lines represent the entrance and exit sites of each protomer. The route of substrate access, binding, and extrusion is shown Copied with permission from Sun et al. (2014)

export was proposed (Murakami et al. 2006). In the first step or the access (or loose, L) state (see Fig. 5.22b, c), a 'vestibule' near the entrance is open to the periplasm and allows potential substrates to access.

Recently, the structure of the AcrAB-TolC multidrug efflux pump was demonstrated (Du et al. 2014). In gram-negative bacteria, many of the transporters form multicomponent 'pumps' that span both inner and outer membranes and are driven energetically by a primary or secondary transporter component (Du et al. 2013; Koronakis et al. 2000; Murakami et al. 2002, 2006; Mikolosko et al. 2006; Seeger et al. 2006; Eicher et al. 2012). A model system for such a pump is the acridine resistance complex of *E. coli* (Du et al. 2013). This pump assembly comprises the outer membrane channel TolC, the secondary transporter AcrB located in the inner membrane, and the periplasmic AcrA, which bridges these two integral MPs. The AcrAB-TolC efflux pump is able to vectorially transport a diverse array of compounds and confers resistance to a broad spectrum of antibiotics. Crystal structures are available for the individual pump components (Koronakis et al. 2000; Murakami et al. 2002, 2006; Mikolosko et al. 2006; Seeger et al. 2006; Eicher et al. 2012). These provided insights into substrate recognition, energy coupling, and the transduction of conformational changes associated with the transport process. But how the subunits are organised in the pump, their stoichiometry, and the details of their interactions are not known clearly. Du et al. (2014) presented the pseudoatomic structure of a complete multidrug efflux pump in complex with a modulatory protein partner (Hobbs et al. 2012). For details, see Figs. 5.23, 5.24 and 5.25. The model

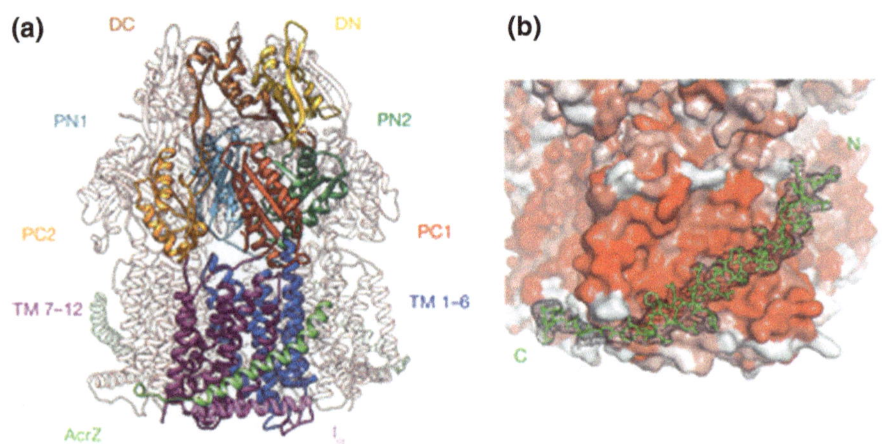


**Fig. 5.23** The crystal structure of the inner-MP AcrB in complex with AcrZ. The crystal structure of the inner-MP AcrB in complex with AcrZ. **a** Overall view of the AcrB/AcrZ trimer. The subdomains of an AcrB protomer are labelled: PN1, PN2, PC1, PC2, DN, and DC; the transmembrane helices are labelled TM1-6, TM7-12, and the linking helix is labelled Iα. AcrB contains a structural repeat, and the structural units comprise PN1-PN2-DN-TM1-6 and PC1-PC2-DC-TM7-12. The protein was crystallized with a DARPin, and one is engaged to the AcrB protomer in the crystallographic asymmetric unit (not shown for clarity). **b** Electron density map with coefficients 2Fo-Fc overlaid on the model for AcrZ (green sticks), the contour level is 1 σ. AcrB is shown in surface representation, with residues colored according to their relative hydrophobicity (hydrophobic in red, hydrophilic in white) (cf. Du et al. 2014) with due permission

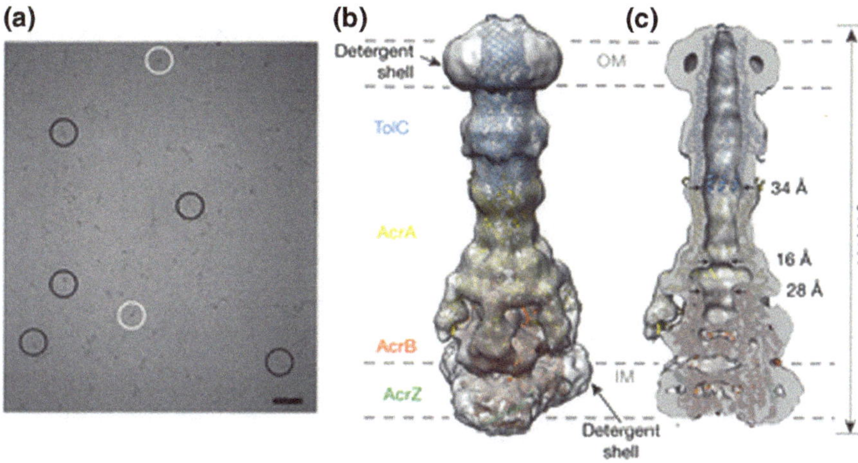


**Fig. 5.24** Cryo-EM image and a pseudoatomic model of the drug efflux pump. **a** A representative raw image of the purified pump. White circles indicate particles with long axis aligned close to the normal to the plane of view; black circles show particles with the long axis parallel to the viewing plane. The scale bar shows 50 nm. **b** The reconstructed map and pseudoatomic model. TolC and AcrBZ are homotrimers that engage six protomers of AcrA to form an assembly with a protein mass of 771 kDa. AcrA has a composite structure of four linearly discontinuous domains connected with flexible linkers. **c** A slice through the reconstruction and model, which shows the continuous conduit that runs from the AcrB funnel through the TolC porin domain (cf . Du et al. 2014) with due permission

defines the quaternary organization of the pump, identifies key domain interactions and suggests a cooperative process for channel assembly and opening. These findings illuminate the basis for drug resistance in numerous pathogenic bacterial species.

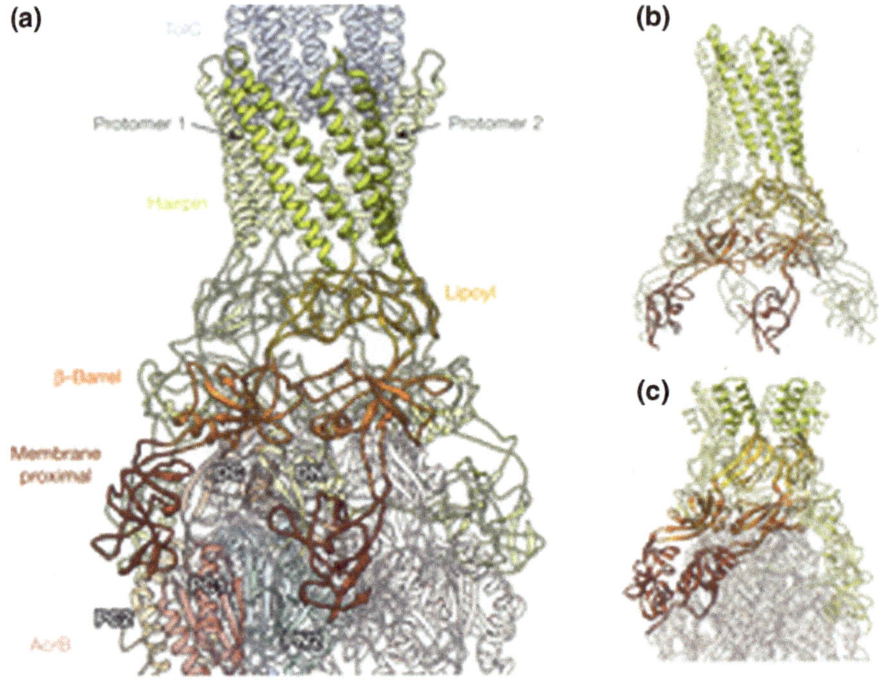


**Fig. 5.25** Modelled interactions of the efflux pump components. **a** Interactions of the AcrA hairpin domain with TolC and the AcrA β-barrel and membrane-proximal domains with AcrB. The lipoyl domains principally interact with each other and make no interactions with AcrB or TolC. Two AcrA protomers in the hexameric ring are shown to illustrate the domain–domain interactions. **b** Homohexameric subunit organization in the structure of the AcrA homologue, *Actinobacillus actinomycetemcomitans* MacA (PDB ID: 4DK0). **c** Interactions of *E. coli* CusA with CusB from the metal-efflux pump15 (cf. Du et al. 2014) with due permission

## References

Wilson, E.B. The Cell in Development and Heredity. 3rd ed. New York: The Macmillan Company; 1925

Taylor, E.W. E.B. Wilson Lecture: The Cell as Molecular Machine. Mol Biol Cell. 2001 Feb; 12 (2): 251–254

Dunkle J.A., Cate J.H. (2010) Ribosome structure and dynamics during translocation and termination. Annu. Rev. Biophys. 39:227–244

Shoji S., Walker S.E., Fredrick K. (2009) Ribosomal translocation: one step closer to the molecular mechanism. ACS Chem. Biol. 4:93–107

Frank J. Jr., Gonzalez R.L. (2010) Structure and dynamics of a processive Brownian motor: the translating ribosome. Annu. Rev. Biochem. 79:381–412

Marina V. Rodnina, Wolfgang Wintermeyer. The ribosome as a molecular machine: the mechanism of tRNA–mRNA movement in translocation. Biochemical Society Transactions Apr 01, 2011, 39 (2) 658–662; https://doi.org/10.1042/bst0390658

Shoji S., Walker S.E., Fredrick K. 2006. Reverse translocation of tRNA in the ribosome. Mol. Cell 24:931–942

Fredrick K., Noller H.F. 2003. Catalysis of ribosomal translocation by sparsomycin. Science 300:1159–1162

Gavrilova L.P., Kostiashkina O.E., Koteliansky V.E., Rutkevitch N.M., Spirin A.S. (1976). Factor-free ('non-enzymic') and factor-dependent systems of translation of polyuridylic acid by Escherichia coli ribosomes. J. Mol. Biol. 101:537–552

Konevega A.L., Fischer N., Semenkov Y.P., Stark H., Wintermeyer W., Rodnina M.V. (2007) Spontaneous reverse movement of mRNA-bound tRNA through the ribosome. Nat. Struct. Mol. Biol. 14:318–324

Cornish P.V., Ermolenko D.N., Noller H.F., Ha T. (2008). Spontaneous intersubunit rotation in single ribosomes. Mol. Cell 30:578–588

Blanchard S.C., Kim H.D., Gonzalez R.L. Jr., Puglisi J.D., Chu S. (2004). tRNA dynamics on the ribosome during translation. Proc. Natl. Acad. Sci. U.S.A. 101:12893–12898

Fischer N., Konevega A.L., Wintermeyer W., Rodnina M.V., Stark H. (2010). Ribosome dynamics and tRNA movement by time-resolved electron cryomicroscopy. Nature 466:329–333

J. Frank. Molecular Machines in Biology. Workshop of the Cell. Editor: Joachim Frank, December, 2011 Cambridge University Press

Bruce Alberts, Alexander Johnson, Julian Lewis, Martin Raff, Keith Roberts, and Peter Walter. Molecular Biology of the Cell. 4th edition. New York: Garland Science; 2002. ISBN-10: 0-8153-3218-1ISBN-10: 0-8153-4072-9

Nobutaka Hirokawa, Yasuko Noda, Yosuke Tanaka & Shinsuke Niwa. Kinesin superfamily motor proteins and intracellular transport. 2009. Nature Reviews Molecular Cell Biology 10, 682–696

O'Connell C, B., M. J. Tyska, and M. S. Mooseker. (2007). Myosin at work: Motor adaptations for a variety of cellular functions. Biochim Biophys Acta 1773(5), 615–630

J.D. Jontes, R.A. Milligan, T.D. Pollard, E.M. Ostap. Kinetic characterization of brush border myosin-I ATPase Proc. Natl. Acad. Sci. U. S. A., 94 (1997), pp. 14332–14337

E.M. Ostap, T.D. Pollard. Biochemical kinetic characterization of the Acanthamoeba myosin-I ATPase. J. Cell Biol., 132 (1996), pp. 1053–1060

J.S. Wolenski, S.M. Hayden, P. Forscher, M.S. Mooseker, Calcium-calmodulin and regulation of brush border myosin-I MgATPase and mechanochemistry, 122 (1993) 613–621

K. Collins, J.R. Sellers, P. Matsudaira. Calmodulin dissociation regulates brush border myosin I (110-kD-calmodulin) mechanochemical activity in vitro. J. Cell Biol., 110 (1990), pp. 1137–1147

M.S. Mooseker, T.R. Coleman. The 110-kD protein–calmodulin complex of the intestinal microvillus (brush border myosin I) is a mechanoenzyme. J. Cell Biol., 108 (1989), pp. 2395–2400

M. El Mezgueldi, N. Tang, S.S. Rosenfeld, E.M. Ostap. The kinetic mechanism of Myo1e (human myosin-IC). J. Biol. Chem., 277 (2002), pp. 21514–21521

T. Lin, N. Tang, E.M. Ostap. Biochemical and motile properties of Myo1b splice isoforms. J. Biol. Chem., 280 (2005), pp. 41562–41567

D.E. Harris, D.M. Warshaw. Smooth and skeletal muscle myosin both exhibit low duty cycles at zero load in vitro. J. Biol. Chem., 268 (1993), pp. 14764–14768

T.Q. Uyeda, S.J. Kron, J.A. Spudich. Myosin step size. Estimation from slow sliding movement of actin over low densities of heavy meromyosin. J. Mol. Biol., 214 (1990), pp. 699–710

D.M. Warshaw, J.M. Desrosiers, S.S. Work, K.M. Trybus. Smooth muscle myosin cross-bridge interactions modulate actin filament sliding velocity in vitro. J. Cell Biol., 111 (1990), pp. 453–463

M. Kovacs, F. Wang, A. Hu, Y. Zhang, J.R. Sellers. Functional divergence of human cytoplasmic myosin II: kinetic characterization of the non-muscle IIA isoform. J. Biol. Chem., 278 (2003), pp. 38132–38140

F. Wang, E.V. Harvey, M.A. Conti, D. Wei, J.R. Sellers. A conserved negatively charged amino acid modulates function in human nonmuscle myosin IIA. Biochemistry, 39 (2000), pp. 5555–5560

S.S. Rosenfeld, J. Xing, L.Q. Chen, H.L. Sweeney. Myosin IIb is unconventionally conventional. J. Biol. Chem., 278 (2003), pp. 27449–27455

F. Wang, M. Kovacs, A. Hu, J. Limouze, E.V. Harvey, J.R. Sellers. Kinetic mechanism of non-muscle myosin IIB: functional adaptations for tension generation and maintenance. J. Biol. Chem., 278 (2003), pp. 27439–27448

M.D. Pato, J.R. Sellers, Y.A. Preston, E.V. Harvey, R.S. Adelstein. Baculovirus expression of chicken nonmuscle heavy meromyosin II-B. Characterization of alternatively spliced isoforms. J. Biol. Chem., 271 (1996), pp. 2689–2695

S. Komaba, A. Inoue, S. Maruta, H. Hosoya, M. Ikebe. Determination of human myosin III as a motor protein having a protein kinase activity. J. Biol. Chem., 278 (2003), pp. 21352–21360

E.M. De La Cruz, A.L. Wells, S.S. Rosenfeld, E.M. Ostap, H.L. Sweeney. The kinetic mechanism of myosin V. Proc. Natl. Acad. Sci. U. S. A., 96 (1999), pp. 13726–13731

A.D. Mehta, R.S. Rock, M. Rief, S.A. Spudich, M.S. Mooseker, R.E. Cheney. Myosin-V is a processive actin-based motor. Nature, 400 (1999), pp. 590–593

E.B. Krementsova, A.R. Hodges, H. Lu, K.M. Trybus. Processivity of chimeric class V myosins. J. Biol. Chem. (2006), 281(9):6079–86

R.E. Cheney, M.K. O'Shea, J.E. Heuser, M.V. Coelho, J.S. Wolenski, E.M. Espreafico, P. Forscher, R.E. Larson, M.S. Mooseker. Brain myosin-V is a two-headed unconventional myosin with motor activity. Cell, 75 (1993), pp. 13–23

S. Watanabe, K. Mabuchi, R. Ikebe, M. Ikebe. Mechanoenzymatic characterization of human Myosin vb. Biochemistry, 45 (2006), pp. 2729–2738

S.L. Reck-Peterson, M.J. Tyska, P.J. Novick, M.S. Mooseker. The yeast class V myosins, Myo2p and Myo4p, are nonprocessive actin-based motors. J. Cell Biol., 153 (2001), pp. 1121–1126

J. Toth, M. Kovacs, F. Wang, L. Nyitray, J.R. Sellers. Myosin V from Drosophila reveals diversity of motor mechanisms within the myosin V family. J. Biol. Chem., 280 (2005), pp. 30594–30603

E.M. De La Cruz, E.M. Ostap, H.L. Sweeney. Kinetic mechanism and regulation of myosin VI. J. Biol. Chem., 276 (2001), pp. 32373–32381

C.A. Morris, A.L. Wells, Z. Yang, L.Q. Chen, C.V. Baldacchino, H.L. Sweeney. Calcium functionally uncouples the heads of myosin VI. J. Biol. Chem., 278 (2003), pp. 23324–23330

A.L. Wells, A.W. Lin, L.Q. Chen, D. Safer, S.M. Cain, T. Hasson, B.O. Carragher, R.A. Milligan, H.L. Sweeney. Myosin VI is an actin-based motor that moves backwards. Nature, 401 (1999), pp. 505–508

M. Yoshimura, K. Homma, J. Saito, A. Inoue, R. Ikebe, M. Ikebe. Dual regulation of mammalian myosin VI motor function. J. Biol. Chem., 276 (2001), pp. 39600–39607

S. Watanabe, R. Ikebe, M. Ikebe. Drosophila myosin VIIA is a high duty ratio motor with a unique kinetic mechanism. J. Biol. Chem., 281 (2006), pp. 7151–7160

A. Inoue, M. Ikebe. Characterization of the motor activity of mammalian myosin VIIA. J. Biol. Chem., 278 (2003), pp. 5478–5487

I.P. Udovichenko, D. Gibbs, D.S. Williams. Actin-based motor properties of native myosin VIIa. J. Cell. Sci., 115 (2002), pp. 445–450

A. Henn, E.M. De La Cruz. Vertebrate myosin VIIb is a high duty ratio motor adapted for generating and maintaining tension. J. Biol. Chem., 280 (2005), pp. 39665–39676

Y. Yang, M. Kovacs, Q. Xu, J.B. Anderson, J.R. Sellers. Myosin VIIB from Drosophila is a high duty ratio motor. J. Biol. Chem., 280 (2005), pp. 32061–32068

P.L. Post, M.J. Tyska, C.B. O'Connell, K. Johung, A. Hayward, M.S. Mooseker. Myosin-IXb is a single-headed and processive motor. J. Biol. Chem., 277 (2002), pp. 11679–11683

A. Inoue, J. Saito, R. Ikebe, M. Ikebe. Myosin IXb is a single-headed minus-end-directed processive motor. Nat. Cell Biol., 4 (2002), pp. 302–306

V. Nalavadi, M. Nyitrai, C. Bertolini, N. Adamek, M. Geeves, M. Bahler. Kinetic mechanism of myosin IXB and the contributions of two class lX specific regions. J Biol Chem. 2005 Nov 25;280(47):38957–68

T. Kambara, M. Ikebe. A unique ATP hydrolysis mechanism of single-headed processive myosin, myosin IX. J Biol Chem. 2006 Feb 24;281(8):4949–57

P.L. Post, G.M. Bokoch, M.S. Mooseker, Human myosin-IXb is a mechanochemically active motor and a GAP for rho, 111 (1998) 941–950

M. Nishikawa, S. Nishikawa, A. Inoue, A.H. Iwane, T. Yanagida, M. Ikebe. A unique mechanism for the processive movement of single-headed myosin-IX. Biochem. Biophys. Res. Commun., 343 (2006), pp. 1159–1164

C.B. O'Connell, M.S. Mooseker. Native myosin-IXb is a plus-, not a minus-end-directed motor. Nat. Cell Biol., 5 (2003), pp. 171–172

M. Kovacs, F. Wang, J.R. Sellers. Mechanism of action of myosin X, a membrane-associated molecular motor. J. Biol. Chem., 280 (2005), pp. 15071–15083

K. Homma, M. Ikebe. Myosin X is a high duty ratio motor. J. Biol. Chem., 280 (2005), pp. 29381–29391

K. Homma, J. Saito, R. Ikebe, M. Ikebe. Motor function and regulation of myosin X. J. Biol. Chem., 276 (2001), pp. 34348–34354

M. Tominaga, H. Kojima, E. Yokota, H. Orii, R. Nakamori, E. Katayama, M. Anson, T. Shimmen, K. Oiwa. Higher plant myosin XI moves processively on actin with 35 nm steps at high velocity. EMBO J., 22 (2003), pp. 1263–1272

A. Herm-Gotz, S. Weiss, R. Stratmann, S. Fujita-Becker, C. Ruff, E. Meyhofer, T. Soldati, D. J. Manstein, M.A. Geeves, D. Soldati. Toxoplasma gondii myosin A and its light chain: a fast, single-headed, plus-end-directed motor. EMBO J., 21 (2002), pp. 2149–2158

S.J. Kron, J.A. Spudich. Fluorescent actin filaments move on myosin fixed to a glass surface. Proc. Natl. Acad. Sci. U. S. A., 83 (1986), pp. 6272–6276

Qing Lu, Jianchao Li, and Mingjie Zhang. Cargo Recognition and Cargo-Mediated Regulation of Unconventional Myosins. Acc. Chem. Res., 2014, 47 (10), pp 3061–3070

Hirano Y, Hatano T, Takahashi A, Toriyama M, Inagaki N, Hakoshima T. Structural basis of cargo recognition by the myosin-X MyTH4–FERM domain. The EMBO Journal (2011) 30, 2734–2747

Richards TA, Cavalier-Smith T (2005) Myosin domain evolution and the primary divergence of eukaryotes. Nature 436: 1113–1118

Yu C, Feng W, Wei Z, Miyanoiri Y, Wen W, Zhao Y, Zhang M (2009) Myosin VI undergoes cargo-mediated dimerization. Cell 138: 537–548

Weber KL, Sokac AM, Berg JS, Cheney RE, Bement WM (2004) A microtubule-binding myosin required for nuclear anchoring and spindle assembly. Nature 431: 325–329

Woolner S, O'Brien LL, Wiese C, Bement WM (2008) Myosin-10 and actin filaments are essential for mitotic spindle function. J Cell Biol 182: 77–88

Toyoshima F, Nishida E (2007) Integrin-mediated adhesion orients the spindle parallel to the substratum in an EB1- and myosin X-dependent manner. EMBO J 26: 1487–1498

Hamada K, Shimizu T, Yonemura S, Tsukita S, Tsukita S, Hakoshima T (2003) Structural basis of adhesion-molecule recognition by ERM proteins revealed by the crystal structure of the radixin-ICAM-2 complex. EMBO J 22: 502–514

Takai Y, Kitano K, Terawaki S, Maesaki R, Hakoshima T (2007) Structural basis of PSGL-1 binding to ERM proteins. Genes Cell 12: 1329–1338

Takai Y, Kitano K, Terawaki S, Maesaki R, Hakoshima T (2008) Structural basis of the cytoplasmic tail of adhesion molecule CD43 and its binding to ERM proteins. J Mol Biol 381: 634–644

Terawaki S, Kitano K, Hakoshima T (2007) Structural basis for type II membrane protein binding by ERM proteins revealed by the radixin-neutral endopeptidase 24.11 (NEP) complex. J Biol Chem 282: 19854–19861

Mori T, Kitano K, Fukami Y, Terawaki S, Hakoshima T (2008) Structural properties of the cytoplasmic tail of adhesion molecule CD44 and its binding to FERM proteins. J Biol Chem 283: 29602–29612

Terawaki S, Maesaki R, Hakoshima T (2006) Structural basis of NHERF recognition by ERM proteins. Structure 14: 777–789

Okada Y, Hirokawa N (2000) Mechanism of the single-headed processivity: diffusional anchoring between the K-loop of kinesin and the C terminus of tubulin. Proc Natl Acad Sci USA 97: 640–645

Hirokawa N, Nitta R, Okada Y (2009) The mechanisms of kinesin motor motility: lessons from the monomeric motor KIF1A. Nat Rev Mol Cell Biol 10: 877–884

Honnappa S, Okhrimenko O, Jaussi R, Jawhari H, Jelesarov I, Winkler FK, Steinmetz MO (2006) Key interaction modes of dynamic +TIP networks. Mol Cell 23: 663–671

Mishima M, Maesaki R, Kasa M, Watanabe T, Fukata M, Kaibuchi K, Hakoshima T (2007) Structural basis for tubulin recognition by cytoplasmic linker protein 170 and its autoinhibition. Proc Natl Acad Sci USA 104: 10346–10351

Lodish H, Berk A, Zipursky SL, et al. Kinesin, Dynein, and Intracellular Transport. Molecular Cell Biology. 4th edition. New York: W. H. Freeman; 2000

E. Lindahl. Unraveling the strokes of ion channel molecular machines in computers. Proc Natl Acad Sci U S A. 2012 Dec 26; 109(52): 21186–21187

Borjesson SI, Elinder F. (2008). Structure, function, and modification of the voltage sensor in voltage-gated ion channels. Cell Biochem Biophys 52:149–74

Schoppa NE, et al. (1992) The size of gating charge in wild-type and mutant Shaker potassium channels. Science 255:1712–1715.

Zagotta WN, Hoshi T, Aldrich RW (1994) Shaker potassium channel gating III: Evaluation of kinetic models for activation. J Gen Physiol 103:321–62

Keynes RD, Elinder F (1998) Modelling the activation, opening, inactivation and reopening of the voltage-gated sodium channel. Proc Biol Sci 265:263–70

Long SB, et al. (2007) Atomic structure of a voltage-dependent K+ channel in a lipid membrane-like environment. Nature 450:376–382

Papazian DM, et al. (1995) Electrostatic interactions of S4 voltage sensor in Shaker K+ channel. Neuron 14:1293–301

DeCaen PG, et al. (2008) Disulfide locking a sodium channel voltage sensor reveals ion pair formation during activation. Proc Natl Acad Sci USA 105:15142–15147

DeCaen PG, et al. (2009) Sequential formation of ion pairs during activation of a sodium channel voltage sensor. Proc Natl Acad Sci USA 106:22498–22503

Villalba-Galea CA, et al. (2008) S4-based voltage sensors have three major conformations. Proc Natl Acad Sci USA 105:17600–17607

Murata Y, et al. (2005) Phosphoinositide phosphatase activity coupled to an intrinsic voltage sensor. Nature 435:1239–1243

Sasaki M, Takagi M, Okamura Y (2006) A voltage sensor-domain protein is a voltage-gated proton channel. Science 312:589–592

Ramsey IS, et al. (2006) A voltage-gated proton-selective channel lacking the pore domain. Nature 440:1213–1216

Schoppa NE, Sigworth FJ (1998) Activation of Shaker potassium channels. III. An activation gating model for wild-type and V2 mutant channels. J Gen Physiol 111:313–342

Larsson H, Elinder F (2000) A conserved glutamate is important for slow inactivation in K+ channels. Neuron 27:573–583

Broomand A, Elinder F (2008) Large-scale movement within the voltage-sensor paddle of a potassium channel-support for a helical-screw motion. Neuron 59:770–777

Webster SM, et al. (2004) Intracellular gate opening in Shaker K+ channels defined by high-affinity metal bridges. Nature 428:864–868

Campos FV, et al. (2007) Two atomic constraints unambiguously position the S4 segment relative to S1 and S2 segments in the closed state of Shaker K channel. Proc Natl Acad Sci USA 104:7904–7909

Li M, et al. (2011) Gating the pore of P2X receptor channels. Nat Neurosci 11:883–887

Lin MC, et al. (2011) R1 in the Shaker S4 occupies the gating charge transfer center in the resting state. J Gen Physiol 138:155–163

Rulisek L, Vondrasek J (1998) Coordination geometries of selected transition metal ions (Co2+, Ni2+, Cu2+, Zn2+, Cd2+, and Hg2+) in metalloproteins. J Inorg Biochem 71:115–127

Ulrike Henriona, Jakob Renhorna, Sara I. Börjessona, Erin M. Nelsona, Christine S. Schwaigerb, Pär Bjelkmarb,c, Björn Wallnerd, Erik Lindahlb,c,1, and Fredrik Elindera. Tracking a complete voltage-sensor cycle with metal-ion bridges. Proc Natl Acad Sci USA. 2012;109(22):8552–8557

Cole KS, Moore JW (1960) Potassium ion current in the squid giant axon: dynamic characteristic. Biophys J 1:1–14

Khalili-Araghi F, et al. (2012) Molecular Dynamics Investigation of the omega-Current in the Kv1.2 Voltage Sensor Domains. Biophys J 102:258–267

Lucie Delemotte, Marina A. Kasimova, Michael L. Klein, Mounir Tarek, and Vincenzo Carnevale. 2015. Free-energy landscape of ion-channel voltage-sensor–domain activation. vol. 112, 124–129

Barducci A, Bussi G, Parrinello M (2008) Well-tempered metadynamics: A smoothly converging and tunable free-energy method. Phys Rev Lett 100(2):020603

Vidyaramanan Ganesan, Timothy Walsh, Kai-Ti Chang, and Marco Colombini. The Dynamics of Bax Channel Formation: Influence of Ionic Strength. Biophys J. 103, 2012, page: 483–491

Unwin, P.N.T., and Ennis, P.D. 1984. Two configurations of a channel-forming membrane protein. Nature 307: 609–613

Perozo, E., Cortes, D.M., and Cuello, L.G. 1999. Structural Rearrangements Underlying K+-Channel Activation Gating. Science 285: 73–78

Perozo, E., Cortes, D.M., Sompornpisut, P., Kloda, A., and Martinac, B. 2002. Open channel structure of MscL and the gating mechanism of mechanosensitive channels. Nature 418: 942–948

Toyoshima, C., and Mizutani, T. 2004. Crystal structure of the calcium pump with a bound ATP analogue. Nature 430: 529–535

Israelachvili, J.N. 1977. Refinement of the fluid-mosaic model of membrane structure. Biochim. Biophys. Acta 469: 221–225

Sackmann, E. 1984. In Biological Membranes, edited by D. Chapman (London: Academic): 105

Gruner, S.M. 1991. Lipid membrane curvature elasticity and protein function in Biologically Inspired Physics, edited by L. Peliti (New York: Plenum): 127–135

Andersen, O.S., Sawyer, D.B., and Koeppe II, R.E. 1992. Biomembrane Structure and Function, edited by K. R. K. Easwaran and B. Gaber (Schenectady, New York: Adenine): 227–244

Brown, M.F. 1994. Modulation of rhodopsin function by properties of the membrane bilayer. Chem. Phys. Lipids 73: 159–180

Helfrich, W. 1973. Elastic properties of lipid bilayers: theory and possible experiments. Z. Naturforsch. 28C: 693–703

Ashrafuzzaman, Md.; Tuszynski, J. Regulation of Channel Function Due to Coupling with a Lipid Bilayer. Journal of Computational and Theoretical Nanoscience, Volume 9, Number 4, April 2012a, pp. 564–570

Ashrafuzzaman, Md, Tuszynski, J. Membrane Biophysics. Springer-Verlag, 2012 (b). ISBN 978-3-642-16105-6

Md. Ashrafuzzaman and J. A. Tuszynski. Ion pore formation in membranes due to complex interactions between lipids and antimicrobial peptides or biomolecules. Handbook of Nanoscience, Engineering, and Technology, Third Edition. Editors: William A Goddard III, Donald W. Brenner, Sergey E. Lyshevski, and Gerald J. Iafrate. CRC Press, 2012 (c); Pages 893–934

Sahar Moradi-Monfareda, Vikram Krishnamurthya, Bruce Cornell. A molecular machine biosensor: Construction, predictive models and experimental studies. Biosensors and Bioelectronics. Volume 34, Issue 1, 15 April 2012, Pages 261–266

Branton, D.; Deamer, D. W.; Marziali, A.; Bayley, H.; Benner, S. A.; Butler, T.; Di Ventra, M.; Garaj, S.; Hibbs, A.; Huang, X.; Jovanovich, S. B.; Krstic, P. S.; Lindsay, S.; Ling, X. S.; Mastrangelo, C. H.; Meller, A.; Oliver, J. S.; Pershin, Y. V.; Ramsey, J. M.; Riehn, R.; Soni, G. V.; Tabard-Cossa, V.; Wanunu, M.; Wiggin, M.; Schloss, J. A. The potential and challenges of nanopore sequencing. Nat. Biotechnol. 2008, 26, 1146–1153

Kocer, A.; Walko, M.; Meijberg, W.; Feringa, B. L.A Light-actuated Nanovalve Derived from a Channel Protein Science 2005, 309, 755–758

Sukharev, S.; Anishkin, A. Mechanosensitive channels: what can we learn from 'simple' model systems? Trends Neurosci. 2004, 27, 345–351

Louhivuori, M.; Risselada, H. J.; van der Giessen, E.; Marrink, S. J. Release of content through mechano-sensitive gates in pressurized liposomes Proc. Natl. Acad. Sci. U. S. A. 2010, 107, 19856–19860

Steinberg-Yfrach, G.; Rigaud, J.-L.; Durantini, E. N.; Moore, A. L.; Gust, D.; Moore, T. A. Light-driven production of ATP catalysed by F0F1-ATP synthase in an artificial photosynthetic membrane Nature 1998, 392, 479–482

Banghart, M.; Borges, K.; Isacoff, E.; Trauner, D.; Kramer, R. H. Light-activated ion channels for remote control of neuronal firing Nat. Neurosci. 2004, 7, 1381–1386

Volgraf, M.; Gorostiza, P.; Numano, R.; Kramer, R. H.; Isacoff, E. Y.; Trauner, D. Allosteric control of an ionotropic glutamate receptor with an optical switch Nat. Chem. Biol. 2006, 2, 47–52

Stein, M.; Breit, A.; Fehrentz, T.; Gudermann, T.; Trauner, D. Optical Control of TRPV1 Channels Angew. Chem., Int. Ed. 2013, 52, 9845–9848

Schoenberger, M.; Damijonaitis, A.; Zhang, Z. N.; Nagel, D.; Trauner, D. Development of a New Photochromic Ion Channel Blocker via Azologization of Fomocaine ACS Chem. Neurosci. 2014, 5, 514–518

Schonberger, M.; Althaus, M.; Fronius, M.; Clauss, W.; Trauner, D. Controlling epithelial sodium channels with light using photoswitchable amilorides Nat. Chem. 2014, 6, 712–719

Kramer, R. H.; Chambers, J. J.; Trauner, D. Photochemical tools for remote control of ion channels in excitable cells Nat. Chem. Biol. 2005, 1, 360–365

Banghart, M. R.; Mourot, A.; Fortin, D. L.; Yao, J. Z.; Kramer, R. H.; Trauner, D. Photochromic Blockers of Voltage-Gated Potassium Channels Angew. Chem., Int. Ed. 2009, 48, 9097–9101

Mourot, A.; Fehrentz, T.; Le Feuvre, Y.; Smith, C. M.; Herold, C.; Dalkara, D.; Nagy, F.; Trauner, D.; Kramer, R. H. Rapid optical control of nociception with an ion-channel photoswitch Nat. Methods 2012, 9, 396–402

Szymanski, W.; Beierle, J. M.; Kistemaker, H. A.; Velema, W. A.; Feringa, B. L. Reversible Photocontrol of Biological Systems by the Incorporation of Molecular Photoswitches Chem. Rev. 2013, 113, 6114–6178 1778

Lougheed, T.; Borisenko, V.; Hennig, T.; Ruck-Braun, K.; Woolley, G. A. Photomodulation of ionic current through hemithioindigo-modified gramicidin channels Org. Biomol. Chem. 2004, 2, 2798–2801

Banghart, M. R.; Volgraf, M.; Trauner, D. Engineering light-gated ion channels Biochemistry 2006, 45, 15129–15141

Sundus Erbas-Cakmak, David A. Leigh, Charlie T. McTernan, and Alina L. Nussbaumer. Artificial Molecular Machines. Chem. Rev., 2015, 115 (18), pp 10081–10206

Arseniev, A.S., Barsukov, I.L., Bystrov, V.F., and Ovchinnikov, Y.A. 1986. Biol. Membr. 3: 437–62

Ketchem, R.R., Roux, B., and Cross, T.A. 1997. High-resolution polypeptide structure in a lamellar phase lipid environment from solid state NMR derived orientational constraints. Structure 5: 1655–69

Townsley, L.E., Tucker, W.A., Sham, S., and Hinton, J.F. 2001. Structures of gramicidins A, B, and C incorporated into sodium dodecyl sulfate micelles. Biochemistry 40: 11676–11686

O'Connell, A.M., Koeppe II, R.E., and Andersen, O.S. 1990. Kinetics of gramicidin channel formation in lipid bilayers: trans-membrane monomer association. Science 250: 1256–1259

Ashrafuzzaman M, Lampson MA, Greathouse DV, Koeppe II RE, Andersen OS. Manipulating lipid bilayer material properties using biologically active amphipathic molecules. J. Phys.: Condens. Matter 2006; 18: S1235–55

Boheim G. Statistical analysis of alamethicin channels in black lipid membranes. J. Mem. Biol. 1974;19: 277–303

He K, Ludtke SJ, Huang HW, Worcester DL. Antimicrobial peptide pores in membranes detected by neutron in-plane scattering. Biochemistry. 1995;34: 15614–18

Matsuzaki K, Murase O, Tokuda H, Fujii N, Miyajima K. An antimicrobial peptide, magainin 2, induced rapid flip-flop of phospholipids coupled with pore formation and peptide translocation. Biochemistry. 1996;35: 11361–68

Ludtke SJ, He K, Heller WT, Harroun TA, Yang L, Huang HW. Membrane pores induced by magainin. Biochemistry. 1996;35:13723–28

Yang L, Harroun T, Weiss TM, Ding L, Huang HW. Barrel-stave model or toroidal model? A case study on melittin pores. Biophys. J. 2001;81: 1475–85

Sobko, A.A., Kotova, E.A., Antonenko, Y.N., Zakharov, S.D., and Cramer, W.A. 2006. Lipid dependence of the channel properties of a colicin E1-lipid toroidal pore. The J. of Biol. Chem. 281: 14408–16

Md Ashrafuzzaman, M Duszyk and J A Tuszynski. Chemotherapy Drugs Thiocolchicoside and Taxol Permeabilize Lipid Bilayer Membranes by Forming Ion Pores. Journal of Physics: Conference Series (2011) 012029

Ashrafuzzaman, M., Tseng, C.-Y., Duszyk, M., Tuszynski, J. Chemotherapy Drugs Form Ion Pores in Membranes Due to Physical Interactions with Lipids. Chem. Biol. Drug Des., Volume 80, Issue 6, December 2012, Pages 992–1002

L. J. Siskind, A. Davoody, N. Lewin, S. Marshall, and M. Colombini. Enlargement and Contracture of C2-Ceramide Channels. Biophysical Journal 85: 1560–1575 (2003)

A. Anishkin, S. Sukharev, M. Colombini. Searching for the molecular arrangement of transmembrane ceramide channels. Biophys. J. 90: 2414–2426 (2006)

S. Samanta, J. Stiban, T.K. Maugel, M. Colombini. Visualization of ceramide channels by transmission electron microscopy. Biochim. Biophys. Acta 1808: 1196–201 (2011)

Abigail Clements, Dejan Bursac, Xenia Gatsos, Andrew J. Perry, Srgjan Civciristov, Nermin Celik, Vladimir A. Likic, Sebastian Poggio, Christine Jacobs-Wagner, Richard A. Strugnell, and Trevor Lithgow. The reducible complexity of a mitochondrial molecular machine. Proc Natl Acad Sci U S A. 2009 Sep 15; 106(37): 15791–15795

Muller M, Martin W (1999) The genome of Rickettsia prowazekii and some thoughts on the origin of mitochondria and hydrogenosomes. Bioessays 21:377–381

Emelyanov VV (2003) Mitochondrial connection to the origin of the eukaryotic cell. Eur J Biochem 270:1599–1618

Gray MW, Burger G, Lang BF (1999) Mitochondrial evolution. Science 283:1476–1481

Dolezal P, Likic V, Tachezy J, Lithgow T (2006) Evolution of the molecular machines for protein import into mitochondria. Science 313:314–318

Murcha MW, et al. (2007) Characterization of the preprotein and amino acid transporter gene family in Arabidopsis. Plant Physiol 143:199–212

Pfanner N, Chacinska A (2002) The mitochondrial import machinery: preprotein-conducting channels with binding sites for presequences. Biochim Biophys Acta 1592:15–24

Neupert W, Herrmann JM (2007) Translocation of proteins into mitochondria. Annu Rev Biochem 76:723–749

van der Laan M, Rissler M, Rehling P (2006) Mitochondrial preprotein translocases as dynamic molecular machines. FEMS Yeast Res 6:849–861

Bolender N, et al. (2008) Multiple pathways for sorting mitochondrial precursor proteins. EMBO Rep 9:42–49

Mokranjac D, Neupert W (2008) Energetics of protein translocation into mitochondria. Biochim Biophys Acta 1777:758–762

Jan Dudek, Peter Rehling, Martin van der Laan. Mitochondrial protein import: Common principles and physiological networks. Biochimica et Biophysica Acta (BBA) - Molecular Cell Research, Volume 1833, Issue 2, February 2013, Pages 274–285

N. Wiedemann, V. Kozjak, A. Chacinska, B. Schönfisch, S. Rospert, M.T. Ryan, N. Pfanner, C. Meisinger. Machinery for protein sorting and assembly in the mitochondrial outer membrane. Nature, 424 (2003), pp. 565–571

S. Kutik, D. Stojanovski, L. Becker, T. Becker, M. Meinecke, V. Krüger, C. Prinz, C. Meisinger, B. Guiard, R. Wagner, N. Pfanner, N. Wiedemann. Dissecting membrane insertion of mitochondrial $\beta$-barrel proteins. Cell, 132 (2008), pp. 1011–1024

S.A. Paschen, T. Waizenegger, T. Stan, M. Preuss, M. Cyrklaff, K. Hell, D. Rapaport, W. Neupert. Evolutionary conservation of biogenesis of β-barrel membrane proteins. Nature, 426 (2003), pp. 862–866

I. Gentle, K. Gabriel, P. Beech, R. Waller, T. Lithgow. The Omp85 family of proteins is essential for outer membrane biogenesis in mitochondria and bacteria. J. Cell Biol., 164 (2004), pp. 19–24

C. Meisinger, M. Rissler, A. Chacinska, L.K. Sanjuan Szklarz, D. Milenkovic, V. Kozjak, B. Schönfisch, C. Lohaus, H.E. Meyer, M.P. Yaffe, B. Guiard, N. Wiedemann, N. Pfanner. The mitochondrial morphology protein Mdm10 functions in assembly of the preprotein translocase of the outer membrane. Dev. Cell, 7 (2004), pp. 61–71

C. Meisinger, S. Pfannschmidt, M. Rissler, D. Milenkovic, T. Becker, D. Stojanovski, M. J. Youngman, R.E. Jensen, A. Chacinska, B. Guiard, N. Pfanner, N. Wiedemann. The morphology proteins Mdm12/Mmm1 function in the major β-barrel assembly pathway of mitochondria. EMBO J., 26 (2007), pp. 2229–2239

M. van der Laan, M. Meinecke, J. Dudek, D.P. Hutu, M. Lind, I. Perschil, B. Guiard, R. Wagner, N. Pfanner, P. Rehling. Motor-free mitochondrial presequence translocase drives membrane integration of preproteins. Nat. Cell Biol., 9 (2007), pp. 1152–1159

K.N. Truscott, P. Kovermann, A. Geissler, A. Merlin, M. Meijer, A.J. Driessen, J. Rassow, N. Pfanner, R. Wagner. A presequence- and voltage-sensitive channel of the mitochondrial preprotein translocase formed by Tim23. Nat. Struct. Biol., 8 (2001), pp. 1074–1082

M. Meinecke, R. Wagner, P. Kovermann, B. Guiard, D.U. Mick, D.P. Hutu, W. Voos, K.N. Truscott, A. Chacinska, N. Pfanner, P. Rehling. Tim50 maintains the permeability barrier of the mitochondrial inner membrane. Science, 312 (2006), pp. 1523–1526

N.N. Alder, R.E. Jensen, A.E. Johnson. Fluorescence mapping of mitochondrial TIM23 complex reveals a water-facing, substrate-interacting helix surface. Cell, 134 (2008), pp. 439–450

T. Komiya, S. Rospert, C. Koehler, R. Looser, G. Schatz, K. Mihara. Interaction of mitochondrial targeting signals with acidic receptor domains along the protein import pathway: evidence for the 'acid chain' hypothesis. EMBO J., 17 (1998), pp. 3886–3898

M.F. Bauer, C. Sirrenberg, W. Neupert, M. Brunner. Role of Tim23 as voltage sensor and presequence receptor in protein import into mitochondria. Cell, 87 (1996), pp. 33–41

A. Chacinska, M. van der Laan, C.S. Mehnert, B. Guiard, D.U. Mick, D.P. Hutu, K.N. Truscott, N. Wiedemann, C. Meisinger, N. Pfanner, P. Rehling. Distinct forms of mitochondrial TOM–TIM supercomplexes define signal-dependent states of preprotein sorting. Mol. Cell. Biol., 30 (2010), pp. 307–318

A. Chacinska, M. Lind, A.E. Frazier, J. Dudek, C. Meisinger, A. Geissler, A. Sickmann, H.E. Meyer, K.N. Truscott, B. Guiard, N. Pfanner, P. Rehling. Mitochondrial presequence translocase: switching between TOM tethering and motor recruitment involves Tim21 and Tim17. Cell, 120 (2005), pp. 817–829

S. Meier, W. Neupert, J.M. Herrmann. Conserved N-terminal negative charges in the Tim17 subunit of the TIM23 translocase play a critical role in the import of preproteins into mitochondria. J. Biol. Chem., 280 (2005), pp. 7777–7785

S. Martinez-Caballero, S.M. Grigoriev, J.M. Herrmann, M.L. Campo, K.W. Kinnally. Tim17p regulates the twin pore structure and voltage gating of the mitochondrial protein import complex TIM23. J. Biol. Chem., 282 (2007), pp. 3584–3593

Jingjing Sun, Ziqing Deng, Aixin Yan. Bacterial multidrug efflux pumps: Mechanisms, physiology and pharmacological exploitations. Biochemical and Biophysical Research Communications, Volume 453, Issue 2, 17 October 2014, Pages 254–267

K. Poole. Efflux pumps as antimicrobial resistance mechanisms. Ann. Med., 39 (2007), pp. 162–176

L.J. Piddock. Multidrug-resistance efflux pumps? Not just for resistance. Nat. Rev. Microbiol., 4 (2006), pp. 629–636

M. Putman, H.W. van Veen, W.N. Konings. Molecular properties of bacterial multidrug transporters. Microbiol. Mol. Biol. Rev., 64 (2000), pp. 672–693

J. Handzlik, A. Matys, K. Kieć-Kononowicz. Recent advances in multi-drug resistance (MDR) efflux pump inhibitors of Gram-positive bacteria S. aureus. Antibiotics, 2 (2013), pp. 28–45

X.-Z. Li, H. Nikaido. Efflux-mediated drug resistance in bacteria. Drugs, 64 (2004), pp. 159–204

L.J. Piddock. Clinically relevant chromosomally encoded multidrug resistance efflux pumps in bacteria. Clin. Microbiol. Rev., 19 (2006), pp. 382–402

S. Murakami, R. Nakashima, E. Yamashita, T. Matsumoto, A. Yamaguchi. Crystal structures of a multidrug transporter reveal a functionally rotating mechanism. Nature, 443 (2006), pp. 173–179

Dijun Du, Zhao Wang, Nathan R. James, Jarrod E. Voss, Ewa Klimont, Thelma Ohene-Agyei, Henrietta Venter, Wah Chiu, and Ben F. Luisi. Structure of the AcrAB-TolC multidrug efflux pump. Nature. 2014 May 22; 509(7501): 512–515

Du, D., Venter, H., Pos, K. M. & Luisi, B. F. in Microbial Efflux Pumps: Current Research Ch. 3 (Caister Academic, 2013)

Koronakis, V., Sharff, A., Koronakis, E., Luisi, B. F. & Hughes, C. Crystal structure of the bacterial membrane protein TolC central to multidrug efflux and protein export. Nature 405, 914–919 (2000)

Murakami, S., Nakashima, R. & Yamashita, E. Cry & Yamaguchi, A. Crystal structure of bacterial multidrug efflux transporter AcrB. Nature 419, 587–593 (2002)

Mikolosko, J., Bobyk, K., Zgurskaya, H. I. & Ghosh, P. Conformational flexibility in the multidrug efflux system protein AcrA. Structure 14, 577–587 (2006)

Seeger, M. A. et al. Structural asymmetry of AcrB trimer suggests a peristaltic pump mechanism. Science 313, 1295–1298 (2006)

Murakami, S., Nakashima, R., Yamashita, E., Matsumoto, T. & Yamaguchi, A. Crystal structures of a multidrug transporter reveal a functionally rotating mechanism. Nature 443, 173–179 (2006)

Eicher, T. et al. Transport of drugs by the multidrug transporter AcrB involves an access and a deep binding pocket that are separated by a switch-loop. Proc. Natl Acad. Sci. USA 109, 5687–5692 (2012)

Hobbs, E. C., Yin, X., Paul, B. J., Astarita, J. L. & Storz, G. Conserved small protein associates with the multidrug efflux pump AcrB and differentially affects antibiotic resistance. Proc. Natl Acad. Sci. USA 109, 16696–16701 (2012)

# Chapter 6
# Cell Transport at Nanoscale Dimensions

A cell has various components. The classified components are composed of various sub-components. These components and sub-components are either static on average but fluctuating around the equilibrium states, or under continuous movement, so are in dynamic state. They may also diffuse from place to place within the kingdom of cell, freely or across certain structures. The movement always happens on a background, so the viscous friction is always present. The friction is also different in different compartments. Cellular components also allow other cell entering elements, materials, proteins, nutrients, drugs, etc., to get diffused or transported throughout the cell compartments. All these together can be referred to as cell transport phenomena. For example, ions like sodium, potassium, chloride, protons move across the cell membrane. Lipids and membrane proteins (MPs) move within the cell membrane. Cell membrane-based ion channels transport materials. All these cellular transports happen mostly within nanometer (nm) length scale what we may brand as nanoscale transport of cell. For example, any transport across cell membrane happens over a length equal to cell membrane thickness ($\sim$3–5 nm) (Lewis and Engelman 1983; Benz et al. 1975; Simon et al. 1982; Harper et al. 2001). Within cell, the MP transport happens in unit scale of lipid cross-sectional dimension (0.6 $nm^2$) (Eze 1991). Cellular transport therefore can be classified within nm dimension length scale. This is also found for slow plant cell transport. In this chapter, we shall provide an in-depth analysis of cell-based nanoscale transport phenomena and mechanisms. Besides addressing existing hypotheses, some novel approaches will be released.

## 6.1 General Cell Transports and Related Theoretical Models

Transport follows some general features and mechanisms. In cell, transport is also no different and follows not very different formalisms. Whenever something moves, the medium in which the movement happens exerts a viscous friction. The friction



M. Ashrafuzzaman, *Nanoscale Biophysics of the Cell*,
https://doi.org/10.1007/978-3-319-77465-7_6

works against the motion. It is just inverse of another popular word 'mobility.' In kinetic theory where movement happens in a thermodynamic condition, we know from the Einstein–Smoluchowski relation that the diffusion constant $D$ determining the easiness of diffusion in a medium is proportional to the absolute temperature. The proportional parameter is the mobility in the medium. The relation follows the equation:

$$D = \mu k_B T \tag{6.1}$$

Here, $k_B$ is the Boltzmann's constant, $T$ is the absolute temperature, and $\mu$ is the mobility.

In Chap. 1, we have explained how cellular compartments may get classified into various physical phase states; namely, the major ones are liquid, plasma, and solid phase states. Every phase state may also contain several sub-phase states within due to mainly the space-dependent fluctuation of density profiles of the constituents creating a specific phase state. The real challenge of addressing diffusion of cellular nutrients across these three major phase states or other sub-phase states within the major phase states lies in determining acceptable values for phase or sub-phase state-specific mobility parameter $\mu$. In most of the cell compartments, multiphase states' coexistence is generally observed. That requires us to address how the diffusion can be coupled between phase states having different values of mobility. We shall develop some analytical expressions in this chapter to address this crucial issue in biological cell. But before going into details on this, let us address on the Einstein diffusion equation and the subsequent related modifications.

The diffusion Eq. (6.1) was found as a result of two independent groundbreaking discoveries that attempted to explain Brownian motion in early twentieth century by Albert Einstein in 1905 (Einstein 1905) and Marian Smoluchowski in 1906 (Smoluchowski 1906). Special forms related to Eq. (6.1) are important from application point of views. They are explained here. When a charged particle moves in a thermodynamic medium, the Einstein–Smoluchowski relation [Eq. (6.1)] takes another form, as follows:

$$D = \frac{\mu_q}{q} k_B T \tag{6.2}$$

Here, $q$ is the charge of the particle, and $\mu_q$ is the mobility which is termed as 'electrical mobility' as the particle bears electrical charges. Eq. (6.2) is popularly known as electrical mobility equation. When spherical particles diffuse in a thermodynamic medium, Einstein–Smoluchowski relation [Eq. (6.1)] takes another distinctive form, popularly known as Stokes–Einstein equation, as follows:

$$D = \frac{1}{6\pi\eta r} k_B T \tag{6.3}$$

Here, $\eta$ is viscosity of the medium, and $r$ is the radius of the spherical particle. Mobility $\mu$ is the inverse of another popularly known generic parameter 'drag coefficient' $\zeta$ where

$$\zeta = 6\pi\eta r \tag{6.4}$$

Here, the form of $\zeta$ [Eq. (6.4)] is in liquid with low Reynolds number. The ratio of the inertial forces to the viscous forces in a fluid medium is known as the Reynolds number. The relative importance of the inertial forces and the viscous forces must be different in different cell phase states. The diffusion constant may also vary as a consequence of possible differences in the form or value of drag coefficient. Depending on the Reynolds number in fluid mechanics, laminar or turbulent flow may occur. At low Reynolds numbers, viscous forces are dominant, so laminar flow occurs. The laminar flow is characterized by relatively smooth and constant fluid motion. Turbulent flow occurs at high Reynolds numbers, and inertial forces dominate causing the turbulent flow to occur. As a result, chaotic, vortices, and various flow instabilities may occur. In cellular phases, both types of flows are possible depending on what nutrients are flowing and in which kind of phase states are they flowing through. We shall avoid going dip into this rather hard core physics not to scare the biological readers with perhaps minimal physics interests. Due to primarily biological interests, we shall try here to see how the diffusion equations addressed here may be found useful in some biological studies. We wish to present an example case here.

Protein aggregation is a crucial biological feature found in disease states. Protein aggregation in the etiology of all amyloid diseases has been implicated partly, or sometimes fully. Understanding the underlying mechanisms and resultant structural effects of the protein aggregation and formation of amyloids is a challenging topic nowadays. Stokes–Einstein equation helps measure the hydrodynamic radii of the aggregate in a thermodynamic fluid state. The main ingredient of such measurements is the viscosity of the fluid. Proteins exist in cell fluid states which are composed of various substances. The classical Stokes–Einstein Eq. (6.3) may be directly used in some studies, but in other cases, the equation requires modifications, most of which are problem specific. Recently, there was a proposal of such a modification in the classical Stokes–Einstein equation, put forward using a mixture theory approach in order to accommodate the changes in viscosity of the solvent due to the changes in solute size and shape, to implement a more realistic model for amyloid beta (Aβ) aggregation involved in Alzheimer's disease (AD) (Achuthan et al. 2011). The kinetics of Aβ aggregation pathway is schematized in Fig. 6.1.

The group has focused on validating this model in protofibril lateral association reactions along the aggregation pathway, which has been experimentally characterized. In this report, this group of researchers has shown that the use of the classical Stokes–Einstein equation may not be found accurate for measuring hydrodynamic radii and consequently molecular weight of an amyloid aggregate. Stokes–Einstein equation helped to show that it is not correct always to assume the viscosity of solvent to be a constant entity especially during reaction processes. In cell, we

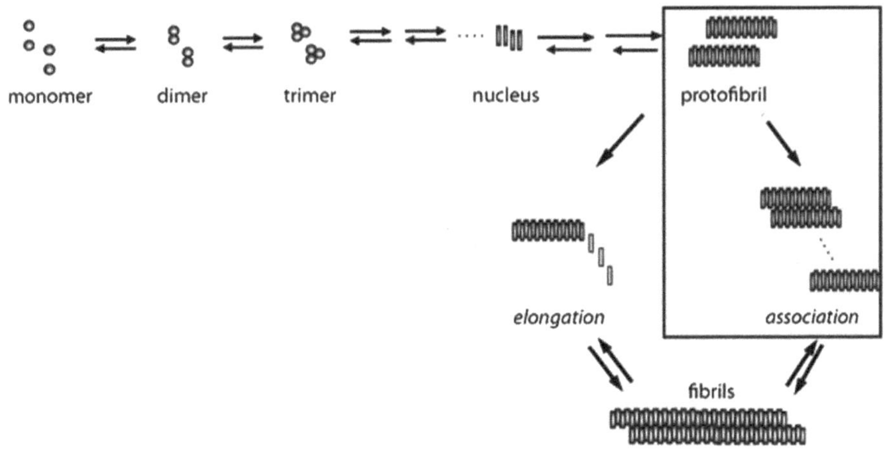


**Fig. 6.1 Kinetics of Aβ aggregation pathway**. Schematic representation of the overall Aβ aggregation toward fibril formation indicating significant steps involved. The boxed area, which represents the conversion of protofibrils to fibrils via lateral association, is the reaction that is being investigated in Achuthan et al. (2011)

propose that there exists various distinguished phase states (see Chap. 1 for details) and sub-phase states within a particular phase (see also Ashrafuzzaman 2015a, b). It is therefore quite obvious that the viscosity of fluid in any major cell phase state may not be constant over a wide diametric distance. The use of Stokes–Einstein equation therefore requires some kind of modifications in the equation itself. Such a proposal has been mentioned in a study (Achuthan et al. 2011). The solvent viscosity being not constant is considered especially in regions of reactions involving Aβ protofibrils. Large size of solute molecules appears to indicate significant amount of effects on viscosity of bulk solvent. An attempt has been made to solve the problem by incorporating modifications to the classical Stokes–Einstein Eq. (6.3).

The proposed modification provides mainly a change in the viscosity of the solvent. As an example, we may address here the mixture theory approach utilized here based on another viscosity model (Quemada 1977, 1998). This Quemada proposed model describes the viscosity of suspensions of hard spheres or of structural units like clusters or aggregates using the following relation:

$$\eta(\emptyset) = \frac{\eta_0}{(1 - \emptyset/\emptyset_m)^2} \tag{6.5}$$

Here, $\eta_0$ is the viscosity of the background solvent, $\emptyset$ is the volume fraction of the suspensions, and $\emptyset_m$ is the maximum of the volume fraction at which the viscosity is found to be singular, i.e., the critical limit of the concentration beyond which the flow tends to cease. The literature estimates $0.58 < \emptyset_m < 0.69$ (Lee et al. 1999; Wolthers et al. 1996). An average value $\emptyset_m \approx 0.6$ can be considered (see Achuthan et al. 2011). In case of the classical mixture theory, the effective viscosity of two liquids with different viscosities is usually taken to be the weighted average viscosity of the participating components. For the case of suspensions, a little caution needs to be exercised. In the model proposed in Achuthan et al. (2011), a mixture approach has been adopted by combining the solvent viscosity appropriately with Eq. (6.5). A schematic display is shown in Fig. 6.2.

Here is presented a recipe for obtaining a suspension with solvent volume $V_{0,}$ and such that, the volume of each constituent particle (molecule) is $V_i$ ($i = 1, 2, \ldots, n$), while requiring that the following three equations are satisfied that ultimately define the volume fraction (Achuthan et al. 2011):

$$\emptyset_k = \frac{V_k}{\sum_{k=0}^{n} V_k} \tag{6.6}$$

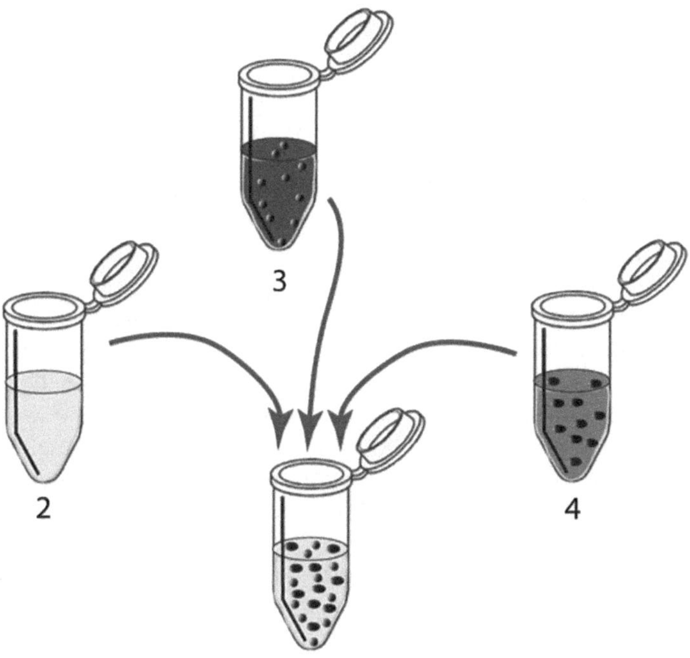


**Fig. 6.2** Mixture theory-based approach to find the effective viscosity. A schematic diagram showing the mixture theory approach of estimating the effective viscosity of a two-species suspension containing the pure solvent (L), species 1 (S1), and species 2 (S2) by means of a weighted average of the individual constituents of the suspension. The tube 2 contains the pure solvent L of viscosity $\eta_0$; tube 3 contains S1 floating in a medium composed of L and S2 and is of viscosity $\eta_1$; tube 4 contains S2 in a medium composed of L and S1 which is of viscosity $\eta_2$. By taking appropriate volumes from all tubes 2, 3, and 4, the final suspension is prepared in tube 1 which now contains L, S1, and S2 which has an effective viscosity $\eta_e$ which is defined in Eq. (6.5)

With the specific constrains,

$$\sum_{k=0}^{n} V_k = \Phi_0, \tag{6.6}$$

where

$$\emptyset_0 + \Phi_0 = 1 \tag{6.7}$$

In a mixture or suspension containing a background liquid and $n$-species of molecules, the volume fraction of each element of this mixture is given by above-mentioned Eq. (6.6).

The effective viscosity, in this scenario, can be defined as weighted average of the individual component taking the following form (Achuthan et al. 2011):

$$\eta_e(\emptyset) = \sum_{j=0}^{n} \emptyset_j \eta_j = \emptyset_0 \eta_0 + \sum_{j=1}^{n} \emptyset_j \eta_0 \left( \prod_{i=j+1}^{j} \frac{1}{\left(1 - \frac{\emptyset_i^{(j)}}{0.6}\right)^2} \right) \tag{6.8}$$

Here, $\eta_i$ $(i \neq 0)$ where $i$ ranges over all cyclic permutations of (1, ..., $j$, $j$ + 1, ..., $n$), represents the relative viscosity of the suspension containing the $j$th solute in a 'solvent' composed of a liquid of viscosity $\eta_0$ and all other solute particles. Note however that Eq. (6.8) is not unique. Various other formulations on the calculation of effective viscosity are possible. Here, we just mentioned this to remind the readers that Stokes–Einstein Eq. (6.3) may offer various diffusion parameters depending on physical parameters characterizing the fluid through which the particle movement happens.

The modification is assumed to be valid for the protofibril lateral association reaction. It is noteworthy that the modifications reflecting the viscosity changes can lead to realistic models not only for Aβ aggregation but also for a majority of amyloid aggregation systems.

In cell, the variation of viscous states may be found frequently while addressing fluidity in a cell phase state due to the presence of various sub-phase states. Diffusion across these phase states and sub-phase states needs to consider dynamics where the explained variations in general diffusion Eqs. (6.1–6.3) may strongly be present. The origins are explained, as example, above and mathematical formalisms also presented. We shall now focus on how the mobility of nutrients can be addressed in cell-based diffusion. As explained in Chap. 1 and also mentioned here, that cell has different compartments which in light of structural phases can be classified into various phase states. The individual phase states may also be further divided into various sub-phase states with varied physical properties. Diffusion through those states certainly follows regulated mobility functions which are different from those in case of free diffusion as explained in this section. These are mathematically modeled here (Ashrafuzzaman 2015a, b).

## 6.2 Diffusion Across Different Cell Phase States

In cell-based diffusion, the frequent appearance of varied viscous states with different fluidity may appear as crucial issue. Viscosity is directly measured by the internal friction of a fluid medium. It is thus measured as the force per unit area resisting a flow in which parallel layers' unit distance apart has unit speed relative to one another. Understanding viscosity in distinguishable cell-based fluid media appears with further complicated formalisms due to the presence of various cell phase states and sub-phase states within any individual phase state. Diffusion across these phase states and sub-phase states needs to consider localized dynamics where such explained variations in general diffusion Eqs. (6.1–6.3) may require scientific considerations. The origins are explained above, as examples, and some mathematical formalisms are also presented there. We shall now focus on how the mobility of nutrients can be addressed in cell-based diffusion. As explained in Chap. 1, that cell has different compartments which in light of structural phases can be classified into various phase and sub-phase states. Diffusion through these states certainly follows regulated mobility functions which are different from those in case of free diffusion as explained earlier. These scenarios are analyzed here using some recently developed mathematical models (see Ashrafuzzaman 2015a, b).

### *6.2.1 Analytic Expressions Explaining Diffusion Across Major Cell-Based Physical Barriers*

In Chap. 1, we have addressed in detail the physical phase states that are detected in cell. The major cell-based structural phase states have been reported as three different distinguishable types, namely liquid, plasma, and solid phase states (Ashrafuzzaman 2015a, b). General cell-based diffusion requires cell constituents to move in and across the mentioned three main phase states (see Fig. 6.3).

While diffusing in a phase state, things diffuse following Einstein's diffusion formula as presented and discussed earlier in this chapter. We shall focus here mainly on the various phase-regulated diffusion properties.

For interphase diffusion that is while diffusing across a boundary separating two phases (see Fig. 6.3), there are two distinguishable probabilities of finding any specific cell constituent in two phase states. If we assume these probabilities to be $P_{s,1}$ and $P_{s,2}$, respectively, in two phases states s,1 and s,2, respectively, then it is obvious that both of these probabilities are proportional to the mobilities $\mu_{s,1}$ and $\mu_{s,2}$, respectively, of the constituents in both phase states. That is,

$$P_{s,1} \sim \mu_{s,1}, \text{ or}, P_{s,1} = k_{s,1}\mu_{s,1} \tag{6.9.1}$$

$$P_{s,2} \sim \mu_{s,2}, \text{ or}, P_{s,2} = k_{s,2}\mu_{s,2} \tag{6.9.2}$$

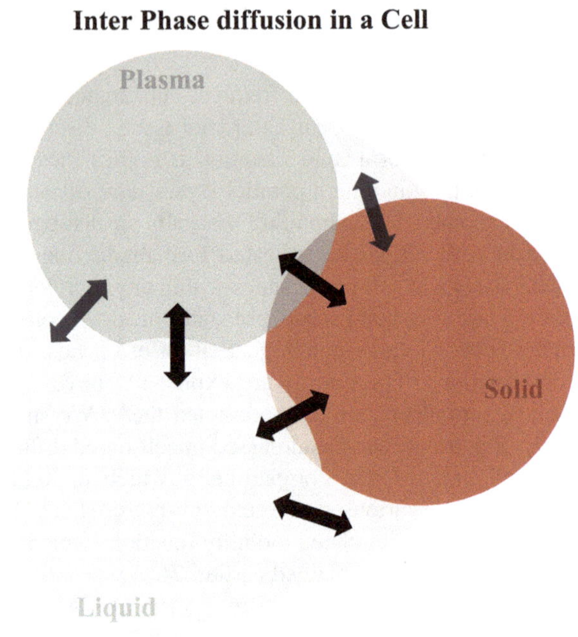


**Fig. 6.3** Diffusion happens between different phases across the interphase boundaries in a cell (this is just a model representation). Here, three equal-sized spheres (size of sphere is chosen arbitrarily) represent three phases, namely liquid-, plasma-, and solid-state structural phases, respectively. The background large sphere represents the cell environment which largely falls within liquid phase

Here, $k_{s,1}$ and $k_{s,2}$ are proportional constants that may be determined by analyzing parameters and properties of the media s,1 and s,2, respectively, that determine their viscosities. In case of no extra interphase barrier except for the differences in properties of various parameters determining the two phases, we can assume the following at the boundary:

$$k_{s,1} = k_{s,2} \tag{6.10}$$

As a result,

$$\frac{\boldsymbol{P}_{s,1}}{\boldsymbol{\mu}_{s,1}} = \frac{\boldsymbol{P}_{s,2}}{\boldsymbol{\mu}_{s,2}} \tag{6.11}$$

This case follows from the diagrammatic representation for the mobility dependence of probability functions as shown in Fig. 6.4. To understand easily, we can compare this with the condition of a frozen sea where the surface contains solid phase (frozen water) above the liquid water phase. At the contact, it is just a line where two phases are separated from each other where each phase exists with viscous properties specific to itself.

In a cell, there are compartments meeting each other but with a buffer zones between them. So at the junction of two phases, there exists a buffer zone. These buffer zones though by usual definition are considered as neutral zones, and they actually create zones of compromises in which distinctive compartments merge. If

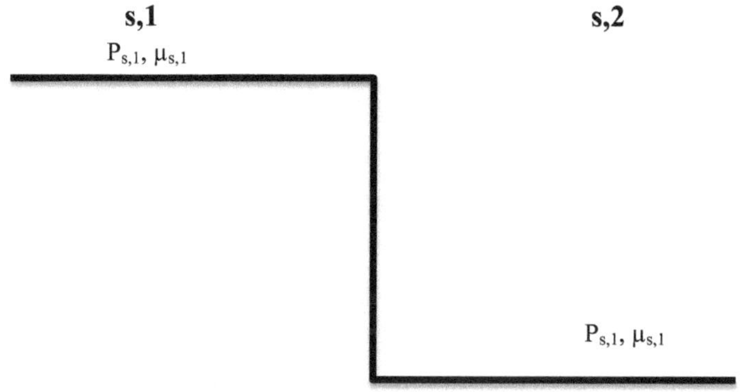


**Fig. 6.4** Two phase states s,1 and s,2 are different by the differences in their viscous properties as a result of the mobilities of the constituents therein. Central step down (left to right) here is just to show the difference in mobilities of the constituents, not to include any (forward or reverse acting) extra interphase barrier

the phases are too distinctive types, the buffer zone plays considerably important role to couple the two zones on its both sides. In general, we can consider a cell membrane to create a buffer zone between cellular exterior and cellular interior liquid states. These two liquid states may not necessarily have exactly identical biochemical and biophysical properties, but both fall in liquid phase state class. Like cell membrane, mitochondrial membrane also separates two near identical phase states. The lipid constructed membranes, without considering the membranes' sole transport properties due to mainly various MPs, make completely insulating barriers for constituents trying pass across the membranes. But as the membrane makes a pretty thick (~3–5 nm) zone and it has already a different phase state (plasma phase state) constituents, while entering into cell, already face a buffer zone existing at the boundary of cellular exterior liquid phase state and membrane's plasma phase state. In this junction or buffer zone, the constituents may not experience the conditions in either cellular exterior liquid phase state or membrane's plasma phase state but a condition that partially holds properties of both phase states and their averages. Therefore, the viscous friction (and as a result the mobility function) may have a different form or value at this buffer zone. This case follows from the diagrammatic representation for the mobility dependence of probability functions as shown in Fig. 6.5.

In case of the presence of buffer zone, we see the previous mathematical formulas addressing the probability functions for 100% diffusion of the constituents to either phase from buffer zone taking the following forms:

$$p_{s,1} = \mu_{s,1}/\left(\mu_{s,1} + \mu_{s,2}\right)p_{s,1/s,2} \tag{6.12.1}$$

$$p_{s,2} = \mu_{s,2}/\left(\mu_{s,1} + \mu_{s,2}\right)p_{s,1/s,2} \tag{6.12.2}$$

**Fig. 6.5** Two phase states s,1 and s,2 are different by only the differences of the mobilities of the constituents there. Central step down with a thicker line, not just a demarcation line like that in previous Fig. 6.4, is to show the presence of a buffer zone that consists of properties (or resultant of properties) of both phases from its left and right sides

Here, $p_{s,1/s,2}$ is the probability of the constituent to be at the boundary of two states. Therefore, the two new probability functions $p_{s,1}$ and $p_{s,2}$ are conditional probabilities of the constituents to get diffused to states s,1 and s,2, respectively, from the buffer zone.

For simplicity, we can assume that the mobility within a phase state is constant. As mentioned earlier, this mobility is nothing but the inverse of the viscous friction of the medium. Here, we consider the probability of the constituent to be at any point within a state is equal. Developing analytic expressions leading to the determination of such mobility factor $\mu_{s,i}$ ($\mu_{s,1}$, $\mu_{s,2}$, …, etc.) depends on the structures of the corresponding state.

## 6.2.2 *Permanent Trap in a Cell Phase State*

In special case of unidirectional flow, the constituents may get trapped permanently inside a phase state. This happens usually due to strong interactions or binding of the diffusive particles with the local structures in a phase state. If we assume that the constituents are subject to transfer from liquid state (LS, s,0) to any of the other two states, plasma state (PS, s,1) or solid state (SS, s,2), we may assume the following:

$$p_{s,1/s,0} = \kappa_{s,1/s,0} P_{s,0} \tag{6.13.1}$$

$$p_{s,2/s,0} = \kappa_{s,2/s,0} P_{s,0} \tag{6.13.2}$$

Here, $P_{s,0}$ is assumed to be the probability function at liquid state. $\kappa_{s,1/s,0}$ and $\kappa_{s,2/s,0}$ are functions that are determined by the partition coefficients active at the buffer zones of boundaries created between states s,1 and s,0 and states s,2 and s,0, respectively. We know that the physical science definition of a partition (or occasionally referred as distribution) coefficient is the ratio of constituent concentration in a mixture of two phases. If the phases are too distinctive types, the buffer zone plays considerable important role to couple the two zones from its both sides. If we consider the concentration of a cell constituent in s,0 to be greater than that in s,1 or s,2, we can assume that the constituent experiences a unidirectional, s,0 → s,1 or s,0 → s2, flow. In that case, the following conditions apply:

$$0 \leq \kappa_{s,1/s,0} \leq 1 \tag{6.14.1}$$

$$0 \leq \kappa_{s,2/s,0} \leq 1 \tag{6.14.2}$$

The previously mentioned state probability functions can be revised, for hypothetical 100% diffusion into either state from the buffer zone, as

$$p_{s,1} = \mu_{s,1} / \left(\mu_{s,1} + \mu_{s,0}\right) \kappa_{s,1/s,0} P_{s,0} \tag{6.15.1}$$

$$p_{s,2} = \mu_{s,2} / \left(\mu_{s,2} + \mu_{s,0}\right) \kappa_{s,2/s,0} P_{s,0} \tag{6.15.2}$$

If the constituent, instead of experiencing regular diffusion, experiences interaction with various interaction causing potential sites in the state, we have to develop an expression for the mobility factor which takes a complicated form.

### *6.2.3 Diffusion Across Sub-phase States within a Cell Phase State*

In a phase state, several sub-states may exist. All these sub-states have generally identical phase state behavior. We have explained this scenario in Chap. 1. In this case like Fig. 6.3, we may have several boundaries for the constituents to cross within the same phase class. We can schematically diagram this case, as shown in Fig. 6.6. Several sub-phases within a classified phase state category may arise due to mainly the presence of different category states defining materials in different regions. For example, in a solid-state phase various solid-state materials may exist and thus various sub-phases within solid-state phase category may emerge. Similarly, due to various other causes like varied chemical compositions, thermodynamic fluctuations, several localized sub-phase states are generally expected to exist although altogether those sub-states are defined by the general biophysical properties of a specific phase state in a cell. Figure 6.6 is schematized to represent such sub-phase states and the intraphase boundaries in any major cell phase state.

For intraphase diffusion that is while diffusing across a boundary separating two sub-phases, there are two distinguishable probabilities of finding any specific cell constituent in two sub-phase states. For simplicity and of course often realistically, we assume that major proportion of the classified phase state is occupied by a

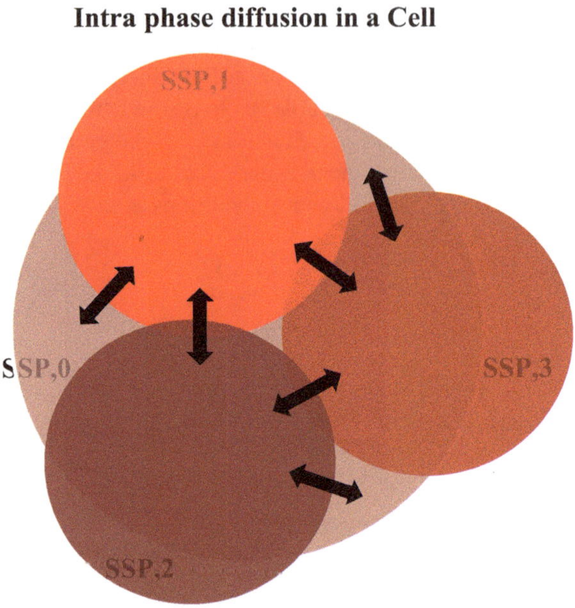


**Fig. 6.6** Diffusion happens between different sub-phases across the intraphase boundaries/ inter-sub-phase boundaries in a classified cell phase like liquid or plasma or solid phase. Here, we have schematized various intrasolid phases or sub-solid phases (SSPs) within classified solid phase of a cell. SSP,0 represents major proportion of SSP, whereas SSP,1, SSP,2, SSP,3, …, etc., represent various intrasolid phases. The intraphase diffusion across the intraphase boundaries follows different quantitative but almost identical qualitative mechanisms as are explained for the diffusion across interphase boundaries. Bidirectional arrows show the directions to which the diffusion may take place

sub-phase state what we may denote as SPS,0 and constituents usually diffuse across the boundaries of SPS,0 (ss,0) and other sub-phase states, e.g., SPS,1 (ss,1), SPS,2 (ss,2). If we assume that the probabilities of constituents to be in SPSs ss,1, ss,2,…, etc., are $p_{\mathrm{ss},1}, p_{\mathrm{ss},2},\ldots$, etc., respectively, then we find the following relation:

$$p_{\mathrm{ss},1} = \mu_{\mathrm{ss},1} / \left(\mu_{\mathrm{ss},1} + \mu_{\mathrm{ss},0}\right) p_{\mathrm{ss},1/\mathrm{ss},0} \tag{6.16.1}$$

$$p_{\mathrm{ss},2} = \mu_{\mathrm{ss},2} / \left(\mu_{\mathrm{ss},2} + \mu_{\mathrm{ss},0}\right) p_{\mathrm{ss},2/\mathrm{ss},0} \tag{6.16.2}$$

etc.

Here, $p_{\mathrm{ss},1/\mathrm{ss},0}, p_{\mathrm{ss},1/\mathrm{ss},0},\ldots$, etc., are the probabilities of the constituents to be at the boundaries of ss,1/ss,0, ss,2/ss,0,…, etc. Therefore, the new probability functions $p_{\mathrm{ss},1}$ and $p_{\mathrm{ss},2},\ldots$, etc., are conditional probabilities of the constituents to get diffused into ss,1, ss,2, …, etc., respectively, from their buffer zones with ss,0. $\mu_{\mathrm{ss},0}$, $\mu_{\mathrm{ss},1}$, $\mu_{\mathrm{ss},2}$, …, etc., are mobilities of the constituent in ss,0, ss,1, ss,2, …, etc., respectively. We may, for simplicity, assume mobility within a specific SPS to be constant.

### 6.2.4 *Permanent Trap in a Sub-phase State within a Cell Phase State*

In a special case where the constituents get trapped permanently inside any SPS and as we assume the constituents to get transferred from ss,0 to any of ss,1, ss,2, …, etc., we can assume the following:

$$p_{ss,1/ss,0} = \kappa_{ss,1/ss,0} p_{ss,0} \tag{6.17.1}$$

$$p_{ss,2/ss,0} = \kappa_{ss,2/ss,0} p_{ss,0} \tag{6.17.2}$$

etc.

Here, as explained earlier, $\kappa_{ss,1/ss,0}$, $\kappa_{ss,1/ss,0}$, …, etc., are functions that are determined by the partition coefficients active at the buffer zones of boundaries created between ss,1 and ss,0, ss,2 and ss,0,…, etc., respectively. The following conditions apply:

$$0 \leq \kappa_{ss,1/ss,0} \leq 1 \tag{6.18.1}$$

$$0 \leq \kappa_{ss,2/ss,0} \leq 1 \tag{6.18.2}$$

Then, the previously mentioned sub-state probability functions can be revised as

$$p_{ss,1} = \mu_{ss,1} / \left(\mu_{ss,1} + \mu_{ss,0}\right) \kappa_{ss,1/ss,0} p_{ss,0} \tag{6.19.1}$$

$$p_{ss,2} = \mu_{ss,2} / \left(\mu_{ss,2} + \mu_{ss,0}\right) \kappa_{ss,2/ss,0} p_{ss,0} \tag{6.19.2}$$

etc.

## 6.3 Numerical Computation and General Trend of Diffusion-Related Probability Functions

The analytical expressions explaining the diffusion-related probability functions that are presented earlier can be computed using numerical computation (NC) methods to address the trend of the diffusion probability function in a phase or sub-phase state. The actual expressions of the probability functions would perhaps be quite complicated while we develop the expressions of the partition coefficients considering the physical properties of the phase states. A general template has been developed for what can be used later, while all the parameters are known.

### 6.3.1 Phase State Probability Functions

We shall present here the NC results and show the trends of the phase state probability functions in case of permanent trap, as mentioned earlier. According to the analytical expressions developed earlier, we get the trapping probability in a state s,i (i = 1, 2, 3, …, etc.) as

$$p_{s,i} = \mu_{s,i}/\left(\mu_{s,i} + \mu_{s,0}\right)\kappa_{s,i/s,0}P_{s,0} \tag{6.20}$$

Therefore, we need to plot the normalized expressions of $p_{s,i}$ that is

$$\left(1/\kappa_{s,i/s,0}\right)p_{s,i}/P_{s,0} = 1/\left(1 + \mu_{s,0}/\mu_{s,i}\right) = 1/\left(1 + 1/\left(\mu_{s,i}/\mu_{s,0}\right)\right) \tag{6.21}$$

The plot is shown in Fig. 6.7.

The plot suggests that $p_{s,i} \leq$ Ps,0. The value of $p_{s,i}$ depends strongly also on the parameter $\kappa_{s,i/s,0}$ which is related to the partition coefficient. Once the expressions of $\mu_{s,0}$ and $\mu_{s,i}$ are analytically developed, the actual trend of the probability functions in any specific phase state can be deducted.

### 6.3.2 Sub-phase State Probability Functions

We shall present the NC results and show the trends of the sub-phase state probability functions in case of permanent trap, as mentioned earlier. We need to plot the normalized expressions of $p_{ss,i}$ which is, according to earlier analysis,

$$\begin{aligned} p_{ss,i} &= \mu_{ss,i}/\left(\mu_{ss,i} + \mu_{ss,0}\right)\kappa_{ss,i/ss,0}p_{ss,0} \\ &= \mu_{ss,i}/\left(\mu_{ss,i} + \mu_{ss,0}\right)\kappa_{ss,i/ss,0}\mu_{s,i}/\left(\mu_{s,i} + \mu_{s,0}\right)\kappa_{s,i/s,0}P_{s,0} \end{aligned} \tag{6.22}$$

Here, $p_{ss,0}$ is the probability for the constituents to be in the most common sub-state ss,0 within state s,i; therefore, $p_{ss,0} = \mu_{s,i}/(\mu_{s,i} + \mu_{s,0})\ \kappa_{s,i/s,0}\ P_{s,0}$ (see analysis presented previously in this chapter).

Fig. 6.7 Plot of $1/\kappa_{s,i/s,0})\ p_{s,i}/P_{s,0}$ as a function of $\mu_{s,i}/\mu_{s,0}$

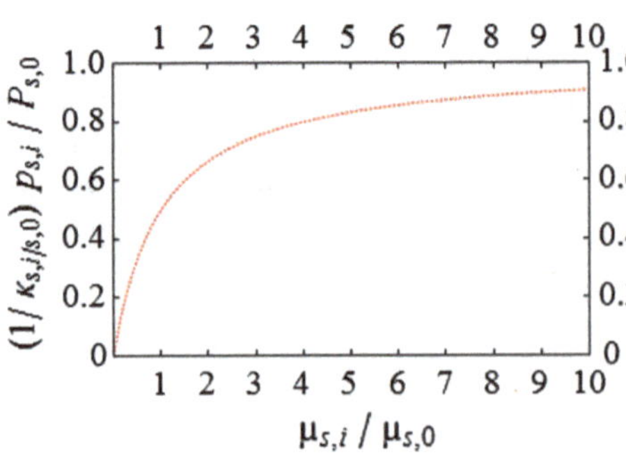

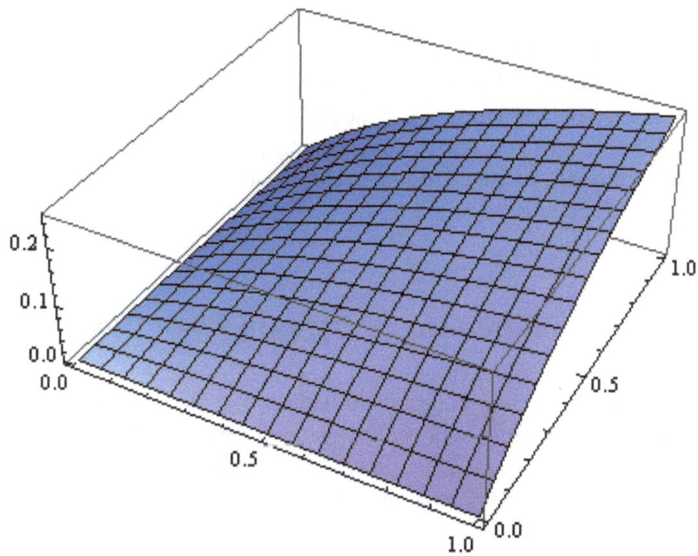


**Fig. 6.8** Plot of $(1/\kappa_{ss,i/ss,0})(1/\kappa_{s,i/s,0})\, p_{ss,i}/P_{s,0}$, along vertical axis, for $\mu_{ss,i}/\mu_{ss,0}$, along $x$-axis, and $\mu_{s,i}/\mu_{s,0}$, along $y$-axis

Therefore, we need to plot

$$\begin{aligned}(1/\kappa_{ss,i/ss,0})(1/\kappa_{s,i/s,0})p_{ss,i}/P_{s,0} &= \mu_{ss,i}/(\mu_{ss,i}+\mu_{ss,0})\kappa_{ss,i/ss,0}\mu_{s,i}/(\mu_{s,i}+\mu_{s,0}) \\ &= \{1/(1+1/(\mu_{ss,i}/\mu_{ss,0})\}\{1/(1+1/(\mu_{s,i}/\mu_{s,0}))\}\end{aligned} \tag{6.23}$$

$(1/\kappa_{ss,i/ss,0})(1/\kappa_{s,i/s,0})\, p_{ss,i}/P_{s,0}$ requires to get plotted in three-dimensional (x, y, z) space where the function is plotted for two factors $\mu_{ss,i}/\mu_{ss,0}$ along $x$-axis and $\mu_{s,i}/\mu_{s,0}$ along $y$-axis (see Fig. 6.8).

If we compare the probability values from two different cases, namely in a classified phase state and in a sub-phase state, it is clear that the value decreases in the latter case. The deeper the constituents penetrate, the lesser is the probability value to be found. 'Deeper' has been used here to mean more sub-states to be crossed by the cell constituents. The values of probabilities need to be normalized further with the values of $\kappa_{ss,i/ss,0}$ and $\kappa_{s,i/s,0}$.

## 6.4 An In Vitro Modeling of Diffusion Across Two Phase States

In vitro modeling of cell has been a popular technique of scientific studies for quite a long time. In this technique, we mimic the biological system, especially one or a few components of the constituents that make a complete system in a test tube made of glass or plastic, etc., or in some commercially available specialized plates, e.g., 96-well plate, Eppendorf tube, etc., to perform some assay studies. In laboratory tests, in which no biological cells or directly extracted cell components are needed to perform this modeling study is planned. Commercially synthesized biocompounds that are active in cell's various structural phases are often used to test their action in this kind of 'in tube' studies. Here, we shall provide a sketch which addresses a list of stepwise doings.

### 6.4.1 Trapping Constituents in Different Phase States

In Fig. 6.9, we have sketched diagrams to address the diffusion of cell constituents or any external agents active in a cell across two representative cell state phases, namely liquid-state and plasma-state phases. To demonstrate this, we shall utilize a recently published strategy used in liposome studies (Tseng et al. 2011; Ashrafuzzaman et al. 2013).

Artificially constructed liposome in solution creates a two phase states coexisting in a tube. The readers are expected to easily understand how constituents may diffuse naturally from one phase to another in the coexisting two phases in a tube.

A lipid solution in methanol is used to get dried in wells, overnight under hood. A dry lipid film is created on the inner surface of the well (see Fig. 6.9) due to the reason that the volatile methanol slowly gets evaporated. Lipids in dry form or in chloroform are commercially available from suppliers, such as Avanti Polar Lipids, Inc. (http://avantilipids.com/), Sigma-Aldrich (http://www.sigmaaldrich.com/). The agent of interest is then dissolved in a solvent at very high concentration. A buffer, usually a solution that mimics cell cytoplasm state (a liquid state), is then made. Usually, a solution of salt like sodium chloride and potassium chloride makes a buffer.

At this stage, we fill the well/tube with this buffer. Then, add aliquots of the agent containing solvent inside the buffer. The tube with solution containing agents is then left in dark for a long time, and course of time is agent specific. Most of the liposome assays use approximately 1 h as incubation time. For details, see Fig. 6.9.

After incubation, we separate the supernatants by gently removing the solution from the well. Here, supernatants refer to agents that are not absorbed in liposomes. The well is then washed with the same buffer for as many as 3–5 times. All the supernatants are collected in a separate container, as shown in Fig. 6.10. After repeated washing and separating, the liposome-containing well is filled with a certain volume of buffer. We now have two separated solution containing agents: (i) liposome dragged agents inside the plasma-state liposome and (ii) the rest of the agents to be left inside liquid-state buffer; they are collected as supernatants. The two samples are now ready to be examined for the relative quantification of agents in two different phase states, as demonstrated in Fig. 6.10.

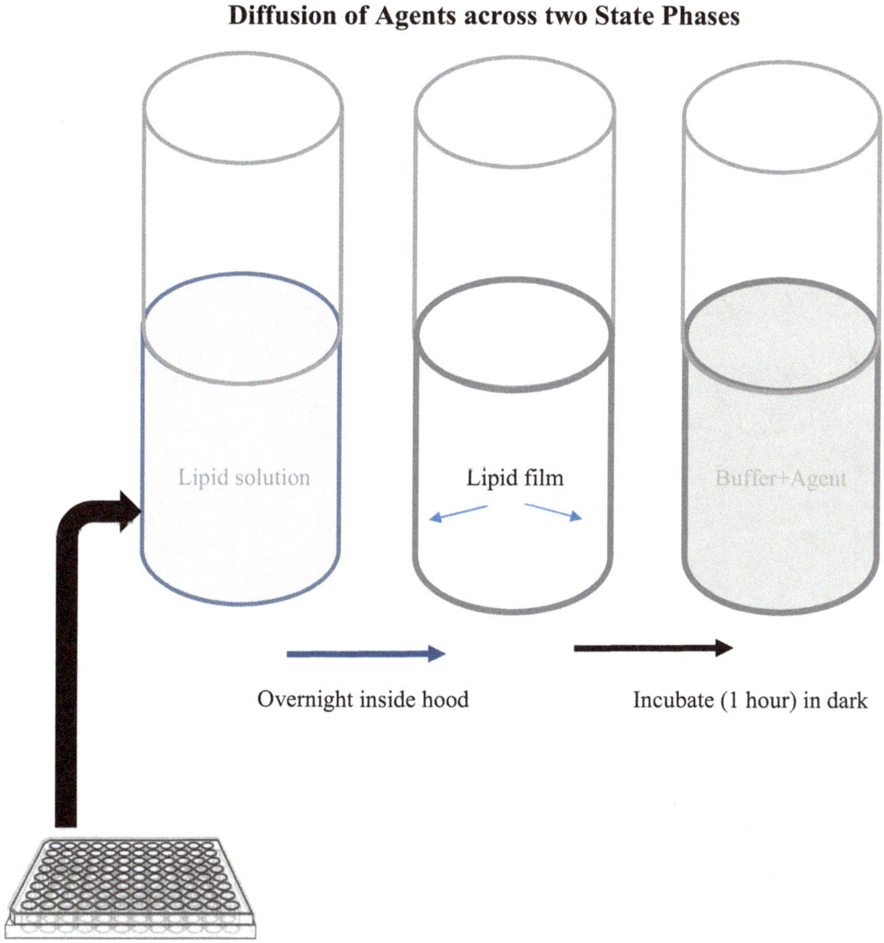


**Fig. 6.9** Sketch of a well from a 96-well plate. We have virtually picked a well from a 96-well plate to address the distribution among various cell state phases

### *6.4.2 Relative Quantification of Agents of Interests in Coexisting Phase States*

In previous section, we have addressed how two coexisting phase states can be separated from each other. Agents partitioned into these different phase states also get separated with the phase state separation. This is why, if we utilize techniques to detect agents in these two separated solutions, we in fact measure the relative number probability of the agents diffused into the states. There are various techniques to do so. The separated solutions can be read or examined by conventional microplate readers, or any other similar machine to quantify the agents in

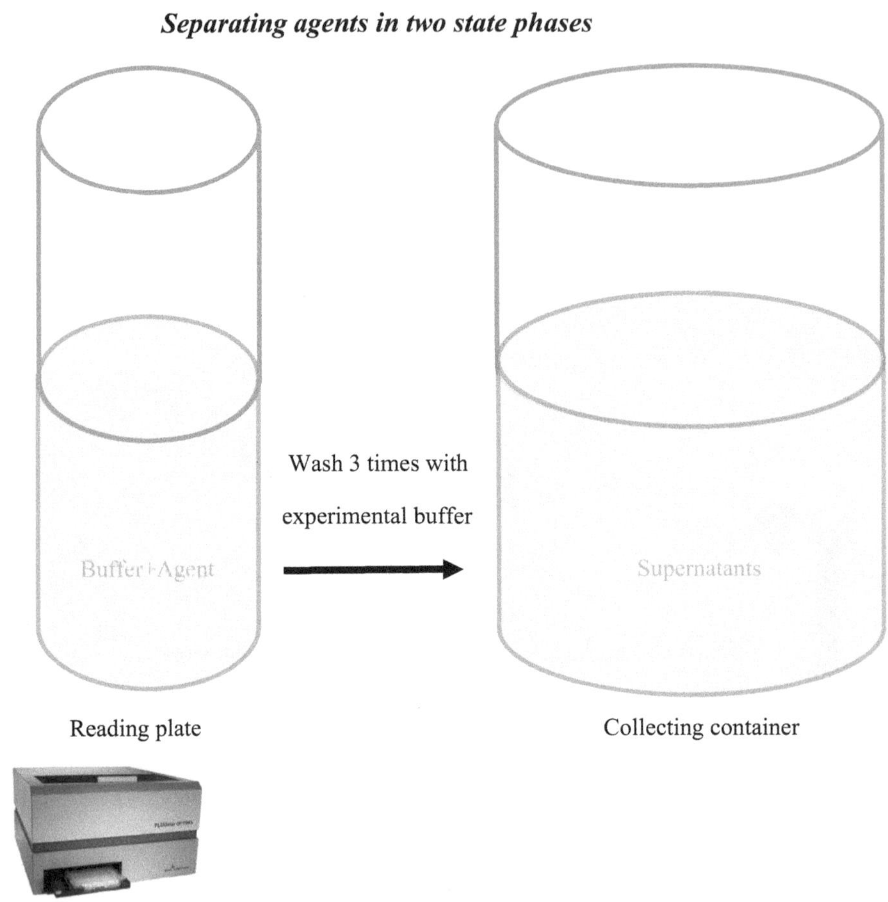


**Fig. 6.10** Wash 3 times with experimental buffer and collect the supernatants. Now add experimental buffer in reading plate and read fluorescence with a microplate reader, e.g., Fluostar Optima (from BMG Labtech)

liposome's plasma phase state and buffer's liquid phase state. A microplate reader that can detect the agent is suitable for the relative quantification of the agents in various state phases. There are chemical detection methods available too. We wish not to detail them as they are readily found in various published articles.

## 6.5 In Silico Diagnosis on Relative Diffusion and Probabilities in Different Cell Phase States

Can we quantify the dynamical properties of agents in a cell?

In biological systems, measurements are always complicated due to the stochastic effects caused by correlated dynamics of many components. Therefore, all of the *in cell* measurements provide mostly the summary or resultant of many effects. Pinpointing at specific phenomenon appears often as a tough job or even sometimes found impossible. You may try to detect a cause or an effect related to an event in the presence of others, but other causes or effects always induce perturbations in your target event. As a result, you end up getting some resultant of many causes or effects. Most of the biological studies have to deal with this kind of scenario. Biophysical ones though may help pinpoint at specific event by mostly isolating that one from others. Technological developments have made some aspects possible, but we are yet far away from achieving the completeness. Specially, when we try to address anything related to interactions or coupling, we end up monitoring many such processes together or the effect that is resultant of many such ones.

Dynamics of agents in a diffusive medium may result due to fluctuations (mostly thermodynamic effects) or interactions (localized or long-range effects), etc., therein or some kind of specific or nonspecific coupling with other agents or medium in or on which the agents move. In computer, we can mimic this kind of environment and observe the dynamical nature of agents. The beauty of performing study in computer is that one can control parameters at one's will. That may not represent the whole biological environment but help address certain aspects inclusively. Molecular dynamic (MD) simulation is such a method (readers will learn on this method later in this chapter and from subsequent chapters). We can virtually create a biological environment in computer and study the dynamics of components within.

### 6.5.1 General Description and In Silico Modeling

As a test case, we shall address the dynamics of a biologically active agent in a virtual biological system. Here, we wish to address how a drug (or an agent), while in the vicinity of or inside the cell, may experience probability to get diffused from one phase state to another and/or be released back to former phase state. Inspection of the dynamics of the agent will help us address these phenomena via the agent-phase state constituent coupling stability measurements. The coupling stability is nothing but a course of time during which the agent will coexist with a phase state. For this study, we shall insert a drug into an in silico designed liquid phase state of a cell and monitor its stability of coupling with this state relative to its coupling stability with a plasma phase state when both liquid and plasma phase states coexist (see Fig. 6.11a). Here, plasma phase state can be constructed by simulating a membrane. But for simplicity, we shall consider a membrane constituent lipid to represent the plasma-state-like cell membrane for in silico dynamical tests (see Fig. 6.11b).

In computational dynamic study, we can mimic a cell environment inside computer as shown in Fig. 6.11. Here, a drug while in cellular cytoplasm or extracellular hydrophilic background can interact with a liposome as sketched here.

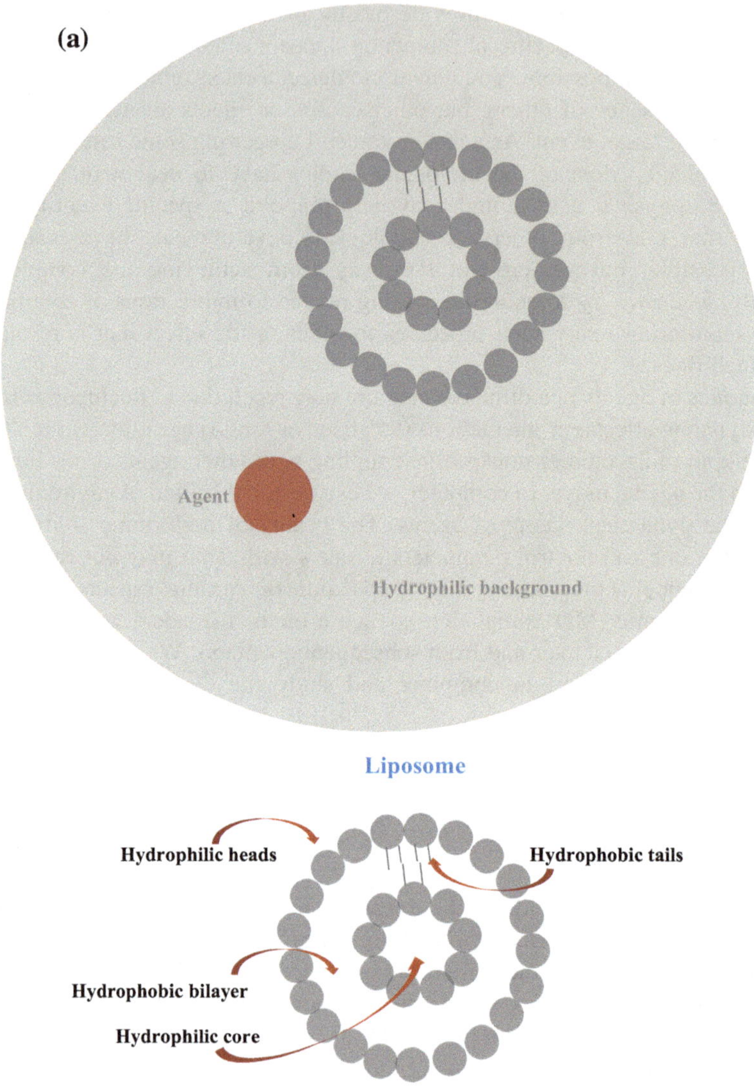


**Fig. 6.11** **a** In upper panel, both agent (red sphere) and liposome are inserted/constructed inside hydrophilic (mostly water environment) bath to address their coupling stability. Lower panel shows a detailed analysis of a liposome. Hydrophobic tails are sketched with only a few lipids though all lipids have them. **b**. Agent–lipid coupling stability assay for in silico dynamical tests. Here, a lipid (main building block of the liposome) is used instead of the whole liposome, for simplicity

(b)

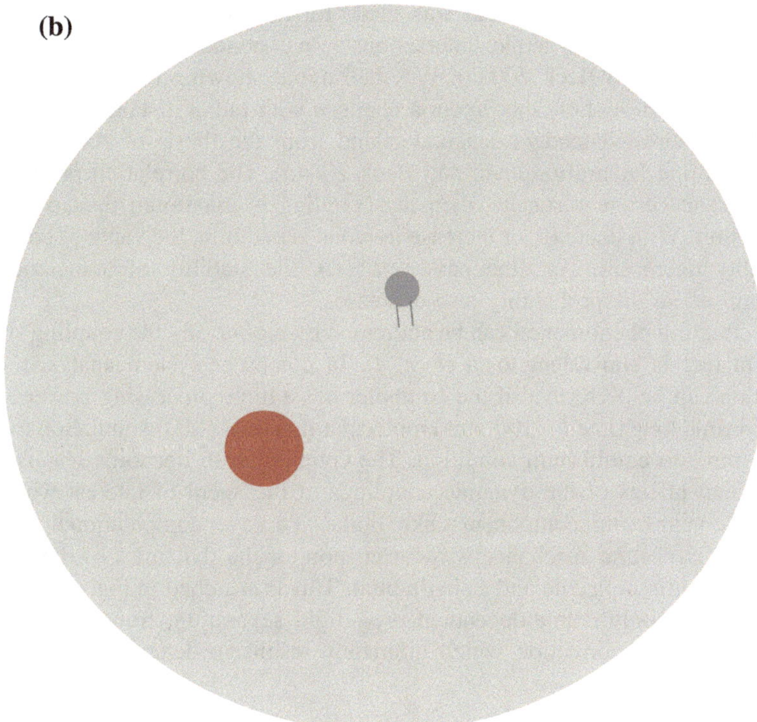

**Fig. 6.11** (continued)

The drug will thus show a diffusive probability into liposome (plasma phase state) and from there back to liquid phase state. In this case, the drug will follow a dynamic correlation with liposome following a drug–liposome coupling phenomenon.

## *6.5.2 Dynamic Correlation Function*

In condensed matter physics, calculation of dynamic correlation function is a popular method to address the time-dependent spatial distribution of interacting particles (Ashrafuzzaman and Beck 2004a, b). In this drug–target structure correlation problem, we can apply the condensed matter physics technique and calculate the dynamic correlation probability/function $\phi\ (r, r', t)$. Here, $r$ is the radial distance between drug and target, $r'$ is the hard core distance, meaning the minimum distance the drug can approach to target, $r \geq r'$, and $t$ is time at which the dynamic correlation is measured. Alternately, we could also consider $\phi\ (r, r', t)$ as the dynamic correlation function where $r$ is the radial distance at time $t$, while in the beginning ($t = 0$) the radial distance was $r'$. We can decouple this function following a detailed analysis that was made for another system, namely vortex–antivortex mutual dynamics/interactions in condensed matter physics (Ashrafuzzaman and Beck 2004a). We shall avoid showing the detailed analysis here but would show how the function changes with radial distance and time (see Fig. 6.12). Interested readers can understand from the thesis of Ashrafuzzaman, uploaded online (Ashrafuzzaman and Beck 2004a). The correlation probability is the highest at certain distance when the coupling is maximum due to complete equilibration. With decrease or increase in radial separation, the value of correlation probability decreases. As time passes ($t > 0$), the stability of association gets compromised so the probability also decreases.

The coupling phenomenon can be addressed by monitoring the coupling stability which in fact is equivalent to $\phi$ $(r, r', t)$. In computer system analysis, such a simulation can be performed if the computer has a huge processing power and we can perform a long time ($\sim$100 s ns) molecular dynamic (MD) simulation to let the system move to equilibrium condition. The coupling with liposome results due to the resultant effects of the dynamic couplings of the agent of interest with major liposome-constructing components like lipids. To save computational time and understand molecular mechanisms, we can monitor the dynamics of the coupling between the drug molecule and a single lipid. This is sketched in Fig. 6.11b. In this case, we can monitor time-dependent drug–lipid separation, from which we can trace the dynamic correction which ultimately results in determining the relative

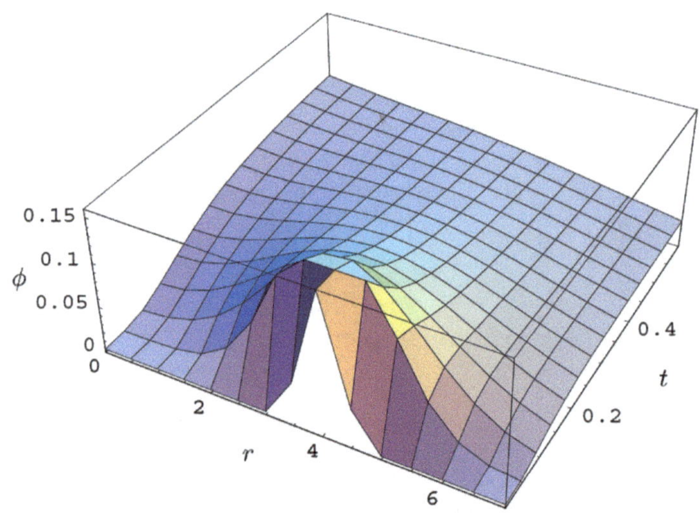


**Fig. 6.12** Plot of $\phi$ $(r, r', t)$ as function of $r$ and $t$ for $r' = 0$. Scales are arbitrary. For details, see Ashrafuzzaman and Beck (2004a). The slopes toward various directions shown here should be considered arbitrary and are, in nature, system specific. This kind of dynamic correlation function needs to be introduced in coupling where the participating agents are in continued movements. In biological system, for example, for the case of mobile agent (any drug) and mobile lipid interactions in cell membrane environment (see Fig. 6.11), the correlation function follows exactly this kind of trend as presented here

liposome adsorption probability (see Ashrafuzzaman et al. 2013, 2014). In these two references, we have presented our novel works to address interactions of peptides [e.g., gramicidin A (gA) and alamethicin (Alm)] (Ashrafuzzaman et al. 2014) and chemotherapy drugs (e.g., capsaicin and taxol derivatives) (Ashrafuzzaman et al. 2013) with major phospholipids (e.g., phosphatidylserine and phosphatidylcholine) in silico MD simulations which got validated in vitro liposome-binding assay experiments. These two studies will surely demonstrate the proof of principles discussed in this chapter.

### 6.5.3 Example Studies: In Silico MD Simulations

Several drugs, namely aptamers (Tseng et al. 2011; Ashrafuzzaman et al. 2013), peptides (Ashrafuzzaman et al. 2014), colchicine, taxol (Ashrafuzzaman et al. 2012), have been tested for their liposome adsorption using the mentioned in silico MD simulation techniques by us. We shall briefly address these studies here.

#### 6.5.3.1 Methods and Materials—In Silico MD Simulation

The binding affinity and related statistical rankings can be addressed using simulation. MD is a technique that helps address these using *in silico* modelling. For theoretical insights into the binding affinity at the molecular level, the MD simulations are utilized. Here, Amber 11 with consideration of latest force field $F$ is utilized (Case et al. 2008, 2010). Following the concept of Monte Carlo method, we consider a number of different relative locations and orientations randomly generated in each agent–target complex as initial structures in a biologically relevant platform. Each location and orientation-specific complex is energy minimized using the steepest descent method for the first few cycles and then followed by a conjugate gradient for another, e.g., 1000 cycles. During the equilibration of the complex using the explicit water TIP3P model, we first apply Langevin dynamics during the process of heating up the system for 200 picosecond (ps) with targets being restrained using a harmonic potentials having a force constant $k$ = 100 N/m to homogeneously distribute water molecules in the system. We then introduce pressure regulation in the simulation to equilibrate solvent density for another 200 ps in addition to temperature regulation. We then perform a 10-ns explicit water MD simulation at 300 K and solution at pH 7 for each complex as a production runs. Note that the targets are gently restrained with a harmonic potential with a force constant $k$ = 10 N/m applied to the phosphorus in PS or PC. Longer time simulation is possible but at the cost of more days, and a higher performing computer facility is required. The whole technique is published in Tseng et al. (2011), Ashrafuzzaman et al. (2013).

#### 6.5.3.2 Aptamer–Lipid Coupling Stability: a Case Study

We modeled a cell environment in computer to perform MD simulation to address the relative probability of drugs to diffuse into hydrophobic lipid membrane environment from hydrophilic water environment. This is equivalent to the diffusion of agents between two phase states, namely liquid phase state and plasma phase states. For simplicity, instead of the whole membrane we used a lipid PC or PS, following the arguments provided earlier. We monitor the distance between two participating agents, namely an aptamer (deoxyribonucleic acid (DNA) short sequence) and PS or PC. Due to dynamic nature, both move in the space but as they have interactions, they occasionally come closer.

Based on MD simulation, results are presented in Fig. 6.13 which plots separation distance $d_{\text{apt–lipid}}$ against time from a set of raw simulation results rather than all cases, in which five randomly generated SIAp4-PS complexes as initial structures, represented in the inset, are considered. Given this result, we evaluated the probability of a designed aptamer and PS being at a specific $d_{\text{apt–lipid}}$ range, which is the time duration both spent at that $d_{\text{apt–lipid}}$ range over the total simulation time. The average probabilities of sequences with PS within $6 < d_{\text{apt–lipid}} < 16$ Å were calculated (the third column of Table 6.1). This time- and distance-dependent probability function is partially equivalent to the previously explained dynamic correlation probability/function $\phi$ $(r, r', t)$.

## 6.5.4 *Universal Correlation Functions: Energetic Scenario in MD Results*

Using MD simulations, we provide computational support for agent–lipid physical interactions which are postulated to be the mechanisms behind the agents' dynamics near the target molecules, e.g., lipids in current case. MD simulations for

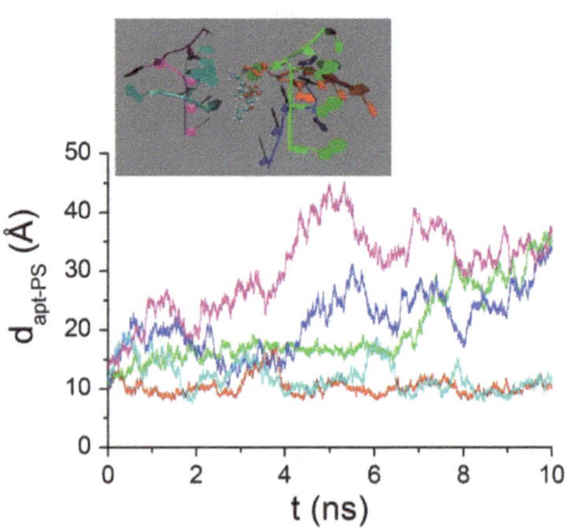


**Fig. 6.13** 10-nanosecond (ns) MD simulation to address the relative distribution between an aptamer and PS, following in silico modeling explained in Fig. 6.11. For details, see Tseng et al. (2011), Ahrafuzzaman et al. (2013). Five traces represent 5 repetitions. Five initial structures have been shown in the inset

**Table 6.1** Two sets of DNA aptamer templates

| Name[a] | Sequence (5′ → 3′) | PS[b] |
|---|---|---|
| SIAp1 | AAAAGA | 0.23 ± 0.11 |
| SIAp2 | AAAGAG | N/A |
| SIAp3 | TAAAGA | 0.57 ± 0.15 |
| SIAp4 | AAAGAC | 0.54 ± 0.19 |
| SIIAp1 | CAGAAAAAAC | 0.4 ± 0.2 |
| SIIAp2 | CAGAAAAAAT | 0.31 ± 0.12 |

[a]SI is designed based on total energy, and SII is designed using interaction energy

[b]Probabilities of aptamers and PS within $6 < d_{\text{PS-lipid}} < 16$ Å. For detail, see Tseng et al. (2011), Ashrafuzzaman et al. (2013)

aptamer–lipid and chemotherapy drug (CD)–lipid have been conducted by us, as mentioned earlier. Using the same settings, similar MD simulations have been carried out for peptides gA–lipid and Alm–lipid pairs (Ashrafuzzaman et al. 2014).

As presented in Tseng et al. (2011), Ashrafuzzaman et al. (2013), three quantities, the separation distance of centers of mass of agent and lipid, $d_{\text{agent–lipid}}$, van der Waals (vdW), and electrostatic (ES) energies, were utilized to analyze simulation results. Note that regarding solvent-accessible area (SA), when both drug and lipid molecules are completely separated, we can expect them to be entirely exposed to solvent, i.e., the corresponding SA areas are at a maximum. The SA areas in all studies are roughly unchanged between the start and the investigated 20 Å length. This suggests that within this range the drug–lipid complexes stay at an equilibrium solvation condition. Therefore, we may focus only on vdW and ES here.

In order to investigate features of physical interactions of all lipid and agents behind their interaction-driven diffusion from MD simulation results, a probabilistic description was considered (developed by Ashrafuzzaman and Tseng in a pending US patent, 2015). Specifically, we first evaluated probability of observing a pair within $d_{\text{agent–lipid}}$ as $P(d_{\text{agent–lipid}}) = \Delta t(d_{\text{agent–lipid}})/T_{\text{sim}}$, where $\Delta t(d_{\text{agent–lipid}})$ is the time duration the agent–lipid pair stays within $d_{\text{agent–lipid}}$ and $T_{\text{sim}}$ is the total simulation time. Second, the probability of having either vdW or ES energy of a lipid and an agent staying at distance $d_{\text{agent–lipid}}$ is given by Boltzmann's distribution, as follows:

$$\mathrm{P}\left(\mathrm{E}\left(\mathrm{d}_{\text{agent–lipid}}\right)\right) = \exp - \beta\mathrm{E}\left(\mathrm{d}_{\text{agent–lipid}}\right)/Z, \tag{6.24}$$

where partition function is $Z = \sum_{d_{\text{agent–lipid}}} \exp -\beta E\left(d_{\text{agent–lipid}}\right)$ and $\beta = 1/k_{\mathrm{B}}T$ with $T = 300$ K.

***Universal footprint shed by MD results***. Figure 6.14 plots $P(E(d_{\text{agent–lipid}}))$ against $P(d_{\text{agent–lipid}})$s, and the corresponding $d_{\text{agent–lipid}}$ values are represented by symbol size, which is illustrated in the bottom panel. Considering all three variables together Fig. 6.14 makes itself a three-dimensional plot. The upper row shows the case with PC and lower row for PS. Furthermore, the left column shows vdW energy and right for ES energy. Several features are revealed in this figure. First, it shows similar trends for all three categories of agents against either PC or PS from vdW interactions point of view. Probabilities of having a pair within $d_{\text{agent–lipid}}$ and vdW energy $E_{\text{vdW}}(d_{\text{agent–lipid}})$ are gradually decreased when a lipid and an agent separation distance $d_{\text{agent–lipid}}$ is increased. Namely, vdW force is likely to play a crucial role in all types of agents to interaction-based drive toward lipids (short $d_{\text{agent–lipid}}$ range). Second, however, from ES energy point of view, two CDs are the only type of agents to show similar trends, and larger the separation distance $d_{\text{agent–lipid}}$ is, the lower the probabilities $P(d_{\text{agent–lipid}})$ and $P(E(d_{\text{agent–lipid}}))$ are, in both PC and PS cases. It suggests that similar to vdW force, ES force will also likely be a mechanism for the driving process for CDs, hence causing diffusion toward lipids. Yet, ES force likely to play a minor role in the driving of lipids and agents such as peptides and aptamers. The cause lies perhaps in the polarities of charges on participating agents, and thus, the ES force can play role either to favor or to disfavor in mutual diffusion toward each other.

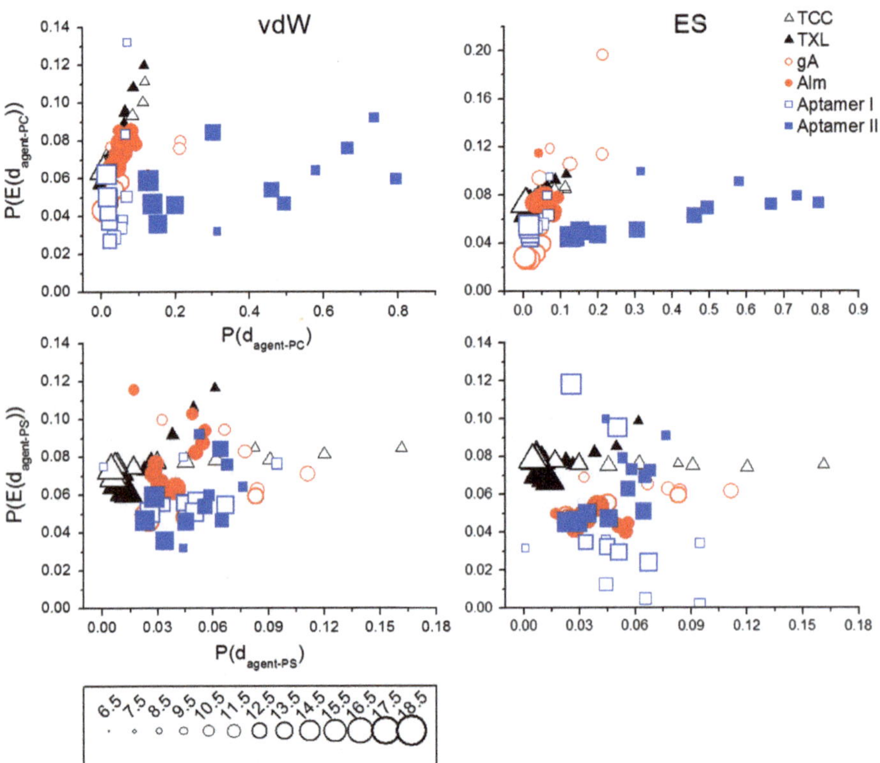


**Fig. 6.14** Analysis of MD results of all lipid–agent pairs. The left column plots van der Waals energy, and the right is electrostatic energy. The upper row is for agent–PC pair, and the lower row is for agent–PS pair. Values of $d_{\text{agent–lipid}}$ are represented by symbol sizes (for simplicity, only circle sizes are shown)

One very important message can be drawn from this novel treatment which is if we can detect all the important energies plying in the molecular-level interactions in a medium (in our case, in a phase state), in the interface of two media we can correctly predict the diffusion phenomena as bourn out of driving forces. In this regard, detection of energies in a single MAA–lipid complex (as done here in our MD simulations) in fact provides better molecular-level understanding than in a usually expected MAA–lipid complex in a membrane. This way we can detect the primary (not the collective) energy values. The overall energies in a MAA–lipid complex in a membrane will just be a combination of such many detected energies. Therefore, our MD simulation strategy and the detected energy-based discovery of the above-mentioned two probabilities appear as strong functions correlating the molecular-level information with the phenomenological observations of MAA–lipid complexes in a phase state as predicted from various in vitro experiments (see Tseng et al. 2011; Ashrafuzzaman et al. 2013, 2014). For more clarification, we may add another statement here. If we compare $d_{\text{agent–lipid}}$ (Fig. 6.14) with the lipid dimension, in the case of our current in silico single-agent single-lipid association assay, we may easily understand the cutoff value of the distance $d_{\text{agent–lipid}}$ beyond which the agent may be considered staying outside the lipid phase state (plasma-state phase). This novel treatment correlates information between in silico and in vitro experiments to address general diffusion and hence interactions between agents in a cell phase state.

The probability calculated in Table 6.1 and generally addressed in Fig. 6.14 is to represent distance between two agents. There is always a cutoff distance of association $d_{\text{cutoff}}$ between these two agents. $d_{\text{cutoff}}$ depends on the nature of the two phases as well as the biophysical properties of agent which has to diffuse between phase states. For the agents to diffuse from liquid phase state (water environment in this in silico simulation) to plasma phase state (lipid constructed membrane plasma phase state), we have to calculate probability functions for all values of $d$ of which the values falling within $d < d_{\text{cutoff}}$ will be the probability for diffusion into the plasma-state phase. This way the interphase and intraphase diffusion of agents can be addressed quite easily. To understand better more in system, validations and corresponding simulations (especially in whole membrane system) are required.

## 6.6 Cellular Transport in Fungal Associated Cells

We shall explain this for the case of a plant cell. A natural symbiotic association of a fungus with the roots of a vascular plant caused by a mycorrhiza is traditional means observed in most of the plants. In these cases, the co-colonization between fungus and the host plant roots is observed either intracellularly for arbuscular mycorrhizal fungi or extracellularly for the fungi known as ectomycorrhizal fungi. The co-localization often causes a seasonally (of the order of timescale equivalent to days-to-weeks-to-months) adjustable geometrical reorganization in the cellular shape, size which result in causing time adjusted in cell transports. Both quantitative and qualitative influences on cells are observed as a result of their fungal association, the mechanism which is supposedly beneficial for both fungi and cells. The transport mechanisms of all nutrients get affected or regulated. Such mechanisms are found to be highly dependent on the change of values of certain parameters related to the seasonal fluctuations like humidity, temperature, salinity, acidity of the cellular environment. The cell-based transport mechanisms therefore need to be understood first biologically. But the nanoscale cellular structural changes draw a clear biophysical perspective in the cell's fungal association process. Mechanical and electrical types of physical cellular processes get affected, and thus, general energetic perspectives that are found to be active in general cell transport mechanisms explained earlier in this chapter are also found to take place in the fungal infected plant cells. We wish to brief about such nanoscale biophysics here which may open up avenues to addressing the cell health condition from physical perspectives.

### 6.6.1 General Behavior of Fungus-Cell Co-localization

We have recently performed a series of experiments to address the biophysics of fungal association with plant cells which are in preparation for publication (Dhar and Ashrafuzzaman 2014). Our main goal is to understand how the transport mechanisms look like in comparison with general cells, which is the topic of this book. The experiment was carried out during the autumn started in October 2013 through the winter and spring up to the beginning of summer in the mid-April 2014 in the net house of King Saud University, Saudi Arabia. The experiment has been designed in completely randomized block design (CRBD). Soil free of mycorrhizal inoculum (pretested for the presence of mycorrhizal inoculum) was used as basic growth media. As indicator, plant Zea mays has been used as it is well established to be the good mycotrophic plant. Prepared inoculum has been used at the rate of 20% (V/V) of the growth medium. In the net house, a thin layer of polyethylene sheet was spread to inhibit the contamination from the net-house floor. In the net-house temperature, humidity and other microclimatic factors were as like as the atmospheric condition. The pots were irrigated to the field capacity of the growth media at one-day interval. A thermometer was hanged in the experimental site to record the everyday temperature of the net house.

Two treatments were maintained in the experiment. Without mycorrhizal inoculum was the control and mycorrhizal inoculum added pots were treated as inoculated. In each pot, two pre-germinated seeds of *Z.* mays were sown, and after germination, pots were thinned to one plant per pot. Pots were supplied with distilled water up to the field capacity at every alternate day to maintain the homogeneous moisture content of the soil. On October 13, 2013, the experiment was set up and at every 30-day interval data were recorded. As a whole, we had six harvestings throughout October 2013 to April 2014. In each harvesting, total pot was used and the roots of the whole plant were taken. Roots were cleaned with tap water and followed by distilled water and cut into 1 cm pieces and stained in trypan blue following the method of Koske and Gemma (1989). Roots were heated in 2.5% KOH solution for 20 min at 80 °C in water bath followed by acidifying in 1% HCl for 24 h. Staining was followed in 0.05% trypan blue solution in acidic glycerol and heated at 80 °C in water bath for 10 min. Randomly, 20 segments were picked up and loaded on the microscopic slides after destaining in 50% acidic glycerol. A cover glass was put on the root segments and gently was pressed to macerate the roots segments facilitating the easy observation of the AM structures inside the roots. Microscopic observation was done with a computerized digital Olympus microscope (Model DP-72). Primarily, roots were observed at 10× magnifications, and final observation and data were recorded at 40× magnification. Length and breadth of the colonized cell and arbuscule within the cell were measured with the CellSans software to calculate the area of the cells colonized by arbuscules and arbuscules within the colonized cell. At least 10 cells were considered randomly in each root segment to measure the area of cells and arbuscules. An average of cell area and arbuscular area of total root segments studied was determined for each harvesting. Arbuscule areas to cell area, cytoplasm area to cell area, arbuscule to cytoplasm ratio were calculated. Obtained results were used for the explanation of the biophysical analysis of the nutrient transfer mechanisms.

We have presented here four figures, one (Fig. 6.15) as control and three (Figs. 6.16, 6.17, and 6.18) to show the seasonally adjusted co-localization between mycorrhiza and plant cells. The readers may read from sole mycorrhiza-focused articles (e.g., see Cavagnaro et al. 2001; Dickson 2004; Pumplin and Harrison

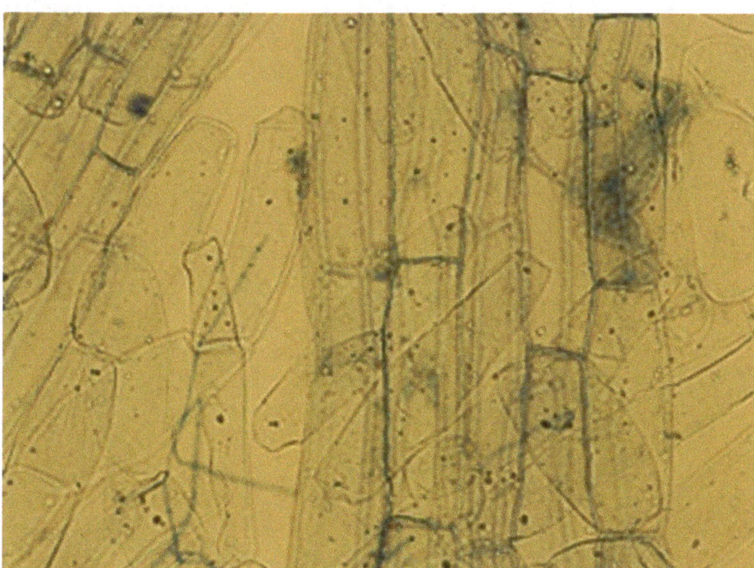

**Fig. 6.15** Root cell in the control treatment without mycorrhizal inoculum (Dhar and Ashrafuzzaman 2014)

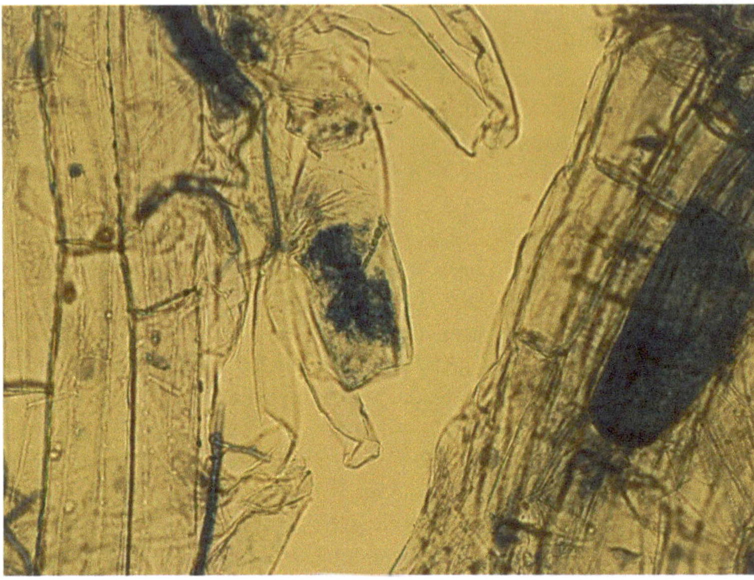

**Fig. 6.16** Root cell with arbuscule at 30 days with mycorrhizal inoculum (Dhar and Ashrafuzzaman 2014)

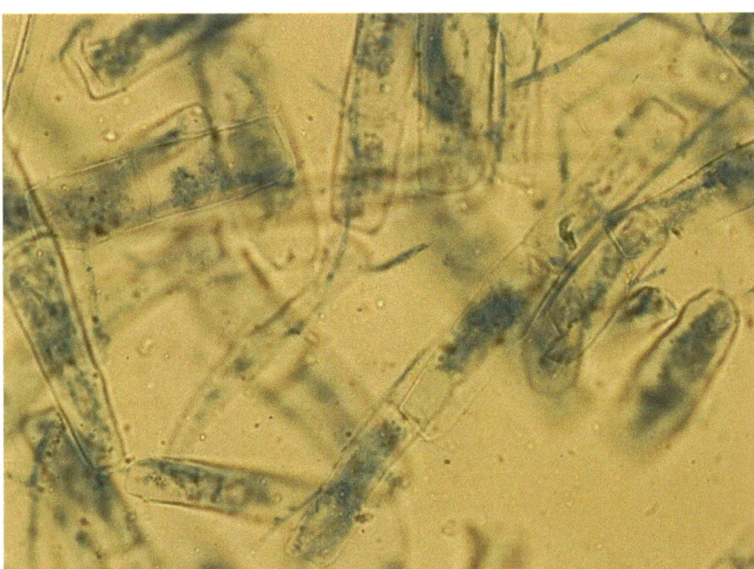

**Fig. 6.17** Root cell with arbuscule at 120 days with mycorrhizal inoculum (Dhar and Ashrafuzzaman 2014)

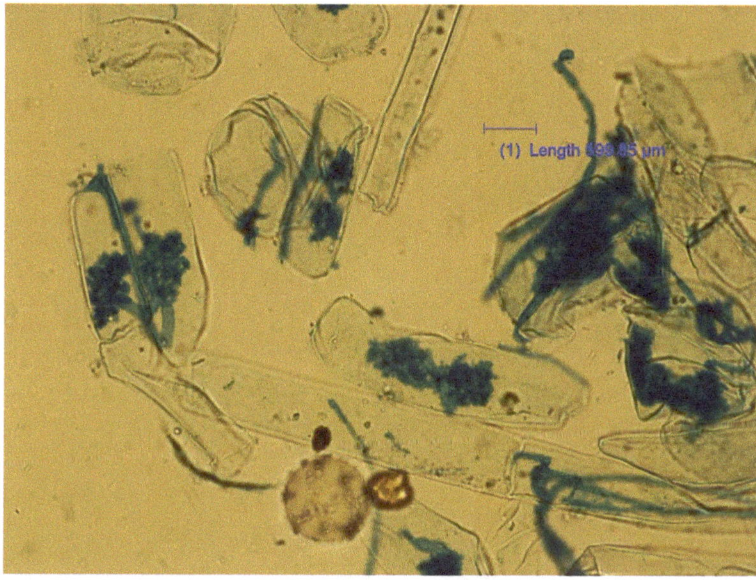


**Fig. 6.18** Root cell with arbuscule at 150 days with mycorrhizal inoculum (Dhar and Ashrafuzzaman 2014)

2009; Ferrol et al. 2000; Morte and Honrubia 2004; Redecker et al. 2013) to understand the in-depth biological perspectives of this co-localization. The images show the geometry and how that geometry changes with the age of the plant. The climate condition and other environment and season-determined parameters have a huge regulatory effect on the relative distribution. We shall focus mainly at the transport across the membrane separating the arbuscule and cell cytoplasm. For that, we wish to address the relative cellular space occupancy as a result of mycorrhizal attack of the cell. To inspect carefully, we have taken the image of a single fungus-cell co-localization here, see Fig. 6.19.

The scenario shown in above-mentioned single co-localized fungus-cell is subject to slow dynamics. A time-dependent change of the relative distribution is observed over weeks-to-months. To monitor such change in a life cycle, we measured the surface areas of both arbuscule occupied region and the rest of the cytoplasm of the cell. A statistical analysis shows the following seasonal changes, as summarized in Fig. 6.20.

### *6.6.2 Model Representation of the Fungus-Cell Co-localization*

By careful inspection of the biological experimental results, a few presented in, e.g., Figures 6.15, 6.16, 6.17, 6.18, and 6.19 and the plots in Fig. 6.20, we may conclude

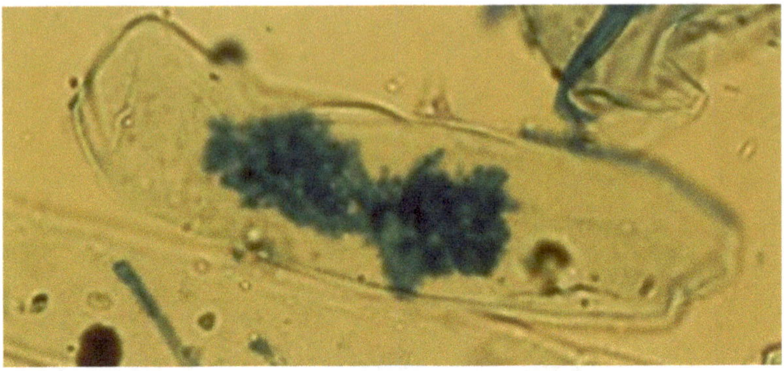

**Fig. 6.19** Image of a single fungus-cell co-localization. The arbuscular penetration into cell's interior region separates the cell cytoplasm. The arbuscular distribution is controlled by a membrane separation from cytoplasm (Dhar and Ashrafuzzaman 2014)

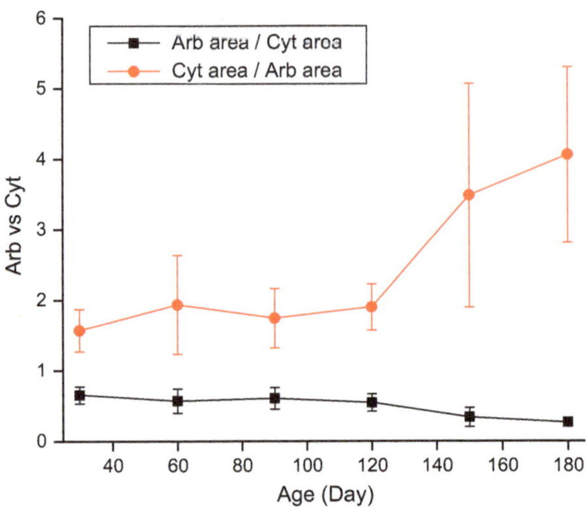


**Fig. 6.20** Arbuscule (Arb area) versus cytoplasm (Cyt area) occupied surface area changes with time [age of the plant, age (day)]. This is what, we wish to call a slow seasonal cell dynamics. The cell surface area is found to be more or less constant although shape/geometry shows somewhat fluctuations

the models on the time-dependent change of the fungus-cell co-localization as presented in Figs. 6.21, 6.22, and 6.23 [manuscript in preparation (Ashrafuzzaman 2016)]. In these three different sketches (Figs. 6.21, 6.22, and 6.23), for modeling purposes only, we have intentionally used different geometry to show their randomness.

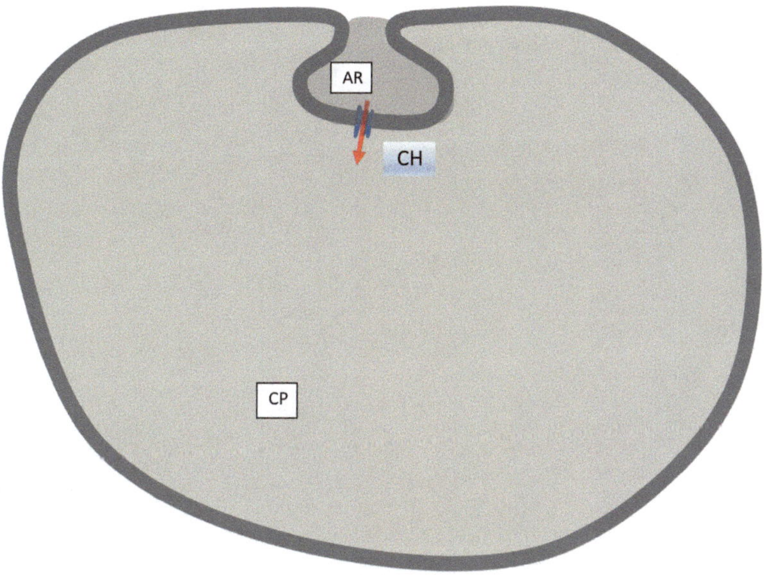


**Fig. 6.21** Fungus-cell co-localized model at plant age = $T_1$ days. Arbuscule (AR) and cellular interior region, mainly occupied by cytoplasm and other constituents is referred as CP, are separated by cell membrane (thick dark line). Ion channels (CHs) transport nutrients, sodium, potassium, phosphorous, etc., from AR to CP region. For simplicity, only one channel (CH) has been sketched, but there are usually many CHs. The number of CHs is expected to vary and probably increases/decreases, with the increase/decrease of the volume occupied by AR, corresponding to simple analogy that the surface area of the volume occupied by AR accounts proportionally for the number of CHs

### *6.6.3 Membrane Transport Across the Membrane Partition in a Fungus-Cell Co-localized State: Analytical Expressions*

Figures 6.21, 6.22, and 6.23 can be generalized into one representative model diagram for biophysical understanding of the transport mechanisms, as shown in Fig. 6.24.

Arbuscular membrane transports nutrients through general membrane diffusion and utilizing specific ion channel transport mechanisms. Higher nutrient concentrations in the arbuscules and lower concentrations in the root cells of host plant lead to creating nutrient concentration gradients. Thus, the efflux of nutrients through arbuscular membrane occurs. Ion channel CHs of various natures experience functional state transitions between opening and closing states, and thus, the nutrient transport through channels happens. The ion channel functions are regulated due to their physical and energetic coupling with arbuscule membrane. Therefore, membrane–ion channel coupling mechanisms are expected to regulate the ion channel transport of nutrients across the membrane separating nutrients on

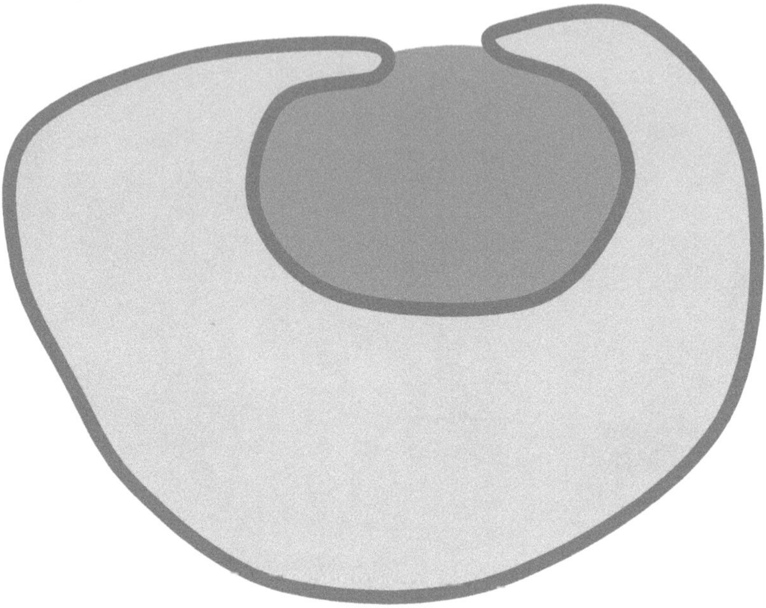

**Fig. 6.22** Fungus-cell co-localized model at age = $T_2$ days ($T_2 > T_1$). For simplicity, no ion channel has been sketched here (see Fig. 6.21). Cell shape, AR and CP, occupied regions' geometry may be varied with time

both sides of the membrane, namely inside arbuscule and cytoplasm. Here, I shall address this biophysical phenomena by developing a clear energetic model on ion channel transport (Ashrafuzzaman 2016) in light of the biological arbuscule transport results recently produced by Dhar and Ashrafuzzaman (2014) (to be submitted elsewhere). The ion channel CH, as sketched in Fig. 6.24, co-localization with membrane can be modeled equivalent to general cell membrane-hosted ion channel case, as shown in Fig. 6.25 (see Ashrafuzzaman and Tuszynski 2012a).

According to the model structure (see Fig. 6.24), let us assume the concentration of nutrients in cytoplasm $C_i$ is due to transported nutrients across the membrane. Therefore,

$$C_i \sim c_i\{a_i/(A_i - a_i)\}[M_{CH} + M_{diff}] \tag{6.25}$$

Here, concentration follows the traditional definition that is the abundance of nutrients divided by the total volume. $M_{CH}$ and $M_{diff}$ are channel and nonchannel diffusion coefficients of nutrients, respectively, with

$$M_{CH} \text{ and } M_{diff} \leq 1 \tag{6.26}$$

Here, we can consider a constant value for the contribution from general diffusion coefficient $M_{diff}$, for simplicity. The contribution of $j$th channel into $M_{CH}$ is $M_{CH,j}$, so

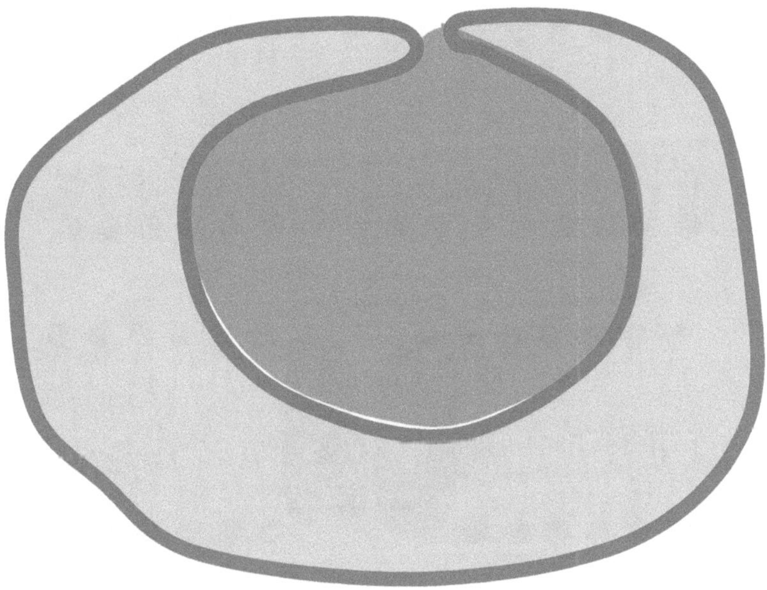

**Fig. 6.23** Fungus-cell co-localized model at age = $T_3$ days ($T_3 > T_2 > T_1$). For simplicity, no ion channel has been sketched here (see Fig. 6.21). Cell shape, AR and CP, occupied regions' geometry may be varied with time

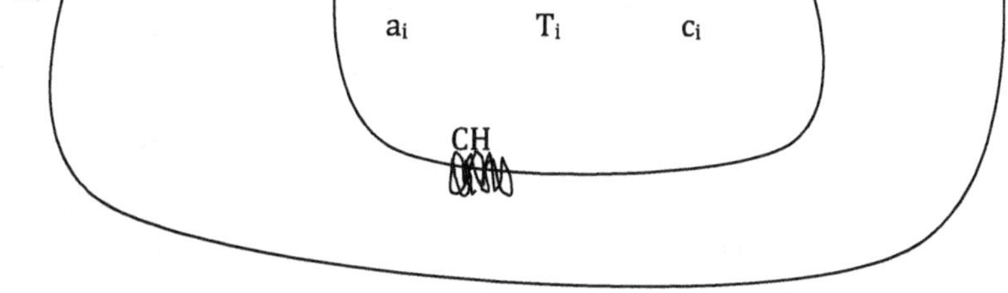


**Fig. 6.24** Fungus-cell co-localized model at an arbitrary age. Arbuscule and cellular interior region, mainly occupied by cytoplasm and other constituents, are separated by cell membrane. Ion channel CHs transport nutrients, sodium, potassium, phosphorous, etc., from arbuscule to cytoplasm region. For simplicity, only one CH has been sketched, but there are usually many CHs. The number of CHs is expected to vary and probably increases/decreases, with the increase/decrease of the volume occupied by arbuscule. Here, $A_i$ is the cell volume, and $a_i$ is the volume occupied by arbuscule. The concentration of nutrients in cytoplasm is $C_i$ and in arbuscule $c_i$

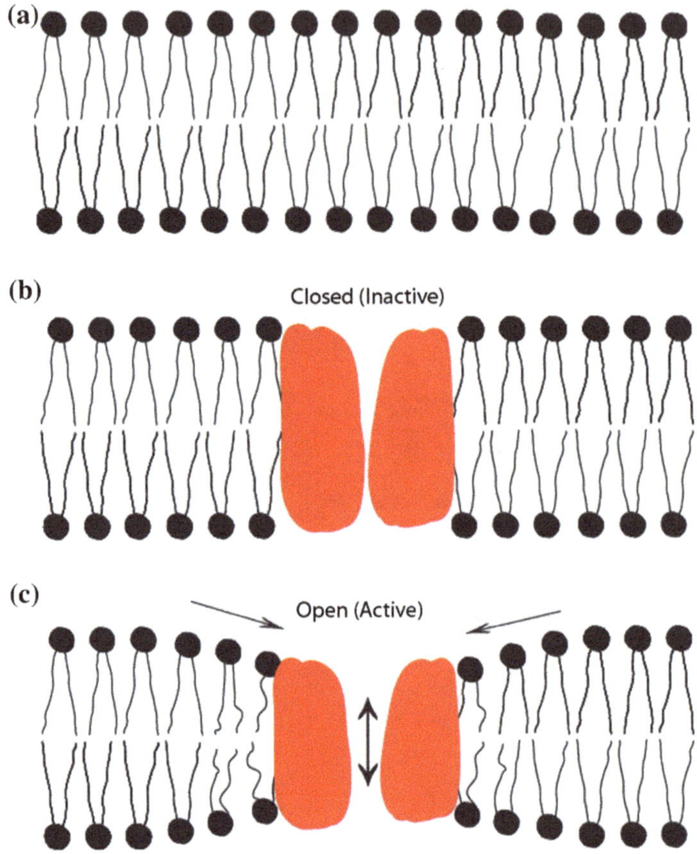


**Fig. 6.25** Lipid bilayer membrane exists with an average constant thickness unless any random MP or ordered channel induces any kind of bilayer deformations. In diagram **a**, a plain lipid bilayer membrane has been schematized. Lipid bilayer behaves like a barrier against ion or nutrient flow across the membrane. In **b**, a lipid bilayer membrane with a possible inactive ion channel (in red color) within it has been schematized. In this case, despite the presence of ion channel lipid bilayer still behaves as a barrier against ion flow across the membrane. The channel in this case is apparently considered to be inactive or nontransporting agent. In **c**, a lipid bilayer membrane with a possible active ion channel has been schematized. In this case unlike the cases represented by (**a**) and (**b**), the bilayer conducts ions, nutrients, etc., across the membrane. Decrease or increase in bilayer thickness near the channel interface as the channel undergoes the conformational transition from closed to open states or vice versa is an important hypothesis in this cartoon representation. A structural change (especially size) in the channel-forming agents is also a prediction. An energetic cost happens while channel undergoes transition from closed (**b**) to open state (**c**). This energetic costs have recently been calculated using the charge-based screened Coulomb interaction model (Ashrafuzzaman and Tuszynski 2012a, b; Ashrafuzzaman et al. 2014). These schematic diagrams have been drawn in light of the gating mechanisms observed in mechanosensitive channels in model membranes (Perozo et al. 2002)

$$M_{\mathrm{CH}} = \sum_j M_{\mathrm{CH},j} = (1/N_{\mathrm{CH}}) \sum_j \mu^r_{\mathrm{CH},j} = (n_{\mathrm{CH}}/N_{\mathrm{CH}})\mu_{\mathrm{CH}} \tag{6.27}$$

$\mu_{\mathrm{CH}}$ is the average nutrient conductance probability across an active channel. $n_{\mathrm{CH}}$ is the number of active channels. $N_{\mathrm{CH}}$ is the total number of channels (active and inactive) that may contribute to the nutrient transport across the arbuscule membrane into the cell cytoplasm. $n_{\mathrm{CH}} \leq N_{\mathrm{CH}}$. $r$ is an exponent dependent on the channel type (see Ashrafuzzaman and Tuszynski 2012b; Ashrafuzzaman et al. 2008; Latorre and Alvarez 1981). For simplicity, let us assume $r = 1$. We get:

$$C_{\mathrm{i}} \sim c_{\mathrm{i}}\{a_{\mathrm{i}}/(A_{\mathrm{i}} - a_{\mathrm{i}})\}[(n_{\mathrm{CH}}/N_{\mathrm{CH}})\mu_{\mathrm{CH}} + M_{\mathrm{diff}}] \tag{6.28}$$

Or,

$$C_{\mathrm{i}} \sim c_{\mathrm{i}}\{a_{\mathrm{i}}/(A_{\mathrm{i}} - a_{\mathrm{i}})\}[(n_{\mathrm{CH}}/N_{\mathrm{CH}})\exp\{-\Delta G_{\mathrm{CH}}/k_{\mathrm{B}}T\} + M_{\mathrm{diff}}] \tag{6.29}$$

Here, we have derived the expression for $\mu_{\mathrm{CH}}$ from information in various publications, e.g., see Lundbæk et al. (2005), Ashrafuzzaman et al. (2006), Ashrafuzzaman and Tuszynski (2012a),

$$\mu_{\mathrm{CH}} \sim \exp\{-\Delta G_{\mathrm{CH}}/k_{\mathrm{B}}T\} \tag{6.30}$$

$\Delta G_{\mathrm{CH}}$ is the energetic changes taking place due to open/close transition mechanism (see Fig. 6.25) that is the amount of energy required to open a channel, often referred to as 'free energy' in channel opening; $k_{\mathrm{B}}$ is Boltzmann's constant; and $T$ is absolute temperature. The back-and-forth channel open/close transition is schematized in cell membrane-hosted ion channel case as shown in Fig. 6.25.

### *6.6.4 Numerical Computation to Address the Theoretical Model of the Arbuscular Membrane Transport*

We may deduct, from Eq. (6.29), the following analytic expression for the arbuscule transport of nutrients into cell cytoplasm:

$$C_{\mathrm{i}}/[c_{\mathrm{i}}\{a_{\mathrm{i}}/(A_{\mathrm{i}} - a_{\mathrm{i}})\}] - M_{\mathrm{diff}} = (n_{\mathrm{CH}}/N_{\mathrm{CH}})\exp\{-\Delta G_{\mathrm{CH}}/k_{\mathrm{B}}T\} \tag{6.31}$$

We can plot the above-mentioned function for constant $M_{\mathrm{diff}}$ at an equilibrium thermodynamic condition. For that, we need to develop an acceptable expression for $\Delta G_{\mathrm{CH}}$. In a recent study on specific protein-lined ion channels, namely gA and Alm channels, a general expression has been proposed for $\Delta G_{\mathrm{CH}}$ in which a membrane–channel mismatch term $\Delta G_{\mathrm{CH}}$ (mismatch) and a membrane curvature (Gruner 1985; Lundbaek et al. 2005; Ashrafuzzaman et al. 2006; Greisen et al. 2011) term $\Delta G_{\mathrm{CH}}$ (curvature) have been discovered using traditional charge-based lipid–channel interactions, following screened Coulomb interaction formalisms (Ashrafuzzaman and Tuszynski 2012a, b; Ashrafuzzaman et al. 2014). Here,

$$\Delta G_{\mathrm{CH}} = \Delta G_{\mathrm{CH}}(\text{mismatch}) + \Delta G_{\mathrm{CH}}(\text{curvature}) \tag{6.32}$$

For a certain type of membrane, the curvature term appears as common term (can be considered for simplicity reason), but the mismatch term regulates the channel functions due to various localized membrane and channel structures especially from the geometric perspectives. The mismatch term appears with the following expressions (Ashrafuzzaman and Tuszynski 2012a, b; Ashrafuzzaman et al. 2014):

$$\Delta G_{\mathrm{CH}}(\text{mismatch}) \sim \exp\{x/\delta_{\mathrm{CH}}\} k_{\mathrm{B}} T \tag{6.33}$$

Here, $x$ accounts for the mismatch between membrane thickness ($d$) and channel length ($l$), that is, $x = d - l$. $\delta_{\mathrm{CH}}$ is a characteristic length or correlation length that determined the length scale unit over which channel–lipid membrane interaction extends. We can assume $\delta_{\mathrm{CH}}$ proportional to the lipid chain lattice constant that is equivalent to a lipid headgroup cross-sectional length. For understanding the deformation of the membrane, one can consider the equivalent case as shown in the channel–bilayer coupling interface in Fig. 6.25c.

For simplicity, we shall use $\Delta G_{\mathrm{CH}}$ (mismatch) term for $\Delta G_{\mathrm{CH}}$ here and perform a NC to understand the trend of channel-induced arbuscule membrane transport mechanism. In reality, various lipid compositions in arbuscule membrane cause the membrane transport to be also regulated but we wish to leave that (e.g., regulation due to lipid curvature) for future studies.

Following exactly the numerical methods addressed in Ashrafuzzaman and Tuszynski (2012b), we used Mathematica 9.1 (Wolfram Mathematica at http://www.wolfram.com) to develop a program for NC/simulation. The results are presented in Fig. 6.26.

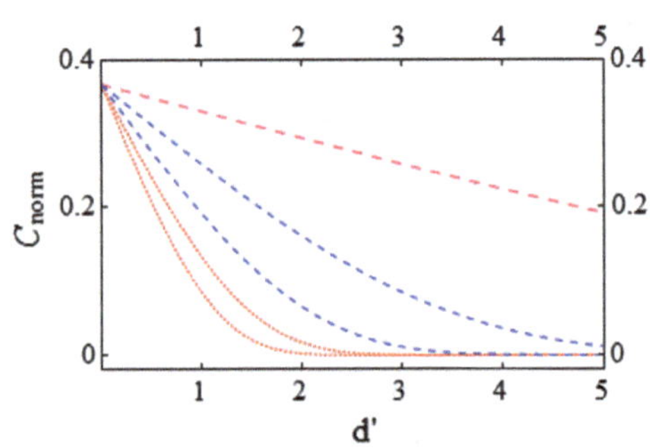


**Fig. 6.26** $C_{\mathrm{norm}} = C_i/[c_i\ \{a_i/(A_i - a_i)\}] - M_{\mathrm{diff}}$ for the case when $n_{\mathrm{CH}}/N_{\mathrm{CH}} = 1$ that is when all channels are active. $d' = d/\delta_{\mathrm{CH}}$. Here, $\delta_{\mathrm{CH}} \approx 0.8$ nm (Eze 1991). Left (or bottom) to right (or top) plots are for channel length occupying 10, 30, 50, 70, 90% of the membrane thickness. The nutrients' transport drastically falls as the mismatch between membrane thickness and channel length increases. $y$-axis values are arbitrary

As $x$ increases, $C_i$ decreases because as observed theoretically the nutrient transport becomes harder in thicker membrane due to higher membrane–ion channel mismatch (see analogy in Ashrafuzzaman and Tuszynski 2012a, b). This gets reflected in cytoplasm/arbuscule area ratio which increases with the increase of age (presented earlier here and see Alexander et al. 1989). This suggests that as cytoplasm area increases, the concentration (nutrients per area) automatically goes down; consider that the supply goes down because the arbuscule area decreases. Here, instead of per volume (theoretical), we have compared the biological data as per area due that we can image the areal not the volume views using the technique we have adapted here.

An especial consideration can also be made. With the increase of arbuscular surface area, the membrane thickness is assumed to be decreasing. This is because the increase in the arbuscular surface area is assumed to result in causing a slow increase in outward (inward to the cell cytoplasm) pressure and thus causes the thinning of the membrane (d decreases). A membrane-thinning mechanism theoretically appears to cause greater inward to cell cytoplasm nutrient flux. This predicted seasonal or age-dependent arbuscule membrane-thinning/-thickening mechanism can be addressed using higher resolution three-dimensional imaging techniques, not covered in our current studies.

Channel length smaller means larger mismatch which also causes arbuscule to cytoplasm transport to get reduced. This gets reflected in our analytical analysis through the numerical data plotted in Fig. 6.26. This was not possible to be addressed biologically using the techniques we used (Dhar and Ashrafuzzaman 2014).

### 6.6.5 Concluding Remarks: Slow Plant Cell Transport Can Be Modeled Biophysically

We performed a biophysical modeling of the traditional biologically observable transport properties of mycorrhizal fungus-colonized plant cell. The biophysical studies are found to explain the symbiotic relationship of mycorrhiza with the plant root system in regard to the nutrient uptake mechanisms quite well.

It has been known quite well that the plant age has an impact on the arbuscular colonization of the cell, arbuscular area, cytoplasmic area, and as a result their areal ratios. With increasing plant age, the arbuscular colonization increases until the fungus occupies a certain highest proportion of the cell volume, e.g., 50–60% (Dhar and Ashrafuzzaman 2014), and in other cases of oats, the fungus occupies 36% of the oat cells (Alexander et al. 1989). At this highest arbuscular occupancy of the cell volume, the frequency and arbuscular area are optimal and their activities are also found at the peak point for quite a long time which is of the order of couple of months (Alexander et al. 1989) when the ratio takes a reverse direction until the end of plant lifetime.

NC (Ashrafuzzaman 2016) addresses a theoretical prediction on the nutrient flow mechanisms across arbuscule membrane through the membrane-hosted channels. A valid theoretical technique developed for a class of certain types of (protein-lined type) channels appears to provide an understandable model representation of the experimentally observable nutrients' conditions in both cell cytoplasm and arbuscule. Cell health condition as addressed here through the state of relative nutrient concentrations of cell cytoplasm and arbuscule is explained biologically. The modeling, presented here, has focused on addressing a theoretical analogy to the biologically observable parameters. NC produces the theoretical trends of the cell nutrients' condition that are found to be able to explain the biological scenario. The membrane physical parameters are found to appear as regulators of membrane transport properties. The membrane physical parameters are reported to get regulated due to various membrane-active agents like various drugs colchicine, taxol, aptamers, peptides, through direct physical interactions or binding (e.g., see Ashrafuzzaman et al. 2012, 2013, 2014). Identical analogy may be applicable for various plant cell active drugs. The current computational formalism as presented here may be extended to testing plant cell active drugs, and thus, perhaps even genetic diversification may also be explained computationally.

In conclusion, this study is expected to help address general biophysical modeling of the biological system 'plant cell' that is fed with a nutrient influx by arbuscular mycorrhizal fungi to maintain the system's health condition (Dhar and Ashrafuzzaman 2014). This novel approach may help address the general nutrient intake-dependent health condition of various biological systems like plant cells. Our modeling may help extend the study of the effects of various cell-targeted drugs (Ashrafuzzaman et al. 2012, 2013, 2014) that might be applied onto the plant cells to test if they also regulate the cell survival period, repair the cellular disorders, enhance the cell health condition via regulating the mechanisms of slow cell transport of nutrients and many similar ones.

## References

Lewis, B.A; Engelman, D.M. Lipid bilayer thickness varies linearly with acyl chain length in fluid phosphatidylcholine vesicles. J. Mol. Biol., 1983, 166, 211–217.

Benz, R.; Fröhlich, Läuger, P.; Montal, M. Electrical capacity of black lipid films and of lipid bilayers made from monolayers. Biochim. Biophys. Acta, 1975, 394, 323–334.

Simon S A, McIntosh T J and Latorre R 1982 Science 216 65–7.

Harper, P.E.; Mannock, D.A.; Lewis, R.N.A.H.; McElhaney, R.N.; Gruner, S.M. X-Ray diffraction structures of some phosphatidylethanolamine lamellar and inverted hexagonal phases. Biophys. J., 2001, 81, 2693–2706.

M. O. Eze. Phase Transitions in Phospholipid Bilayers: Lateral Phase Separations Play Vital Roles in Biomembranes. BIOCHEMICAL EDUCATION 19(4) 1991.

Einstein, A. (1905). "Über die von der molekularkinetischen Theorie der Wärme geforderte Bewegung von in ruhenden Flüssigkeiten suspendierten Teilchen". Annalen der Physik (in German) 322 (8): 549–560. Bibcode:1905AnP...322..549E. https://doi.org/10.1002/andp.19053220806.

von Smoluchowski, M. (1906). "Zur kinetischen Theorie der Brownschen Molekularbewegung und der Suspensionen". Annalen der Physik (in German) 326 (14): 756–780. Bibcode:1906AnP...326..756 V. https://doi.org/10.1002/andp.19063261405.

S. Achuthan, B.J. Chung, P. Ghosh, V. Rangachari, and A. Vaidya. A modified Stokes-Einstein equation for Aβ aggregation. BMC Bioinformatics 2011, 12(Suppl 10):S13.

Ashrafuzzaman, M., 2015a. Diffusion across cell phase states. Biomedical Sci. Today. 1:e4.

Ashrafuzzaman, M., 2015b. Phenomenology and energetics of diffusion across cell phase states. Saudi J. of Biol. Sci., 22: 666–673.

Quemada D: Rheology of concentrated disperse systems. I. Minimum energy dissipation principle and viscosity-concentration relationship. Rheologica Acta 1977, 16(1):82–94.

Quemada D: Rheological modelling of complex fluids. I. The concept of effective volume fraction revisited. European Physical Journal of Applied Physics 1998, 1(1):119–127.

Lee JD, So JH, Yang SM: Rheological behavior and stability of concentrated silica suspensions. Journal of Rheology 1999, 43(5):1117–1140.

Wolthers W, Van den Ende D, Duits MHG, Mellema J: The viscosity and sedimentation of aggregating colloidal dispersions in a Couette flow. Journal of Rheology 1996, 40(1):55–67.

C-Y. Tseng, Md. Ashrafuzzaman, J. Mane, J. Kapty, J. Mercer, J. Tuszynski, Entropic fragment based approach to aptamer design, Chem Biol Drug Des (2011) 78, 1–13.

Ashrafuzzaman M., Tseng C.Y., Kapty J., Mercer J.R., Tuszynski J.A. A computationally designed DNA aptamer template with specific binding to phosphatidylserine. Nucleic Acid Ther. 2013 Dec; 23(6):418–26.

Ashrafuzzaman M., Beck H. 2004a. Vortex dynamics in two-dimensional Josephson junction arrays, (University of Neuchatel, http://doc.rero.ch/record/2894?ln=fr), ch 5, p 85.

M Ashrafuzzaman, H Beck. 2004b. Vortex dynamics in dilute two-dimensional Josephson junction arrays. J. Magnetism Magnetic Materials 272, 284–285.

Ashrafuzzaman, M., Tseng, C.-Y., Duszyk, M., Tuszynski, J. Chemotherapy Drugs Form Ion Pores in Membranes Due to Physical Interactions with Lipids. Chem. Biol. Drug Des., Volume 80, Issue 6, December 2012, Pages 992–1002.

Ashrafuzzaman, M., Tseng C.Y., Tuszynski J.A. Regulation of channel function due to physical energetic coupling with a lipid bilayer. Biochem. Biophys. Res. Commun. 2014 Mar 7;445 (2):463–8.

CASE, D.A., DARDEN, T.A., CHEATHAM, T.E. III, SIMMERLING, C.L., WANG, J., DUKE, R.E., LUO, R., CROWLEY, M., WALKER, R.C., ZHANG, W., et al. (2008). AMBER 10. (University of California, San Francisco).

CASE, D.A., DARDEN, T.A., CHEATHAM, T.E. III, SIMMERLING, C.L., WANG, J., DUKE, R.E., LUO, R., CROWLEY, M., WALKER, R.C., ZHANG, W., et al. (2010). AMBER 11. (University of California, San Francisco).

P. P. Dhar and Md. Ashrafuzzaman. CELL MEMBRANE REGULATES THE MYCORRHIZAL FUNGAL TRANSPORT OF NUTRIENTS. 2014 (to be submitted).

Koske, R.E. & Gemma, J. N. 1989. A modified procedure for staining roots to detect VA mycorrhizas. **Mycological Research 92**: 486–505.

Cavagnaro, T.R., Gao, L.-L., Smith, A.F. and Smith, S.E. (2001). Morphology of arbuscular mycorrhizas is influenced by fungal identity. *New Phytol.* 151: 469–475.

Dickson, S. (2004) The *Arum–Paris* continuum of mycorrhizalsymbioses. *New Phytol.* 163: 187–200.

Pumplin, N., and Harrison, M.J. (2009). Live-cell imaging reveals periarbuscular membrane domains and organelle location in Medicago truncatula roots during arbuscular mycorrhizal symbiosis. Plant Physiol. 151: 809–819.

Ferrol, N.; J.M. Barea & C. Azcon-Aguilar. 2000. Molecular approaches to study plasma membrane HC-ATPases inarbuscular mycorrhizas, *Plant and Soil* 226: 219–225.

Asuncion Morte and Mario Honrubia, 2004. Ultrastructure of the mycorrhiza formed by *Tetraclinisarticulata* (Vahl) Masters (*Cupressaceae*)Anales de Biologia 26: 179–190.

Redecker, D.; Schüßler, A.; Stockinger, H.; Stürmer, S.L.; Morton, J. B. & Walker, C. 2013. An evidence-based consensus for the classification of arbuscular mycorrhizal fungi (*Glomeromycota*). **Mycorrhiza 23**:515–531.

M. Ashrafuzzaman. Theoretical and computational understanding of the cell membrane regulation of the mycorrhizal fungal transport of nutrients. 2016 (to be submitted).

Ashrafuzzaman, M., Tuszynski, J., Membrane Biophysics, Springer (Heidelberg), 2012a, ISSN 1618-7210, ISBN 978-3-642-16104-9 ISBN 978-3-642-16105-6 (eBook), https://doi.org/10.1007/978-3-642-16105-6.

Ashrafuzzaman, M., and J. A. Tuszynski. 2012b. Regulation of channel function due to coupling with a lipid bilayer, J. Comput. Theor. Nanosci. 9: 564–570.

Perozo, E., Cortes, D.M., Sompornpisut, P., Kloda, A., and Martinac, B. 2002. Open channel structure of MscL and the gating mechanism of mechanosensitive channels. Nature 418: 942–948.

Md. Ashrafuzzaman, O.S. Andersen, R.N. McElhaney. The antimicrobial peptide gramicidin S permeabilizes phospholipid bilayer membranes without forming discrete ion channels. Biochimica et Biophysica Acta 1778 (2008) 2814–2822.

R. Latorre, R. Alvarez, Voltage-dependent channels in planar lipid bilayer membranes, Physiol. Rev. 61 (1981) 77–150.

Lundbæk J A, Birn P, Tape S E, Toombes G E, Sogaard R, Koeppe R E II, Gruner S M, Hansen A J and Andersen O S. 2005. Mol. Pharmacol. 68 680–9.

Md Ashrafuzzaman, M A Lampson, D V. Greathouse, R E. Koeppe II, and O S Andersen. Manipulating lipid bilayer material properties using biologically active amphipathic molecules. J. Phys.: Condens. Matter 18 (2006) S1235–S1255.

Gruner S M 1985 *Proc. Natl Acad. Sci. USA* **82** 3665–9.

Per Greisen, Jr., Kevin Lum, Md. Ashrafuzzaman, Denise V. Greathouse, Olaf S. Andersen, and Jens A. Lundbæk. Linear rate-equilibrium relations arising from ion channel-bilayer energetic coupling. PNAS, August 2, 2011, vol. 108, 12717–12722.

Alexander, T., Toth, R., Meier, R. and Weber, H.C. (1989) Dynamics of arbuscule development and degeneration in onion, bean, and tomato with reference to vesicular-arbuscular mycorrhizae in grasses. *C an. J. Bot.* 67: 2505–2513.

# Chapter 7
# Cell-Based Nanotechnology—Interactions and Energetics

Cells have natural nanometer (nm) scale structures and nanotechnologies active inside. In earlier chapters, we have addressed enough of the nm-dimensional aspects of cells. Cell membrane thickness is of the order of 3–5 nm (for details, see Ashrafuzzaman and Tuszynski 2012a), membrane constituent lipid cross section is of the order of 0.6 $nm^2$ (Eze 1991), membrane-transporting integral ion channels maintain nm dimensions in lengths and cross sections as their dimensions are limited within membrane dimensions, microtubule filaments are of the order of 25 nm diameters, etc. There is a huge number of cellular processes that structurally fall in nanoscale class. Similarly, there are many such processes that function in nanoscale dimensions, for example lipid curvature geometry profile regulation, migration of lipids on the membrane surface with the change of the mechanical property of the cell membrane, change of interlipid separation to cause thermodynamic membrane phase transitions between various phases like cubic, liquid crystal, hexagonal, etc., membrane protein–lipid membrane association/dissociation as a result of electrical charge-based interactions (Ashrafuzzaman and Tuszynski 2012b), microtubule polymerization through formation of $\alpha\beta$ heterodimers of tubulins (Sekulić and Satarić 2015), and many such processes. Above all, most of the interactions happen at low nanoscale dimensions or several angstroms scales. Nanoscale physics and engineering are cellular reality in both normal functional and abnormal conditions.

Nanotechnology exists in many naturally occurring cellular processes. They are generally found active in nanomachines of cells. In Chap. 5, we have explained about a few such molecular machines. As the cell holds many nanomachines and nanotechnology processes it is natural to consider the application of artificial nanotechnology in these cellular processes. Imaging of cells to understand the cell components' structures and behavior requires machineries to apply nanotechnology. Manipulating cellular processes, altering cell components' structural disorders, performing artificially induced interactions, migrating external agents into cellular sections, etc., can easily be performed with the application of nanoscale physics and technology. This chapter will address on these aspects quite in detail.



M. Ashrafuzzaman, *Nanoscale Biophysics of the Cell*,
https://doi.org/10.1007/978-3-319-77465-7_7

We shall address on the structural aspects of nanoscale cellular components where the nanotechnology approaches may naturally or artificially be found applicable. We shall extend the analysis by providing some details on the theoretical aspects of the low-scale interactions and energetics active in cell environment.

## 7.1 Nanoscale-Dimension Building Blocks Constructing Large and Small Biomolecules

DNA, RNA, and protein have various structures. The structures are also found to undergo transitions through various geometric structural states with state-specific energy conditions. For example, the 20 amino acids that are primarily responsible to make the polymer protein may have different arrangements among them. A primary protein structure which is nothing but a chain of amino acid is shown in Fig. 7.1.

Amino acids are building blocks of proteins and therefore play central roles both in construction of proteins and in determining metabolism. All of these 20 amino acids of proteins play crucial roles in biological actions. Only 10 of 20 amino acids are produced in human body, and the rest are supplied through daily food intake. Besides the 10 naturally produced in human body, a regular supply of the rest 10 has to be ensured to avoid degradation of protein structures in body. Quality food choice is, therefore, an essence for sustainable strength of body muscle and all protein-related organs.

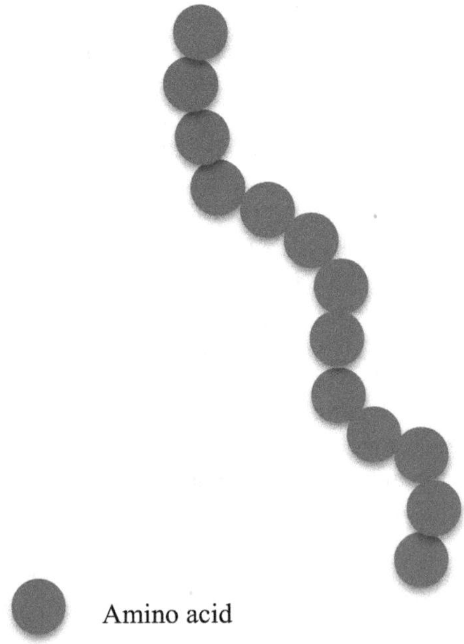


**Fig. 7.1** Amino acid chain makes the primary structure of a protein. Primary protein structure is, therefore, just a sequence of amino acids making a chain as shown here, schematic diagram. Twenty amino acids through variety in their combination can create various primary structures

The amino acid dimension falls in fraction of nm scale. Modeling studies are able to predict various aspects of size, charge, hydrophobicity, etc., of amino acids (Biro 2006). In solution, the dimension can alternatively be deducted from various structural studies, e.g., ion channel in membrane. Gramicidin A channel, dimer of gramicidin A monomers where each monomer contains 12–18 amino acids, covers the whole lipid bilayer of thickness about 4–5 nm (for details, see Mitra et al. 2004; Andersen and Koeppe II 2007; Ashrafuzzaman et al. 2006, 2008; Ashrafuzzaman and Tuszynski 2012a, b). This suggests that each amino acid covers about a distance which is of the order of value around 0.2 nm. Certainly there is a range around this order, and the value changes modestly with the change of type of amino acids. But it is clear that the primary structure of protein consists of amino acids that are arranged at lower nm dimension.

The protein structure takes various forms. Here is a flowchart shown in Fig. 7.2. The amino acid sequence that creates protein's primary structures leads to various other structures in solution like secondary, tertiary, quaternary structures. These structures depend on environment in which they take place. That is why these structures may also get changed as the environmental components get changed. The structure in water solution may get another form from that in plasma membrane and vice versa. But the primary composition of the amino acids usually is kept intact although the mutual arrangements may get modulated or distorted. The average

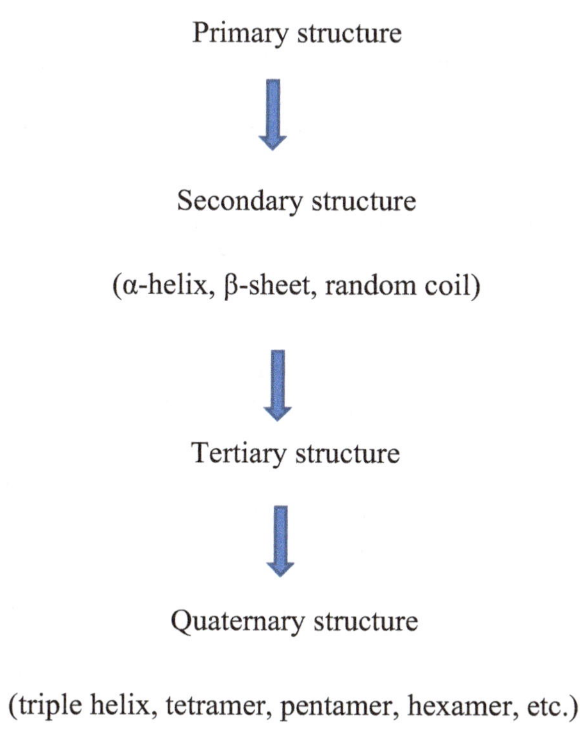


**Fig. 7.2** Proteins with various structures get originated from primary structures

amino acid separation, angular successive arrangement, etc., usually get perturbed as the protein structural transitions happen. This means the basis of the protein structural changes may be found in the building block arrangement/structure where nm-dimension geometric properties exist.

**Nanoscale-Dimension Structural Building Blocks to Construct a Small Structure**
Earlier we have briefly discussed mainly the structural units, subsequent bonding, and transition mechanisms through which a complete molecular structure undergoes and finally settles to create a big molecule. We have described a protein in this case. Now we shall focus at units that create smaller molecules which can be utilized to construct agents that may work as biomarkers to detect targets or as regulators, inhibitors, etc., to act on other small or big structures in biological system.

There is a considerable number of molecule classes that make the nanoscale-dimension structural building blocks to create small structures. Aptamers made of DNA, RNA, peptide, etc., make a class of molecules. Various biomolecules, natural products also make such classes. We shall focus here at aptamers as an example case to investigate how we can utilize them, being constructed from nanoscale building blocks that work within nanoscale functioning mechanisms.

DNA primary structure is made of four nucleotides adenine (A), guanine (G), thymine (T), and cytosine (C). Similarly, RNA primary structure is made of four nucleotides adenine (A), guanine (G), uracil (U), and cytosine (C). Altogether out of DNA and RNA, we find five nucleotide building blocks (NBBs), namely A, G, T, C, U. These NBBs can be utilized to construct aptamers that are found useful as drugs to target cellular biomolecules (see two of our important papers in Tseng et al. 2011; Ashrafuzzaman et al. 2013). Later in this chapter, we shall explain how earlier mentioned two techniques EgM and EpM can be applied to design aptamers as drugs out of these five NBBs using the drug–target molecule binding energetics and entropy information, respectively. Before that let us try to understand how the drugs may generally target biomolecules in cellular conditions and what happens as a result of drug association with target structures.

## 7.2 Drugs Target Biomolecules to Construct Drug–Biomolecule Complex Structures

The objective of drug discovery is often to discover or select a drug molecule that fits in the disordered section of a target biomolecule structure. A protein, nucleic acid RNA, nucleic acid DNA, lipid membrane, mitochondria, etc., can be target structures. The target biomolecule structure is generally very big. A section or various sections of the target structure usually get distorted or damaged which we call a disease state. A drug's therapeutic role is to reach at this specific distorted section of the target molecule and interact with that section to perform any of the following activities, namely:

(i) repair the damaged section of the structure,
(ii) inhibit the functionality of the damaged section of the structure,
Or, if both are not realistic then,
(iii) create further damage in the damaged section to destroy it's functionality in the target molecule.

As the molecule approaches the target structure, it creates a complex in the attachment region. The complex's geometry is sometimes an important determinant of the action of the molecule at its target structure region. We can schematize the scenario as shown in Fig. 7.3. Here we have schematized the target structure and drug candidate molecule structure quite simple. The real structures do not look so

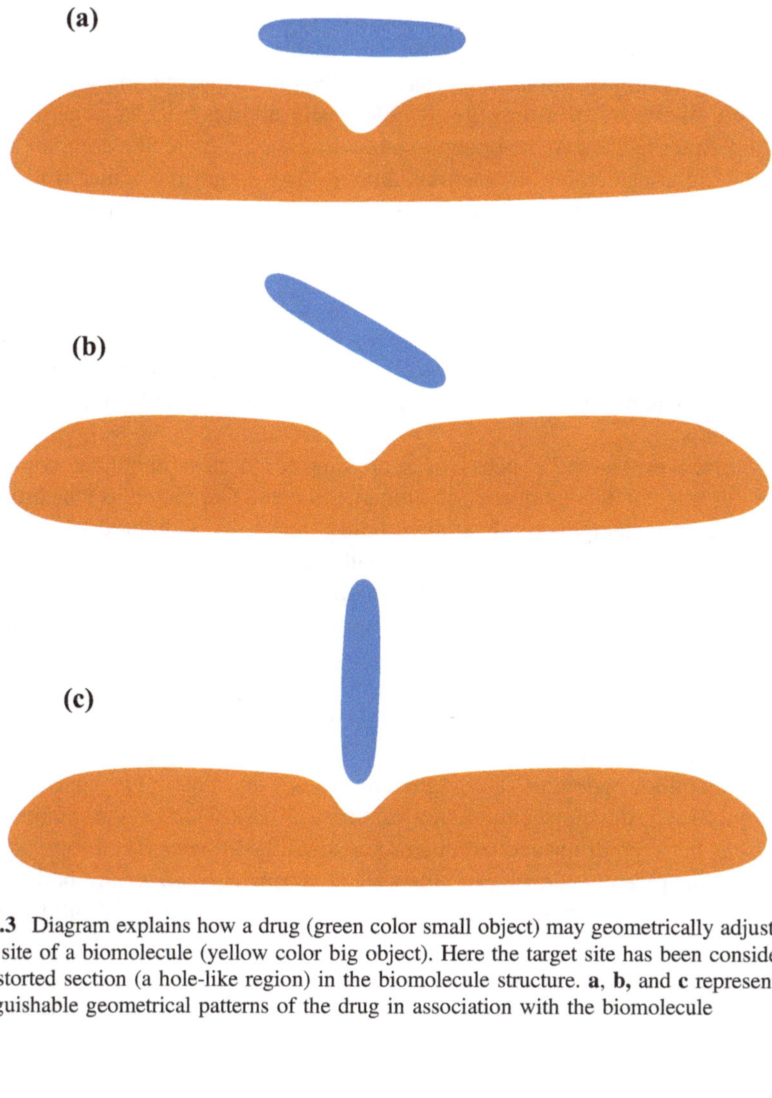


**Fig. 7.3** Diagram explains how a drug (green color small object) may geometrically adjust at the target site of a biomolecule (yellow color big object). Here the target site has been considered at the distorted section (a hole-like region) in the biomolecule structure. **a**, **b,** and **c** represent three distinguishable geometrical patterns of the drug in association with the biomolecule

simple. We shall provide a few real scenario later. Here what we have attempted to address is that the aspect of drug molecule approaching the target structure is important. The size of the drug molecule relative to the target region, the orientation of the drug candidate relative to the target structural geometry, the viscous condition of the region in which the drug–target complex gets created, etc., are clear determinants to create the condition of a background in which drug–target interactions take place.

## 7.3 Interactions in Drug–Biomolecule Complex Structures

Drug discovery is a rising field in which lots of scientists worldwide are now working to mainly design drugs for specific diseases. Biologists, biochemists, chemists, biophysicists, pharmacologist, and medical scientists are working and utilizing various subject-oriented approaches. The ultimate target of every approach is to ensure discovery of a drug that would easily approach to its target structure site(s) and either make a stable or semi-stable drug–target molecule complex and/or ensure having considerable interactions among drug–target molecule. But often a substantial amount of drug molecules end up being distributed to others regions or off-target biological structures. Thus, the process causes severe cytotoxicity. To mimic such off-target drug distribution, we shall develop here a novel drug discovery strategy in which a template will be created. This template will utilize drug–target structure binding-related properties as inputs while discovering drugs. We shall mainly use charge-based drug–target structure interaction energetics-related properties to design novel compounds out of a set of biomolecules falling within a chemical class.

Drug discovery is the process of searching for a new chemical compound suitable for use for the treatment of human diseases. Computer use in drug discovery and the emergence of a new field known as Computer-Aided Drug Discovery which uses the power of computer helps to discover new drugs. The emergence of this area helps the researcher to develop drug discovery tools quite easily. Computer modelling of physical laws to be applied on a system or process can perform computational simulations which produces almost valid *in system* data. A strategic *in vitro* assay further validates the model data in experimental laboratories. We shall utilize the power of computer and program using, e.g., Mathematica 9.1 (Wolfram) to address phosphatidylserine (PS) binding of aptamers, created from DNA nucleotides adenine (A), guanine (G), thymine (T), and cytosine (C) (unpublished, developed by Ashrafuzzaman 2015). Mainly, the charge-based interactions between drugs and target structures will be utilized. Therefore, we shall use here our recently developed theoretical formalism on many-body/charge interactions in which charges interact in presence of other charges (Ashrafuzzaman and Tuszynski 2012b).

### 7.3.1 Charge-Based Interactions of Drug–Target Molecule Complex Structures

In both target molecule and chemicals to be utilized for drug design, there are charges that we can use to interact with each other. At the end, the problem of interactions in a drug–target complex is equivalent to the many-charge interaction scenario in a biological system. We shall briefly address on the issue in this section. Utilizing the theoretical knowledge on the many-charge interactions and using the power of computational techniques, we can pinpoint biophysical properties to design certain drugs. For that, we shall use some example targets and some chosen chemicals.

In a many-charge system, each charge interacts with another following screened Coulomb interaction (SCI) formula, recently developed by us (Ashrafuzzaman and Tuszynski 2012b) in Fourier space:

$$V_{\mathrm{sc}}(k) = \frac{V(k)}{1 + \left(\frac{1}{2\pi k_{\mathrm{B}} T} n\right) V(k)} \tag{7.1}$$

Here $V(k)$ is the direct Coulomb interaction (D.CI) between two charges $q_i$ and $q_j$ and follows the relation:

$$V(k) = (1/\varepsilon_0 \varepsilon_r) q_i q_j / k^2 \tag{7.2}$$

Here $k_{\mathrm{B}}$ is the Boltzmann's constant, $n$ is number density of the participating particles (in this case charges), $k$ is wavenumber, and $T$ is absolute temperature providing the thermodynamic condition of the system.

Let us assume,

$$f(T, n) = \frac{1}{2\pi k_B T} n \tag{7.3}$$

We then get the following pattern:

$$V_{\mathrm{sc}}(k) = V(k) \left(\frac{1}{1 + x}\right) \tag{7.4}$$

where we have assumed

$$x = \left(\frac{1}{2\pi k_{\mathrm{B}} T} n\right) V(k) = f(T, n) V(k) \tag{7.5}$$

Using Taylor's series expansion we get

$$
\begin{aligned}
V_{sc}(k) &= V(k)\left(\frac{1}{1+x}\right) \\
&= V(k)\left(1+x+\frac{x^2}{2!}+\frac{x^3}{3!}+\cdots\right) \\
&= V(k)+V(k)x+V(k)\frac{x^2}{2!}+V(k)\frac{x^3}{3!}+\cdots \\
&= V(k)+V(k)x+V(k)\frac{x^2}{2!}+V(k)\frac{x^3}{3!}+\cdots \\
&= V(k)+V(k)f(T,n)V(k)+V(k)\frac{(f(T,n)V(k))^2}{2!}+V(k)\frac{(f(T,n)V(k))^3}{3!}+\cdots \\
&= V(k)+f(T,n)V(k)V(k)+\frac{(f(T,n))^2}{2}V(k)V(k)V(k) \\
&\quad+\frac{(f(T,n))^3}{6}V(k)V(k)V(k)V(k)+\cdots
\end{aligned}
\tag{7.6}
$$

Therefore, we can conclude the following:

$$
\begin{aligned}
V_{sc}(k) &= \mathrm{D.CI}+1\mathrm{st order\ scr\ CI}+2\mathrm{nd\ order\ scr\ CI} \\
&\quad+3\mathrm{rd\ order\ scr\ CI}+4\mathrm{th\ order\ scr\ CI}+\cdots
\end{aligned}
\tag{7.7}
$$

Here CI is Coulomb interaction, D for direct, scr for screened.

Let us consider an aptamer–lipid interaction scenario as example case. Lipid can be phosphatidylserine (PS), phosphatidylcholine (PC), etc., and aptamer can be a DNA sequence like AGGGTT (telomere specifically), AAAAAA, AAA, GGGGG, GGG, ATGCTT, etc., any combination specific for specific target structures (Tseng et al. 2011; Ashrafuzzaman et al. 2013). We have chosen PS as our target structure due to the important roles of PS in apoptosis which are explained here.

Cancer is split disorganized cell division and continues without interruption caused by a defect in the control of cell divisions. Cancer treatment is difficult, due to mainly the inability to target only cancer cells without the normal cells. Drugs targeting and modulating apoptosis have a recognized potential in cancer diagnosis and therapy. There are two main apoptotic pathways:

(i) the death receptor (extrinsic) pathway, and
(ii) the mitochondrial (intrinsic) pathway.

The death receptor pathway (MAPK) is activated by the binding of FAS or TRAIL ligands to their receptors (DR4/5), stimulating receptor aggregation.

In normal cells, apoptosis is a biomolecular process in which PS externalization happens. PS goes out across cell membrane through apoptotic region. In cancerous cells, apoptosis gets stopped so PS migration gets distorted or stopped.

In anticancer treatment, apoptosis is induced by certain therapeutic drugs. This means PS externalization in cancerous cells under treatment is a marker for treatment's efficacy. A designed drug that binds to the externalized PS can raise the possibility to check the anticancer drug efficacy. Here the PS-binding drug is used to quantify the externalized PS and thus helps us diagnose the treatment stages. We shall attempt to design such a set of 'diagnostic drugs.' We shall utilize above-mentioned electrostatic interaction scenario to design our desired PS-binding diagnostic drugs.

### *7.3.2 Mechanical Coupling of Drug–Target Molecule Complex Structures*

The mechanical coupling between a drug and a target molecule usually happens due to the elastic properties of the participating agents. A detailed explanation on this issue for a peptide–lipid membrane interaction has been made by us recently and can be found in our publications (see Ashrafuzzaman and Tuszynski 2012a, b). In this section, we wish to focus on issues that are important for designing drugs. We, therefore, avoid detailing the mechanical aspects of drug–target coupling as that becomes important after the drugs get designed. This will be presented in another platform. Instead, we better keep ourselves focused on the charge-based interactions as these phenomena will be utilized to design drugs.

## 7.4 Energetics of Drug–Biomolecule Complex Structures. Aptamer PS Interaction Energetics

Let us determine the interaction energetics between PS and AAAAAA. PS has a resultant negative charge. In Fig. 7.4, PS is seen to have three charges: two negatives and a positive. As a resultant, it may show one charge which is one negative charge in solution. Similarly, in Fig. 7.5, we see A, T, G, C to have charges in various places in their structures. Both A and T have three charges each, whereas G and C have four charges each in solution.

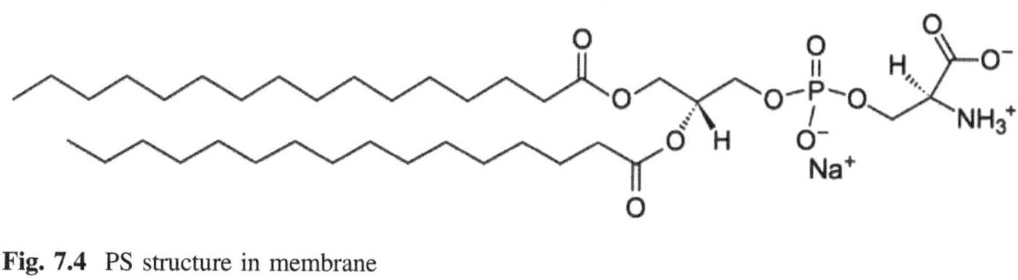


**Fig. 7.4** PS structure in membrane

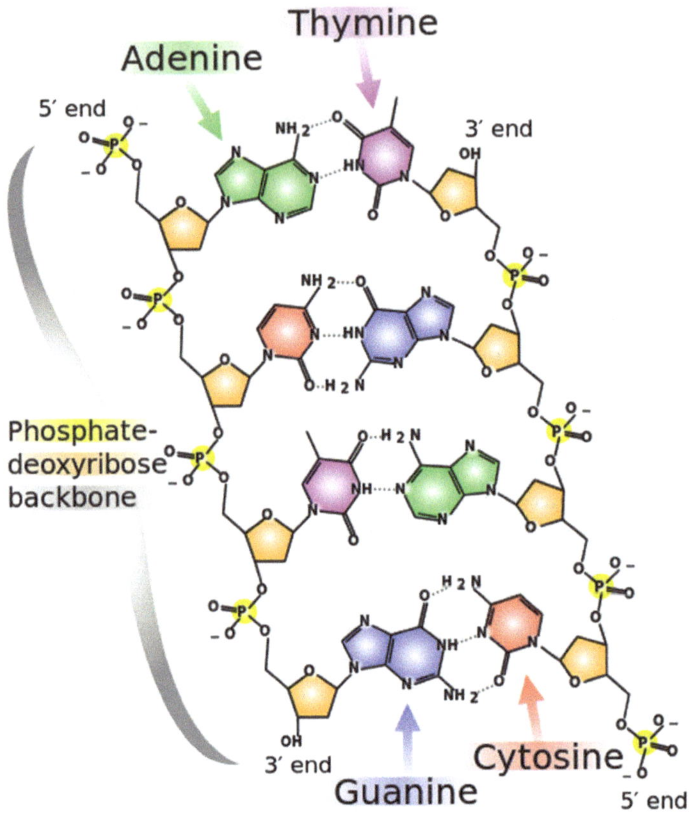


**Fig. 7.5** DNA structure in solution showing charge groups in nucleotides adenine (A), guanine (G), thymine (T), and cytosine (C). The DNA nucleotides exist in solution with active (charge) sites. This is shown in figure here

When PS and A or T or G or C come closer, their charges interact with each other. When a sequence of nucleotides, which is considered as aptamer in drug design science, comes closer to PS many charges on nucleotides come closer to PS. For simplicity, we can then assume each nucleotide to have one effective charge exactly following the assumption used for PS to have one effective charge despite having three. We can now consider a chain of charges in which charges interact following many-charge interaction formalism as explained earlier. From the charge-based interaction scenario, we can calculate the energies which will be explained here.

Let us try to address the interaction energetics of PS–AAAAA complex, for example.

In a PS–AAAAA (PS-A-A-A-A-A) interaction we can assume each molecule (PS or A) to have an effective charge, namely $q_{PS}$ for PS and $q_A$ for A. This will make the system to have a line of charges as shown in Fig. 7.6. The related interaction orders are shown in Figs. 7.6a–e.

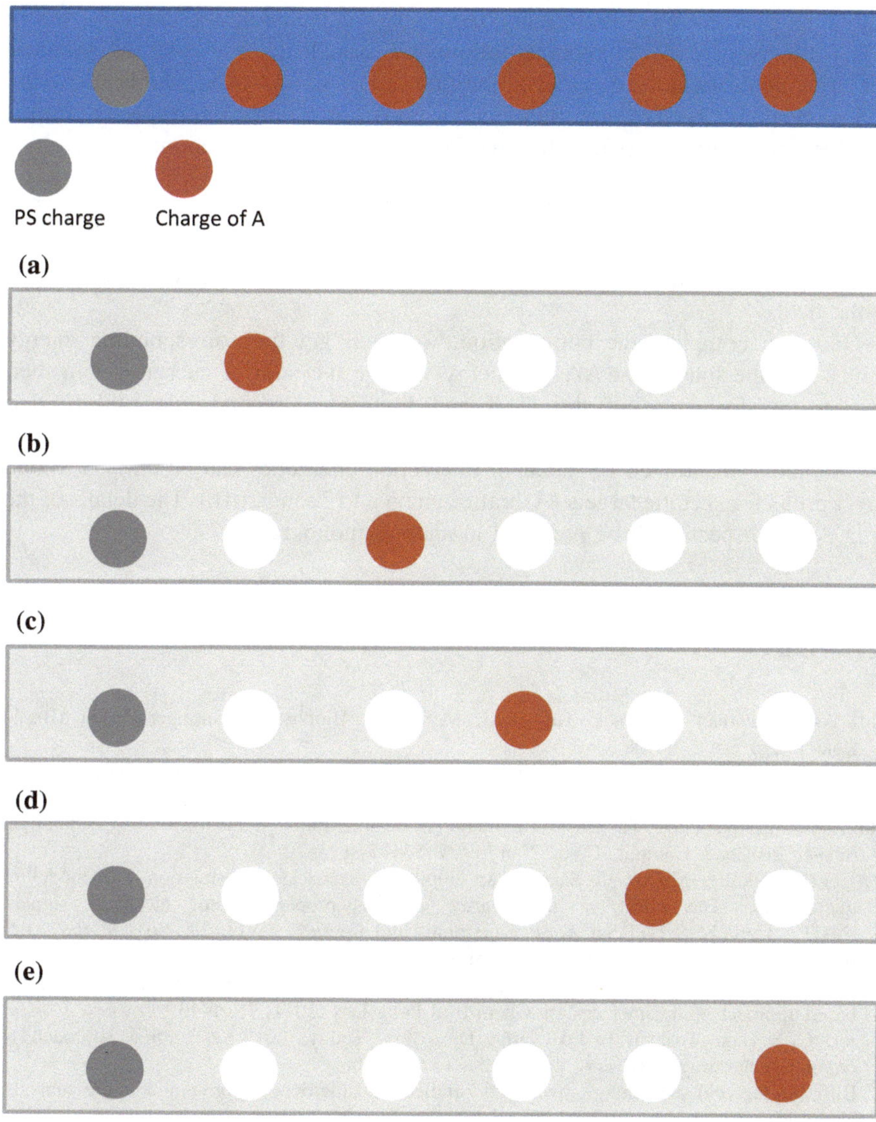


**Fig. 7.6** Charge arrangement in a PS–AAAAA interaction **a** D.CI between PS and first A (nearest-neighbor A), **b** first-order SCI between PS and second A (next nearest-neighbor A), **c** second-order SCI between PS and third A, **d** third-order SCI between PS and fourth A, and **e** fourth-order SCI between PS and fifth A

Now we shall address on how the interaction energies for the scenarios explained in Figs. 7.6a–e can be calculated.

In Fig. 7.6a, the interaction energy seems to be a D.CI energy. The calculation is straightforward following standard Coulomb formula which we studied in high school physics courses. Performing the energy calculation for cases shown in Figs. 7.6b–e where all the energies follow SCI formalisms, as explained earlier, is not so straightforward. To calculate the first-order, second-order, third-order, etc., SCI energies, we need to perform a Fourier transformation on the respective forms. The form of Fourier transform follows from the following equation:

$$V_{sc}(r) = \int d^3 k e^{(ik.r)} V_{sc}(k) \tag{7.8}$$

The production of these results is left for enthusiastic scientists who are working in the field.

Once we complete the computation, we shall get the corresponding energy values and the total PS–AAAAA energy will be the sum of energies from their Fourier transformation of the D.CI and first-order, second-order, third-order, fourth-order SCIs presented earlier.

The above-mentioned PS–aptamer interaction energetics can be utilized to discover drugs for specific targets (Ashrafuzzaman and Tseng 2016). The details of the drug design aspects will be provided in another monograph.

## References


Md. Ashrafuzzaman and Jack Tuszynski, Membrane Biophysics, Springer-Verlag (Berlin Heidelberg), Sept., 2012a.

MICHAEL O EZE. Phase Transitions in Phospholipid Bilayers: Lateral Phase Separations Play Vital Roles in Biomembranes. BIOCHEMICAL EDUCATION 19(4) 1991.

Md. Ashrafuzzaman and J. Tuszynski, Regulation of channel function due to coupling with a lipid bilayer. 2012b. J. Comput. Theor. Nanosci. 9, 564–570.

Dalibor L. Sekulić, Miljko V. Satarić. An improved nanoscale transmission line model of microtubule: The effect of nonlinearity on the propagation of electrical signals. FACTA UNIVERSITATIS, Series: Electronics and Energetics Vol. 28, No1, March 2015, pp. 133–142, https://doi.org/10.2298/fuee1501133s.

C-Y. Tseng, Md. Ashrafuzzaman, J. Mane, J. Kapty, J. Mercer, J. Tuszynski, Entropic fragment based approach to aptamer design. Chem Biol Drug Des (2011) 78, 1–13.

Tseng, C.-Y. and Tuszynski, J.A. Drug Discovery Today. 2015. A unified approach to computational drug discovery. Volume 20, 1328–1336.

JC Biro. Amino acid size, charge, hydropathy indices and matrices for protein structure analysis. Theor Biol Med Model. 2006; 3: 15.

Kakoli Mitra, Iban Ubarretxena-Belandia, Tomohiko Taguchi, Graham Warren, and Donald M. Engelma. Modulation of the bilayer thickness of exocytic pathway membranes by membrane proteins rather than cholesterol. PNAS March 23, 2004, vol. 101, no. 12, 4083–4088.

Andersen OS, Koeppe RE II. Bilayer thickness and membrane protein function: an energetic perspective. Annu Rev Biophys Biomol Struct. 2007;36:107–30.

Md. Ashrafuzzaman, O. S. Andersen, R. N. McElhaney, The antimicrobial peptide gramicidin S permeabilizes phospholipid bilayer membranes without forming discrete ion channels, Biochim. Biophys. Acta—Biomembranes (2008) 1778, 2814–2822.

Ashrafuzzaman M, Tseng CY, Kapty J, Mercer JR, Tuszynski JA. A computationally designed DNA aptamer template with specific binding to phosphatidylserine. Nucleic Acid Ther. 2013 Dec;23(6):418–26.

Cox, R.T., ed. (1961) The Algebra of Probable Inference, The Johns Hopkins Press.
Jaynes, E.T. 1957b. Information theory and statistical mechanics II. Phys. Rev. 108, 171–190.
Shore, J.E. and Johnson, R.W. (1981) Properties of cross-entropy minimization. IEEE Trans. Inf. Theory 27, 472–482.
Berg, E.L. (2014) Systems biology in drug discovery and development. Drug Discov. Today 19, 113–125.
CHEN, C.-C., TSENG, C.-Y., and DONG, J.-J. (2007). New entropy-based method for variables selection and its application to the debris flow hazard assessment. Eng. Geol. 94, 19–24.
Md Ashrafuzzaman, M A Lampson, D V Greathouse, R E Koeppe, O S Andersen, (2006) Manipulating lipid bilayer material properties using biologically active amphipathic molecules. Journal of Physics: Condensed Matter 18 (28):S1235–S1255.
M. Ashrafuzzaman and C.-Y. Tseng. 2016. Method for direct detection of lipid binding agents in membrane. US 9529006 B1. https://www.google.com/patents/US9529006.

# Chapter 8
# Ion Channels in Cancer

Ion channels in a cell are responsible for most of the cellular transport mechanisms. They play crucial roles in determination of cell membrane structure and cell function. In cancer condition, different ion channels play various kinds of critical roles in regard to cell proliferation, malignant angiogenesis, migration, and metastasis. The perturbed physiology of the cancer cell may regulate ion channel structures, energetics, and functions.

During apoptosis, in general both the extrinsic and intrinsic pathways in cells lead to analogous events including cell shrinkage, internucleosomal DNA fragmentation, and apoptotic body formation (Bortner and Cidlowski 2014). The cell volume loss or cell shrinkage is hallmark of apoptosis (Bortner and Cidlowski 2007, 2011; Lang et al. 2007), occasionally termed as apoptotic volume decrease (Maeno et al. 2000). The changes in intracellular ions are found to be linked with the loss of cell volume during apoptosis. The loss of intracellular potassium is found to play critical role in the downstream activation of the apoptotic machinery (Cain et al. 2001; Thompson et al. 2001; Bortner et al. 1997; Hughes et al. 1997). Various studies suggest that voltage-gated, delayed rectifier and inward rectifier potassium channels play a role during the activation of apoptosis (Bortner and Cidlowski 2007; Lang et al. 2005, 2007). From other studies, we see that the direct inhibition of potassium channels using various agents can help cells to avert death (Gantner et al. 1995; Wang et al. 2000; Heimlich et al. 2004). In the cell death process, an increase in intracellular sodium ion concentrations has been reported (Bortner and Cidlowski 2003; Yurinskaya et al. 2005; Franco et al. 2006; Bortner et al. 2008). By inhibiting the intracellular sodium increase, the apoptosis can be prevented. Conversely, by removing sodium from the extracellular region, apoptosis has been found to occur in absence of cell shrinkage (Bortner and Cidlowski 2003; Yurinskaya et al. 2005). Chloride ion in apoptosis is also found to play quite important role (see Okada et al. 2004, 2006; Heimlich and Cidlowski 2006). Chloride flux modulation impairs the intrinsic apoptotic induction pathways, but has no effect on the extrinsic pathways (Heimlich and Cidlowski 2006).



M. Ashrafuzzaman, *Nanoscale Biophysics of the Cell*,
https://doi.org/10.1007/978-3-319-77465-7_8

A volume-sensitive outwardly rectifying chloride channel creates a primary anionic conductance during the apoptosis (Wang et al. 2005).

Ion channels' roles in cell proliferation are immense, and they are found to have important roles in cancer development too (Kunzelmann 2005). Besides contributing into membrane conductance, channels appear to control the membrane voltage and $Ca^{2+}$ signaling in proliferating cells. Cellular $Ca^{2+}$ overload or perturbation of intracellular $Ca^{2+}$ compartmentalization is reported to cause cytotoxicity and trigger either apoptotic or necrotic (the premature autolysis of cell) cell death (Orrenius et al. 2003). Homeostatic parameters, such as the intracellular ion concentrations, pH of the cytosol, and volume of the cell, are also governed by ion channel activity. In normal cell, pH gradient is tightly maintained by the action of pH-regulatory membrane proteins such as the isoform protein of the Na(+)/H(+) exchanger (NHE1). The activity of NHE1 in neoplastic microenvironment dysregulates pH homeostasis and alters the survival, differentiation, and proliferation mechanisms of cancer cells. Thus, the cells become tumorigenic (Amith and Fliegel 2013; Cardone et al. 2005; Malo and Fliegel 2006). Cytoplasmic alkalinization in breast cancer cells occurs as a result of increased activity of NHE1. The membrane domain of NHE1 is active for ion exchange. The details of this topic go beyond the specific interest and scope of our book. But to mention that the pH dysregulation in cancer cell has a general role in cancer cell-based ion exchange mechanisms. The main role of ion channels is nothing but to deal with the ion exchange or migration across them. pH certainly has huge direct or indirect roles in influencing the channel activity (Ashrafuzzaman et al. 2008).

Quite earlier finding suggests that tumor necrosis factor-alpha (TNF, cachectin) formed pH-dependent, voltage-dependent, ion-permeable channels in lipid bilayer membranes and increased the sodium permeability of human U937 histiocytic lymphoma cells (Kagan et al. 1992). Thus, some of the physiological effects of TNF may be elicited through its intrinsic ion channel-forming activity.

Here we have discussed how in cancer state various ion channels carrying various types of electrolytes may determine various cell physiological conditions. Due to rise of cancer condition, various channel abnormalities are observed. We have created a table (see Table 8.1) to list various ion channels that are expressed in various types of cancers (Ismail et al. 2009; Li and Xiong 2011; Rao et al. 2015; Spugninia et al. 2015).

Ion channels that are expressed or regulated in cancer are listed in Table 8.1. We have addressed the details about general structure and behavior of normal ion channels in earlier Chap. 5 and in our earlier book 'Membrane Biophysics' (Ashrafuzzaman et al. 2012). Here we wish to address briefly on some of the example channel physiological conditions in cancer state.

We have discussed the physiological conditions of various ions in potassium, sodium, calcium, chloride, etc., in cancer conditions. Specific channels in cancer cells undergo perturbations in their physiological conditions. Sometimes both channel structures and functions, or either of them, get altered due to altered

**Table 8.1** Ion channels in specific cancer cells—a brief list presented here for comparison reason

| Ion channel | Cancer type where expressed | Effect |
|---|---|---|
| Voltage-gated $K^+$ channel Kv 10.1 kV 11.1 | Cervical cancer, breast cancer, ovarian cancer, osteosarcoma, glioma, melanoma, small-cell lung cancer, myeloid leukemia | Reduced proliferation by increasing apoptosis or cell cycle arrest at G0/G1, G1/S, or G2/M phase<br>Diochot et al. (2003), Garcia-Ferreiro et al. (2004), Weber et al. (2006), Restano-Cassulini et al. (2006), Gomez-Varela et al. (2007), Zhao et al. (2008), Fiore et al. (2013), Lansu et al. (2013), Perez-Neut et al. (2015) |
| Voltage-gated $Ca^{++}$ channel Cav 1/Cav 1.3 Cav 3/Cav 3.1 | Adrenal adenomas, prostate cancer, melanoma, glioblastoma | Reduced cell proliferation by inducing apoptosis<br>Das et al. (2012), Yu et al. (2014), Chen et al. (2014) |
| Voltage-gated $Na^+$ channel Nav 1.5, 1.6, 1.7, 1.9 | Prostate cancer, ovarian cancer, cervical cancer, breast cancer | Reduced cell proliferation via cell cycle arrest observed with the small-molecule inhibitors<br>Sikes et al. (2003), Fraser et al. (2014), Nelson et al. (2015) |
| Voltage-gated $Cl^-$ channel | Gliomas | Reduced proliferation by cell cycle inhibition at G1, G2/M phase via cell volume regulation. Channel activators induce apoptosis<br>Zhang et al. (1994), Deshane et al. (2003), Li et al. (2008), Sontheimer (2008), Zhu et al. (2012), Liu et al. (2013), Dardevet et al. (2015) |
| TRP channels<br>(i) TRPM1,<br>(ii) TRPM7,<br>(iii) TRPM8,<br>(iv) TRPV6 | (i) Melanoma<br>(ii) FaDu cell line, glioma, A172 cell line<br>(iii) Prostate, breast, colon, lung, skin cancers<br>(iv) Prostate adenocarcinoma, colorectal cancer | (i) Cell growth<br>(ii) Cell proliferation<br>(iii) Proliferation and apoptosis<br>(iv) Cell proliferation<br>Li and Xiong (2011) |
| Epithelial $Na^+$ channel/degenerin family (ENaC/DEG) | Glioblastoma cells; leukemic cell line | Cell swelling<br>Li and Xiong (2011) |

(continued)

**Table 8.1** (continued)

| Ion channel | Cancer type where expressed | Effect |
|---|---|---|
| Proton channel proton exchangers and transporters, including V-ATPase, $Na^+/H^+$ exchanger (NHE), monocarboxylate transporters (MCTs), and carbonic anhydrases (CAs) | General tumor | Aberrant pH gradient due to acidification of cancer microenvironment Spugninia et al. (2015) |
| Aquaporin water channel | Prostate cancer | Inhibition of aquaporin in plasma membrane increases the sensitivity of prostate cancer cells to freezing Ismail et al. (2009) |

Details are listed and explained in channel-specific subsections later in this chapter

physiological cancer conditions. We shall provide a detailed analysis of a few example cases. Although there are other channels that might not be mentioned here due to limited space, we shall try to focus on the most important ones or categories.

## 8.1 Voltage-Gated Ion Channels in Cancer Cells

Voltage-gated ion channels (VGICs) are widely observed to be expressed in excitable and nonexcitable cells. Besides being involved in a variety of physiological processes, VGICs work in developing various diseases. Among them, cardiac arrhythmias, epilepsy, hyperkalemia, some hereditary diseases (Fiske et al. 2006) are especially mentionable. The role of VGICs in cancer cell is now quite evident. VGICs are found to be engaged in the onset, proliferation, migration, and survival of various types of cancers and cancer cell lines (Fiske et al. 2006).

VGICs are subdivided considering the types of ions they transport, namely voltage-gated potassium channels (VGKCs), voltage-gated sodium channels (VGNaCs), voltage-gated calcium channels (VGCaCs), and voltage-gated chloride channels (VGClCs). But all these sub-types share similar structural topology. The main feature of the structure consists of a tetramic association of a series of six transmembrane α-helical segments, numbered S1–S6, connected by both intracellular and extracellular loops, the interlinkers of α-helical segments (Lehmann-Horn et al. 1999). Besides α-sub-unit, voltage-gated channels also have multifunctional β-sub-units, which modulate gating properties and assist in associations between different ion channels (Fiske et al. 2006). The structure-specific cancer cell effects are now quite known. We shall brief on them later in this section. Among many articles, a recent review is especially mentionable which has presented a brief description about them (Li and Xiong 2011).

VGICs play important roles in cancer cell progression, proliferation, and migration. VGICs play major roles in cell-specific differentiation program for neuronal and non-neuronal cell types. In some cancer types, the peptide hormones and other growth factors in the control of VGICs are also very important. Nav in prostate cancer cells are regulated by steroid hormone (Fiske et al. 2006). On the other hand, VGICs may act dependently in the physiological conditions. Cancer cells survive in modulated proton ($H^+$) concentrations. In revised pH conditions, the VGICs may regulate the transport of ions, may get regulated due to membrane depolarization, which could then repolarize the membrane, effectively blocking activation of voltage-gated sodium, calcium, etc., channels. Thus, disturbed synchronization among VGICs may cause production of cancer cells. Therefore, VGICs may also be considered as important therapeutic targets to repair cancer cells or stop cancer cells from being further produced.

In a recent review article, a beautiful schematic representation of possible involvement of different VGIC activity during the cell cycle phases of cancer cells has been put forward (Rao et al. 2015) (see Fig. 8.1). A cell cycle can be distinguished into various phases as shown in Fig. 8.1, namely the G0 phase, comprised mainly of the nonproliferating cells, G1 phase with cells getting primed for DNA replication, followed by the S phase with cells undergoing DNA replication, leading to the G2 phase, where the cells get ready to undergo mitosis/cell division. Finally, the mitosis (M) phase results in complete division of the cells with daughter cells ensuing that individually continue the process of cell cycle (Rao et al. 2015; Garcia et al. 2004). The roles of membrane potential and ion channels in all these phases are quite known. A number of studies have reported that cells with a much hyperpolarized resting potential, such as muscle cells and neurons, typically show little or no mitotic activity, while proliferating cells, particularly cancer cells, have a depolarized membrane potential in comparison to normal cells (Garcia-Ferreir et al. 2004; Pardo et al. 2004; Weber et al. 2006; Stuhmer et al. 2006; Hemmerlein et al. 2006; Mello et al. 2006).

VGICs, depending on their ion transport specificity, act in various cancer cells with different mechanisms or mode of actions. In Table 8.2, we have listed a few of the VGICs along with their host cancer cells and possible mechanisms. Presenting the details is beyond the scope of this book. But enough references are quoted here for more focussed academic readerships.

## 8.2 Transient Receptor Potential Channels in Cancer Cells

Considerably a large number of the transient receptor potential (TRP) family members of $Ca^{2+}$ and $Na^+$ permeable channels show altered physiological conditions with modified expressions in cancer cells (Ramsey et al. 2006; Nilius et al. 2007; Prevarskaya et al. 2007; Pla and Gkika 2013). The most studied early

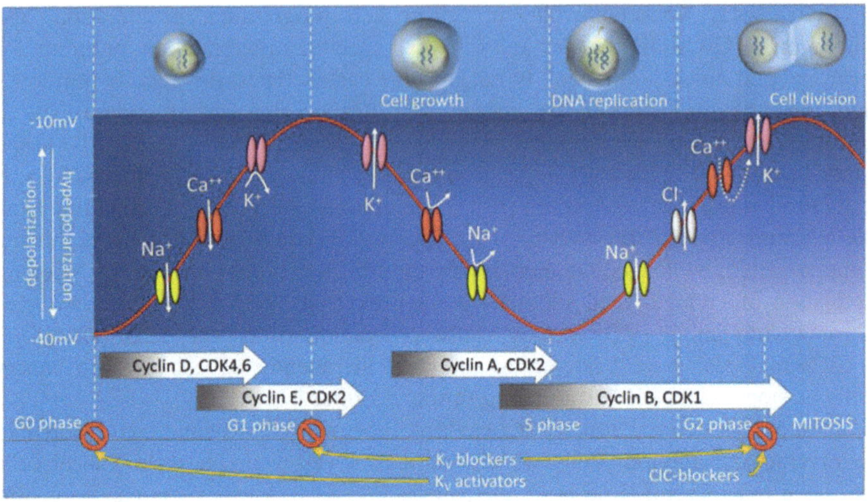


**Fig. 8.1** Schematic representation of possible involvement of different VGIC activities during the cell cycle of cancer cells. According to their expression level, several VGICs have been found playing important roles during the cell cycle. Opening of the voltage-gated $Na^+$ and/or $Ca^{++}$ channel moves positive charges from the extracellular space to the cytoplasm causing depolarization of the membrane. This event appears to be essential to promote transition from the G0/G1 phase to the S phase of the cell cycle. In contrast, membrane potential during the S phase tends to repolarize due to the opening of $K^+$ channels and/or the closing of $Ca^{++}$ and/or $Na^+$ channels. Mitosis is associated with more depolarized membrane potential compared to cells in the rest of the S phase. This is possible due to increased cytosolic $Na^+$ and/or $Ca^{++}$. Depolarization and augmented $Ca^{++}$ entry will allow stimulation of $Ca^{++}$-activated $K^+$ channels and activation of $Cl^-$ channels. Both $K^+$ and $Cl^-$ outward currents are responsible for water leaving the cytoplasm, which leads to a cell shrinkage before cell division. Chronic application of $K^+$ channel blockers (e.g., Kv11.1 blocker E4031) or openers (Kv11.1 opener NS1634) leads to changes of the membrane potential in the opposite direction, but both type of drugs cause arrests of the cell cycle in the G0/G1 phase, while blockade of the VGClC arrests the cell cycle in the G2 phase. This suggests that oscillation of the membrane potential is a fundamental event that promotes progression of the cell cycle. ↓ = inward ionic flux; ↑ = outward ionic flux; ✓ = no ionic flux. Copied from Rao et al. (2015) with due permission

members were TRPM8, TRPV6, TRPM1, and TRPV1 with whom a few more have been subsequently added (see Table 8.3). Most changes in TRP protein involve increased or decreased expression levels of the normal/wild-type TRP protein, depending on the stage of the cancer. It is not yet possible to conclude whether these changes in TRP expression are central dogma in cancer progression or are secondary to other changes, although the latter is most likely (Prevarskaya et al. 2007). Several TRP proteins are also proven to be valuable markers in predicting the progress of cancers and are recently considered as potential targets for anti-cancer drug design and possible pharmaceutical treatment (Duncan et al. 1998; Tsavaler et al. 2001; Fuessel et al. 2003; Fixemer et al. 2003; Bodding 2007). Since a number of TRP family channels are found to be involved in sensory neurons, TRP

**Table 8.2** Involvement of VGICs in the pathology of cancer/tumor cells

| Voltage-gated ion channels | Cancer type/cell line | Channel function |
|---|---|---|
| Kv 1.1 | MCF-7 human breast cancer cell line | Regulate tumor cell proliferation, cell cycle progression, and apoptosis<br>Ouadid-Ahidouch and Ahidouch (2008) |
| Kv 2.1–2.9 | Uterine cancer cells | Influences proliferation of cancer cells<br>Suzuki et al. (2009) |
| Kv 1.3 | Breast cancer cells | Regulate tumor cell proliferation, cell cycle progression, and apoptosis<br>Jang et al. (2009), Khaitan et al. (2009) |
| Kv 4.1 | Human breast epithelial cells | Inhibits cell proliferation of tumorigenic cells<br>Jang et al. (2009a) |
| KATP channel | Glioblastoma multiforme;<br>Allogeneic brain tumor model;<br>Uterine leiomyoma | Expression level increases, increased tumor permeability<br>Black et al. (2008), Park et al. (2008) |
| G-protein-coupled inwardly rectifying potassium channel 1 (GIRK1) | Breast carcinoma | Overexpressed in breast carcinoma, involved in proliferation and oncogenesis<br>Ouadid-Ahidouch and Ahidouch (2008), Brevet et al. (2008) |
| HERG K+ channels | Leukemia cell lines K562, HL-60,<br>Primary leukemia cells;<br>Gastric cancer cells | Overexpressed in cancer cells, regulate leukemia cell proliferation, and cell cycle<br>Li et al. (2008), Shao et al. (2008) |
| BKCa | Osteosarcoma;<br>Human glioma cells;<br>Human ovarian cancer cells;<br>Breast cancer (KCNMA1);<br>MCF-7 Human breast cancer cell line;<br>LNCaP prostate cancer cells | High sensitivity to calcium. Expression of BK channel auxiliary beta-sub-units by glioma cells<br>Wonderlin et al. (1996), Ransom et al. (2002), Khaitan et al. (2009), Han et al. (2008), Black et al. (2008), Sontheimer (2008), Cambien et al. (2008), Gessner et al. (2005) |
| KCa3.1 | Leukocytes, mitogen-induced endothelial cells, vascular smooth muscle cells;<br>Human colon cancer cells | Inflammatory and oncology diseases<br>Chou et al. (2008), De Marchi et al. (2009) |
| TREK-1 (two-pore domain (K (2P)) potassium channel) | Prostate cancer | Overexpression in cancer and is associated with abnormal cell proliferation<br>Voloshyna et al. (2008) |

(continued)

**Table 8.2** (continued)

| Voltage-gated ion channels | Cancer type/cell line | Channel function |
|---|---|---|
| TASK3 (TWIK-related acid-sensitive $K^+$ channel, KCNK9) | Human glioma cell lines; Glioma specimens; Breast tumors | Effects on membrane topology, involved in cell death, reduce cell survival in different glioma cell lines. Ouadid-Ahidouch and Ahidouch (2008), Meuth et al. (2008) |
| Kir 4.1 | Human astrocytic tumors | Mislocalization of Kir 4.1 in various human brain tumors (low- and high-grade astrocytomas and oligodendrogliomas), buffering capacity of glial cells may be compromised, leading to water influx (cytotoxic edema) Tan et al. (2008), Warth et al. (2005) |
| Voltage-gated $Ca^{++}$ channel Cav 1/Cav 1.3 Cav 3/Cav 3.1 | Adrenal adenomas, prostate cancer, melanoma, glioblastoma | Reduced cell proliferation by inducing apoptosis Das et al. (2012), Yu et al. (2014) |
| Voltage-gated $Na^+$ channel Nav 1.5, 1.6, 1.7, 1.9 | Prostate cancer, ovarian cancer, cervical cancer, breast cancer | Reduced cell proliferation via cell cycle arrest observed with the small-molecule inhibitors. Sikes et al. (2003), Fraser et al. (2014), Nelson et al. (2015) |
| Voltage-gated $Cl^-$ channel | Gliomas | Reduced proliferation by cell cycle inhibition at G1, G2/M phase via cell volume regulation Channel activators induce apoptosis. Zhang et al. (1994), Deshane et al. (2003), Li et al. (2008), Zhao et al. (2008), Sontheimer (2008), Zhu et al. (2012), Liu et al. (2013), Cheng et al. (2014), Dardevet et al. (2015) |

channels play a role in sensing pain in cancer (Nilius et al. 2007); therefore, these channels are considered prime targets in strategies to alleviate the cancer-related pain (Menendez et al. 2006; Ghilardi et al. 2005) which is also an important remedy cancer patients require especially during chemotherapy and the late stages of cancer.

TRPs are six transmembrane cation-permeable channels. They are widely expressed in mammalian tissues. They regulate intracellular $Ca^{2+}$, $Na^+$ ion concentrations and membrane voltages in both excitable and nonexcitable cells. TRP channels, as signal integrators, can be activated by multiple distinct stimuli and localized physiological perturbations. With the opening of TRP channels, $Ca^{2+}$ and

**Table 8.3** TRP channels/proteins that exhibit altered expression in cancer cells

| TRP channel/ protein | Cancer type where expressed | Effects and/or changes in TRP expression |
|---|---|---|
| TRPM1 | Melanoma | Cell growth gets affected. Decreased expression in tumorigenic melanocytes.<br>Duncan et al. (1998), Fang and Setaluri (2000); Deeds et al. (2000), Duncan et al. (2001), Prevarskaya et al. (2007) |
| TRPM7 | FaDu cell line, glioma, A172 cell line | Cell proliferation<br>Park et al. (2014) |
| TRPM8 | Prostate, breast, colon, lung, skin cancers<br>(iv) Prostate | Proliferation and apoptosis. Increased expression in androgen-insensitive prostate cancer cells.<br>Tsavaler et al. (2001), Fussel et al. (2003), Prevarskaya et al. (2007) |
| TRPV1 | Prostate, bladder, colon, pancreas | Increased expression in prostate cancer, colon cancer, and pancreatic cancer<br>Dömötör et al. (2005), Hartel et al. (2006), Lazzeri et al. (2005), Sánchez et al. (2005) |
| TRPV6 | Rostate adenocarcinoma, colorectal cancer | Cell proliferation<br>Li et al. (2011) |
| TRPC | Ovarian cancer | Promote human ovarian cancer cell proliferation and tumorigenesis<br>Zeng et al. (2013) |

$Na^+$ ions flow into the cell, which couple their activity to downstream cellular signal amplification via calcium permeation and membrane depolarization (Clapham 2003).

Based on amino acid sequence homology, mammalian TRP channels can be divided into seven subfamilies, such as TRPA, TRPC, TRPM, TRPML, TRPN, TRPP, and TRPV (Montell et al. 2002; Clapham et al. 2003). A brief report has been presented in Li et al. (2011). As $Ca^{2+}$ is an essential regulator for cell cycle and proliferation (Means 1994), TRP channels are observed to play important roles in many physiological processes. They are known to be involved in a variety of pathological processes including tumor formation and metastasis. The main objective here is to briefly analyze the TRP channels' engagement in cancer promotion or regulation. Table 8.3 summarizes such a brief report on TRP channels in cancer.

Studies from the last two decades suggest that the TRP channels and the transport proteins play important roles in cell migration. Various recent literatures reveal a critical role for TRP channels in the migration process, both in cancer cells and in tumor vascularization. A recent review mainly focused on this important issue (Pla and Gkika 2013). It provided an overview of advances in this field describing TRP channels' contributions to the vascular and cancer cell migration process and relevant molecular mechanisms involved. Table 8.4 presents the summary of TRP/Orai1 functions in cancer and endothelial cell migration.

**Table 8.4** TRP/Orai1 functions in cancer and endothelial cell migration [Copied from Pla and Gkika (2013)]

| Channel | Cell type | Cell migration | | Techniques used | Proposed mechanism | References |
|---|---|---|---|---|---|---|
| | | Epithelial | Endothelial | | | |
| TRPC1 | Glioma cell lines; zebra fish; HUVEC | + | + | Boyden chamber; morpholinos on zebra fish; tubulogenesis in vitro | EGF-mediated migration, involvement of lipid raft. ClC-3 interaction/filopodia extension | Yu et al.(2010), Bomben et al. (2011), Antigny et al. (2012), Cuddapah et al. (2013) |
| TRPC3 | MCF-7 (breast cancer) | + | ND | Boyden chamber | Indirect link | Zhang et al. (2012) |
| TRPC6 | Head and neck carcinoma cell line; glioblastoma multiple; HMEC; HPAEC; HUVEC | + | + | Wound healing; Matrigel invasion assays on transwell; soft agar clonogenic assay; tubulogenesis in vitro | Notch activation under hypoxia in turn promotes TRPC6 expression; in EC PTEN regulates TRPC6 expression | Hamdollah Zadeh et al. (2008), Ge et al. (2009), Chigurupati et al. (2010), Kini et al. (2010), Bernaldo de Quirbs et al. (2013) |
| TRPC5/ TRPC6 | BAEC, MAEC | ND | – | Wound healing | Lysophosphatidylcholine activates TRPC6 which in turn promotes TRPC5 membrane expression | Chaudhuri et al. (2008) |
| TRPV1 | Human hepatoblastoma cells (HepG2); cervical and bladder cancer cell | ± | – | Random cell migration, Boyden chamber, Matrigel invasion assays, in vivo xenografts on nude mice | TRPV1 antagonist capsazepine inhibits both cannabidiol-induced tissue inhibitors of the matrix metalloproteinase 1 (TIMP-1) expression and activation of the MAPKs p38 and p42/44; capsaicin promotes IGF (insulin-like | Waning et al. (2007), Ramer and Hinz (2008), Ramer et al. (2010), Caprodossi et al. (2011) |

(continued)

**Table 8.4** (continued)

| Channel | Cell type | Cell migration | | Techniques used | Proposed mechanism | References |
|---|---|---|---|---|---|---|
| | | Epithelial | Endothelial | | | |
| | | | | | growth factor)-1 release, GZMA and MMP9 activation, a-tubulin disassembly and cytoskeleton degradation. The effect is reverted by TRPV1 overexpression | |
| TRPV2 | Prostate cancer cells (PC3), urothelial cancer veils (T24/83) | + | ND | Migration assays on transwell, Matrigel invasion assays on transwell, in vivo xenografts on nude mice | Lysophospholipids and adrenomedullin activate TRPV2 via PI3 K pathway. TRPV2 activation induces MMP2, MMP9, and cathepsin B52 expression | Monet et al. (2010), Oulidi et al. (2013) |
| TRPV4 | Hepatoblastoma cell line (HepG2); bovine capillary endothelial (BCE) cells and human dermal microvascular endothelial (HMVE); BHMEC; TEC | + | + | Random motility; wound healing; live cell microscopy after mechanical shear stress application | Ultrarapid activation by bl-integrin; activation by AA via actin remodeling | Thodeti et al. (2009), Matthews et al. (2010), Fiorio Pla et al. (2012b) |
| TRPM7 | Breast cancer, lung cancer, nasopharyngeal cancer, and pancreatic ductal adeno | + | ND | Matrigel invasion assays on transwell | TRPM7 activation of Src-MAPK signaling pathway, focal adhesion number; EGF-mediated TRPM7 membrane expression | Gao et al. (2011), Middelbeek et al. (2012), Meng et al. (2013) |

(continued)

**Table 8.4** (continued)

| Channel | Cell type | Cell migration | | Techniques used | Proposed mechanism | References |
|---|---|---|---|---|---|---|
| | | Epithelial | Endothelial | | | |
| TRPM8 | Prostate metastatic cancer; glioblastoma; squamous carcinoma cell lines | – | – | Wound healing; transwell; random motility; in vivo xenografts on nude mice | Activation by icilin and PSA; TRPM8 diminish PFAK levels | Wondergem et al. (2008), Yang et al. (2009b), Gkika et al. (2010), Zhu et al. (2011), Okamoto et al. (2012), Valero et al. (2012) |
| ORAI1/ STIM1 | Breast cancer; cervical cancer; HUVEC; EA. hy926 cells; EPC | + | + | Transwell; Matrigel invasion assays on transwell random migration; in vivo xenografts on nude mice; in vitro tubulogenesis; wound healing | Stimulation of focal adhesion turnover via ras and rac GTPases; downstream to VEGF | Abdullaev et al. (2008), Yang et al. (2009a), Chen et al. (2011), Dragoni et al. (2011), Li et al. (2011), Beech (2012) |

Orai is calcium release-activated calcium modulator 1

The listed references in table: Yu et al. (2010), Bomben et al. (2011), Antigny et al. (2012), Cuddapah et al. (2013), Zhang et al. (2012), Hamdollah et al. (2008), Ge et al. (2009), Chigurupati et al. (2010), Kini et al. (2010), Bernaldo de Quiros et al. (2013) Chaudhuri et al. (2008), Waining et al. (2008), Ramer and Hinz (2008), Ramer et al. (2010), Caprodossi et al. (2011), Monet et al. (2010), Oulidi et al. (2013), Thodeti et al. (2009), Matthews et al. (2010), Fiorio Pla et al. (2012b), Gao et al. (2011), Middelbeek et al. (2012), Meng et al. (2013), Wondergem et al. (2008), Yang et al. (2009b), Gkika et al. (2010), Zhu et al. (2011), Okamoto et al. (2012), Valero et al. (2012), Abdullaev et al. (2008), Yang et al. (2009a), Chen et al. (2011), Dragoni et al. (2011), Li et al. (2011), Beech (2012)

*HMEC,* human microvascular EC; *HPAEC,* human pulmonary artery EC; *HUVEC,* human umbilical vein EC; *EA.hy926,* EC line derived from HUVECs fused with human lung adenocarcinoma cell line A549; *BTEC,* tumor derived EC from breast carcinoma; *MAEC,* mouse aortic EC; *BHMEC,* brain microvascular EC; *EPC,* endothelial precursors cells; *RCC-EPC,* EPC isolated from renal carcinoma patients; *EGF,* epithelial growth factor; *ClC-3,* chloride channel; *PTEN,* phosphatase and tensin homolog protein; *TIMP1,* metallopeptidase inhibitor 1; *MAPK,* mitogen-activated protein kinase; *IGF,* insulin-like growth factor; *GZMA,* granzyme A; *MMP9,* matrix metalloproteinase 9; *PI3 K,* phosphatidylinositol 3-kinase; *MMP2,* matrix metalloproteinase 2; *AA,* arachidonic acid

Figure 8.2 presents the mechanisms in representative cancer cells and endothelial cells without any tissue specification (Pla and Gkika 2013). AA, arachidonic acid; ClC-3, chloride channel-3; EC, endothelial cells; ER, endoplasmic reticulum; bFGF, basic fibroblast growth factor; GZMA, granzyme A; GPCR, G-protein-coupled receptor; IGF, insulin-like growth factor; LPL, lysophospholipids; LPC,

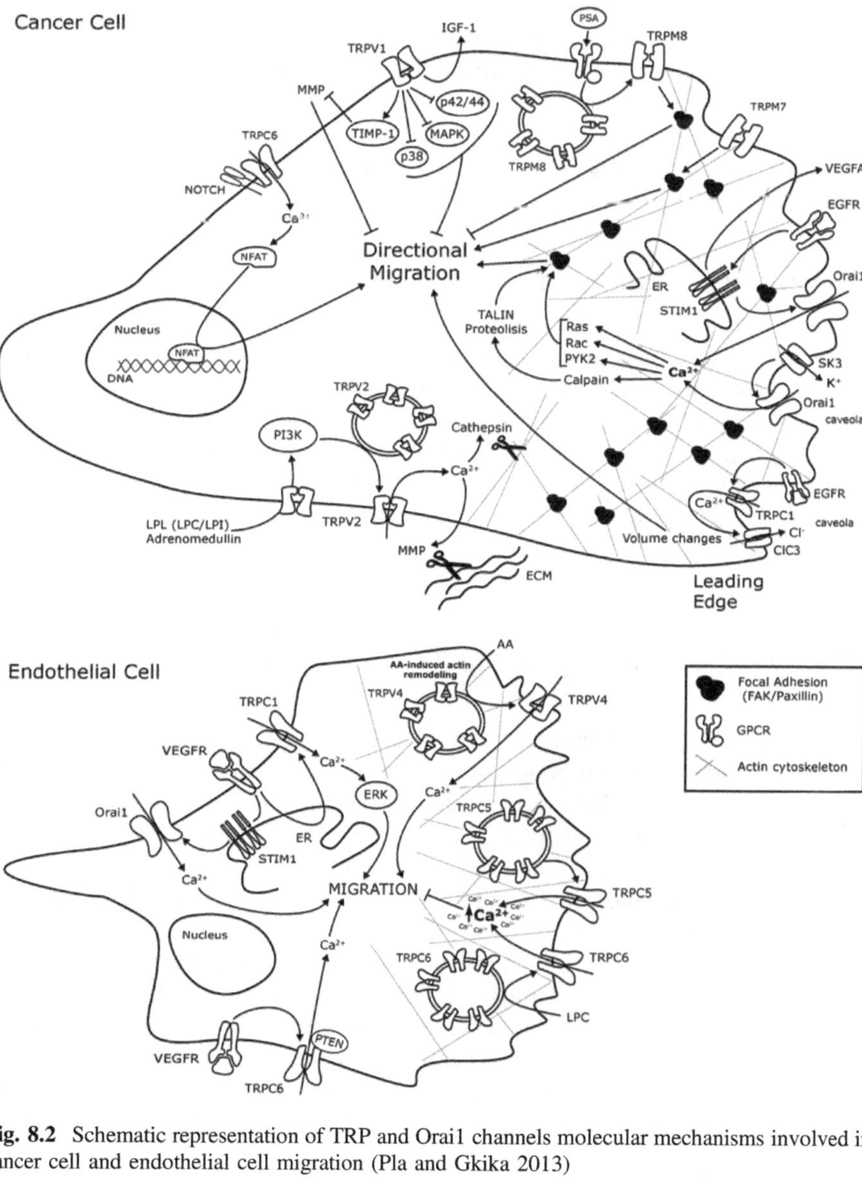


**Fig. 8.2** Schematic representation of TRP and Orai1 channels molecular mechanisms involved in cancer cell and endothelial cell migration (Pla and Gkika 2013)

lysophosphatidylcholine; LPI, lysophosphatidylinositol; MAPK, mitogen-activated protein kinase; MMP, matrix metalloproteinase; MMP2, matrix metalloproteinase 2; NFAT, nuclear factor of activated T-cells; PI3 K, phosphatidylinositol 3-kinase; PTEN, phosphatase and tensin homolog; Pyk2, protein tyrosine kinase Pyk2; SK3, K + channel; TIMP1, metallopeptidase inhibitor 1; VEGF, vascular endothelium growth factor; VEGFR, VEGF receptor.

In this section, we have addressed the diversified roles of TRP channels in cancer cells' physiological functioning as well as cell migration. We have avoided presenting the details as there are various reviews, as listed here, available where the readers may find most of the information in topic specific aspects. Brief report presented here may help understand the general behavior of TRP channels in cancer cells and their possible use as target structures for therapeutic purposes while developing anticancer drugs and therapy templates.

## 8.3 Proton Channels in Cancer Cells

Cancer cells and tissues generally experience an aberrant regulation of hydrogen ion dynamics. This leads to rise of a reversal of the intracellular to extracellular pH gradient. Such deranged pH control is a common feature of most cancer histotypes. The perturbation in pH dynamics is known to rise very early in carcinogenesis which makes it one of the most common pathophysiological hallmarks of tumors. Here we shall brief about a few proton channels and address on how proton channels are active in cancer cells.

### 8.3.1 *Proton Exchangers and Transporters in Cancer*

The reversed proton gradient is driven by a series of proton export mechanisms that underlie the initiation and progression of the neoplastic process. The $Na^+/H^+$ exchanger isoform 1 (NHE1) is an integral membrane transport protein involved in regulating intracellular and extracellular pH level. In tumor cells, NHE1 is a major contributor to the production and maintenance of their reversed proton gradient. A recent review presented the latest progress in understanding the role of NHE1 in determining tumor progression and invadopodia-guided invasion/metastasis and recent patents for NHE1 inhibitors including novel therapeutic protocols for anti-NHE1 pharmacological approaches (Reshkin and Rosa 2013).

Cancer becomes isolated in biological system due to mainly the activity of major determinants of aberrant pH gradient in cancer which are proton exchangers and transporters, including V-ATPase, $Na^+/H^+$ exchanger (NHE), monocarboxylate transporters (MCTs), carbonic anhydrases (CAs). This isolation, often also known

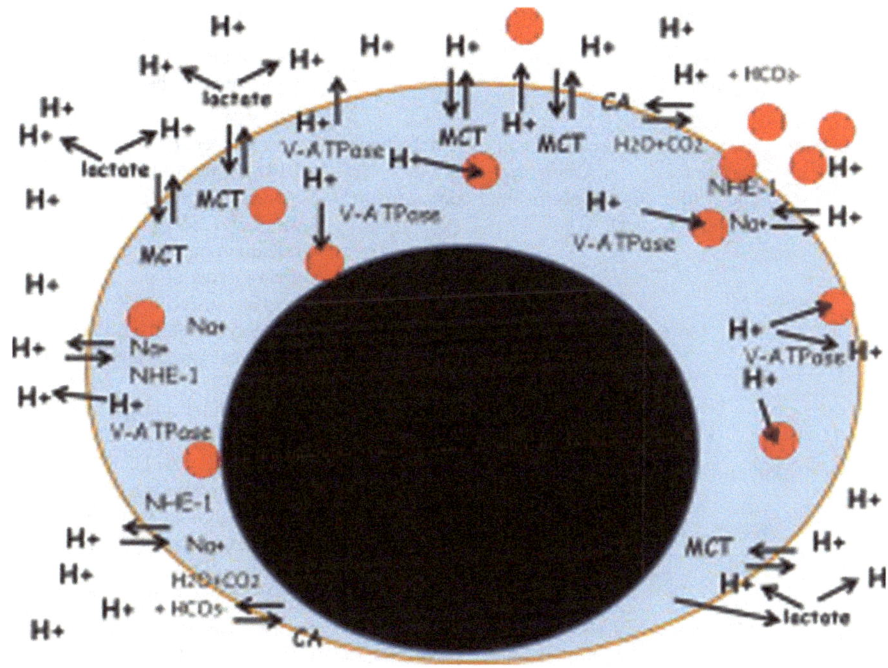


**Fig. 8.3** Proton channels. This figure is copied from Spugninia et al. (2015)

as protection, is not only from the body reaction against the tumor growth, but also from the vast majority of drugs that when protonated into the acidic tumor microenvironment do not enter into physiologically isolated cancer cells. Apparently, some key characters of proton transporters and exchangers present in tumor cells help them survive in hostile microenvironments that the cells create around them. A recent review which is part of a special issue entitled 'Membrane channels and transporters in cancers,' critically presents a novel and efficient approach to fighting cancer through blocking or inhibiting proton exchangers and transporters that are active in human cancer cells (Spugninia et al. 2015). A model (see Fig. 8.3) and list of proton exchangers and their roles in cancer have been given in Table 8.5 (Spugninia et al. 2015).

Balance between acid and base conditions needs to be maintained to ensure cellular homeostasis. A narrow pH range of the cell physiology can only ensure the living processes working normally. The mentioned proton exchangers in Table 8.5 have direct roles to play in regard to development of cancer. We shall avoid detailing there structures and functions here, but interested readers are encouraged to understand them from quoted references. In cancer cells, these proton exchangers are found to be prime targets to design therapeutic agents for various types of cancers (Spugninia et al. 2015).

**Table 8.5** Proton exchangers and their role in cancer

| Pump type | Cellular localization | Function | Cancer type where found active (major cancer types are listed here) |
|---|---|---|---|
| $H^+$ ATPase | Plasma membrane and acidic organelles | Acidification of extracellular microenvironment and endolysosomal compartment | Breast cancer; Feng et al. (2014), Feng et al. (2013), Schwarzenberg et al. (2014), You et al. (2009)<br>Oral squamous cell carcinoma; Garcia-Garcia et al. (2012), Perez-Sayans et al. (2010), Perez-Sayans et al. (2010a),<br>Esophageal carcinoma; Huang et al. (2012),<br>Hepatocellular and pancreatic carcinoma; Xu et al. (2012), Chung et al. (2011),<br>Lung carcinoma; Lu et al. (2013), Sarcoma Perut et al. (2014), Avnet et al. (2013),<br>General solid tumors; Perez-Sayans et al. (2009) |
| $Na^+/H^+$ exchangers (NHE) | Plasma membrane | Alkalinization of cytosol and acidification of extracellular microenvironment | Breast cancer; Cardone et al. (2005, 2007), Lauritzen et al. (2012),<br>Melanoma; Vahle et al. (2014), Stuwe et al. (2007), Ludwig et al. (2013),<br>Lung cancer; Provost et al. (2012),<br>Acute myeloid leukemia; Jin et al. (2012), Rentsch et al. (2007) |
| MCT1 ($H^+$/lactate symporter) | Plasma membrane | Elimination of lactate from glucose catabolism, and acidification of extracellular milieu | Breast cancer; Busk et al. (2011), Kennedy et al. (2013),<br>Melanoma; Ho et al. (2012),<br>Pancreatic cancer; Guillaumond et al. (2013) |
| Carbonic anhydrase 9 | Plasma membrane | Regulation of intracellular pH and pH gradients | Lung; Giatromanolaki et al. (2001),<br>Cervical carcinoma; Brewer et al. (1996),<br>Esophagus; Turner et al. (1997),<br>Bladder; Klatte et al. (2008),<br>Breast; Chia et al. (2001),<br>Colorectal cancers; Saarnio et al. (1998)<br>Head and neck cancer; Choi et al. (2008), Roh et al. (2009), Oliveira et al. (2011); Eckert et al. (2011), Kondo et al. (2011) |

(continued)

**Table 8.5** (continued)

| Pump type | Cellular localization | Function | Cancer type where found active (major cancer types are listed here) |
|---|---|---|---|
| $H^+/K^+$ ATPase | Gastric epithelial cell line | Regulation of extracellular pH. Gastric $H^+/K^+$-ATPase inhibitors can inhibit V-ATPase via binding to sub-unit A of the nucleotide binding domain | Major types of cancers Bowman et al. (1998), Huss et al. (2011), Muroi et al. (1998), Wiedmann et al. (2012), Mattsson et al. (1991) |

### *8.3.2 Human Voltage-Gated Proton Channels (Hv Channels)*

Specific expressions of the human voltage-gated proton channel Hv1 in highly metastatic breast cancer cells are found to promote tumor progression and metastasis (Wang et al. 2011). This is quite important novel finding. About three decades ago, the voltage-gated proton channels (Hv channels) were first documented in snail neurons (Thomas and Meech 1982). They were later identified at the molecular level, called Hv1 for human and mVSOP for mouse (Ramsey et al. 2006; Sasaki et al. 2006). Hv1/mVSOP is extremely selective for proton $H^+$, with no detectable permeability to other popular cell-based transported cations. Here the voltage activation relationship depends strongly on pH of both intracellular and extracellular regions. Increasing extracellular pH or decreasing intracellular pH is found to promote $H^+$ channel opening via shifting the channel activation threshold to more negative values of the potentials.

Wang et al. (2011) have recently shown, apparently for the first time ever, that Hv1 is expressed at a high level in highly metastatic human breast tissues and cell lines, and at a very low level in poorly metastatic human breast tissues and cell lines. The expressed Hv1 was found to regulate breast cancer cytosolic pH, to acidify extracellular milieu and to exacerbate the migratory ability of metastatic cells. Figure 8.4 explains the expression of Hv1 in human breast cancer tissues and cell lines. Down-regulation of Hv1 expression significantly inhibits the ability of the highly metastatic cancer cell invasion, migration, and proton secretion. The present data suggest that Hv1 regulates breast cancer intracellular pH and exacerbates the migratory ability of metastatic cells. Hv channels are activated by depolarization and intracellular acidification, whose activities maintain intracellular pH to be neutral to keep reactive oxygen species (ROS) generation (Clark et al. 1987; Morgan et al. 2005). Hv channels not only regulate pH in cytoplasm, but could also provide protons in phagosome, a closed membrane compartment for killing and digesting the pathogen (DeCoursey 2003). pH condition is one of the major concerns in cancer cells. Hv's role in controlling or regulating pH may be considered important for therapeutics.

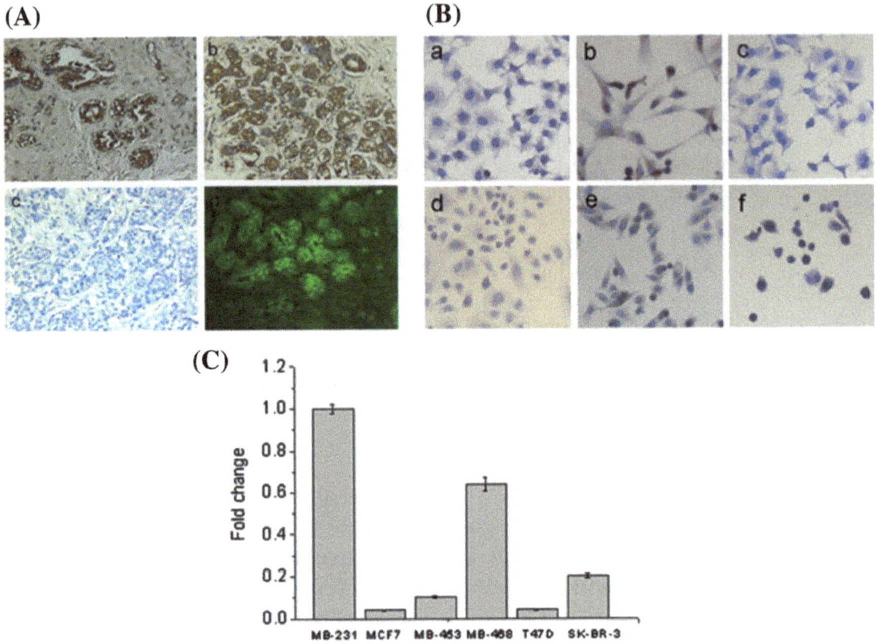


**Fig. 8.4** Expression of Hv1 in human breast cancer tissues and cell lines (Wang et al. 2011). (**A**) Distribution of Hv1 in highly (a, b, and d) and poorly (c) metastatic breast cancer tissues detected by immunohistochemistry (a, b, and c) and immunofluorescence (d). In the highly metastatic breast tumor tissues, intense immunoreactivity to Hv1 is shown by brownish staining (a and b) and green fluorescence (d). In contrast, the poorly metastatic breast tumor tissue (c) hardly shows the immunoreactivity. The background nuclei are in blue. Original magnification 40×. (**B**) Expression of Hv1 in breast cancer cell lines, MCF-7 (a), MDA-MB-231 (b), T-47D (c), SK-BR-3 (d), MDA-MB-468 (e), and MDA-MB-453 (f) detected by immunocytochemistry. (**C**) mRNA expression levels of Hv1 showed significant difference among these breast cancer cell lines. Values are means ± SD ($n = 3$). B and C clearly show that Hv1 is expressed at a high level in the highly metastatic MDA-MB-231 cells, but at a low level in the lowly metastatic MCF-7 cells

The role of Hv1 in cancer cell migration is addressed in Fig. 8.5. Hv1 is found to increase cell migration and invasion (Wang et al. 2011). Invasion and migration are two important hallmarks of tumor malignancy. Figure 8.5A shows the migration kinetics of the (MDA-MB-231) and poorly (MCF-7) metastatic cells. A wounded monolayer of MDA-MB-231 cells allows wound closure after 48 h (Fig. 8.5A, b, e and h). MDA-MB-231 cells closed the wound faster than MCF-7 cells (Fig. 8.5A, a, d and g). Suppressed Hv1 by siRNA3-Hv1 clearly decreased the migration in the highly metastatic cells (Fig. 8.5A, c, f, and i), but almost without affecting the poorly metastatic cells [data not shown Wang et al. (2011)]. This group then studied the invasion and migration of MDA-MB-231 and MCF-7 cells using transwell inserts. The expression of Hv1 in the highly metastatic MDA-MB-231 cells was inhibited by the three siRNA duplexes targeting Hv1 (siRNA1-Hv1, siRNA2-Hv1, and siRNA3-Hv1). The inhibition of these three siRNA duplexes targeting Hv1 on

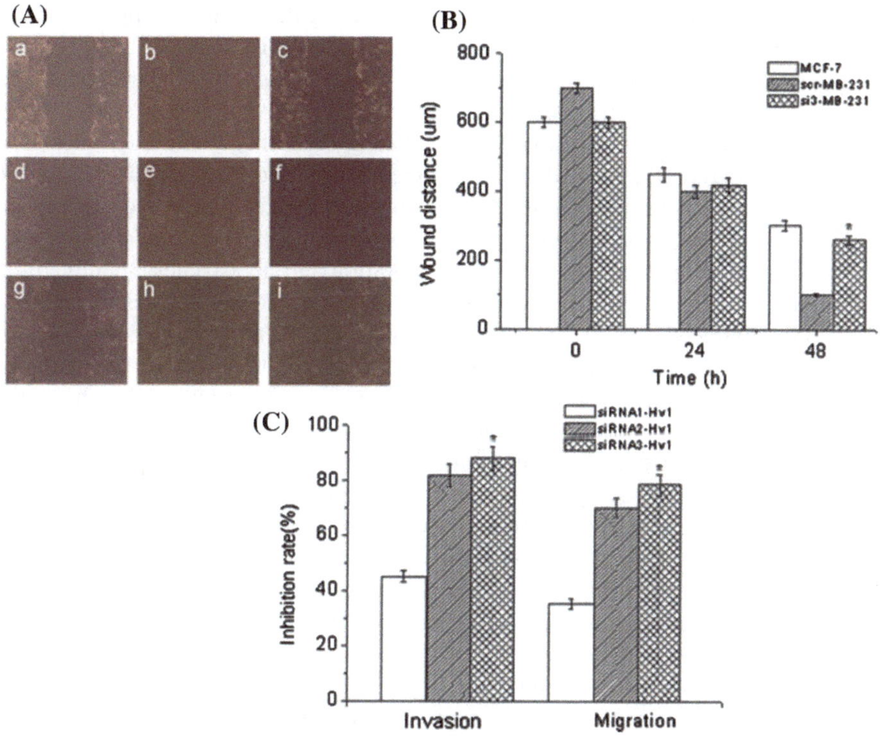


**Fig. 8.5** Hv1 increases cell migration and invasion (Wang et al. 2011). (**A** and **B**) Migration kinetics of MCF-7 (a, d, and g), scr-MB-231 (b, e, and h), and si3-MB-231 (c, f, and i) assayed by wounded monolayer model. a, d, and g, migration of MCF-7 cells for 0, 24, and 48 h, respectively; (b, e, and h), migration of si-MB-231 cells for 0, 24, and 48 h, respectively; (c, f, and i), migration of si-MB-231 cells for 0, 24, and 48 h, respectively. Values are means ± SD ($n = 3$). $^{*}P < 0.05$, compared with scr-MB-231 cells. (C) Invasion and migration are significantly inhibited by siRNA targeting Hv1 in the highly (MDA-MB-231) metastatic cells. The inhibition rate of invasion/migration was calculated as (1 − invasion cell No. of test/invasion cell No. of control) × 100%. Values are means ± SD ($n = 3$). $^{*}P < 0.05$, compared with scr-MB-231 cells

the invasion and migration of MDA-MB-231 cells was studied. MDA-MB-231 cells grown at confluence were transfected with these three siRNA duplexes and one negative control with scrambled random sequence for 24 h, respectively. And MCF-7 and the transfected cells were placed into transwell inserts coated with Matrigel (invasion) or uncoated (migration) and assessed the inhibition rates of the invasion and migration of the highly metastatic MDA-MB-231 cells after time interval 24 h. The inhibition rates of the invasion and migration of MDA-MB-231 cells by the three siRNAs are shown in Fig. 8.5C. The siRNA duplex siRNA3-Hv1 remarkably decreased the invasion and migration of MDA-MB-231 cells. The data indicate that the suppression of Hv1 expression could inhibit the invasion and migration of the highly metastatic breast cancer cells in vitro, and suggest that Hv1 is involved in the invasion and migration of the metastatic cells of human breast cancer.

Wang et al. were the first to show that the voltage-gated proton channel Hv1 was specifically expressed in highly metastatic human breast tumor tissues and cell lines (Wang et al. 2011). In subsequent work (Wang et al. 2012), they addressed on the contribution of Hv1 to breast carcinogenesis to some extent. In this study, Wang et al. showed that Hv1 expression was significantly correlated with the tumor size, tumor classification, lymph node status, clinical stage, and Her-2 status. High Hv1 expression was associated significantly with shorter overall and recurrence-free survival. In vitro, knockdown of Hv1 expression in the highly metastatic MDA-MB-231 cells decreased the cell proliferation and invasiveness, inhibited the cell proton secretion and intracellular pH recovery, and blocked the cell capacity of acidifying extracellular milieu. Furthermore, the activity of gelatinase in MDA-MB-231 cells that suppressed Hv1 was reduced. In vivo, the breast tumor size of the implantation of the MDA-MB-231 xenografts in nude mice that were knocked down by Hv1 was dramatically smaller than that in the control groups. The results demonstrated that the inhibition of Hv1 function via knockdown of Hv1 expression could effectively retard the cancer growth and suppress the cancer metastasis by the decrease of proton extrusion and the down-regulation of gelatinase activity. We shall avoid going to the details here, but readers with more interests may consider reading the original article (Wang et al. 2012).

Voltage-gated proton channels have been implicated in late-stage breast cancer (Wang et al. 2012), where Hv1 expression correlated with tumor size and poor prognosis. Hv1 knockdown reduced cell proliferation and migration as well as matrix metalloprotease release. However, the presence of functional Hv1 in the membrane of cancer cells has not been demonstrated, and the mechanism by which they affect the function of these cells has not been elucidated. Just recently, Morgan et al. have shown the presence of functional Hv1 on the membranes of MDA-MB-231 cells (Morgan et al. 2015), a highly metastatic triple negative cell line. They performed patch-clamp experiments on these cells and were able to detect bona fide voltage- and pH-gated proton selective channels. The membrane density of the channels in these cells was recorded as 3.5 pA/pF which is roughly fivefold lower than the 15 pA/pF displayed in neutrophils. In order to show that Hv1 expresses at a level sufficient to impact pH regulation within these cells, the investigators acid loaded the cells using the ammonium pre-pulse technique and monitored internal pH recovery utilizing SEER with SNARF-1. Inhibiting Hv1 with 1 mM $Zn^{2+}$ slowed recovery from an acid load by threefold, demonstrating that the expression of Hv1 on these cells affects pH regulation in the cells. The ultimate conclusion came out of this latest work that one mechanism by which Hv1 may influence the pathophysiology of breast cancer is by improving the ability of breast cancer cells to regulate their internal pH (Morgan et al. 2015). Membrane patch-clamp experiments as addressed here came out as a traditional technique that is usually applied to address the membrane hosted ion channels' electrophysiology conditions. We wish not to detail the results and other related aspects here, but the interested readers may consider reading the original article (Morgan et al. 2015).

## 8.4 Water Channels in Cancer Cells

Aquaporins (*AQPs*) are intrinsic membrane proteins that facilitate selective water and small solute movement across the plasma membrane (Verkman and Mitra 2000). Optimal concentration of water and solutes is required for survival of cells. This balance is largely maintained by water channels created by *AQPs*. The expression of *AQPs* has been found in more than 20 human cancer types (Papadopoulos and Saadoun 2015). The pore expression is found to be significantly correlated with the severity of histological tumors and prognosis of patients with cancer. Recent evidence showed that *AQPs* could also play a role in tumor-associated edema, proliferation and migration of tumor cells, and tumor angiogenesis in solid and hematological tumors. Inhibitors of *AQPs* in tumor cells and microvessels have been suggested as new therapeutic strategies. A recent review provided important information on *AQPs* in cancer cells and their importance as anticancer therapeutic targets (Wang et al. 2015). Ismail et al. investigated the role of inhibiting *AQPs* in sensitizing prostate cancer cells to cryotherapy and reports suggest quite remarkable therapeutic effects on cancer (Ismail et al. 2009).

The aquaporin membrane proteins were initially characterized in human red blood and renal cells (Preston et al. 1992; Nielsen et al. 1993). Subsequently, they were identified in mammals, plants, yeasts, and arthropods. Thirteen members of the *AQP* family have so far been identified in mammals (*AQP0–12*). Mammalian *AQPs* have two subgroups. First subgroup members *AQP1*, *AQP2*, *AQP4*, *AQP5*, *AQP6,* and *AQP8* are highly selective for the passage of water across the plasma membrane. The second subgroup is the aquaglyceroporins which are *AQP3*, *AQP7*, *AQP9,* and *AQP10* and permit the transport of small nonionic molecules, e.g., glycerol, in addition to water molecules (Ishibashi et al. 1994). Recently, aquaporin 3 was found to be expressed in normal and malignant prostate tissue and is assumed generally to be involved in tumor initiation and development (Wang et al. 2007).

Prostate cancer represents the second most common cause among all sorts of cancer deaths (Jemal et al. 2008). Prostate cryotherapy is localized application of freezing temperature resulting in in situ tissue ablation. Cryotherapy still represents a minimally invasive alternative treatment for locally advanced prostate cancer. Clinical case series studies confirmed the feasibility of cryotherapy as a primary and salvage treatment for patients with localized prostate cancer (Ismail et al. 2007; Cohen et al. 2008). The main aim of prostate cryotherapy is to selectively destroy neoplastic tissue and preserve vital prostate regional structures like rectum and urinary bladder. A standard freezing process is required to achieve this mentioned goal. The tissue at target has to be exposed to a temperature lower than −40 °C to ensure eradication of the cancer tissue completely (Larson et al. 2000) which is complicated temperature cooling therapy, not achievable without huge side effects. Cellular-level cold injury starts when temperature falls to sub-zero levels and extracellular ice starts to form at temperature range between −7 and −20 °C. Lower temperatures ($<-15$ °C) cause intracellular ice formation, which is quite often lethal to the living cells (Gage and Baust 1998). One of the most important aspects perhaps is to consider how aquaporins respond to such cooling therapy in regard to their capability of water molecule transport across plasma membranes.

As an example, we shall provide the brief report of a group that investigated the effect of cryotherapy on aquaporin channels in cancer cells (Ismail et al. 2009). The Ismail et al. group (Ismail et al. 2009) cooled PC-3 and DU145 prostate cancer cells down to 0, −5, and −10 °C. The *AQP3* expression in response to freezing was determined using real-time quantitative polymerase chain reaction (RT–qPCR) and Western blot analysis. Aquaporins were inhibited using mercuric chloride ($HgCl_2$) and small-interfering RNA (siRNA) duplex. Cell survival was assessed using a colorimetric assay. There was a significant increase observed in *AQP3* expression in response to freezing. *AQP3* siRNA-treated cells were observed to be more sensitive to cryoinjury compared with control cells. $HgCl_2$ inhibition of the *AQPs* increased the sensitivity of both cell lines to cryoinjury, and a complete loss of cell viability was observed at −10 °C. Figure 8.6 demonstrates how *AQP3* is expressed on the plasma membrane of prostate cancer PC-3 cells as a result of cryotherapy. The Ismail et al. group was thus quite successful to demonstrate the involvement of membrane protein *AQP3* directly in cryoinjury (Ismail et al. 2009). The bottom line message is that inhibition of *AQP3* increases the sensitivity of prostate cancer cells to freezing.

Latest progresses have added further details into our understanding of the *AQPs* in cancer biology and cellular activity (Papadopoulos and Saadoun 2014; Wang

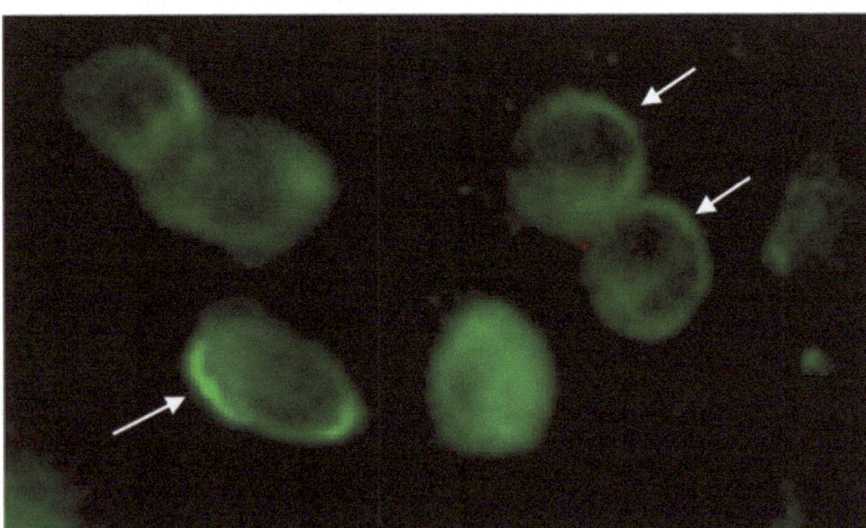

**Fig. 8.6** Immunofluorescence expression and localization of aquaporin 3 (*AQP3*) water channel post-cryotherapy (copied from Ismail et al. 2009). It is clear that *AQP3* is expressed on the plasma membrane of prostate cancer PC-3 cells. A similar finding was observed by the same group in prostate cancer DU145 cells (Ismail et al. 2009). Original magnification × 600. Inhibition of aquaporin in plasma membrane increases the sensitivity of prostate cancer cells to freezing

et al. 2015). Expression of *AQPs* is found positively correlated with tumor types, grades, proliferation, migration, angiogenesis, and tumor-associated edema (Saadoun et al. 2002; Guan et al. 2009; Nico and Ribatti 2011), which can be considered as diagnostic and therapeutic targets in cancer. For example, *AQP1* and *AQP4* were found to be massively up-regulated in high-grade astrocytomas, as compared with the low-grade ones and normal brain tissues (Saadoun et al. 2002; Saadoun et al. 2002a). The high *AQP4* expression was correlated with the formation of more prominent type brain edema (Warth et al. 2007). *AQP1* was found up-regulated in lung adenocarcinoma, and inhibition of *AQP1* expression can cause inhibition of tumor cell invasion, which thereby were proposed as the prognostic index and therapeutic targets for lung cancer (see Hoque et al. 2006).

*AQPs* are evident to be expressed in several solid and hematological cancer cells. Table 8.6 summarizes a detail of *AQPs* expression in different kinds of cancer tissues. One type of *AQP* protein can be expressed in different cancer tissues, and there is a lack of the tumor-specific prosperity. The reverse is also true meaning one type of cancer cell may host expression of several *AQPs*. *AQP1* is reported to be overexpressed in brain, breast, lung, colorectal, and ovarian cancers (Saadoun et al. 2002a; Moon et al. 2003; Yin et al. 2008; Mobasheri et al. 2005; Yang et al. 2006). *AQP3* expression is up-regulated in as many as cutaneous, esophageal and oral squamous, pulmonary, renal, or hepatocellular cancer cells (Hara-Chikuma and Verkman 2008; Liu et al. 2013; Guo et al. 2013; Kafe et al. 2004; Kusayama et al. 2011). The expression of *AQP4* is found to be increased in brain, lung, or thyroid cancers (Saadoun et al. 2002a; Xie et al. 2012; Niu et al. 2012). *AQP5* is over-expressed in lung, chronic myelogenous leukemia (CML), stomach, ovarian, or colorectal cancers (Kang et al. 2008; Chae et al. 2008; Zhang et al. 2010; Yan et al. 2012; Watanabe et al. 2009). *AQP7* expression is found to be increased in thyroid cancer. *AQP9* expressed mainly in brain or ovarian cancers (Lacroix et al. 2005; Tan et al. 2008; Yang et al. 2011). However, *AQP8* is down-expressed in hepatocellular or colorectal cancers (Jablonski et al. 2007; Wang et al. 2012).

*AQP5* are exclusively for water while *AQP3* has ability to transport water and other small solutes such as glycerol. Figure 8.7 shows some of the important features, *AQP*-associated cancer cell pathways in model demonstration. *AQP3* increases intracellular glycerol content which is transported to mitochondria to form ATP. The ATP formation provides energy for tumor cell proliferation. The cascaded diagram shows various other regulatory function provided by *AQP3*; see Fig. 8.7. *AQP3* can directly or indirectly reduce the natural degradation of HIF-1$\alpha$/HIF-2$\alpha$ to cause an increase in the expression of vascular endothelial growth factor (VEGF), which is a critical regulator in vascular maturation and tumor angiogenesis. *AQP3* may directly or indirectly activate AKT (which behaves as oncoprotein) to cause an increase in MMPs, resulting in tumor invasion. *AQP5* induces activation of the epidermal growth factor receptor (EGFR) and extracellular receptor kinase (ERK1/2) pathway to facilitate tumor cell proliferation and metastasis. *AQP5* is phosphorylated on Ser156 and binds the SH3 domain of Src to promote EMT activity in tumor/cancer cells. The roles of *AQP3* and *AQP5* in these pathways are yet to be fully elucidated.

**Table 8.6** *AQPs* expression in different kinds of human tumors

| Aquaporin type | Tumor or cancer type where specific AQP is expressed | Correlated functions in cancer cells |
|---|---|---|
| *AQP0* | Shi et al. (2013) | Unknown |
| *AQP1* | Brain cancer (Saadoun et al. 2002), breast cancer (Yin et al. 2008), colorectal cancer (Moon et al. 2003), cervical cancer, laryngeal cancer, lung cancer (Xie et al. 2012), nasopharyngeal cancer, ovarian cancer (Yang et al. 2006) | Tumor grade, prognosis, tumor angiogenesis, tumor necrosis, tumor cell migration, tumor invasion, and metastasis |
| *AQP2* | | |
| *AQP3* | Colorectal cancer (Moon et al. 2003), cervical cancer, liver cancer (Guo et al. 2013), lung cancer (Papadopoulos and Saadoun 2015), esophageal cancer (Liu et al. 2013), renal cancer (Kafe et al. 2004), skin cancer (Hara-Chikuma and Verkman 2008), stomach cancer, tongue cancer (Kusayama et al. 2011) | Tumor grade, prognosis, tumor angiogenesis, tumor invasion, tumor cell migration, and tumor energy metabolism |
| *AQP4* | Brain cancer (Saadoun et al. 2002), lung cancer (Xie et al. 2012), thyroid cancer (Niu et al. 2012) | Tumor grade, migration, tumor-associated edema, adhesion, invasion, and tumor apoptosis |
| *AQP5* | Breast cancer (Papadopoulos and Saadoun 2015), colorectal cancer (Moon et al. 2003; Kang et al. 2008), cervical cancer, leukemia (Chae et al. 2008), liver cancer (Guo et al. 2013), lung cancer (Zhang et al. 2010), esophageal cancer (Liu et al. 2013), ovarian cancer (Yan et al. 2012), stomach cancer (Watanabe et al. 2009), tongue cancer (Kusayama et al. 2011) | Tumor prognosis, tumor proliferation, tumor invasion, tumor cell migration, and drug resistance |
| *AQP6* | Renal cell carcinoma (GeneCards) | Unclear |
| *AQP7* | Thyroid cancer (Lacroix et al. 2005) | Unknown |
| *AQP8* | Cervical cancer (Shi et al. 2013) | Tumor cell migration, tumor invasion, tumor metastasis, and anti-apoptosis. |
| *AQP9* | Brain cancer (Tan et al. 2008), ovarian cancer (Yang et al. 2011) | Tumor grade, drug resistance, and energy metabolism |
| *AQP10* | Breast cancer (Shi et al. 2012) | Unclear |
| *AQP11* | Breast cancer (Shi et al. 2012) | Unclear |
| *AQP12* | Breast cancer (Shi et al. 2012) | Unclear |

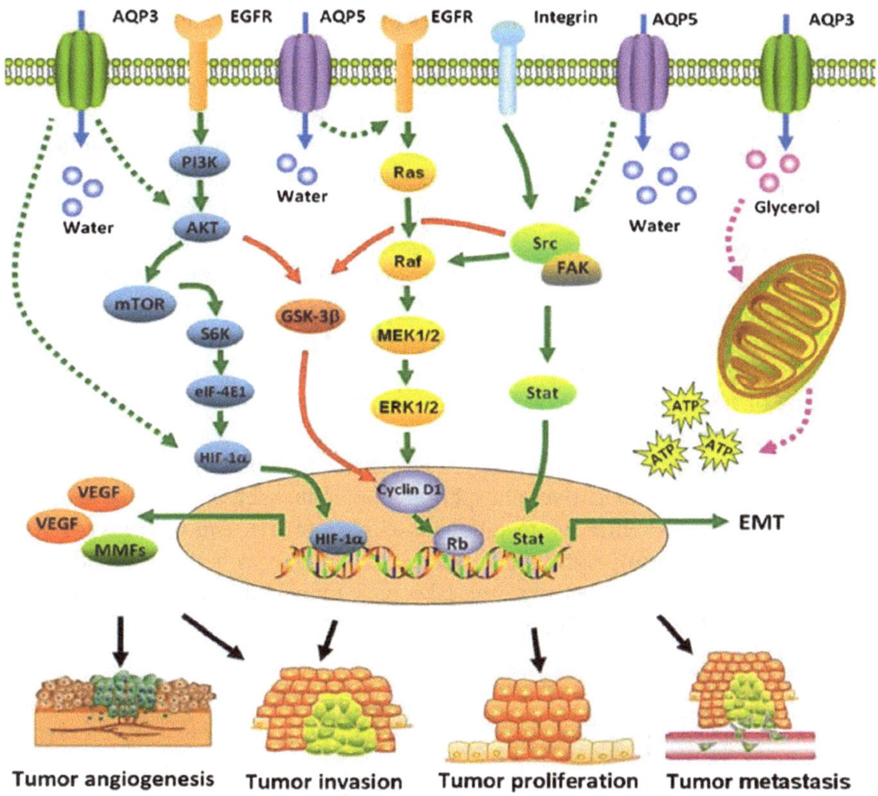


**Fig. 8.7** Proposed model of novel roles of *AQP3* and *AQP5* in tumor biology. Copied from Wang et al. (2015)

## 8.5 A Mitochondria-K$^+$ Channel Axis Suppression in Cancer Cells

In several human cancer cells, high mitochondrial membrane potential ($\Delta\Psi_m$) and low K$^+$ channel Kv1.5 expressions have been observed. This unique metabolic profile of cancer might contribute to apoptosis resistance (Bonnet et al. 2007). In cancer, mitochondrial glucose oxidation is known to be inhibited and energy production relies on cytoplasmic glycolysis. This mitochondrial inactivity likely induces some apoptosis resistance.

Dichloroacetate (DCA) is found to show substantial therapeutic values by inhibiting mitochondrial pyruvate dehydrogenase kinase (PDK), thus shifts metabolism from glycolysis to glucose oxidation, decreases mitochondrial membrane potential $\Delta\Psi_m$, increases the amount of mitochondrial $H_2O_2$, and activates potassium channels in all cancer. DCA acts by up-regulating Kv1.5 by a nuclear factor of activated T-cell NFAT1-dependent mechanism. DCA induces apoptosis, decreases proliferation, and inhibits tumor growth, without apparent toxicity.

Glycolytic phenotype is reported to be associated with a state of apoptosis resistance (Plas and Thompson 2002). Many glycolytic enzymes have been recognized to also regulate apoptosis, and several oncoproteins are found to induce the expression of glycolytic enzymes (Kim and Dang 2005). For example, AKT, which stimulates glycolysis and induces resistance to apoptosis (Elstrom et al. 2004), activates hexokinase, an enzyme catalyzing the first and irreversible step in glycolysis. Via its downstream mediator glycogen synthase kinase 3 (GSK3), AKT induces the translocation of hexokinase to the mitochondrial membrane where it binds to the voltage-dependent anion channel (VDAC), suppressing apoptosis (Kim and Dang 2005; Pastorino et al. 2005). Inhibition of GSK3 in cancer cells causes unbinding of hexokinase from VDAC, induces apoptosis, and increases sensitivity to chemotherapy (Pastorino et al. 2005). This suggests that perhaps the metabolic phenotype in cancer is due to a potentially plastic mitochondrial remodeling that results in suppressed oxidative phosphorylation, enhanced glycolysis, and suppressed apoptosis. Whether the metabolism of glucose will end with glycolysis in the cytoplasm (converting pyruvate to lactate) or continue with glucose oxidation in the mitochondria is controlled by a gate-keeping mitochondrial enzyme, pyruvate dehydrogenase (PDH) (see Fig. 8.8). PDH converts pyruvate to acetyl-CoA which,

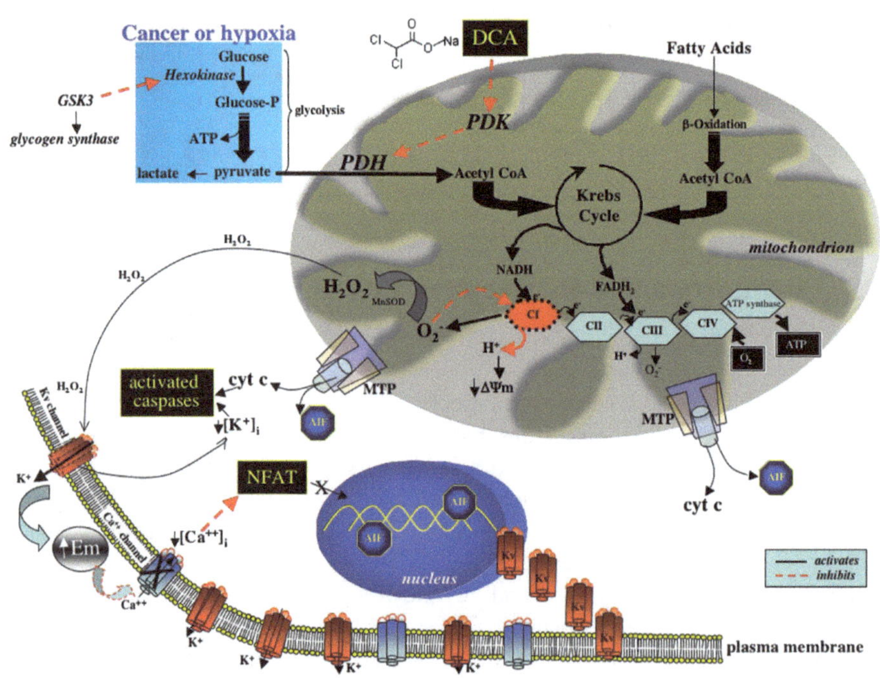


**Fig. 8.8** A reversible metabolic-electrical remodeling in cancer is shown to contribute to resistance to apoptosis [with due copyright permission taken from Bonnet et al. (2007)]

along with the acetyl-CoA from the fatty acid β-oxidation, is fed to the Krebs cycle, producing the electron donors NADH and $FADH_2$. NADH donates electrons to complex I of the electron transport chain (ETC) (and $FADH_2$ to complex III). The flux of electrons down the ETC is associated with the production of ROS and with the efflux of $H^+$, which causes a negative mitochondrial membrane potential ($\Delta\Psi_m$). The F1F0-ATP synthase uses the stored energy of the $\Delta\Psi_m$ to synthesize ATP; thus, the $\Delta\Psi_m$ reflects ETC activity and mitochondrial function. PDH is inhibited by phosphorylation by PDH kinase (PDK). The role of PDH and PDK in cancer is unknown.

Figure 8.8 demonstrates important physiological features (Bonnet et al. 2007). Activation of PDH by DCA increases glucose oxidation by promoting the influx of acetyl-CoA into the mitochondria and the Krebs cycle, thus increasing NADH delivery to complex I of the electron transport chain, increasing the production of superoxide, which in the presence of MnSOD is dismutated to the more stable $H_2O_2$. Sustained increase in ROS generation can damage the redox-sensitive complex I, inhibiting $H^+$ efflux and decreasing $\Delta\Psi_m$. Opening of the $\Delta\Psi_m$-sensitive mitochondrial transition pore (MTP) allows the efflux of cytochrome *c* and apoptosis-inducing factor (AIF). Both cytochrome *c* and $H_2O_2$ open the redox-sensitive $K^+$ channel Kv1.5 in the plasma membrane and hyperpolarize the cell (increased $E_m$), inhibiting a voltage-dependent $Ca^{2+}$ entry. The decreased $(Ca^{2+})_i$ suppresses a tonic activation of NFAT, resulting in its removal from the nucleus, thus increasing Kv1.5 expression. The increased efflux of $K^+$ from the cell decreases the tonic inhibition of $(K^+)_i$ on caspases, further enhancing apoptosis. DCA's selectivity is based on its ability to target the unique metabolic profile that characterizes most cancers, and its effectiveness is explained by its dual mechanism of apoptosis induction, both by depolarizing mitochondria (proximal pathway) and activating/up-regulating Kv1.5 (distal pathway).

## 8.6 Mitochondrial Membrane Channels in Cancer Cells

Abnormalities in the mitochondrial membrane transport properties may be indirectly responsible for turning normal cells into cancerous. The mitochondrial membrane transport mainly happens through various types of channels. ATP and other energy metabolites are delivered from mitochondria to the cell's interior region through a voltage-dependent anion channel (VDAC) (for details see, e.g., Maldonado et al. 2014). But mitochondrial apoptosis-induced channel (MAC) releases cell death signals regulating cytochrome c (Dejean et al. 2006) (see Fig. 8.9).

Bcl-2 family proteins control the apoptosis or cell death program by regulating the permeabilization of the mitochondrial outer membrane (MOM) and, hence, the release of cytochrome c and other pro-apoptotic factors. MAC is responsible for cytochrome c release early in apoptosis. MAC activity is usually detected by patch-clamping mitochondria at the time of release of cytochrome c. MAC was first

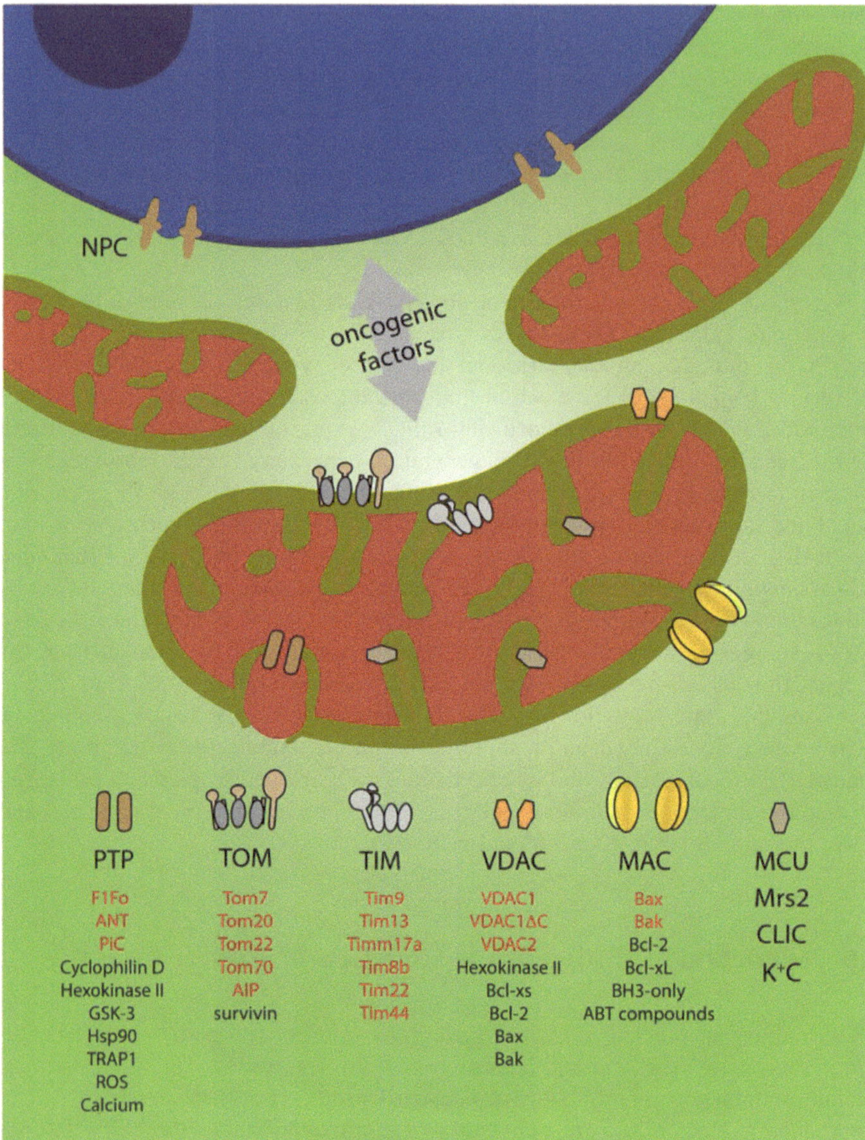


**Fig. 8.9** Mitochondrial ion channels might mediate exchange of oncogenic factors with the nucleus. Nuclear oncogenic factors either directly (via nuclear pore complexes, NPC) (Simon and Rout 2014) or indirectly (via oncogenic expression) are transmitted to mitochondria to reprogram metabolism and cell death mechanisms. Channels in the outer and the inner membranes might provide the pathways for production and/or bidirectional transport of oncogenic factors. Alternatively, they might be direct targets (see text). The legend indicates channel components (red) or modulators (black) associated with cancer (see text for references). Components and regulators of TIM22 and TIM23 channels were listed in the same category. $K^+C$ represents the potassium channels $BK_{Ca}$ and $IK_{Ca}$, ATP-dependent $K_{ATP}$, Kv1.3, two-pore TWIK-related acid-sensitive $K^+$ channel-3 (TASK-3). Inspired by Odra Noel's 'Mitochondrial Dawn.' For details, see Madamba et al. (2015)

detected in patch-clamp experiments on mitochondria isolated from apoptotic FL5.12 cells 12 h after interleukin-3 (IL-3) withdrawal (Pavlov et al. 2001). Electrophysiology record of MAC suggests that MAC is a high-conductance channel of the outer membrane permeable to polymers (see Fig. 8.10).

The Bcl-2 family proteins mainly regulate apoptosis by controlling the formation of MAC. Depending on the type of cells and apoptotic inducer, Bax and/or Bak are structural component(s) of MAC. The negative action on MAC activity occurs through overexpression of anti-apoptotic protein Bcl-2 that eliminates MAC activity. Dejean et al. (2006) produced a nice review article to address the biophysical characterization of MAC activity and its regulation by Bcl-2 family proteins. A model showing the Bcl-2 family members to play a critical role in regulating MAC formation is presented in Fig. 8.11 (for details, see Dejean et al. 2006).

Some cancer cells reprogram the local power plant to adjust with revised physiological conditions by decreasing the open probability of mitochondrial channels involved in the transport of energy metabolites (Goldin et al. 2008; Rostovtseva and Bezrukov 2012; Maldonado et al. 2013) and relying on glycolytic metabolism to fuel competitive reproduction (Maldonado et al. 2014). The payoff for relinquishing energy conversion efficiency is that cancer cells suffer lower oxidative stress which seems to be one of the determinants of tumor cell transformation (Gupta et al. 2012; Yang et al. 2013). Exacerbated ROS emission causes the opening of pores in the inner membrane that breach the membrane permeability barrier to solutes and the ensuing uncontrolled matrix swelling, membrane rupture, and spillage of mitochondrial proteins into the cytosol. Although the exact permeability mechanism of the inner membrane is still being deciphered, cancer cells expertly avert them by equally mysterious ways, not known yet. Cancer cell mitochondria are coated with sentinel proteins, which prevent the formation of MAC in the outer membrane, rendering the power plant oblivious to cancer transformation and unrestrained growth. Cancer cells are known to overexpress the sentinel protein Bcl-2. It may also be considered counterintuitive that the MAC components Bax and Bak are structurally related and belong to the Bcl-2 family of proteins. The truth is that this multifaceted family of proteins controls permeabilization of the outer membrane by either inducing or preventing it. In cancer cells, the second mode prevails and might even go beyond the outer membrane, as some anti-apoptotic members (Bcl-xL and Mcl-1) have been shown to interact with and regulate inner membrane proteins, including putative components of the permeability transition pore complex (Madamba et al. 2015; Vento et al. 2010; Quinn et al. 2011; Arbel et al. 2012). Mitochondrial dysfunction in cancer cells may also be linked to the characteristic genomic instability of many human cancer cells (Delsite et al. 2002; Rasmussen et al. 2003; Miceli et al. 2005; Singh et al. 2005; Kulawiec et al. 2008; Porporato et al. 2014). Additional mitochondrial channels may also be responsible for controlling calcium, potassium and magnesium flux, organelle volume, nucleic acid transport, etc., which are directly or indirectly associated with the rise of cancer condition (Baughman et al. 2011; De Stefani et al. 2011; Peixoto et al. 2012; Leanza et al. 2013, 2014). Table 8.7 addresses a brief

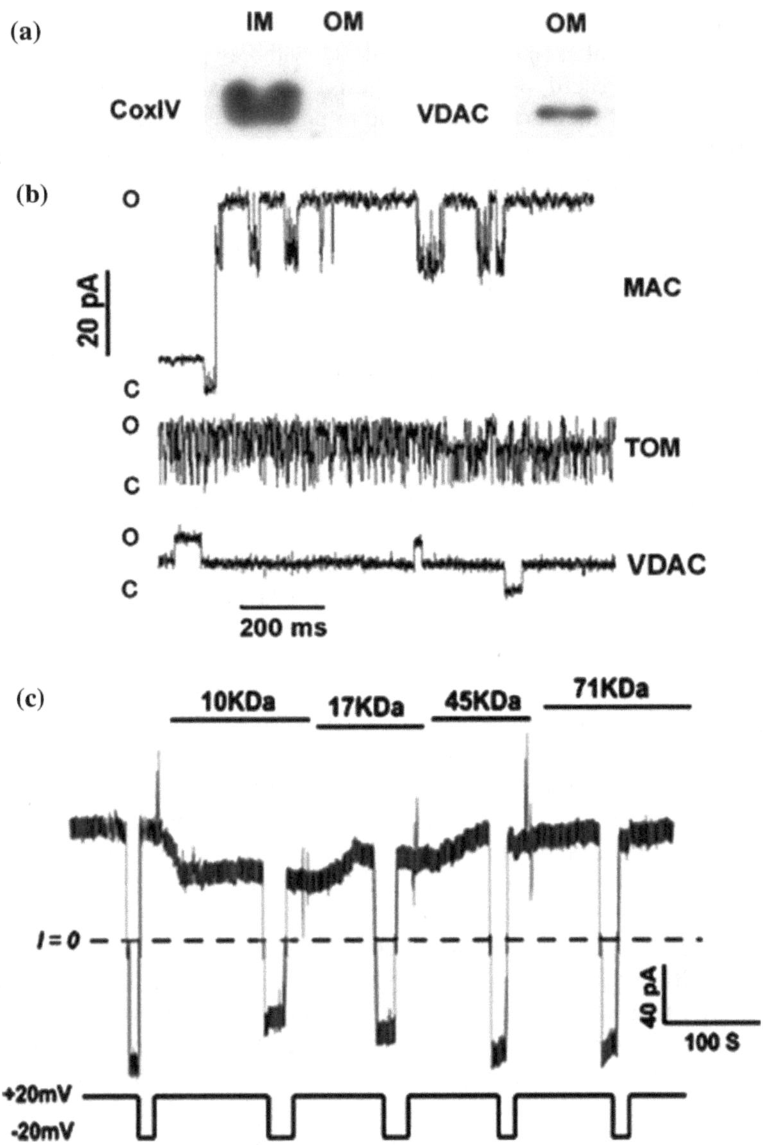


report on a few important mitochondrial ion channels as their involvement is identified in various cancer cells. We shall also provide here some details about two important channel classes, namely VDAC and MAC in cancer cells.

◂**Fig. 8.10** MAC is a high-conductance channel of the outer membrane permeable to up to 17 kDa polymers. **a** Immunoblots show the presence of the outer membrane protein VDAC but not the inner membrane protein cytochrome oxidase sub-unit IV (CoxIV) in the outer membranes (OM, 2 μg) purified from mitochondria of apoptotic FL5.12 cells. Inner membranes (IM, 2 μg) are the positive control for CoxIV. (**b** and **c**) MAC activity was reconstituted by incorporating mitochondrial outer membranes of apoptotic FL5.12 cells into proteoliposomes. **b** Current traces at 20 mV allow comparisons of the single-channel behavior of MAC, VDAC, and TOM channels. O and C indicate open- and closed-state current levels, respectively. **c** The pore size of MAC was estimated by the polymer exclusion method. Current traces are shown of a 4 nS MAC after sequential perfusion of the bath with media containing 5% w/vol of indicated MW dextrans as the voltage was switched between ±20 mV. 10 and 17 kDa MW dextrans are permeant as they induce a decrease in the current, but 45 and 71 kDa polymers are not. Parts of this figure were reprinted from Pavlov et al. (2001), Guo et al. (2004), Martinez-Caballero et al. (2004)

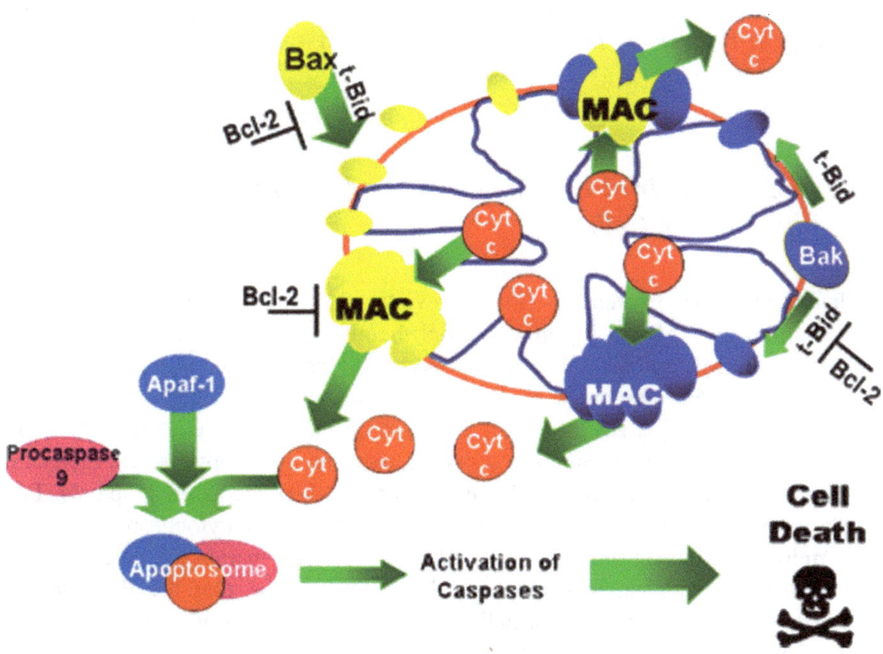


**Fig. 8.11** MAC provides a pathway for cytochrome c to exit the mitochondria during apoptosis (Dejean et al. 2006). Model shows that Bcl-2 family members play a critical role regulating MAC formation. Induction of the intrinsic pathway results in Bax translocation to mitochondria and oligomerization. The BH3 domain-only protein t-Bid facilitates activation and oligomerization of Bax and Bak to form MAC. The anti-apoptotic protein Bcl-2 suppresses MAC formation. MAC could be predominantly composed of either Bax or Bak. There may be other components, and Bax and Bak may or may not form hetero-oligomers. MAC formation results in the release of cytochrome c and ultimately apoptosome formation, caspase activation, and cell death

### 8.6.1 *Voltage-Dependent Anion Channels in Mitochondrial Membrane of Cancer Cells*

Reference to a nice report by Tajeddine et al. (2008), the group studied A549 NSCLC cells (carcinomic human alveolar basal epithelial cell line) being transfected with a battery of small-interfering RNAs (siRNAs) that knock down distinct proteins involved in the regulation of apoptosis-related MMP (de La Motte Rouge et al. 2007). It provides explanation on the role of VDAC channels. Figure 8.12 shows the effects of VDAC1 knockdown on apoptosis-associated mitochondrial transmembrane potential ($\Delta\Psi_m$) dissipation and plasma membrane permeabilization. Subsequent Fig. 8.13 shows the influence of VDAC1 on the release of mitochondrial intermembrane space (IMS) proteins and caspase activation induced by CDDP, cisplatin which induces cell death (Tajeddine et al. 2008). Providing details on VDAC channels' roles in cancer cells goes beyond the scope of this book, but readers may read the details from various references quoted in Table 8.7.

### 8.6.2 *Mitochondrial Apoptosis-Induced Channels in Cancer Cells*

Madamba et al. have recently written a review where the group has attempted to address the role of MACs in mitochondrial membrane in cancer cells in brief, but quite in style (Madamba et al. 2015). MAC is a dynamic channel formed in the outer mitochondrial membrane by at least two pro-apoptotic Bcl-2 protein family members, Bak and Bax (Dejean et al. 2006). A decade ago, Hetz et al. (2005) first demonstrated that Bax channel activity is required in apoptosis. Bax channel inhibitors were found to protect cells in vitro against Bax-mediated apoptosis. Two small molecular weight Bax channel inhibitors prevented cytochrome c release from mitochondria, inhibited decrease of mitochondrial membrane potential, and protected HeLa cells against apoptosis. Here to mention that the Bax channel inhibitors did not affect the conformational activation of Bax or its translocation and insertion into the mitochondrial membrane in cells undergoing apoptosis (Hetz et al. 2005). MOM's localized physiological condition is another important feature that might regulate apoptosis. Recently, Colombini group (Ganesan et al. 2012) investigated the dynamics of Bax-mediated mitochondrial outer membrane permeabilization (MOMP), disorder of which creates direct regulation in apoptosis. Here, the rate of Bax-mediated MOMP in isolated mitochondria was investigated considering ionic strength as a tool. Higher ionic strength was found to increase the sensitivity of Bax-mediated MOMP to inhibition by Bcl-xL. Decreasing ionic strength resulted in the formation of larger channels that allow high flux rates of cytochrome c and synchronously release both small and large proteins. High ionic strength favored the initial formation of small channels, and these develop more

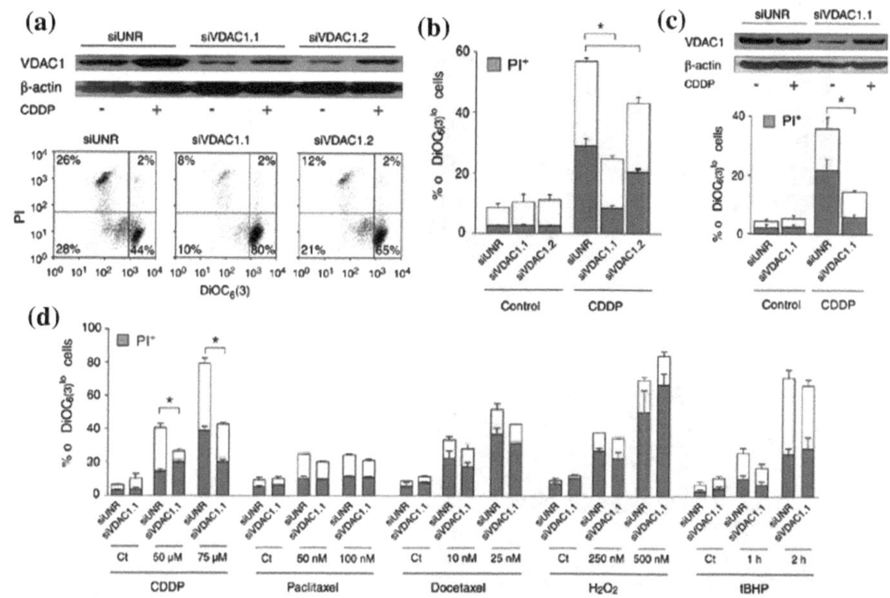


**Fig. 8.12** Effects of VDAC1 knockdown on apoptosis-associated mitochondrial transmembrane potential ($\Delta\Psi_m$) dissipation and plasma membrane permeabilization (Tajeddine et al. 2008). After transient transfection (for 48 h) with either of the two siRNAs down-regulating VDAC1 (i.e., siVDAC1.1 and siVDAC1.2) or with a control siRNA (siUNR), A549 (**a**, **b**, **d**), and HeLa (**c**) cells were left untreated or were treated for additional 24 h with 50 μM CDDP or with the indicated concentrations of other cytotoxic molecules, and finally labeled for the cytofluorometric assessment of $\Delta\Psi_m$ [with the $\Delta\Psi_m$-sensitive probe $DiOC_6(3)$] and plasma membrane integrity (with propidium iodide, that is, PI). Due to its rapid effects, *tert*-butylhydroperoxide (tBHP) was administered at the fixed concentration of 100 μM for 1–2 h. White and gray columns represent the percentage of cells exhibiting $\Delta\Psi_m$ loss alone ($DiOC_6(3)^{low}$) or in association with plasma membrane breakdown ($PI^+$), respectively (mean ± s.e.m., $n = 3$, $^*P < 0.05$). (**a–c**) VDAC1 depletion inhibits CDDP-induced apoptosis-associated modifications. (**a**) Representative dot plots recorded upon CDDP treatment. Numbers in each quadrant indicate the percentage of cells. In (**b**) and (**c**), quantitative data are presented for A549 and HeLa cells, respectively. Insets in (**a**) and (**c**) report immunoblots for the detection of VDAC1 following siRNA-mediated down-regulation in A549 and HeLa cells, respectively. (**d**) Knockdown of VDAC1 fails to inhibit cell death induced by various toxic stimuli other than CDDP. CDDP, cisplatin; $DiOC_6(3)$, 3,3′dihexyloxalocarbocyanine iodide; $H_2O_2$, hydrogen peroxide; siRNA, small-interfering RNA; VDAC1, voltage-dependent anion channel 1

slowly into the larger structures that were observed much sooner at low ionic strength. More in cancer cell detection of parameters related to Bax channel's regulated dynamics needed to address this phenomenological aspect.

The MAC pores may not be entirely of protein nature. A few quite important studies showed that sphingolipid ceramides or intermediates of ceramide biosynthesis pathway had the potential to stimulate cytochrome c release in a Bak- and/or Bax-dependent manner, indicating that these lipids are integral part of MAC conductance (Siskind et al. 2010; Chipuk et al. 2012). In theory, MAC function mainly

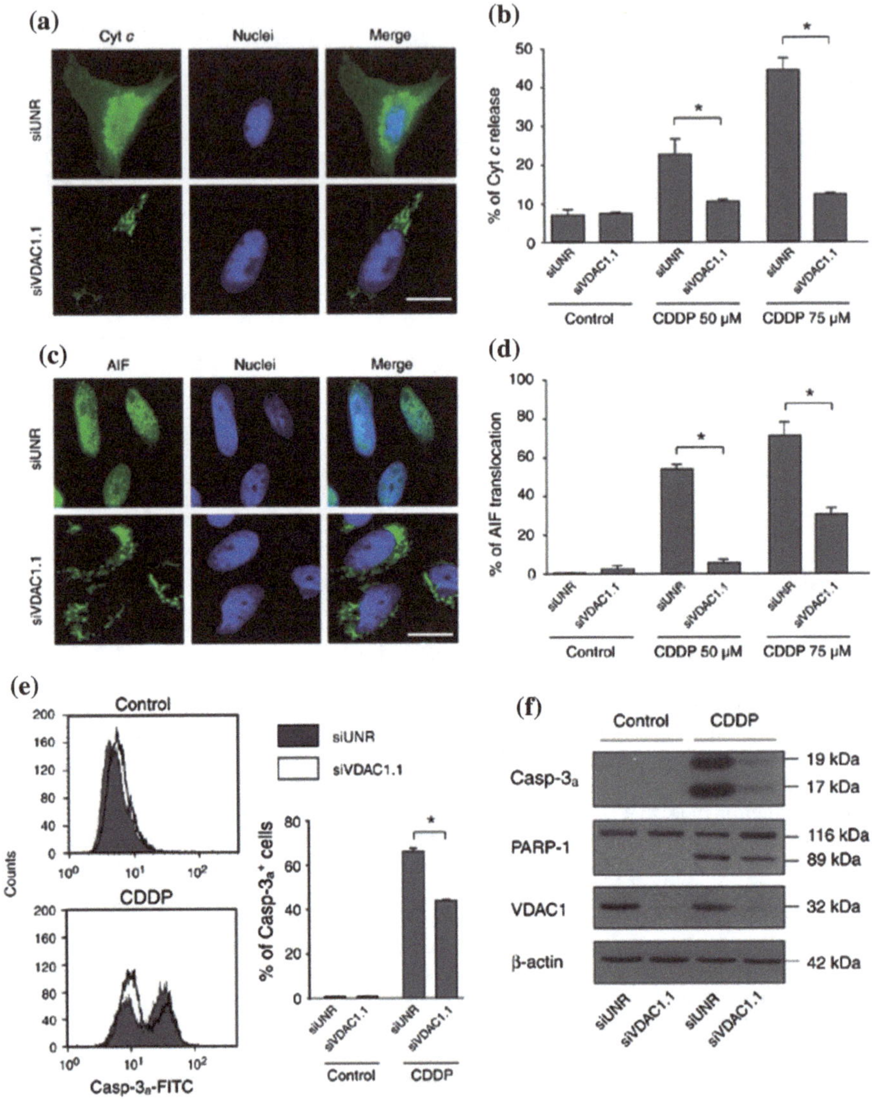


antagonizes cancer cell transformation and is, in fact, kept at bay by oncogenic gene expression. Mechanism of regulation of the Bcl-2 or Bcl-xL proto-oncogenes is associated with formation of tumor and more particularly with B-cell non-Hodgkin's lymphoma (Egle et al. 2004; Swanson et al. 2004; Meijerink et al. 2005). Prolymphocytic cell lines overexpressing Bcl-2 are known to exhibit resistance to MAC formation and the ability to induce lymphoma in mice (Pavlov et al. 2001; Meijerink et al. 2005; Goping et al. 1998; Gross et al. 1998).

◂**Fig. 8.13** Influence of VDAC1 on the release of mitochondrial intermembrane space (IMS) proteins and caspase activation induced by CDDP (Tajeddine et al. 2008). (**a–d**) IMS proteins are retained in mitochondria of VDAC1-depleted cells following CDDP treatment. A549 cells transiently transfected (48 h) with an siRNA targeting VDAC1 (siVDAC1.1) or with an unrelated, control siRNA (siUNR) were exposed or not to the indicated doses of CDDP for 24 h followed by immunofluorescence staining to visualize nuclei (with Hoechst 33342) and cytochrome *c* (Cyt *c*; (**a**) and (**b**)) or apoptosis-inducing factor (AIF; (**c**) and (**d**)). In (**a**) and (**c**), representative immunofluorescence pictures of CDDP-treated cells are shown (white scale bar = 10 μm), whereas the percentage of cells exhibiting diffuse Cyt *c* and nuclear AIF staining is depicted in (**b**) and (**d**), respectively (mean ± s.e.m., $n = 3$, $^{*}P < 0.05$). (**e** and **f**) The knockdown of VDAC1 inhibits CDDP-induced caspase-3 activation. A549 cells transfected as in (**a**) were left untreated or incubated with 50 μM CDDP for additional 24 h, then subjected to cytofluorometric assessments (**e**) or immunoblotting (**f**) for the analysis of caspase-3 activation. In (**e**), left panels depict representative histograms recorded following the administration of CDPP to control or VDAC1-depleted cells. In the right panel, the percentage of cells (mean ± s.e.m., $n = 3$, $^{*}P < 0.05$) staining positive for active caspase-3 (Casp-$3_a$ $^{+}$) is reported. (**f**) Total protein extracts were separated on SDS–PAGE and analyzed by immunoblotting with antibodies specifically recognizing the active fragments of caspase-3 (Casp-$3_a$), poly (ADP-ribose) polymerase 1 (PARP-1), VDAC1 and β-actin (loading control). The activation of caspase-3 results in the accumulation of its active fragments of 19 and 17 kDa, paralleled by the cleavage of PARP-1 (116 kDa) to a degradation product of 89 kDa. CDDP, cisplatin; SDS–PAGE, sodium dodecyl sulfate-polyacrylamide gel electrophoresis; siRNA, small-interfering RNA; VDAC1, voltage-dependent anion channel 1

Paradoxically, it was shown that an overabundance of Bcl-2 not only caused mitochondrial sequestration of BH3-only proteins, which are activators of the formation of MAC, but also that of the MAC core component Bax (Certo et al. 2006; Letai 2008; Teijdo and Dejean 2010). This phenomenon seems to be a common feature of anti-apoptotic Bcl-2 family members as an overabundance of Bcl-xL was also recently shown to induce both Bax accumulation and functional activation at the mitochondria (Gautier et al. 2011; Renault et al. 2013). This accumulation has been termed 'priming for death,' and this concept is currently being used to develop new anticancer drugs and strategies of cancer therapies (Certo et al. 2006; Letai 2008).

In a decade-old study by Tarek et al. (2006), the role of the most studied ceramide was investigated. Ceramide is a pro-apoptotic molecule. The accumulation of ceramide following treatment of cells with various apoptotic agents has implicated this lipid in its biological responses of these agents (Radin 2003). Because of the apoptosis-inducing effects in cancer cells, ceramide has been considered as 'tumor suppressor lipid' (Hanun 1997). Ceramide (Cer) has been found linked to most of the tumor suppressor protein p53, the Bcl-2 family of proteins and several protease classes that are all key components of the tumor response to stress insults (see Fig. 8.14, copied from Tarek et al. 2006). Ceramide is therefore found a promising target for ceramide-based therapeutics in cancer treatment (Morad and Cabot 2013). We are also in progress to design ceramide binding agents, mostly aptamer based, to address the ceramide-targeted anticancer drug discovery (Ashrafuzzaman and co-workers).

**Table 8.7** Mitochondrial ion channels in cancer cells or tissue

| Ion channel | Cancer type where role is proven | Function in cancer cell |
|---|---|---|
| VDAC1<br>*In mammals 3 isoforms of VDAC: VDAC1, VDAC2, VDAC3* | Breast<br>Colon<br>Liver<br>Lung<br>Pancreatic<br>Thyroid cancer | VDAC1 gene is up-regulated. Loss of growth control, role in tumor microenvironments, which are often hypoxic |
| VDAC2 | Prostate<br>Melanoma | Microtubule formation regulates apoptosis through its interactions with Bak, activates Bax to efficiently mediate cell death, facilitates recruitment of both Bak and Bax to the mitochondria. Overexpression of VDAC2 inhibits inhibition of Bak and decreases apoptosis by Bcl-x(S) in melanoma cells Plotz et al. (2012),<br>Asmarinah et al. (2014) |
| VDAC3 | Prostate<br>Liver (HepG2 cells) | No significance effect on VDAC3 observed in prostate cancer tissue. VDAC3 is a regulator of sperm motility, male mice missing VDAC3 result in infertility Reina et al. (2010), VDAC3 knockdown decreases mitochondrial membrane potential in HepG2 cells Maldonado et al. (2013), mutations in VDAC3 are also associated with Parkinson's disease. |
| MAC-Bax<br>*MAC: formed by proteins Bax and Bak and lipid ceramide.* | Hematological cancer<br>Human colon cancer cells (HCT116) Chandra et al.( 2005)<br>Human prostate cancer cell lines (DU145) Chandra et al. (2005) | Bax deficiency delayed and inhibited apoptosis Chandra et al. (2005). |
| MAC-Bak | Human colon cancer cells (HCT116) Chandra et al. (2005)<br>Human prostate cancer cell lines (DU145) Chandra et al. (2005) | Down-regulation of Bak in Bax-expressing cells only slightly inhibits mitochondrial protein release, caspase activation, and apoptosis Chandra et al. (2005) |
| MAC-ceramide | | Ceramide is popularly known as the 'tumor suppressor lipid' Taha et al. (2006) |

(continued)

**Table 8.7** (continued)

| Ion channel | Cancer type where role is proven | Function in cancer cell |
|---|---|---|
| Protein import channels, pathway through channel complexes: translocase of the outer membrane (TOM); proteins Tom20, Tom70, Tom40, Tom22, Tom7, Tom6, Tom5 Rapaport (2005) | Cervical carcinoma HeLa cells<br>Breast adenocarcinoma MCF-7 cells<br>Epithelial cancer cells | Tom20 abolished mitochondrial import of survivin and sensitized tumor cells to apoptosis, whereas silencing of Tom70 had no effect. Rapaport (2005), Sotgia et al. (2012) |
| Translocases of the inner membrane (TIM); TIM22, TIM23 | Thyroid carcinomas | Tim22 forms channel<br>Tim44 observed in oncotic thyroid carcinomas<br>TIM23 peripheral component might have a role in cancer transformation possibly by induction of ROS production or impairment of protein import Handa et al. (2007) |
| Permeability transition pore (PTP) in the inner mitochondrial membrane | General tumor cell<br>Skin cancer | Cancer cells down-regulate PTP inducers (i.e., ROS and calcium), alter expression of chaperones that regulate PTP opening, desensitize the channel through kinase signaling pathways (GSK-3 and hexokinase II), leading to reduced mitochondrial oxidative phosphorylation regimes characterized by high ATP/ADP ratios, low-inorganic phosphate, and high-ROS levels Bonora and Pinton (2014), Rasola and Bernard (2015)<br>Inhibition of the PTP promotes skin cancer Norman et al. (2010) |

This is just a brief list.
There are many other but we wish not to go for the whole database here.

### *8.6.3 Mitochondrial Inner Membrane Permeability Transition Pore in Cancer Cells*

Induction of permeability transition pore (PTP) in inner mitochondrial membrane is a key effector of death elicited by oxidative stress (Rasola and Bernardi 2011). Prolonged opening of PTP constitutes the point of no return toward the demise of the cell, irreversibly committing it either to apoptosis, provided that there is enough ATP present to activate caspases, or to necrosis, elicited by massive release of $Ca^{2+}$ in the cytosol and through impairment of various mitochondrial functions. Various

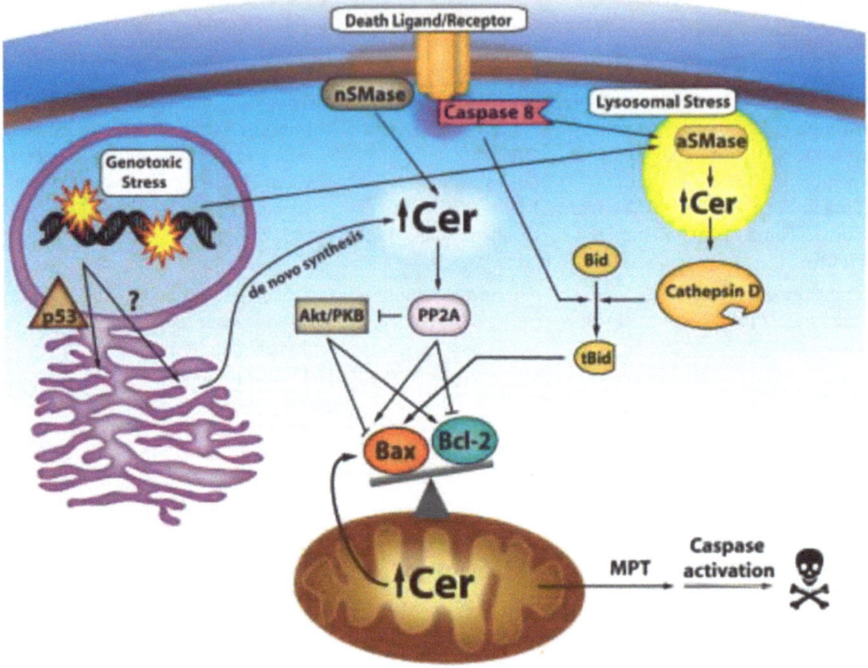


**Fig. 8.14** Ceramide is a central mediator of many apoptotic pathways. Following genotoxic stress, ceramide accumulates, but it is still unclear whether the increase is dependent or independent of p53. Signaling through the lysosomes involves increases in ceramide via ASMase. Ceramide can directly activate the lysosomal protease cathepsin D which can subsequently cleave and activate Bid. Another direct target of ceramide is protein phosphatase 2A, which is known to negatively regulate the prosurvival kinase AKT/PKB. Both PP2A and AKT/PKB regulate the Bax/Bcl-2 rheostat; if the PP2A prevails, as is the case in ceramide signaling, Bax prevails over Bcl-2 and the mitochondrion is permeabilized. Death receptor activation can lead to ceramide formation via nSMase and caspase 8-mediated cleavage of Bid; lysosomes may also be activated downstream of death receptors, with both signals converging on MMP. Additionally, generation of ceramide from sphingomyelin hydrolysis in mitochondria results in Bax translocation and MMP

pro-oxidant agents are able to induce PTP opening. Oxidants increase intracellular release of $Ca^{2+}$ from the endoplasmic reticulum and inhibit $Ca^{2+}$ extrusion from plasma membrane (Camello-Almaraz et al. 2006). Mitochondrial $Ca^{2+}$ enhances oxidative phosphorylation (OXPHOS), and therefore generation of ROS as by-products. Above a certain level, mitochondrial $Ca^{2+}$ or membrane hyperpolarization might also produce ROS through inhibition of respiratory complexes and eventually by PTP induction, which completely stops electron flow along the respiratory chain and causes release of mitochondrial glutathione (GSH) (Kowaltowski et al. 2009; Leanza et al. 2013). A transient PTP induction may in turn induce short bursts of ROS in the so-called ROS-induced ROS release (RIRR) (Zorov et al. 2006; Zorov et al. 2014).

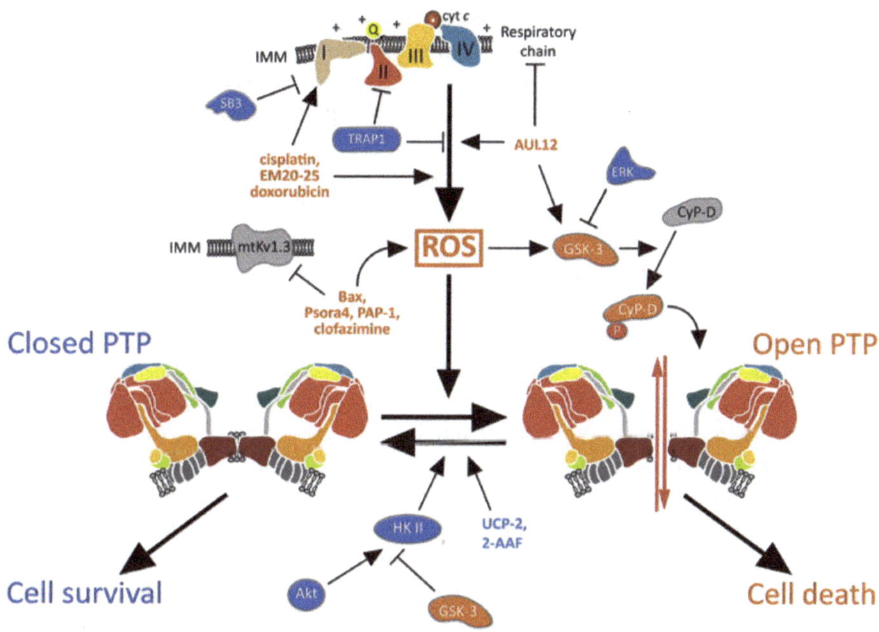


**Fig. 8.15** Mechanisms of PTP regulation in tumor cells. A variety of factors control tumor cell viability by modulating PTP opening, mainly by modulating ROS levels. A ROS surge elicits PTP opening and cell death, whereas ROS inhibition keeps the pore locked and protects cells from noxious stimuli (Rasola and Bernardi 2015). PTP inducers are indicated in orange, PTP inhibitors in blue. I-IV: respiratory complexes; 2-AAF: 2-acetylaminofluorene; Q: coenzyme Q; cyt c: cytochrome c; HK II: hexokinase II; IMM: inner mitochondrial membrane; SB3: serpin B3; UCP-2: uncoupling protein 2

Various molecules may induce ROS-dependent PTP opening (see Fig. 8.15) (Rasola and Bernardi 2015). Inhibition of a mitochondrial voltage-dependent $K^+$ channel, termed mtKV1.3, induces production of ROS and a CsA-sensitive mitochondrial depolarization, indicating opening of PTP. Inhibition can be elicited by specific residue on the pro-apoptotic protein Bax (Szabo et al. 2008) or by the membrane-permeant, selective inhibitors Psora-4, PAP-1, and clofazimine, which at the end induce death in several human and mouse cancer cell lines and in primary B-chronic lymphocytic leukemia cells (Leanza et al. 2012; Leanza et al. 2013a).

## 8.7 Targeting Ion Channels in Cancer Cells

This chapter is dedicated to addressing the state of ion channels in cancer cells. We have so far discussed about the altered expression of ion channels, some of the channels' structural and functional features, and their specific molecular actions in cancer cells. A clear understanding of these structural and functional phase state alterations is key to success in developing therapeutics for cancer cure. Altered channel expressions can be exploited for diagnostic purposes or for addressing traceable or cytotoxic compounds to specific neoplastic tissue (Arcangeli et al. 2009).

Evidences indicate that blocking ion channel activity impairs the growth of some tumors, both in vitro and in vivo. This opens up a new field for medicinal chemistry studies. This can avail many available tools, such as blocking antibodies, antisense oligonucleotides, small-interfering RNAs, peptide toxins and a large variety of small organic compounds, designed small-molecule inhibitors, molecular regulators. Ion channels are generally overexpressed in numerous types of tumor cells. Their altered activity plays a significant role in apoptosis resistance, proliferation, and metastasis of cancer cells. Controlling channel growth is an obvious strategy in therapy. Blocking the activity of ion channels to impair cancer growth is an obvious scientific strategy (Bose et al. 2015). This is associated with altered levels of ion channel expression and their activity in the mutated cancer cells (Lang and Stournaras 2014). Some ion channel blockers, however, produce serious side effects, often considered as cytotoxicity, in the vicinity of on or off target areas, such as cardiac arrhythmias. Therefore, rigorous multidimensional cytotoxicity assays need to be concomitantly explored, besides finding agents for regulating altered ion channel states, structures, and functions. We present here a Table 8.8 [modified from another table in Bose et al. (2015)] listing a few most updated ion channel blockers and regulators and their roles in cancer cell-based ion channel functions. We are also in process of designing and validating nucleic acid aptamers (Ashrafuzzaman 2016) based on patented techniques entropy fragment-based approach (Tseng et al. 2011; Ashrafuzzaman et al. 2013) and screened Coulomb interaction approach (Ashrafuzzaman and Tuszynski 2012). We cannot list the aptamer sequences here because we are still in process of patenting them.

Beside conventional therapy, ion channels in cancer treatment are sometimes targeted using some dietary compounds (Frede et al. 2013). Curcumin, resveratrol (grape polyphenol), docosahexaenoic acid (omega-3), and epigallocatechin gallate (catechin from green tea) extract were investigated and found to modulate ion channels activity by suppressing migration and growth of cells in breast and ovarian cancers (Ji et al. 2008; Lee et al. 2009; Isbilen et al. 2006; Yan et al. 2012).

We have substantial amount of knowledge about the altered ion channel conditions in cancer cells, as explained in this chapter. But the progress in channel-targeted therapeutics and diagnostic discoveries is not satisfactory. Our book does not aim at covering the discovery part here, but our explained knowledge on the channels' alteration conditions in this chapter will certainly help the discovery scientists to work for further developments in this area.

**Table 8.8** Ion channel blockers in immune and cancer cells

| Ion channel blocker | Ion channel | Cell type | Comments | References |
|---|---|---|---|---|
| Margatoxin (MgTX) Charybdotoxin (CTX) | Kv1.3 | T-lymphocytes, Jurkat cells | Antiproliferative effect in T-lymphocytes, regulation of immunoresponsiveness | Garcia-Calvo et al. (1993), Helms et al. (1997) |
| TRAM-34, NS6180, ShK-186 | Kv1.3, KCa3.1 | NK cells, leukemia cells | Inhibition of KCa3.1 increased the degranulation of adherent NK cells and their ability to kill K562 leukemia cells | Koshy et al. (2013) |
| R-roscovitine | Kv1.3, Kv2.1, Kv4.2, HERG (Kv11.1) | Leukemia | Roscovitine is well-known cyclin-dependent kinase inhibitor | Buraei et al. (2007), Meijer et al. (1997) |
| mAb56 | EAG1 (Kv10.1) | Pancreas carcinoma, breast cancer | Inhibition of tumor cell growth both in vitro and in vivo | Gomez-Varela et al. (2007) |
| Way 123,398 | HERG (Kv11.1) | Colorectal cancer | Reduced cell migration of H630, HCT, and HCT8 cells; unaffected growth of HEK 293 cells | Lastraioli et al. (2004) |
| Way 123,398; CsCl; E4031 | HERG (Kv11.1) | Acute myeloid leukemia | Impaired cell proliferation | Pillozzi et al. (2002, 2007) |
| Cisapride | HERG (Kv11.1) | Gastric cancer | Inhibition of cells entering S phase from G1 phase of the cell cycle. | Shao et al. (2005) |
| Verapamil | ERG (Kv11.1) | Lung cancer, melanoma, colon cancer | Increased survival rate for patients treated with verapamil + chemotherapy | Millward et al. (1993), Yohem et al. (1991) |
| UNBS0 (Cardenolide) | $Na^+/K^+$ ATPase | Glioblastoma | Decrease in intracellular ATP concentration leads to autophagy in glioma cells UNBS0 shows antiproliferative activity in vitro in 58 human cancer cell lines | Lefranc and Kiss (2008), Van Quaquebeke et al. (2005) |

(continued)

**Table 8.8** (continued)

| Ion channel blocker | Ion channel | Cell type | Comments | References |
|---|---|---|---|---|
| Tetrodotoxin (TTX) | Nav1.5, Nav1.6 Voltage-gated $Na^+$ channels | Human melanoma, macrophages, breast cancer | TTX and shRNA knockdown of Nav1.6 has inhibitory effects on both cellular invasion of macrophages and melanoma cells | Carrithers et al. (2009), Fraser et al. (2005) |
| Charybdotoxin (CTX) | $K_{ir}$ (IK1) | Human melanoma | Reduced migration of melanoma cells treated with CTX | Schwab et al. (1999) |
| Zinc, methanandamide | $K_{2P}9.1$ (TASK-3) | Ovarian cancer | Reduction in cell proliferation and increase in apoptosis | Innamaa et al. (2013) |
| DNA aptamers | MAC released Bcl-2 protein | | Anti-apoptotic cytosol protein | Ashrafuzzaman (2016a) |
| DNA aptamers | MAC released Bax protein | | Pro-apoptotic cytol protein | Ashrafuzzaman (2016a) |
| DNA aptamers | Ceramide channel | | MAC channel releasing apoptotic proteins | Ashrafuzzaman (2016b) |

## 8.8 Conclusions

States of ion channels especially structures and functions in cancer cells are often observed way different from that of normal cells. Understanding of pinpointed aspects and targeting that for chemotherapy is a promising field now. As anticancer drug discovery is also on a rise, certain ion channel structure may be at the target to design novel molecules. The geometric locations of most of the ion channels that are analyzed here can be sketched in a single model diagram (see Fig. 8.16) as presented in 2016 (Peruzzo et al. 2016).

Two decades ago, ion channels were proposed as potential targets for cancer treatment as the channels are often aberrantly expressed in cancer tissues compared to healthy ones. In a nutshell, several hallmarks of cancer cells rely on the physiological conditions of various of these cell-based ion channels (Hanahan and Weinberg 2011), such as cell migration and invasiveness (Litan and Langhans 2015; Djamgoz and Onkal 2013), metabolic reprogramming, e.g., (Andersen et al. 2014; Leanza et al. 2013, 2014), limitless proliferative potential (Urrego et al. 2014;

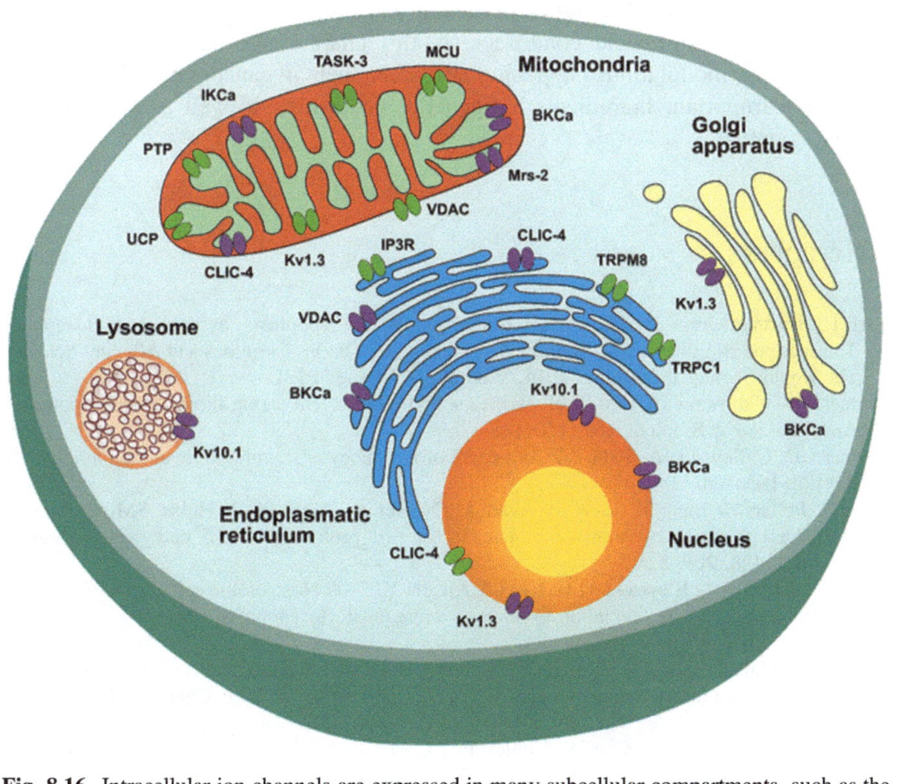


**Fig. 8.16** Intracellular ion channels are expressed in many subcellular compartments, such as the nucleus, endoplasmic reticulum, golgi apparatus, lysosome, mitochondria. Ion channels with roles in cancer development and/or progression are stained in *green*

Pardo and Stühmer 2014), apoptosis resistance (Wang 2004; Hoffmann and Lambert 2014), stimulation of neo-angiogenesis (Munaron 2015).

As summarized in Fig. 8.16, inside the cell, various ion channels are expressed in several organelles (Xu et al. 2015) including mitochondria, endoplasmic reticulum, nucleus, lysosome, Golgi apparatus, peroxisomes, endosomes. Most of the identified 'oncogenic' intracellular channels are located in the outer or inner mitochondrial membrane, since this organelle is a key point of control of important hallmarks of cancer, such as ATP-linked metabolism and apoptotic cell death (for details, see the analysis in Peruzzo et al. (2016). Another quite relevant intracellular location is the endoplasmic reticulum (ER). Ion channels and transporters of the ER modulate the cytosolic $Ca^{2+}$ concentration by controlling the uptake and release of calcium from this intracellular store and thus impact several cellular pathways implicated in the determination of cell fate as well in cell metabolism (Peruzzo et al. 2016). Modulated gene expressions in several cancer tissues linked to various cancer cell ion channels have also been evident (for details, see Peruzzo et al. 2016).

The ion channels require more in-depth investigations to understand their cascaded roles in regard to their structural aspects as well as structures in the vicinity, namely the ion channel hosting membrane modulation, modulation in the dielectric environment, physiological conditions. In this chapter, we have tried to briefly pinpoint at the modulations happening at ion channels in cancer cells. This chapter may serve important ingredients for general readerships as well as for drug discovery scientists.

## References


Carl D. Bortner, John A. Cidlowski. Ion channels and apoptosis in cancer. Philosophical Transactions B. Edited by Mustafa B. A. Djamgoz, R. Charles Coombes and Albrecht Schwab. The Royal Society Publishing. 2014. Volume: 369 Issue: 1638.

Bortner CD, Cidlowski JA. 2007 Cell shrinkage and monovalent cation fluxes: role in apoptosis. Arch. Biochem. Biophys. 462, 176–188.

Bortner CD, Cidlowski JA. 2011 Life and death of lymphocytes: a volume regulation affair. Cell. Physiol. Biochem. 28, 1079–1088.

Lang F, Foller M, Lang K, Lang P, Ritter M, Vereninov A, Szabo I, Huber SM, Gulbins E. 2007 Cell volume regulatory ion channels in cell proliferation and cell death. Methods Enzymol. 428, 209–225.

Maeno E, Ishizaki Y, Kanaseki T, Hazama A, Okada Y. 2000 Normotonic cell shrinkage because of disordered volume regulation is an early prerequisite to apoptosis. Proc. Natl Acad. Sci. USA 97, 9487–9492.

Cain K, Langlais C, Sun XM, Brown DG, Cohen GM. 2001 Physiological concentrations of K + inhibit cytochrome c-dependent formation of the apoptosome. J. Biol. Chem. 276, 41 985–41 990.

Thompson GJ, Langlais C, Cain K, Conley EC, Cohen GM. 2001 Elevated extracellular (K +) inhibits death-receptor and chemical-mediated apoptosis prior to caspase activation and cytochrome c release. Biochem. J. 357, 137–145.

Bortner CD, Hughes FM Jr., Cidlowski JA. 1997 A primary role for K + and Na + efflux in the activation of apoptosis. J. Biol. Chem. 272, 32 436–32 442.
Hughes FM Jr., Bortner CD, Purdy GD, Cidlowski JA. 1997 Intracellular K + suppresses the activation of apoptosis in lymphocytes. J. Biol. Chem. 272, 30 567–30 576.
Lang F, Foller M, Lang KS, Lang PA, Ritter M, Gulbins E, Vereninov A, Huber SM. 2005 Ion channels in cell proliferation and apoptotic cell death. J. Membr. Biol. 205, 147–157.
Gantner F, Uhlig S, Wendel A. 1995 Quinine inhibits release of tumor necrosis factor, apoptosis, necrosis and mortality in a murine model of septic liver failure. Eur. J. Pharmacol. 294, 353–355.
Wang X, Xiao AY, Ichinose T, Yu SP. 2000 Effects of tetraethylammonium analogs on apoptosis and membrane currents in cultured cortical neurons. J. Pharmacol. Exp. Ther. 295, 524–530.
Heimlich G, Bortner CD, Cidlowski JA. 2004 Apoptosis and cell volume regulation: the importance of ions and ion channels. Adv. Exp. Med. Biol. 559, 189–203.
Bortner CD, Cidlowski JA. 2003 Uncoupling cell shrinkage from apoptosis reveals that Na + influx is required for volume loss during programmed cell death. J. Biol. Chem. 278, 39 176–39 184.
Yurinskaya VA, et al. 2005 Potassium and sodium balance in U937 cells during apoptosis with and without cell shrinkage. Cell. Physiol. Biochem. 16, 155–162.
Franco R, Bortner CD, Cidlowski JA. 2006 Potential roles of electrogenic ion transport and plasma membrane depolarization in apoptosis. J. Membr. Biol. 209, 43–58.
Bortner CD, Sifre MI, Cidlowski JA. 2008 Cationic gradient reversal and cytoskeleton-independent volume regulatory pathways define an early stage of apoptosis. J. Biol. Chem. 283, 7219–7229.
Okada Y, Maeno E, Shimizu T, Manabe K, Mori S, Nabekura T. 2004 Dual roles of plasmalemmal chloride channels in induction of cell death. Pflugers Arch. 448, 287–295.
Okada Y, Shimizu T, Maeno E, Tanabe S, Wang X, Takahashi N. 2006 Volume-sensitive chloride channels involved in apoptotic volume decrease and cell death. J. Membr. Biol. 209, 21–29.
Heimlich G, Cidlowski JA. 2006 Selective role of intracellular chloride in the regulation of the intrinsic but not extrinsic pathway of apoptosis in Jurkat T-cells. J. Biol. Chem. 281, 2232–2241.
Wang X, Takahashi N, Uramoto H, Okada Y. 2005 Chloride channel inhibition prevents ROS-dependent apoptosis induced by ischemia-reperfusion in mouse cardiomyocytes. Cell. Physiol. Biochem. 16, 147–154.
Karl Kunzelmann. Ion Channels and Cancer. The Journal of Membrane Biology, 2005, Volume 205, Issue 3, pp 159–173.
Sten Orrenius, Boris Zhivotovsky, Pierluigi Nicotera. Regulation of cell death: the calcium–apoptosis link. Nature Reviews Molecular Cell Biology 4, 552–565 (July 2003).
Amith SR1, Fliegel L. Regulation of the Na +/H + Exchanger (NHE1) in Breast Cancer Metastasis. Cancer Res. 2013 Feb 15;73(4):1259–64. https://doi.org/10.1158/0008-5472.can-12-4031. Epub 2013 Feb 7.
Cardone RA, Casavola V, Reshkin SJ. The role of disturbed pH dynamics and the Na +/H + exchanger in metastasis. Nat Rev 2005;5: 786–95.
Malo ME, Fliegel L. Physiological role and regulation of the Na +/H + exchanger. Can J Physiol Pharmacol 2006;84:1081–95.
Ashrafuzzaman M, Andersen OS, McElhaney RN. (2008) The antimicrobial peptide gramicidin S permeabilizes phospholipid bilayer membranes without forming discrete ion channels. Biochim. Biophys. Acta.; 1778: 2814–22.
BL Kagan, RL Baldwin, D Munoz, BJ Wisnieski. Formation of ion-permeable channels by tumor necrosis factor-alpha. Science 13 March 1992, Vol. 255 no. 5050 pp. 1427–1430.
M Ismail, S Bokaee, R Morgan, J Davies, K J Harrington, and H Pandha. Inhibition of the aquaporin 3 water channel increases the sensitivity of prostate cancer cells to cryotherapy. Br J Cancer. 2009 Jun 16; 100(12): 1889–1895. (Corrected later: Br J Cancer. 2009 Aug 4; 101(3): 549.

Minghua Li, Zhi-Gang Xiong. Ion channels as targets for cancer therapy. Int J Physiol Pathophysiol Pharmacol 2011;3(2):156–166.

Vidhya R. Rao, Mathew Perez-Neut, Simon Kaja, and Saverio Gentile. Voltage-Gated Ion Channels in Cancer Cell Proliferation. Cancers (Basel). 2015 Jun; 7(2): 849–875.

Enrico Pierluigi Spugninia, Pierre Sonveauxb, Christian Stockc, Mario Perez-Sayansd, Angelo De Militoe, Sofia Avnetf, Abel Garcìa Garcìad, Salvador Harguindeyg, Stefano Faisa. Proton channels and exchangers in cancer. Biochimica et Biophysica Acta (BBA)—Biomembranes 1848, Issue 10, Part B, October 2015, Pages 2715–2726 (Special issue: Membrane Channels and Transporters in Cancers).

Diochot S., Loret E., Bruhn T., Beress L., Lazdunski M. APETx1, a new toxin from the sea anemone Anthopleura elegantissima, blocks voltage-gated human ether-a-go-go-related gene potassium channels. Mol. Pharmacol. 2003;64:59–69.

A. Garcia-Garcia, M. Perez-Sayans Garcia, M.J. Rodriguez, J. Antunez-Lopez, F. Barros Angueira, M. Somoza-Martin, et al., Immunohistochemical localization of C1 subunit of V ATPase (ATPase C1) in oral squamous cell cancer and normal oral mucosa, Biotech. Histochem. 87 (2) (Feb 2012) 133–139.

Weber C, Mello DQ, Downie BR, Suckow A, Stuhmer W, Pardo LA. Silencing the activity and proliferative properties of the human EagI Potassium Channel by RNA Interference. J.Biol Chem. 2006;281:13030–13037.

Restano-Cassulini R., Korolkova Y.V., Diochot S., Gurrola G., Guasti L., Possani L.D., Lazdunski M., Grishin E.V., Arcangeli A., Wanke E. Species diversity and peptide toxins blocking selectivity of ether-a-go go-related gene subfamily K + channels in the central nervous system. Mol. Pharmacol. 2006;69:1673–1683.

Gomez-Varela D, Zwick-Wallasch E, Knotgen H, Sanchez A, Hettmann T, Ossipov D et al. Monoclonal antibody blockade of the human Eag1 potassium channel function exerts antitumor activity. Cancer Res 2007; 67: 7343–7349.

Zhao J., Wei X.L., Jia Y.S., Zheng J.Q. Silencing of herg gene by shRNA inhibits SH-SY5Y cell growth in vitro and in vivo. Eur. J. Pharmacol. 2008;579:50–57.

Fiore A., Carraresi L., Morabito A., Polvani S., Fortunato A., Lastraioli E., Femia A.P., De Lorenzo E., Caderni G., Arcangeli A. Characterization of hERG1 channel role in mouse colorectal carcinogenesis. Cancer Med. 2013;2:583–594.

Lansu K., Gentile S. Potassium channel activation inhibits proliferation of breast cancer cells by activating a senescence program. Cell. Death Dis. 2013;4:e652.

Perez-Neut M., Rao V.R., Gentile S. hERG1/Kv11.1 activation stimulates transcription of p 21waf/cip in breast cancer cells via a calcineurin-dependent mechanism. Oncotarget. 2015 in press.

Das A., Pushparaj C., Bahi N., Sorolla A., Herreros J., Pamplona R., Vilella R., Matias-Guiu X., Marti R.M., Canti C. Functional expression of voltage-gated calcium channels in human melanoma. Pigment. Cell. Melanoma Res. 2012;25:200–212.

Yu W., Wang P., Ma H., Zhang G., Yulin Z., Lu B., Wang H., Dong M. Suppression of T-type Ca2 + channels inhibited human laryngeal squamous cell carcinoma cell proliferation running title: roles of T-type Ca2 + channels in LSCC cell proliferation. Clin. Lab. 2014;60:621–628.

Cheng Y., Zhao J., Qiao W., Chen K. Recent advances in diagnosis and treatment of gliomas using chlorotoxin-based bioconjugates. Am. J. Nucl. Med. Mol. Imaging. 2014;4:385–405.

Sikes R.A., Walls A.M., Brennen W.N., Anderson J.D., Choudhury-Mukherjee I., Schenck H.A., Brown M.L. Therapeutic approaches targeting prostate cancer progression using novel voltage-gated ion channel blockers. Clin. Prostate Cancer. 2003;2:181–187.

Fraser S.P., Ozerlat-Gunduz I., Brackenbury W.J., Fitzgerald E.M., Campbell T.M., Coombes R. C., Djamgoz M.B. Philos. Regulation of voltage-gated sodium channel expression in cancer: Hormones, growth factors and auto-regulation. Trans. R Soc. Lond. B Biol. Sci. 2014;369.

Nelson M., Yang M., Dowle A.A., Thomas J.R., Brackenbury W.J. The sodium channel-blocking antiepileptic drug phenytoin inhibits breast tumour growth and metastasis. Mol. Cancer. 2015;14:13.

Zhang J.J., Jacob T.J., Valverde M.A., Hardy S.P., Mintenig G.M., Sepulveda F.V., Gill D.R., Hyde S.C., Trezise A.E., Higgins C.F. Tamoxifen blocks chloride channels. A possible mechanism for cataract formation. J. Clin. Invest. 1994;94:1690–1697.

Deshane J., Garner C.C., Sontheimer H. Chlorotoxin inhibits glioma cell invasion via matrix metalloproteinase-2. J. Biol. Chem. 2003;278:4135–4144.

Li X, Link JM, Stekhova S, Yagle KJ, Smith C, Krohn KA and Tait JF. (2008).. Site specific labeling of annexin V with F-18 for apoptosis imaging. Bioconjug Chem. 19:1684–1688.

Sontheimer H. An unexpected role for ion channels in brain tumor metastasis. Exp Biol Med (Maywood) 2008;233:779–791.

Zhu L., Yang H., Zuo W., Yang L., Zhang H., Ye W., Mao J., Chen L., Wang L. Differential expression and roles of volume-activated chloride channels in control of growth of normal and cancerous nasopharyngeal epithelial cells. Biochem. Pharmacol. 2012;83:324–334.

Liu J., Zhang D., Li Y., Chen W., Ruan Z., Deng L., Wang L., Tian H., Yiu A., Fan C. Discovery of bufadienolides as a novel class of ClC-3 chloride channel activators with antitumor activities. J. Med. Chem. 2013;56:5734–5743.

Dardevet L., Rani D., Aziz T.A., Bazin I., Sabatier J.M., Fadl M., Brambilla E., de Waard M. Chlorotoxin: A helpful natural scorpion peptide to diagnose glioma and fight tumor invasion. Toxins (Basel) 2015;7:1079–1101.

Ashrafuzzaman, Md, Tuszynski, J. Membrane Biophysics. Springer-Verlag, 2012. ISBN: 978-3-642-16104-9 (Print), 978-3-642-16105-6. (Online).

Fiske JL, Fomin VP, Brown ML, Duncan RL, Sikes RA. Voltage-sensitive ion channels and cancer. Cancer Metastasis Rev. 2006 Sep; 25(3):493–500.

Lehmann-Horn F, Jurkat-Rott K. Voltage-gated ion channels and hereditary disease. Physiol Rev. 1999 Oct; 79(4):1317–72.

Garcia-Ferreiro R.E., Kerschensteiner D., Major F., Monje F., Stuhmer W., Pardo L.A. Mechanism of block of hEag1 K + channels by imipramine and astemizole. J. Gen. Physiol. 2004;124:301–317.

Pardo LA. Voltage-gated potassium channels in cell proliferation. Physiology (Bethesda) 2004;19:285–292. (PubMed).

Stuhmer W, Alves F, Hartung F, Zientkowska M, Pardo LA. Potassium channels as tumour markers. FEBS Lett. 2006;580:2850–2852. (PubMed).

Hemmerlein B, Weseloh RM, Mello DQ, Knotgen H, Sanchez A, Rubio ME, Martin S, Schliephacke T, Jenke M, Heinz JR, Stuhmer W, Pardo LA. Overexpression of Eag1 potassium channels in clinical tumours. Mol Cancer. 2006;5:41. (PMC free article), (PubMed).

Mello DQ, Suarez-Kurtz G, Stuhmer W, Pardo LA. Ether a go-go potassium channel expression in soft tissue sarcoma patients. Mol Cancer. 2006;5:42.

Ouadid-Ahidouch H. and Ahidouch, AK + channel expression in human breast cancer cells: involvement in cell cycle regulation and carcinogenesis. J Membr Biol. 2008;221:1–6.

Suzuki T, Takimoto K. Selective expression of HERG and Kv2 channels influences proliferation of uterine cancer cells. Int J Oncol. 2004;25:153–159.

Jang SH, Kang KS, Ryu PD, Lee SY. Kv1.3 voltage-gated K(+) channel subunit as a potential diagnostic marker and therapeutic target for breast cancer. BMB Rep. 2009;42:535–539.

Khaitan D, Sankpal UT, Weksler B, Meister EA, Romero IA, Couraud PO, Ningaraj NS. Role of KCNMA1 gene in breast cancer invasion and metastasis to brain. BMC Cancer. 2009;9:258.

Jang SH, Choi C, Hong SG, Yarishkin OV, Bae YM, Kim JG, O'Grady SM, Yoon KA, Kang KS, Ryu PD, Lee SY. Silencing of Kv4.1 potassium channels inhibits cell proliferation of tumorigenic human mammary epithelial cells. Biochem Biophys Res Commun. 2009a;384:180–186.

Black KL, Yin D, Konda BM, Wang X, Hu J, Ko MK, Bayan JA, Sacapano MR, Espinoza AJ, Ong JM, Irvin D, Shu Y. Different effects of KCa and KATP agonists on brain tumor permeability between syngeneic and allogeneic rat models. Brain Res. 2008;1227:198–206. (PMC free article).

Park SH, Ramachandran S, Kwon SH, Cha SD, Seo EW, Bae I, Cho C, Song DK. Upregulation of ATP-sensitive potassium channels for estrogen-mediated cell proliferation in human uterine leiomyoma cells. Gynecol Endocrinol. 2008;24:250–256.

Brevet M, Ahidouch A, Sevestre H, Merviel P, El Hiani Y, Robbe M, Ouadid-Ahidouch H. Expression of K + channels in normal and cancerous human breast. Histol Histopathol. 2008;23:965–972.

Li H, Liu L, Guo L, Zhang J, Du W, Li X, Liu W, Chen X, Huang S. HERG K + channel expression in CD34 +/CD38-/CD123(high) cells and primary leukemia cells and analysis of its regulation in leukemia cells. Int J Hematol. 2008;87:387–392.

Shao XD, Wu KC, Guo XZ, Xie MJ, Zhang J, Fan DM. Expression and significance of HERG protein in gastric cancer. Cancer Biol Ther. 2008;7:45–50.

Wonderlin WF. and Strobl JS. Potassium channels, proliferation and G1 progression. J Membr Biol. 1996;154:91–107.

Ransom CB, Liu X, Sontheimer H. BK channels in human glioma cells have enhanced calcium sensitivity. Glia. 2002;38:81–291.

Han X, Xi L, Wang H, Huang X, Ma X, Han Z, Wu P, Ma X, Lu Y, Wang G, Zhou J, Ma D. The potassium ion channel opener NS1619 inhibits proliferation and induces apoptosis in A2780 ovarian cancer cells. Biochem Biophys Res Commun. 2008;375:205–209.

Sontheimer H. An unexpected role for ion channels in brain tumor metastasis. Exp. Biol. Med. (Maywood) 2008;233:779–791. https://doi.org/10.3181/0711-mr-308. (PMC free article).

Cambien B, Rezzonico R, Vitale S, Rouzaire-Dubois B, Dubois JM, Barthel R, Karimdjee BS, Mograbi B, Schmid-Alliana A, Schmid-Antomarchi H. Silencing of hSlo potassium channels in human osteosarcoma cells promotes tumorigenesis. Int J Cancer. 2008;123:365–371.

Gessner G, Schonherr K, Soom M, Hansel A, Asim M, Baniahmad A, Derst C, Hoshi T, Heinemann SH. BKCa channels activating at resting potential without calcium in LNCaP prostate cancer cells. J Membr Biol. 2005;208:229–240.

Chou CC, Lunn CA, Murgolo NJ. (KCa3.1: target and marker for cancer, autoimmune disorder and vascular inflammation? Expert Rev Mol Diagn. 2008;8:179–187.

De Marchi U, Sassi N, Fioretti B, Catacuzzeno L, Cereghetti GM, Szabo I, Zoratti M. Intermediate conductance Ca2 + -activated potassium channel (KCa3.1) in the inner mitochondrial membrane of human colon cancer cells. Cell Calcium. 2009;45:509–516.

Voloshyna I, Besana A, Castillo M, Matos T, Weinstein IB, Mansukhani M, Robinson RB, Cordon-Cardo C, Feinmark SJ. TREK-1 is a novel molecular target in prostate cancer. Cancer Res. 2008;68:1197–1203.

Meuth SG, Herrmann AM, Ip CW, Kanyshkova T, Bittner S, Weishaupt A, Budde T, Wiendl H. The two-pore domain potassium channel TASK3 functionally impacts glioma cell death. J Neurooncol. 2008;87:263–270.

Tan G, Sun SQ, Yuan DL. Expression of Kir 4.1 in human astrocytic tumors: correlation with pathologic grade. Biochem Biophys Res Commun. 2008;367:743–747.

Warth A, Mittelbronn M, Wolburg H. Redistribution of the water channel protein aquaporin-4 and the K + channel protein Kir4.1 differs in low- and high-grade human brain tumors. Acta Neuropathol. 2005;109:418–426.

S.I. Ramsey, M. Delling, D.E. Clapham. An introduction to TRP channels. Annu. Rev. Physiol., 68 (2006), pp. 18.1–18.29.

B. Nilius, G. Owsianik, T. Voets, J.A. Peters. Transient receptor cation channels in disease. Physiol. Rev., 87 (2007), pp. 165–217.

Natalia Prevarskaya, Lei Zhang, Greg Barritt. TRP channels in cancer. Biochimica et Biophysica Acta (BBA) - Molecular Basis of Disease. V1772, 2007, 937–946.

Alessandra Fiorio Pla and Dimitra Gkika. Emerging role of TRP channels in cell migration: from tumor vascularization to metastasis. Front. Physiol. 2013. V4, Art. 311.

L.M. Duncan, J. Deeds, J. Hunter, J. Shao, L.M. Holmgren, E.A. Woolf, R.I. Tepper, A.W. Shyjan Down regulation of the novel gene melastatin correlates with potential for melanoma metastasis. Cancer Res., 58 (1998), pp. 1515–1520.

L. Tsavaler, M.H. Shapero, S. Morkowski, R. Laus. Trp-p 8, a novel prostate-specific gene, is up-regulated in prostate cancer and other malignancies and shares high homology with transient receptor potential calcium channel proteins. Cancer Res., 61 (2001), pp. 3760–3769.
S. Fuessel, D. Sickert, A. Meye, U. Klenk, U. Schmidt, M. Schmitz, et al. Multiple tumour marker analyses (PSA, hK2, PSCA, trp-p 8) in primary prostate cancers using quantitative RT-PCR. Int. J. Oncol., 23 (2003), pp. 221–228.
T. Fixemer, U. Wissenbach, V. Flockerzi, H. Bonkhoff. Expression of the Ca2 + -selective cation channel TRPV6 in human prostate cancer: a novel prognostic marker for tumour progression. Oncogene, 22 (2003), pp. 7858–7861.
M. Bodding. TRP proteins and cancer. Cell. Signal., 19 (2007), pp. 617–624.
L. Menendez, L. Juarez, E. Garcia, O. Garcia-Suarez, A. Hidalgo, A. Baamonde. Analgesic effects of capsazepine and resiniferatoxin on bone cancer pain in mice. Neurosci. Lett., 393 (2006), pp. 70–73.
J.R. Ghilardi, H. Rohrich, T.H. Lindsay, M.A. Sevcik, M.J. Schwei, K. Kubota, K.G. Halvorson, J. Poblete, S.R. Chaplan, A.E. Dubin, N.I. Carruthers, D. Swanson, M. Kuskowski, C.M. Flores, D. Julius, P.W. Mantyh. Selective blockade of the capsaicin receptor TRPV1 attenuates bone cancer pain. J. Neurosci., 25 (2005), pp. 3126–3131.
Clapham DE. TRP channels as cellular sensors Nature. 2003;426:517–524.
Montell C, Birnbaumer L, Flockerzi V, Bindels RJ, Bruford EA, Caterina MJ, Clapham DE, Harteneck C, Heller S, Julius D, Kojima I, Mori Y, Penner R, Prawitt D, Scharenberg AM, Schultz G, Shimizu N, Zhu MX. A unified nomenclature for the superfamily of TRP cation channels. Mol Cell. 2002;9:229–231.
Clapham DE, Montell C, Schultz G, Julius D. International Union of Pharmacology. XLIII. Compendium of voltage-gated ion channels: transient receptor potential channels. Pharmacol Rev. 2003;55:591–596.
Li M, Yu Y, Yang J. Structural biology of TRP channels. Adv Exp Med Biol. 2011;704:1–23.
Means AR. Calcium, calmodulin and cell cycle regulation. FEBS Lett. 1994;347:1–4.
D. Fang, V. Setaluri. Expression and up-regulation of alternatively spliced transcripts of melastatin, a melanoma metastasis-related gene, in human melanoma cells. Biochem. Biophys. Res. Commun., 279 (2000), pp. 53–61.
J. Deeds, F. Cronin, L.M. Duncan. Patterns of melastatin mRNA expression in melanocytic tumors. Hum. Pathol., 31 (2000), pp. 1346–1356.
L.M. Duncan, J. Deeds, F.E. Cronin, M. Donovan, A.J. Sober, M. Kauffman, J.J. McCarthy. Melastatin expression and prognosis in cutaneous malignant melanoma. J. Clin. Oncol., 19 (2001), pp. 568–576.
Hyun Soo Park, Chansik Hong, Byung Joo Kim, and Insuk So. The Pathophysiologic Roles of TRPM7 Channel. Korean J Physiol Pharmacol. 2014 Feb; 18(1): 15–23.
Zeng B, Yuan C, Yang X, Atkin SL, Xu SZ. TRPC channels and their splice variants are essential for promoting human ovarian cancer cell proliferation and tumorigenesis. Curr Cancer Drug Targets. 2013 Jan;13(1):103–16.
Yu, P., Gu, S., Bu, J., and Du, J. (2010). TRPC1 is essential for in vivo angiogenesis in zebrafish. Circ. Res. 106, 1221–1232. https://doi.org/10.1161/circresaha.109.207670.
Bomben, V. C., Turner, K. L., Barclay, T.-T. C., and Sontheimer, H. (2011). Transient receptor potential canonical channels are essential for chemotactic migration of human malignant gliomas. J. Cell. Physiol. 226, 1879–1888. https://doi.org/10.1002/jcp.22518.
Antigny, F., Girardin, N., and Frieden, M. (2012). Transient receptor potential canonical channels are required for in vitro endothelial tube formation. J. Biol. Chem. 287, 5917–5927.
Cuddapah, V. A., Turner, K. L., and Sontheimer, H. (2013). Calcium entry via TRPC1 channels activates chloride currents in human glioma cells. Cell Calcium 53, 187–194.
Zhang, H., Zhou, L., Shi, W., Song, N., Yu, K., and Gu, Y. (2012). A mechanism underlying the effects of polyunsaturated fatty acids on breast cancer. Int. J. Mol. Med. 30, 487–494.
Hamdollah Zadeh, M. A., Glass, C. A., Magnussen, A., Hancox, J. C., and Bates, D. O. (2008). VEGF-mediated elevated intracellular calcium and angiogenesis in human microvascular

endothelial cells in vitro are inhibited by dominant negative TRPC6. Microcirculation 15, 605–614.

Ge, R., Tai, Y., Sun, Y., Zhou, K., Yang, S., Cheng, T., et al. (2009). Critical role of TRPC6 channels in VEGF-mediated angiogenesis. Cancer Lett. 283, 43–51.

Chigurupati, S., Venkataraman, R., Barrera, D., Naganathan, A., Madan, M., Paul, L.,et al. (2010). Receptor channel TRPC6 is a key mediator of notch-driven glioblastoma growth and invasiveness. Cancer Res. 70, 418–427.

Kini, V., Chavez, A., and Mehta, D. (2010). A new role for PTEN in regulating transient receptor potential canonical channel 6-mediated Ca2 + entry, endothelial permeability, and angiogenesis. J. Biol. Chem. 285, 33082–33091.

Bernaldo de Quirós, S., Merlo, A., Secades, P., Zambrano, I., de Santa María, I. S., Ugidos, N., et al. (2013). Identification of TRPC6 as a possible candidate target gene within an amplicon at 11q21-q22.2 for migratory capacity in head and neck squamous cell carcinomas. BMC Cancer 13:116.

Chaudhuri, P., Colles, S. M., Bhat, M., Van Wagoner, D. R., Birnbaumer, L., and Graham, L. M. (2008). Elucidation of a TRPC6-TRPC5 channel cascade that restricts endothelial cell movement. Mol. Biol. Cell 19, 3203–3211.

Waning, J., Vriens, J., Owsianik, G., Stüwe, L., Mally, S., Fabian, A., et al. (2007). A novel function of capsaicin-sensitive TRPV1 channels: involvement in cell migration. Cell Calcium 42, 17–25.

Ramer, R., and Hinz, B. (2008). Inhibition of cancer cell invasion by cannabinoids via increased expression of tissue inhibitor of matrix metalloproteinases-1. J. Natl. Cancer Inst. 100, 59–69.

Ramer, R., Merkord, J., Rohde, H., and Hinz, B. (2010). Cannabidiol inhibits cancer cell invasion via upregulation of tissue inhibitor of matrix metalloproteinases-1. Biochem. Pharmacol. 79, 955–966.

Caprodossi, S., Amantini, C., Nabissi, M., Morelli, M. B., Farfariello, V., Santoni, M., et al. (2011). Capsaicin promotes a more aggressive gene expression phenotype and invasiveness in null-TRPV1 urothelial cancer cells. Carcinogenesis 32, 686–694.

Monet, M., Lehen'kyi, V., Gackiere, F., Firlej, V., Vandenberghe, M., Roudbaraki, M., et al. (2010). Role of cationic channel TRPV2 in promoting prostate cancer migration and progression to androgen resistance. Cancer Res. 70, 1225–1235.

Oulidi, A., Bokhobza, A., Gkika, D., Vanden Abeele, F., Lehen'kyi, V., Ouafik, L., et al. (2013). TRPV2 Mediates adrenomedullin stimulation of prostate and urothelial cancer cell adhesion, migration and invasion. PLoS ONE 8:e64885.

Thodeti, C. K., Matthews, B., Ravi, A., Mammoto, A., Ghosh, K., Bracha, A. L., et al. (2009). TRPV4 channels mediate cyclic strain-induced endothelial cell reorientation through integrin-to-integrin signaling. Circ. Res. 104, 1123–1130.

Matthews, B. D., Thodeti, C. K., Tytell, J. D., Mammoto, A., Overby, D. R., and Ingber, D. E. (2010). Ultra-rapid activation of TRPV4 ion channels by mechanical forces applied to cell surface beta1 integrins. Integr. Biol. (Camb). 2, 435–442.

Fiorio Pla, A., Ong, H. L., Cheng, K. T., Brossa, A., Bussolati, B., Lockwich, T., et al. (2012b). TRPV4 mediates tumor-derived endothelial cell migration via arachidonic acid-activated actin remodeling. Oncogene 31, 200–212.

Gao, H., Chen, X., Du, X., Guan, B., Liu, Y., and Zhang, H. (2011). EGF enhances the migration of cancer cells by up-regulation of TRPM7. Cell Calcium 50, 559–568.

Middelbeek, J., Kuipers, A. J., Henneman, L., Visser, D., Eidhof, I., van Horssen, R., et al. (2012). TRPM7 is required for breast tumor cell metastasis. Cancer Res. 72, 4250–4261.

Meng, X., Cai, C., Wu, J., Cai, S., Ye, C., Chen, H., et al. (2013). TRPM7 mediates breast cancer cell migration and invasion through the MAPK pathway. Cancer Lett. 333, 96–102.

Wondergem, R., Ecay, T. W., Mahieu, F., Owsianik, G., and Nilius, B. (2008). HGF/SF and menthol increase human glioblastoma cell calcium and migration. Biochem. Biophys. Res. Commun. 372, 210–215.

Yang, Z.-H., Wang, X.-H., Wang, H.-P., and Hu, L.-Q. (2009b). Effects of TRPM8 on the proliferation and motility of prostate cancer PC-3 cells. Asian J. Androl. 11, 157–165.

Gkika, D., Flourakis, M., Lemonnier, L., and Prevarskaya, N. (2010). PSA reduces prostate cancer cell motility by stimulating TRPM8 activity and plasma membrane expression. Oncogene 29, 4611–4616.

Zhu, G., Wang, X., Yang, Z., Cao, H., Meng, Z., Wang, Y., et al. (2011). Effects of TRPM8 on the proliferation and angiogenesis of prostate cancer PC-3 cells in vivo. Oncol. Lett. 2, 1213–1217.

Okamoto, Y., Ohkubo, T., Ikebe, T., and Yamazaki, J. (2012). Blockade of TRPM8 activity reduces the invasion potential of oral squamous carcinoma cell lines. Int. J. Oncol. 40, 1431–1440.

Valero, M. L., Mello de Queiroz, F., Stühmer, W., Viana, F., and Pardo, L. A. (2012). TRPM8 ion channels differentially modulate proliferation and cell cycle distribution of normal and cancer prostate cells. PLoS ONE 7:e51825.

Abdullaev, I. F., Bisaillon, J. M., Potier, M., Gonzalez, J. C., Motiani, R. K., and Trebak, M. (2008). Stim1 and Orai1 mediate CRAC currents and store-operated calcium entry important for endothelial cell proliferation. Circ. Res. 103, 1289–1299.

Yang, S., Zhang, J. J., and Huang, X.-Y. (2009a). Orai1 and STIM1 are critical for breast tumor cell migration and metastasis. Cancer Cell 15, 124–134.

Chen, Y.-F., Chiu, W.-T., Chen, Y.-T., Lin, P.-Y., Huang, H. J., Chou, C.-Y., et al. (2011). Calcium store sensor stromal-interaction molecule 1-dependent signaling plays an important role in cervical cancer growth, migration, and angiogenesis. Proc. Natl. Acad. Sci. U.S.A. 108, 15225–15230.

Dragoni, S., Laforenza, U., Bonetti, E., Lodola, F., Bottino, C., Berra-Romani, R., et al. (2011). Vascular endothelial growth factor stimulates endothelial colony forming cells proliferation and tubulogenesis by inducing oscillations in intracellular Ca2 + concentration. Stem Cells 29, 1898–1907.

Li, J., Cubbon, R. M., Wilson, L. A., Amer, M. S., McKeown, L., Hou, B., et al. (2011a). Orai1 and CRAC channel dependence of VEGF-activated Ca2 + entry and endothelial tube formation. Circ. Res. 108, 1190–1198.

Beech, D. J. (2012). Orai1 calcium channels in the vasculature. Pflugers Arch. 463, 635–647.

Stephan J. Reshkin, Rosa A. Cardone1 and Salvador Harguindey. Na + -H + Exchanger, pH Regulation and Cancer. Recent Patents on Anti-Cancer Drug Discovery, 2013, 8, 85–99 85.

S. Feng, M. Cai, P. Liu, L. Wei, J. Wang, J. Qi, et al., Atp6v1c1 may regulate filament actin arrangement in breast cancer cells, PLoS One 9 (1) (Jan 15 2014) e84833.

S. Feng, G. Zhu, M. McConnell, L.Deng, Q. Zhao,M.Wu, et al., Silencing of atp6v1c1 prevents breast cancer growth and bone metastasis, Int. J. Biol. Sci. 9 (8) (Sep 5 2013) 853–862.

K. von Schwarzenberg, T. Lajtos, L. Simon, R. Muller, G. Vereb, A.M. Vollmar, V-ATPase inhibition overcomes trastuzumab resistance in breast cancer, Mol. Oncol. 8 (1) (Feb 2014) 9–19.

H. You, J. Jin, H. Shu, B. Yu, A. De Milito, F. Lozupone, et al., Small interfering RNA targeting the subunit ATP6L of proton pump V-ATPase overcomes chemoresistance of breast cancer cells, Cancer Lett. 280 (1) (Jul 18 2009) 110–119.

M. Perez-Sayans, J.M. Somoza-Martin, F. Barros-Angueira, P.G. Diz, J.M. Rey, A. Garcia-Garcia, Multidrug resistance in oral squamous cell carcinoma: the role of vacuolar ATPases, Cancer Lett. 295 (2) (Sep 28 2010) 135–143.

M. Perez-Sayans, M.D. Reboiras-Lopez, J.M. Somoza-Martin, F. Barros-Angueira, P.G. Diz, J.M. Rey, et al., Measurement of ATP6V1C1 expression in brush cytology samples as a diagnostic and prognostic marker in oral squamous cell carcinoma, Cancer Biol. Ther. 9 (12) (Jun 15 2010a) 1057–1064.

L. Huang, Q. Lu, Y. Han, Z. Li, Z. Zhang, X. Li, ABCG2/V-ATPase was associated with the drug resistance and tumor metastasis of esophageal squamous cancer cells, Diagn. Pathol. 7 (Dec 17 2012) 180 (ORJ-toolbar" src = "./pubmed15_files/toolbar).

J. Xu, R. Xie, X. Liu, G.Wen, H. Jin, Z. Yu, et al., Expression and functional role of vacuolar H (+)-ATPase in human hepatocellular carcinoma, Carcinogenesis 33 (12) (Dec 2012) 2432–2440.

C. Chung, C.C. Mader, J.C. Schmitz, J. Atladottir, P. Fitchev, M.L. Cornwell, et al., The vacuolar-ATPase modulates matrix metalloproteinase isoforms in human pancreatic cancer, Lab. Invest. 91 (5) (May 2011) 732–743.

Q. Lu, S. Lu, L. Huang, T.Wang, Y.Wan, C.X. Zhou, et al., The expression of V-ATPase is associated with drug resistance and pathology of non-small-cell lung cancer, Diagn. Pathol. 8 (Aug 28 2013) 145 (ORJ-toolbar" src = "./pubmed18_files/toolbar).

F. Perut, S. Avnet, C. Fotia, S.R. Baglio, M. Salerno, S. Hosogi, et al., V-ATPase as an effective therapeutic target for sarcomas, Exp. Cell Res. 320 (1) (Jan 1 2014) 21–32.

S. Avnet, G. Di Pompo, S. Lemma, M. Salerno, F. Perut, G. Bonuccelli, et al., V-ATPase is a candidate therapeutic target for Ewing sarcoma, Biochim. Biophys. Acta 1832 (8) (Aug 2013) 1105–1116.

M. Perez-Sayans, A. Garcia-Garcia, M.D. Reboiras-Lopez, P. Gandara-Vila, Role of V-ATPases in solid tumors: importance of the subunit C (review), Int. J. Oncol. 34 (6) (Jun 2009) 1513–1520.

R.A. Cardone, A. Bellizzi, G. Busco, E.J.Weinman, M.E. Dell'Aquila, V. Casavola, et al., The NHERF1 PDZ2 domain regulates PKA-RhoA-p 38-mediated NHE1 activation and invasion in breast tumor cells, Mol. Biol. Cell 18 (5) (May 2007) 1768–1780.

R.A. Cardone, V. Casavola, S.J. Reshkin, The role of disturbed pH dynamics and the Na +/H + exchanger in metastasis, Nat. Rev. Cancer 5 (10) (Oct 2005) 786–795.

G. Lauritzen, C.M. Stock, J. Lemaire, S.F. Lund, M.F. Jensen, B. Damsgaard, et al., The Na +/H + exchanger NHE1, but not the Na + , HCO3(−) cotransporter NBCn1, regulates motility of MCF7 breast cancer cells expressing constitutively active ErbB2, Cancer Lett. 317 (2) (Apr 28 2012) 172–183.

A.K. Vahle, B. Domikowsky, C. Schwoppe, H. Krahling, S. Mally, M. Schafers, et al., Extracellular matrix composition and interstitial pH modulate NHE1-mediated melanoma cell motility, Int. J. Oncol. 44 (1) (Jan 2014) 78–90.

L. Stuwe, M. Muller, A. Fabian, J.Waning, S. Mally, J. Noel, et al., pH dependence of melanoma cell migration: protons extruded by NHE1 dominate protons of the bulk solution, J. Physiol. 585 (Pt 2) (Dec 1 2007) 351–360.

F.T. Ludwig, A. Schwab, C. Stock, The Na +/H + -exchanger (NHE1) generates pH nanodomains at focal adhesions, J. Cell. Physiol. 228 (6) (Jun 2013) 1351–1358.

J.J. Provost, D. Rastedt, J. Canine, T. Ngyuen, A. Haak, C. Kutz, et al., Urokinase plasminogen activator receptor induced non-small cell lung cancer invasion and metastasis requires NHE1 transporter expression and transport activity, Cell. Oncol. (Dordr.) (Jan 31 2012) (Epub ahead of print).

W. Jin,Q. Li, J.Wang,G. Chang, Y. Lin, H. Li, et al., Na +/H + exchanger 1 inhibition contributes to K562 leukaemic cell differentiation, Cell Biol. Int. 36 (8) (Aug 1 2012) 739–745.

M.L. Rentsch, C.G. Ossum, E.K. Hoffmann, S.F. Pedersen, Roles of Na +/H + exchange in regulation of p 38 mitogen-activated protein kinase activity and cell death after chemical anoxia in NIH3T3 fibroblasts, Pflugers Arch. 454 (4) (Jul 2007) 649–662.

M. Busk, S. Walenta, W. Mueller-Klieser, T. Steiniche, S. Jakobsen, M.R. Horsman, et al., Inhibition of tumor lactate oxidation: consequences for the tumor microenvironment, Radiother. Oncol. 99 (3) (Jun 2011) 404–411.

K.M. Kennedy, P.M. Scarbrough, A. Ribeiro, R. Richardson, H. Yuan, P. Sonveaux, et al., Catabolism of exogenous lactate reveals it as a legitimatemetabolic substrate in breast cancer, PLoS One 8 (9) (Sep 12 2013) e75154.

J. Ho, M.B. deMoura, Y. Lin, G. Vincent, S. Thorne, L.M. Duncan, et al., Importance of glycolysis and oxidative phosphorylation in advanced melanoma, Mol. Cancer 11 (Oct 9 2012) 76 (4598-11-76).

F. Guillaumond, J. Leca, O. Olivares, M.N. Lavaut, N. Vidal, P. Berthezene, et al., Strengthened glycolysis under hypoxia supports tumor symbiosis and hexosamine biosynthesis in pancreatic adenocarcinoma, Proc. Natl. Acad. Sci. U. S. A. 110 (10) (Mar 5 2013) 3919–3924.

A. Giatromanolaki, M.I. Koukourakis, E. Sivridis, J. Pastorek, C.C.Wykoff, K.C. Gatter, et al., Expression of hypoxia-inducible carbonic anhydrase-9 relates to angiogenic pathways and

independently to poor outcome in non-small cell lung cancer, Cancer Res. 61 (21) (Nov 1 2001) 7992–7998.
C.A. Brewer, S.Y. Liao, S.P.Wilczynski, S. Pastorekova, J. Pastorek, J. Zavada, et al., A study of biomarkers in cervical carcinoma and clinical correlation of the novel biomarker MN, Gynecol. Oncol. 63 (3) (Dec 1996) 337–344.
J.R. Turner, R.D. Odze, C.P. Crum, M.B. Resnick, MN antigen expression in normal, preneoplastic, and neoplastic esophagus: a clinicopathological study of a new cancer-associated biomarker, Hum. Pathol. 28 (6) (Jun 1997) 740–744.
T. Klatte, A.S. Belldegrun, A.J. Pantuck, The role of carbonic anhydrase IX as amolecular marker for transitional cell carcinoma of the bladder, BJU Int. 101 (Suppl. 4) (Jun 2008) 45–48.
S.K. Chia, C.C. Wykoff, P.H. Watson, C. Han, R.D. Leek, J. Pastorek, et al., Prognostic significance of a novel hypoxia-regulated marker, carbonic anhydrase IX, in invasive breast carcinoma, J. Clin. Oncol. 19 (16) (Aug 15 2001) 3660–3668.
J. Saarnio, S. Parkkila, A.K. Parkkila, K. Haukipuro, S. Pastorekova, J. Pastorek, et al., Immunohistochemical study of colorectal tumors for expression of a novel transmembrane carbonic anhydrase, MN/CA IX, with potential value as a marker of cell proliferation, Am. J. Pathol. 153 (1) (Jul 1998) 279–285.
S.W. Choi, J.Y. Kim, J.Y. Park, I.H. Cha, J. Kim, S. Lee, Expression of carbonic anhydrase IX is associated with postoperative recurrence and poor prognosis in surgically treated oral squamous cell carcinoma, Hum. Pathol. 39 (9) (Sep 2008) 1317–1322.
J.L. Roh, K.J. Cho, G.Y. Kwon, C.H. Ryu, H.W. Chang, S.H. Choi, et al., The prognostic value of hypoxiamarkers in T2-staged oral tongue cancer, Oral Oncol. 45 (1) (Jan 2009) 63–68.
L.R. Oliveira, A. Ribeiro-Silva, Prognostic significance of immunohistochemical biomarkers in oral squamous cell carcinoma, Int. J. Oral Maxillofac. Surg. 40 (3) (Mar 2011) 298–307.
A.W. Eckert, M.H. Lautner, A. Schutze, H. Taubert, J. Schubert, U. Bilkenroth, Coexpression of hypoxia-inducible factor-1alpha and glucose transporter-1 is associated with poor prognosis in oral squamous cell carcinoma patients, Histopathology 58 (7) (Jun 2011) 1136–1147.
Y. Kondo, K. Yoshikawa, Y. Omura, A. Shinohara, Y. Kazaoka, J. Sano, et al., Clinicopathological significance of carbonic anhydrase 9, glucose transporter-1, ki-67 and p 53 expression in oral squamous cell carcinoma, Oncol. Rep. 25 (5) (May 2011) 1227–1233.
E.J. Bowman, A. Siebers, K. Altendorf, Bafilomycins: a class of inhibitors of membrane ATPases from microorganisms, animal cells, and plant cells, Proc. Natl. Acad. Sci. U. S. A. 85 (21) (Nov 1988) 7972–7976.
M. Huss, O. Vitavska, A. Albertmelcher, S. Bockelmann, C.Nardmann, K. Tabke, et al., Vacuolar H(+)-ATPases: intra- and intermolecular interactions, Eur. J. Cell Biol. 90 (9) (Sep 2011) 688–695.
M. Muroi, N. Shiragami, K. Nagao, M. Yamasaki, A. Takatsuki, Folimycin (concanamycin A), a specific inhibitor of V-ATPase, blocks intracellular translocation of the glycoprotein of vesicular stomatitis virus before arrival to the Golgi apparatus, Cell Struct. Funct. 18 (3) (Jun 1993) 139–149.
R.M. Wiedmann, K. von Schwarzenberg, A. Palamidessi, L. Schreiner, R. Kubisch, J. Liebl, et al., The V-ATPase-inhibitor archazolid abrogates tumor metastasis via inhibition of endocytic activation of the rho-GTPase Rac1, Cancer Res. 72 (22) (Nov 15 2012) 5976–5987.
J.P. Mattsson, K. Vaananen, B. Wallmark, P. Lorentzon, Omeprazole and bafilomycin, two proton pump inhibitors: differentiation of their effects on gastric, kidney and bone H(+)-translocating ATPases, Biochim. Biophys. Acta 1065 (2) (Jun 18 1991) 261–268.
Y. Wang, S.J. Li, J. Pan, Y. Che, J. Yin, Q. Zhao. Specific expression of the human voltage-gated proton channel Hv1 in highly metastatic breast cancer cells, promotes tumor progression and metastasis. Biochemical and Biophysical Research Communications, Volume 412, Issue 2, 26 August 2011, Pages 353–359.
R.C. Thomas, R.W. Meech. Hydrogen ion currents and intracellular pH in depolarized voltage-clamped snail neurons. Nature, 299 (1982), pp. 826–828.
I.S. Ramsey, M.M. Moran, J.A. Chong, D.E. Clapham. A voltage-gated proton-selective channel lacking the pore domain. Nature, 440 (2006), pp. 1213–1216.

M. Sasaki, M. Takagi, Y. Okamur. A voltage sensor-domain protein is a voltage-gated proton channel. Science, 312 (2006), pp. 589–592.

R.A. Clark, K.G. Leidal, D.W. Pearson, W.M. Nauseef. NADPH oxidase of human neutrophils. Subcellular localization and characterization of an arachidonate-activatable superoxide-generating system. J. Biol. Chem., 262 (1987), pp. 4065–4074.

D. Morgan, V.V. Cherny, R. Murphy, B.Z. Katz, T.E. DeCoursey. The pH dependence of NADPH oxidase in human eosinophils. J. Physiol., 569 (2005), pp. 419–431.

T.E. DeCoursey. Voltage-gated proton channels and other proton transfer pathways. Physiol. Rev., 83 (2003), pp. 475–579.

Yifan Wang, Shu Jie Li, Xingye Wu, Yongzhe Che, and Qiang Li. Clinicopathological and Biological Significance of Human Voltage-gated Proton Channel Hv1 Protein Overexpression in Breast Cancer. JOURNAL OF BIOLOGICAL CHEMISTRY. APRIL 20, 2012•VOLUME 287•NUMBER 17. PP 13877–13888.

Deri Morgan, Patrick McIntire, Vladimir Cherny, Susan Smith, Boris Musset, Thomas DeCoursey. Proton Channels are Present in Cell Membranes of the Breast Cancer Cell Line MDA MB 231 and Affect Recovery from an Acid Load. Biophys. J. 108, Issue 2, Supplement 1, p 587a, 27 January 2015.

Verkman AS, Mitra AK (2000) Structure and function of aquaporin water channels. Am J Physiol Renal Physiol 278: F13–F28.

Papadopoulos MC, Saadoun S. Key roles of aquaporins in tumor biology. Biochim Biophys Acta. Volume 1848, Issue 10, Part B, October 2015, Pages 2576–2583. (Special issue: Membrane Channels and Transporters in Cancers).

Jian Wang, Li Feng, Zhitu Zhu, Minghuan Zheng, Diane Wang, Zhihong Chen and Hongzhi Sun. Aquaporins as diagnostic and therapeutic targets in cancer: How far we are? Journal of Translational Medicine (2015) 13:96.

Saadoun S, Papadopoulos MC, Davies DC, Bell BA, Krishna S. Increased aquaporin 1 water channel expression in human brain tumours. Br J Cancer. 2002; 87:621–3.

Guan G, Dong Z, Sun K. Correlation between the expression of aquaporin 1 and the micro-angiogenesis in laryngeal carcinoma. Lin Chung Er Bi Yan Hou Tou Jing Wai Ke Za Zhi. 2009; 23:219-21.

Nico B, Ribatti D. Role of aquaporins in cell migration and edema formation in human brain tumors. Exp Cell Res. 2011; 317:2391-6.

Saadoun S, Papadopoulos MC, Davies DC, Krishna S, Bell BA. Aquaporin-4 expression is increased in oedematous human brain tumours. J Neurol Neurosurg Psychiatry. 2002a; 72:262-5.

Warth A, Simon P, Capper D, Goeppert B, Tabatabai G, Herzog H et al.. Expression pattern of the water channel aquaporin-4 in human gliomas is associated with blood-brain barrier disturbance but not with patient survival. J Neurosci Res. 2007; 85:1336-46.

Hoque MO, Soria JC, Woo J, Lee T, Lee J, Jang SJ et al.. Aquaporin 1 is overexpressed in lung cancer and stimulates NIH-3 T3 cell proliferation and anchorage-independent growth. Am J Pathol. 2006; 168:1345-53.

Moon C, Soria JC, Jang SJ, Lee J, Obaidul HM, Sibony M et al.. Involvement of aquaporins in colorectal carcinogenesis. Oncogene. 2003; 22:6699-703.

Yin T, Yu S, Xiao L, Zhang J, Liu C, Lu Y et al.. Correlation between the expression of aquaporin 1 and hypoxia-inducible factor 1 in breast cancer tissues. J Huazhong Univ Sci Technolog Med Sci. 2008; 28:346-8. PubMed Abstract| Publisher Full Text OpenURL.

Mobasheri A, Airley R, Hewitt SM, Marples D. Heterogeneous expression of the aquaporin 1 (AQP1) water channel in tumors of the prostate, breast, ovary, colon and lung: a study using high density multiple human tumor tissue microarrays. Int J Oncol. 2005; 26:1149-58. PubMed Abstract| Publisher Full Text OpenURL.

Yang JH, Shi YF, Chen XD, Qi WJ. The influence of aquaporin-1 and microvessel density on ovarian carcinogenesis and ascites formation. Int J Gynecol Cancer. 2006; 16 Suppl 1:400-5. PubMed Abstract| Publisher Full Text OpenURL.

Hara-Chikuma M, Verkman AS. Prevention of skin tumorigenesis and impairment of epidermal cell proliferation by targeted aquaporin-3 gene disruption. Mol Cell Biol. 2008; 28:326-32. PubMed Abstract| Publisher Full Text OpenURL.

Liu S, Zhang S, Jiang H, Yang Y, Jiang Y. Co-expression of AQP3 and AQP5 in esophageal squamous cell carcinoma correlates with aggressive tumor progression and poor prognosis. Med Oncol. 2013; 30:636. PubMed Abstract| Publisher Full Text OpenURL.

Guo X, Sun T, Yang M, Li Z, Li Z, Gao Y. Prognostic value of combined aquaporin 3 and aquaporin 5 overexpression in hepatocellular carcinoma. Biomed Res Int., 2013:206525.

Kafe H, Verbavatz JM, Cochand-Priollet B, Castagnet P, Vieillefond A. Collecting duct carcinoma: an entity to be redefined? Virchows Arch. 2004; 445:637-40. PubMed Abstract| Publisher Full Text OpenURL.

Kusayama M, Wada K, Nagata M, Ishimoto S, Takahashi H, Yoneda M et al.. Critical role of aquaporin 3 on growth of human esophageal and oral squamous cell carcinoma. Cancer Sci. 2011; 102:1128-36.

Xie Y, Wen X, Jiang Z, Fu HQ, Han H, Dai L. Aquaporin 1 and aquaporin 4 are involved in invasion of lung cancer cells. Clin Lab. 2012; 58:75-80. PubMed Abstract| Publisher Full Text OpenURL.

Niu D, Kondo T, Nakazawa T, Kawasaki T, Yamane T, Mochizuki K et al.. Differential expression of aquaporins and its diagnostic utility in thyroid cancer. PLoS One. 2012; 7:e40770.

Kang SK, Chae YK, Woo J, Kim MS, Park JC, Lee J et al.. Role of human aquaporin 5 in colorectal carcinogenesis. Am J Pathol. 2008; 173:518-25. PubMed Abstract| Publisher Full Text OpenURL.

Chae YK, Kang SK, Kim MS, Woo J, Lee J, Chang S et al.. Human AQP5 plays a role in the progression of chronic myelogenous leukemia (CML). PLoS One. 2008; 3:e2594. PubMed Abstract| Publisher Full Text OpenURL.

Zhang Z, Chen Z, Song Y, Zhang P, Hu J, Bai C. Expression of aquaporin 5 increases proliferation and metastasis potential of lung cancer. J Pathol. 2010; 221:210-20. PubMed Abstract| Publisher Full Text OpenURL.

Yan C, Yang J, Shen L, Chen X. Inhibitory effect of Epigallocatechin gallate on ovarian cancer cell proliferation associated with aquaporin 5 expression. Arch Gynecol Obstet. 2012; 285:459-67. PubMed Abstract| Publisher Full Text OpenURL.

Watanabe T, Fujii T, Oya T, Horikawa N, Tabuchi Y, Takahashi Y et al.. Involvement of aquaporin-5 in differentiation of human gastric cancer cells. J Physiol Sci. 2009; 59:113-22. PubMed Abstract| Publisher Full Text OpenURL.

Lacroix L, Lazar V, Michiels S, Ripoche H, Dessen P, Talbot M et al.. Follicular thyroid tumors with the PAX8-PPARgamma1 rearrangement display characteristic genetic alterations. Am J Pathol. 2005; 167:223-31. PubMed Abstract| Publisher Full Text OpenURL.

Tan G, Sun SQ, Yuan DL. Expression of the water channel protein aquaporin-9 in human astrocytic tumours: correlation with pathological grade. J Int Med Res. 2008; 36:777-82. PubMed Abstract| Publisher Full Text OpenURL.

Yang JH, Yan CX, Chen XJ, Zhu YS. Expression of aquaglyceroporins in epithelial ovarian tumours and their clinical significance. J Int Med Res. 2011; 39:702-11. PubMed Abstract| Publisher Full Text OpenURL.

Jablonski EM, Mattocks MA, Sokolov E, Koniaris LG, Hughes FJ, Fausto N et al.. Decreased aquaporin expression leads to increased resistance to apoptosis in hepatocellular carcinoma. Cancer Lett. 2007; 250:36-46.

Wang W, Li Q, Yang T, Bai G, Li D, Li Q et al.. Expression of AQP5 and AQP8 in human colorectal carcinoma and their clinical significance. World J Surg Oncol. 2012; 10:242.

Shi YH, Rehemu N, Ma H, Tuokan T, Chen R, Suzuke L. Increased migration and local invasion potential of SiHa cervical cancer cells expressing Aquaporin 8. Asian Pac J Cancer Prev. 2013;14(3):1825-8.

Shi Z, Zhang T, Luo L, Zhao H, Cheng J, Xiang J, Zhao C. Aquaporins in human breast cancer: identification and involvement in carcinogenesis of breast cancer. J Surg Oncol. 2012 Sep 1;106(3):267-72.

Sébastien Bonnet, Stephen L. Archer, Joan Allalunis-Turner, Alois Haromy, Christian Beaulieu, Richard Thompson, Christopher T. Lee, Gary D. Lopaschuk, Lakshmi Puttagunta, Sandra Bonnet, Gwyneth Harry, Kyoko Hashimoto, Christopher J. Porter, Miguel A. Andrade, Bernard Thebaud, Evangelos D. Michelakis. A Mitochondria-K + Channel Axis Is Suppressed in Cancer and Its Normalization Promotes Apoptosis and Inhibits Cancer Growth. Cancer Cell, Volume 11, Issue 1, January 2007, Pages 37–51.

D.R. Plas, C.B. Thompson. Cell metabolism in the regulation of programmed cell death. Trends Endocrinol. Metab., 13 (2002), pp. 75–78.

J.W. Kim, C.V. Dang. Multifaceted roles of glycolytic enzymes. Trends Biochem. Sci., 30 (2005), pp. 142–150.

R.L. Elstrom, D.E. Bauer, M. Buzzai, R. Karnauskas, M.H. Harris, D.R. Plas, H. Zhuang, R.M. Cinalli, A. Alavi, C.M. Rudin, C.B. Thompson. Akt stimulates aerobic glycolysis in cancer cells. Cancer Res., 64 (2004), pp. 3892–3899.

J.G. Pastorino, J.B. Hoek, N. Shulga. Activation of glycogen synthase kinase 3β disrupts the binding of hexokinase II to mitochondria by phosphorylating voltage-dependent anion channel and potentiates chemotherapy-induced cytotoxicity. Cancer Res., 65 (2005), pp. 10545–10554.

Maldonado EN, Lemasters JJ.. 2014 ATP/ADP ratio, the missed connection between mitochondria and the Warburg effect. Mitochondrion (2014) 19(Pt A):78–84.

Dejean LM, Martinez-Caballero S, Kinnally KW.. Is MAC the knife that cuts cytochrome c from mitochondria during apoptosis? Cell Death Differ (2006) 13:1387–95.

Simon DN, Rout MP. Cancer and the nuclear pore complex. Adv Exp Med Biol (2014) 773:285–307.

Stephen M. Madamba, Kevin N. Damri, Laurent M. Dejean, and Pablo M. Peixoto. Mitochondrial Ion Channels in Cancer Transformation. Front Oncol. 2015; 5: 120.

Pavlov EV, Priault M, Pietkiewicz D, Cheng EH, Antonsson B, Manon S, Korsmeyer SJ, Mannella CA and Kinnally KW (2001). A novel, high conductance channel of mitochondria linked to apoptosis in mammalian cells and Bax expression in yeast. J. Cell. Biol. 155: 725–731.

Guo L, Pietkiewicz D, Pavlov EV, Grigoriev SM, Kasianowicz JJ, Dejean LM, Korsmeyer SJ, Antonsson B and Kinnally KW (2004) Effects of cytochrome c on the mitochondrial apoptosis-induced channel MAC. Am.J. Physiol. Cell Physiol. 286: C1109–C1117.

Martinez-Caballero S, Dejean LM and Kinnally KW (2004) Some amphiphilic cations block the mitochondrial apoptosis-induced channel, MAC. FEBS Lett. 568: 35–38.

Dejean LM, Martinez-Caballero S, Kinnally KW. Is MAC the knife that cuts cytochrome c from mitochondria during apoptosis? Cell Death Differ. 2006 Aug;13(8):1387-95. Epub 2006 May 5.

Goldin N, Arzoine L, Heyfets A, Israelson A, Zaslavsky Z, Bravman T, et al. Methyl jasmonate binds to and detaches mitochondria-bound hexokinase. Oncogene (2008) 27:4636–43.10.1038/onc.2008.108.

Rostovtseva TK, Bezrukov SM.. VDAC inhibition by tubulin and its physiological implications. Biochim Biophys Acta (2012) 1818:1526–35.10.1016/j.bbamem.2011.11.004.

Maldonado EN, Sheldon KL, Dehart DN, Patnaik J, Manevich Y, Townsend DM, et al. Voltage-dependent anion channels modulate mitochondrial metabolism in cancer cells: regulation by free tubulin and erastin. J Biol Chem (2013) 288:11920–9.10.1074/jbc.M112.433847.

Gupta SC, Hevia D, Patchva S, Park B, Koh W, Aggarwal BB.. Upsides and downsides of reactive oxygen species for cancer: the roles of reactive oxygen species in tumorigenesis, prevention, and therapy. Antioxid Redox Signal (2012) 16:1295–322.10.1089/ars.2011.4414.

Yang Y, Karakhanova S, Werner J, Bazhin AV. Reactive oxygen species in cancer biology and anticancer therapy. Curr Med Chem (2013) 20:3677–92.10.2174/0929867311320999165.

Vento MT, Zazzu V, Loffreda A, Cross JR, Downward J, Stoppelli MP, et al. Praf2 is a novel Bcl-xL/Bcl-2 interacting protein with the ability to modulate survival of cancer cells. PLoS One (2010) 5:e15636.10.1371/journal.pone.0015636.

Quinn BA, Dash R, Azab B, Sarkar S, Das SK, Kumar S, et al. Targeting Mcl-1 for the therapy of cancer. Expert Opin Investig Drugs (2011) 20:1397–411.10.1517/13543784.2011.609167.

Arbel N, Ben-Hail D, Shoshan-Barmatz V.. Mediation of the antiapoptotic activity of Bcl-xL protein upon interaction with VDAC1 protein. J Biol Chem (2012) 287:23152–61.10.1074/jbc.M112.345918.

Delsite R, Kachhap S, Anbazhagan R, Gabrielson E, Singh KK.. Nuclear genes involved in mitochondria-to-nucleus communication in breast cancer cells. Mol Cancer (2002) 1:6.10.1186/1476-4598-1-6.

Rasmussen AK, Chatterjee A, Rasmussen LJ, Singh KK.. Mitochondria-mediated nuclear mutator phenotype in Saccharomyces cerevisiae. Nucleic Acids Res (2003) 31:3909–17.10.1093/nar/gkg446.

Miceli MV, Jazwinski SM.. Nuclear gene expression changes due to mitochondrial dysfunction in ARPE-19 cells: implications for age-related macular degeneration. Invest Ophthalmol Vis Sci (2005) 46:1765–73.10.1167/iovs.04-1327.

Singh KK, Kulawiec M, Still I, Desouki MM, Geradts J, Matsui S.. Inter-genomic cross talk between mitochondria and the nucleus plays an important role in tumorigenesis. Gene (2005) 354:140–6.10.1016/j.gene.2005.03.027.

Kulawiec M, Safina A, Desouki MM, Still I, Matsui S, Bakin A, et al. Tumorigenic transformation of human breast epithelial cells induced by mitochondrial DNA depletion. Cancer Biol Ther (2008) 7:1732–43.10.4161/cbt.7.11.6729.

Porporato PE, Payen VL, Perez-Escuredo J, De Saedeleer CJ, Danhier P, Copetti T, et al. A mitochondrial switch promotes tumor metastasis. Cell Rep (2014) 8:754–66.10.1016/j.celrep.2014.06.043.

Baughman JM, Perocchi F, Girgis HS, Plovanich M, Belcher-Timme CA, Sancak Y, et al. Integrative genomics identifies MCU as an essential component of the mitochondrial calcium uniporter. Nature (2011) 476:341–5.10.1038/nature10234.

De Stefani D, Raffaello A, Teardo E, Szabo I, Rizzuto R.. A forty-kilodalton protein of the inner membrane is the mitochondrial calcium uniporter. Nature (2011) 476:336–40.10.1038/nature10230.

Peixoto PM, Dejean LM, Kinnally KW. The therapeutic potential of mitochondrial channels in cancer, ischemia-reperfusion injury, and neurodegeneration. Mitochondrion (2012) 12:14–23.10.1016/j.mito.2011.03.003.

Leanza L, Biasutto L, Manago A, Gulbins E, Zoratti M, Szabo I. Intracellular ion channels and cancer. Front Physiol (2013) 4:227.10.3389/fphys.2013.00227.

Leanza L, Zoratti M, Gulbins E, Szabo I.. Mitochondrial ion channels as oncological targets. Oncogene (2014) 33:5569–81.10.1038/onc.2013.578.

Plötz M1, Gillissen B, Hossini AM, Daniel PT, Eberle J. Disruption of the VDAC2-Bak interaction by Bcl-x(S) mediates efficient induction of apoptosis in melanoma cells. Cell Death Differ. 2012 Dec;19(12):1928-38.

Asmarinah Asmarinah Agnieszka Paradowska-Dogan Ria Kodariah Budiana Tanuhardja Przemyslaw Waliszewski Chaidir Arif Mochtar Yuli Budiningsih Wolfgang Weidner Elvira Hinsch. Expression of the Bcl-2 family genes and complexes involved in the mitochondrial transport in prostate cancer cells. Int. J. Oncology, 2014, 45: 1489-1496.

Reina S, Palermo V, Guarnera A, Guarino F, Messina A, Mazzoni C, De Pinto V (Jul 2010). "Swapping of the N-terminus of VDAC1 with VDAC3 restores full activity of the channel and confers anti-aging features to the cell". FEBS Letters 584 (13): 2837–44.

Eduardo N. Maldonado, Kely L. Sheldon‖, David N. DeHart, Jyoti Patnaik, Yefim Manevich, Danyelle M. Townsend, Sergey M. Bezrukov, Tatiana K. Rostovtseva‖ and John J. Lemasters. Voltage-dependent Anion Channels Modulate Mitochondrial Metabolism in Cancer Cells. REGULATION BY FREE TUBULIN AND ERASTIN*. J. Biol. Chem., 2013, 288: 11920-11929.

Chandra D, Choy G, Daniel PT, Tang DG. Bax-dependent regulation of Bak by voltage-dependent anion channel 2. J Biol Chem. 2005;280:19051–19061.

Taha, T.A., Mullen, T.D. and Obeid, L.M. (2006). "A house divided: ceramide, sphingosine, and sphingosine-1-phosphate in programmed cell death". Biochim. Biophys. Acta 1758 (12): 2027–36.

Rapaport D. (2005). How does the TOM complex mediate insertion of precursor proteins into the mitochondrial outer membrane? J. Cell Biol. 171, 419–423.

Sotgia F, Whitaker-Menezes D, Martinez-Outschoorn UE, Salem AF, Tsirigos A, Lamb R, et al. Mitochondria "fuel" breast cancer metabolism: fifteen markers of mitochondrial biogenesis label epithelial cancer cells, but are excluded from adjacent stromal cells. Cell Cycle (2012) 11:4390–401. https://doi.org/10.4161/cc.22777.

Handa N, Kishishita S, Morita S, Akasaka R, Jin Z, Chrzas J, et al. Structure of the human Tim44 C-terminal domain in complex with pentaethylene glycol: ligand-bound form. Acta Crystallogr D Biol Crystallogr (2007) 63:1225–34. https://doi.org/10.1107/s0907444907051463.

Bonora M, Pinton P. The mitochondrial permeability transition pore and cancer: molecular mechanisms involved in cell death. Front Oncol (2014) 4:302. https://doi.org/10.3389/fonc.2014.00302.

Rasola A, Bernardi P. The mitochondrial permeability transition pore and its adaptive responses in tumor cells. Cell Calcium (2015). https://doi.org/10.1016/j.ceca.2015.03.004.

Kimberly G. Norman, Jeffrey A. Canterc, Mingjian Shia, Ginger L. Milned, Jason D. Morrowd, James E. Sligh. Cyclosporine A suppresses keratinocyte cell death through MPTP inhibition in a model for skin cancer in organ transplant recipients. Mitochondrion, Volume 10, Issue 2, March 2010, Pages 94–101.

N Tajeddine, L Galluzzi, O Kepp, E Hangen, E Morselli, L Senovilla, N Araujo, G Pinna, N Larochette, N Zamzami, N Modjtahedi, A Harel-Bellan and G Kroemer. Hierarchical involvement of Bak, VDAC1 and Bax in cisplatin-induced cell death. Oncogene (2008) 27, 4221–4232; https://doi.org/10.1038/onc.2008.63.

de La Motte Rouge T, Galluzzi L, Olaussen KA, Zermati Y, Tasdemir E, Robert T et al. (2007). A novel epidermal growth factor receptor inhibitor promotes apoptosis in non-small cell lung cancer cells resistant to erlotinib. Cancer Res 67: 6253–6262.

Claudio Hetz, Pierre-Alain Vitte, Agnes Bombrun, Tatiana K. Rostovtsevall, Sylvie Montessuit, Agnes Hiver, Matthias K. Schwarz, Dennis J. Church, Stanley J. Korsmeyer, Jean-Claude Martinou, and Bruno Antonsson. Bax Channel Inhibitors Prevent Mitochondrion-mediated Apoptosis and Protect Neurons in a Model of Global Brain Ischemia. The JOURNAL OF BIOLOGICAL CHEMISTRY, 2005, VOL. 280, NO. 52, pp. 42960–42970.

Vidyaramanan Ganesan, Timothy Walsh, Kai-Ti Chang, and Marco Colombini. The Dynamics of Bax Channel Formation: Influence of Ionic Strength. Biophysical J., Volume 103, August 2012, 483–491.

Siskind LJ, Mullen TD, Romero Rosales K, Clarke CJ, Hernandez-Corbacho MJ, Edinger AL, et al. The BCL-2 protein BAK is required for long-chain ceramide generation during apoptosis. J Biol Chem (2010) 285:11818–26. https://doi.org/10.1074/jbc.m109.078121.

Chipuk JE, McStay GP, Bharti A, Kuwana T, Clarke CJ, Siskind LJ, et al. Sphingolipid metabolism cooperates with BAK and BAX to promote the mitochondrial pathway of apoptosis. Cell (2012) 148:988–1000. https://doi.org/10.1016/j.cell.2012.01.038.

Egle A, Harris AW, Bath ML, O'Reilly L, Cory S. VavP-Bcl2 transgenic mice develop follicular lymphoma preceded by germinal center hyperplasia. Blood (2004) 103:2276–83. https://doi.org/10.1182/blood-2003-07-2469.

Swanson PJ, Kuslak SL, Fang W, Tze L, Gaffney P, Selby S, et al. Fatal acute lymphoblastic leukemia in mice transgenic for B cell-restricted bcl-xL and c-myc. J Immunol (2004) 172:6684–91. https://doi.org/10.4049/jimmunol.172.11.6684.

Meijerink JP, Van Lieshout EM, Beverloo HB, Van Drunen E, Mensink EJ, Macville M, et al. Novel murine B-cell lymphoma/leukemia model to study BCL2-driven oncogenesis. Int J Cancer (2005) 114:917–25.

Pavlov EV, Priault M, Pietkiewicz D, Cheng EH, Antonsson B, Manon S, et al. A novel, high conductance channel of mitochondria linked to apoptosis in mammalian cells and Bax expression in yeast. J Cell Biol (2001) 155:725–31. https://doi.org/10.1083/jcb.200107057.

Goping IS, Gross A, Lavoie JN, Nguyen M, Jemmerson R, Roth K, et al. Regulated targeting of BAX to mitochondria. J Cell Biol (1998) 143:207–15. https://doi.org/10.1083/jcb.143.1.207.

Gross A, Jockel J, Wei MC, Korsmeyer SJ. Enforced dimerization of BAX results in its translocation, mitochondrial dysfunction and apoptosis. EMBO J (1998) 17:3878–85. https://doi.org/10.1093/emboj/17.14.3878.

Certo M, Del Gaizo Moore V, Nishino M, Wei G, Korsmeyer S, Armstrong SA, et al. Mitochondria primed by death signals determine cellular addiction to antiapoptotic BCL-2 family members. Cancer Cell (2006) 9:351–65. https://doi.org/10.1016/j.ccr.2006.03.027.

Letai AG. Diagnosing and exploiting cancer's addiction to blocks in apoptosis. Nat Rev Cancer (2008) 8:121–32. https://doi.org/10.1038/nrc2297.

Teijido O, Dejean L. Upregulation of Bcl2 inhibits apoptosis-driven BAX insertion but favors BAX relocalization in mitochondria. FEBS Lett (2010) 584:3305–10. https://doi.org/10.1016/j.febslet.2010.07.002.

Gautier F, Guillemin Y, Cartron PF, Gallenne T, Cauquil N, Le Diguarher T, et al. Bax activation by engagement with, then release from, the BH3 binding site of Bcl-xL. Mol Cell Biol (2011) 31:832–44. https://doi.org/10.1128/mcb.00161-10.

Renault TT, Teijido O, Antonsson B, Dejean LM, Manon S. Regulation of Bax mitochondrial localization by Bcl-2 and Bcl-x(L): keep your friends close but your enemies closer. Int J Biochem Cell Biol (2013) 45:64–7. https://doi.org/10.1016/j.biocel.2012.09.022.

Tarek A. Taha, Thomas D. Mullen, and Lina M. Obeid. A house divided: ceramide, sphingosine, and sphingosine-1-phosphate in programmed cell death. Biochim Biophys Acta. 2006 Dec; 1758(12): 2027–2036.

Radin NS. Killing tumours by ceramide-induced apoptosis: a critique of available drugs. Biochem J. 2003;371:243–56.

Hannun YA. Sphingolipid second messengers: tumor suppressor lipids. Adv Exp Med Biol. 1997;400A:305–12.

Samy A. F. Morad & Myles C. Cabot. Ceramide-orchestrated signalling in cancer cells. Nature Reviews Cancer 13, 51-65 (January 2013)| https://doi.org/10.1038/nrc3398.

Ashrafuzzaman, Md. Designing aptamer based drug molecules to bind to various apoptotic proteins and construction of an Aptamer Drug Bank. Biochemistry Department, King Saud University, Saudi Arabia, 2016 (ongoing project).

A. Rasola, P. Bernardi. Mitochondrial permeability transition in Ca(2 +)-dependent apoptosis and necrosis. Cell Calcium, 50 (2011), pp. 222–233.

C. Camello-Almaraz, P.J. Gomez-Pinilla, M.J. Pozo, P.J. Camello. Mitochondrial reactive oxygen species and Ca2 + signaling. Am. J. Physiol. Cell Physiol., 291 (2006), pp. C1082–C1088.

A.J. Kowaltowski, N.C. de Souza-Pinto, R.F. Castilho, A.E. Vercesi. Mitochondria and reactive oxygen species. Free Radic. Biol. Med., 47 (2009), pp. 333–343.

L. Leanza, L. Biasutto, A. Manago, E. Gulbins, M. Zoratti, I. Szabo. Intracellular ion channels and cancer. Front. Physiol., 4 (2013), p. 227.

D.B. Zorov, M. Juhaszova, S.J. Sollott. Mitochondrial ROS-induced ROS release: an update and review. Biochim. Biophys. Acta, 1757 (2006), pp. 509–517.

D.B. Zorov, M. Juhaszova, S.J. Sollott. Mitochondrial reactive oxygen species (ROS) and ROS-induced ROS release. Physiol. Rev., 94 (2014), pp. 909–950.

I. Szabo, J. Bock, H. Grassme, M. Soddemann, B. Wilker, F. Lang, M. Zoratti, E. Gulbins. Mitochondrial potassium channel Kv1.3 mediates Bax-induced apoptosis in lymphocytes. Proc. Natl. Acad. Sci. U. S. A., 105 (2008), pp. 14861–14866.

L. Leanza, B. Henry, N. Sassi, M. Zoratti, K.G. Chandy, E. Gulbins, I. Szabo. Inhibitors of mitochondrial Kv1.3 channels induce Bax/Bak-independent death of cancer cells. EMBO Mol. Med., 4 (2012), pp. 577–593.

L. Leanza, L. Trentin, K.A. Becker, F. Frezzato, M. Zoratti, G. Semenzato, E. Gulbins, I. Szabo. Clofazimine, Psora-4 and PAP-1, inhibitors of the potassium channel Kv1.3, as a new and selective therapeutic strategy in chronic lymphocytic leukemia. Leukemia, 27 (2013a), pp. 1782–1785.

Arcangeli, A.; Crociani, O.; Lastraioli, E.; Masi, A.; Pillozzi, S.; Becchetti, A. Targeting Ion Channels in Cancer: A Novel Frontier in Antineoplastic Therapy. Current Medicinal Chemistry, Volume 16, Number 1, January 2009, pp. 66-93(28).

T Bose, A Cieślar-Pobuda and E Wiechec. Role of ion channels in regulating Ca2 + homeostasis during the interplay between immune and cancer cells. Cell Death and Disease (2015) 6, e1648; https://doi.org/10.1038/cddis.2015.23.

Lang F, Stournaras C. Ion channels in cancer: future perspectives and clinical potential. Philos Trans R Soc Lond B Biol Sci 2014; 369: 20130108.

TSENG C.-Y., ASHRAFUZZAMAN MD., MANE J., KAPTY J., MERCER J., and TUSZYNSKI J. (2011). Entropic fragment based approach to aptamer design. Chem. Biol. Drug Design 78,1–13.

Md Ashrafuzzaman, Chih-Yuan Tseng, Janice Kapty, John R. Mercer, and Jack A. Tuszynski. A Computationally Designed DNA Aptamer Template with Specific Binding to Phosphatidylserine. Nucleic Acid Ther. 2013 Dec 1; 23(6): 418–426.

Ashrafuzzaman, Md.; Tuszynski, J. Regulation of Channel Function Due to Coupling with a Lipid Bilayer. Journal of Computational and Theoretical Nanoscience, Volume 9, Number 4, April 2012a, pp. 564-570.

Garcia-Calvo M, Leonard RJ, Novick J, Stevens SP, Schmalhofer W, Kaczorowski GJ et al. Purification, characterization, and biosynthesis of margatoxin, a component of Centruroides margaritatus venom that selectively inhibits voltage-dependent potassium channels. J Biol Chem 1993; 268: 18866–18874.| PubMed| CAS|.

Helms LM, Felix JP, Bugianesi RM, Garcia ML, Stevens S, Leonard RJ et al. Margatoxin binds to a homomultimer of K(V)1.3 channels in Jurkat cells. Comparison with K(V)1.3 expressed in CHO cells. Biochemistry 1997; 36: 3737–3744.| Article| PubMed| ISI|.

Koshy S, Wu D, Hu X, Tajhya RB, Huq R, Khan FS et al. Blocking KCa3.1 channels increases tumor cell killing by a subpopulation of human natural killer lymphocytes. PloS One 2013; 8: e76740.| Article| PubMed|.

Buraei Z, Schofield G, Elmslie KS. Roscovitine differentially affects CaV2 and Kv channels by binding to the open state. Neuropharmacology 2007; 52: 883–894.| Article| PubMed| ISI| CAS|.

Meijer L, Borgne A, Mulner O, Chong JP, Blow JJ, Inagaki N et al. Biochemical and cellular effects of roscovitine, a potent and selective inhibitor of the cyclin-dependent kinases cdc2, cdk2 and cdk5. Eur J Biochem 1997; 243: 527–536.| Article| PubMed| ISI| CAS|.

Lastraioli E, Guasti L, Crociani O, Polvani S, Hofmann G, Witchel H et al. herg1 gene and HERG1 protein are overexpressed in colorectal cancers and regulate cell invasion of tumor cells. Cancer Res 2004; 64: 606–611.| Article| PubMed| ISI| CAS|.

Pillozzi S, Brizzi MF, Balzi M, Crociani O, Cherubini A, Guasti L et al. HERG potassium channels are constitutively expressed in primary human acute myeloid leukemias and regulate cell proliferation of normal and leukemic hemopoietic progenitors. Leukemia 2002; 16: 1791–1798.| Article| PubMed| ISI| CAS|.

Pillozzi S, Brizzi MF, Bernabei PA, Bartolozzi B, Caporale R, Basile V et al. VEGFR-1 (FLT-1), beta1 integrin, and hERG K + channel for a macromolecular signaling complex in acute myeloid leukemia: role in cell migration and clinical outcome. Blood 2007; 110: 8–0.| Article|.

Shao XD, Wu KC, Hao ZM, Hong L, Zhang J, Fan DM. The potent inhibitory effects of cisapride, a specific blocker for human ether-a-go-go-related gene (HERG) channel, on gastric cancer cells. Cancer Biol Ther 2005; 4: 295–301.| Article| PubMed| ISI| CAS|.

Millward MJ, Cantwell BM, Munro NC, Robinson A, Corris PA, Harris AL. Oral verapamil with chemotherapy for advanced non-small cell lung cancer: a randomised study. Br J Cancer 1993; 67: 1031–1035.| Article| PubMed| ISI| CAS|.

Yohem KH, Clothier JL, Montague SL, Geary RJ, Winters AL 3rd, Hendrix MJ et al. Inhibition of tumor cell invasion by verapamil. Pigment Cell Res 1991; 4: 225–233.

Lefranc F, Kiss R. The sodium pump alpha1 subunit as a potential target to combat apoptosis-resistant glioblastomas. Neoplasia 2008; 10: 198–206.

Van Quaquebeke E, Simon G, Andre A, Dewelle J, El Yazidi M, Bruyneel F et al. Identification of a novel cardenolide (2"-oxovoruscharin) from Calotropis procera and the hemisynthesis of novel derivatives displaying potent in vitro antitumor activities and high in vivo tolerance: structure-activity relationship analyses. J Med Chem 2005; 48: 849–856.| Article| PubMed| ISI|.

Carrithers MD, Chatterjee G, Carrithers LM, Offoha R, Iheagwara U, Rahner C et al. Regulation of podosome formation in macrophages by a splice variant of the sodium channel SCN8A. J Biol Chem 2009; 284: 8114–8118.| Article| PubMed| ISI|.

Fraser SP, Diss JK, Chioni AM, Mycielska ME, Pan H, Yamaci RF et al. Voltage-gated sodium channel expression and potentiation of human breast cancer metastasis. Clin Cancer Res 2005; 11: 5381–5389.| Article| PubMed| ISI| CAS|.

Schwab A, Reinhardt J, Schneider SW, Gassner B, Schuricht B. K (+) channel-dependent migration of fibroblasts and human melanoma cells. Cell Physiol Biochem 1999; 9: 126–132.| Article| PubMed| ISI| CAS|.

Innamaa A, Jackson L, Asher V, Van Shalkwyk G, Warren A, Hay D et al. Expression and prognostic significance of the oncogenic K2P potassium channel KCNK9 (TASK-3) in ovarian carcinoma. Anticancer Res 2013; 33: 1–8.

Ashrafuzzaman, Md., Tseng, C.-Y. DNA aptamers for Bcl 2 and BAX proteins, 2016a, unpublished.

Ashrafuzzaman, Md. DNA aptamers for ceramides, 2016b, unpublished.

Frede J, Fraser SP, Oskay-Ozcelik G, Hong Y, Ioana Braicu E, Sehouli J et al. Ovarian cancer: Ion channel and aquaporin expression as novel targets of clinical potential. Eur J Cancer 2013; 49: 2331–2344.| Article| PubMed| ISI|.

Ji C, Cao C, Lu S, Kivlin R, Amaral A, Kouttab N et al. Curcumin attenuates EGF-induced AQP3 up-regulation and cell migration in human ovarian cancer cells. Cancer Chemother Pharmacol 2008; 62: 857–865.| Article| PubMed| ISI| CAS|.

Lee MH, Choi BY, Kundu JK, Shin YK, Na HK, Surh YJ. Resveratrol suppresses growth of human ovarian cancer cells in culture and in a murine xenograft model: eukaryotic elongation factor 1A2 as a potential target. Cancer Res 2009; 69: 7449–7458.| Article| PubMed| ISI|.

Isbilen B, Fraser SP, Djamgoz MB. Docosahexaenoic acid (omega-3) blocks voltage-gated sodium channel activity and migration of MDA-MB-231 human breast cancer cells. Int J Biochem Cell Biol 2006; 38: 2173–2182.| Article| PubMed| ISI|.

Yan C, Yang J, Shen L, Chen X. Inhibitory effect of Epigallocatechin gallate on ovarian cancer cell proliferation associated with aquaporin 5 expression. Arch Gynecol Obstet 2012; 285: 459–467.

Roberta Peruzzo, Lucia Biasutto, Ildikò Szabò, Luigi Leanza. (2016) Impact of intracellular ion channels on cancer development and progression. European Biophysics Journal 45 (7):685–707.

Hanahan D, Weinberg RA (2011). Hallmarks of cancer: the next generation. Cell 144:646–674.

Litan A, Langhans SA (2015) Cancer as a channelopathy: ion channels and pumps in tumor development and progression. Front Cell Neurosci 9:86.

Djamgoz MB, Onkal R (2013) Persistent current blockers of voltage-gated sodium channels: a clinical opportunity for controlling metastatic disease. Recent Pat Anticancer Drug Discov 8:66–84.

Andersen AP, Moreira JM, Pedersen SF (2014) Interactions of ion transporters and channels with cancer cell metabolism and the tumour microenvironment. Philos Trans R Soc Lond B Biol Sci 369:20130098.

Leanza L, Zoratti M, Gulbins E, Szabo I (2014) Mitochondrial ion channels as oncological targets. Oncogene 33:5569–5581.

Urrego D et al (2014) Potassium channels in cell cycle and cell proliferation. Philos Trans R Soc Lond B Biol Sci 369:20130094.

Pardo LA, Stühmer W (2014) The roles of K + channels in cancer. Nature Rev Cancer 14:39–48.

Wang Z (2004) Roles of K + channels in regulating tumour cell proliferation and apoptosis. Pflugers Arch 448:274–286.

Hoffmann EK, Lambert IH (2014) Ion channels and transporters in the development of drug resistance in cancer cells. Philos Trans R Soc Lond B Biol Sci 369:20130109.

Munaron L (2015) Systems biology of ion channels and transporters in tumor angiogenesis: An omics view. Biochim Biophys Acta 1848(10 Pt B):2647–56.

Xu H, Martinoia E, Szabo I (2015) Organellar channels and transporters. Cell Calcium 58:1–10.

Chen R., Zeng X., Zhang R., Huang J., Kuang X., Yang J., Liu J., Tawfik O., Thrasher J.B., Li B. Cav1.3 channel alpha1D protein is overexpressed and modulates androgen receptor transactivation in prostate cancers. Urol. Oncol. 2014; 32:524–536.

Rafael E. García-Ferreiro, Daniel Kerschensteiner, Felix Major, Francisco Monje, Walter Stühmer, Luis A. Pardo, (2004) Mechanism of Block of hEag1 K Channels by Imipramine and Astemizole . The Journal of General Physiology 124 (4):301–317.

András Dömötör, Zsanett Peidl, Áron Vincze, Béla Hunyady, János Szolcsányi, László Kereskay, György Szekeres, Gyula Mózsik, (2005) Immunohistochemical distribution of vanilloid receptor, calcitonin-gene related peptide and substance P in gastrointestinal mucosa of patients with different gastrointestinal disorders. InflammoPharmacology 13 (1-3):161–177.

M Hartel, (2006) Vanilloids in pancreatic cancer: potential for chemotherapy and pain management. Gut 55 (4):519–528.

Massimo Lazzeri, Maria Giuliana Vannucchi, Michele Spinelli, Elisa Bizzoco, Patrizia Beneforti, Damiano Turini, Maria-Simonetta Faussone-Pellegrini, (2005) Transient Receptor Potential Vanilloid Type 1 (TRPV1) Expression Changes from Normal Urothelium to Transitional Cell Carcinoma of Human Bladder. European Urology 48 (4):691–698.

María G. Sánchez, Ana M. Sánchez, Beatriz Collado, Sophie Malagarie-Cazenave, Nuria Olea, María J. Carmena, Juan C. Prieto, Inés Díaz-Laviada, (2005) Expression of the transient receptor potential vanilloid 1 (TRPV1) in LNCaP and PC-3 prostate cancer cells and in human prostate tissue. European Journal of Pharmacology 515 (1-3):20–27.

Jianbo Wang, Nozomu Tanji, Tadahiko Kikugawa, Masachika Shudou, Xishuang Song, Masayoshi Yokoyama, (2007) Expression of aquaporin 3 in the human prostate. International Journal of Urology 14 (12):1088–1092.

A. Jemal, R. Siegel, E. Ward, Y. Hao, J. Xu, T. Murray, M. J. Thun, (2008) Cancer Statistics, 2008. CA: A Cancer Journal for Clinicians 58 (2):71–96.

Mohamed Ismail, Shwan Ahmed, Christof Kastner, John Davies, (2007) Salvage cryotherapy for recurrent prostate cancer after radiation failure: a prospective case series of the first 100 patients. BJU International 100 (4):760–764.

Jeffrey K. Cohen, Ralph J. Miller, Sharmila Ahmed, Meredith J. Lotz, John Baust, (2008) Ten-Year Biochemical Disease Control for Patients with Prostate Cancer Treated with Cryosurgery as Primary Therapy. Urology 71 (3):515–518.

Andrew A. Gage, John Baust, (1998) Mechanisms of Tissue Injury in Cryosurgery. Cryobiology 37 (3):171–186.

G. M. Preston, T. P. Carroll, W. B. Guggino, P. Agre, (1992) Appearance of Water Channels in Xenopus Oocytes Expressing Red Cell CHIP28 Protein. Science 256 (5055):385–387.

S. Nielsen, B. L. Smith, E. I. Christensen, P. Agre, (1993) Distribution of the aquaporin CHIP in secretory and resorptive epithelia and capillary endothelia. Proceedings of the National Academy of Sciences 90 (15):7275–7279.

K. Ishibashi, S. Sasaki, K. Fushimi, S. Uchida, M. Kuwahara, H. Saito, T. Furukawa, K. Nakajima, Y. Yamaguchi, T. Gojobori, (1994) Molecular cloning and expression of a member of the aquaporin family with permeability to glycerol and urea in addition to water expressed at the basolateral membrane of kidney collecting duct cells. Proceedings of the National Academy of Sciences 91 (14):6269–6273.

Thayne R Larson, David W Rrobertson, Alberto Corica, David G Bostwick, (2000) In vivo interstitial temperature mapping of the human prostate during cryosurgery with correlation to histopathologic outcomes. Urology 55 (4):547–552.

# Chapter 9
# Quantum Mechanics of the Cell: An Emerging Field

Cell is generally considered a classical system. The molecular structures inside it appear with ultra-level complexities. General physics concepts help construct popular biophysics techniques to understand the energy states and physiological functions of various cellular structures. Besides using statistical mechanics, classical mechanics, and other general physics rules, it is also found recently that quantum mechanics may be utilized to understand some of the crucial cellular aspects. Even about six decades ago, the quantum effects of the electron tunneling of cytochrome were explained (see Chance and Nishimura (1960); Vredenberg and Duysens (1964); Devault et al. (1967). In this chapter, we shall briefly address the application of quantum mechanics to understand cell biology. We shall provide hints on the future outlook of cellular research that may connect quantum mechanics techniques to address both fundamental cellular aspects and related diseases dealt in biomedical research.

## 9.1 Introduction

Quantum effects are usually observed in nanometer (nm) or lower scale size objects that are surrounded by vacuum at ultra-low controlled laboratory temperatures. Whereas, biological cells exist in messy warm environment at microscopic scale. Some reactions between the cell-approaching agents, e.g., drugs and their cell-based targets, are often found to happen at low nm scale. These cell targets or components may be individual plasma and mitochondrial membrane-based lipids, cholesterol, hydrocarbons, small membrane proteins, cytosol proteins, etc., and the compartments such as cell membrane, nuclear membrane, mitochondrial membrane, and various other structures. Moreover, the fluctuations, dynamics, movement of these small components, and their physical interactions that take place inside a cell happen also within nm scale dimensions. We have explained some of these aspects



M. Ashrafuzzaman, *Nanoscale Biophysics of the Cell*,
https://doi.org/10.1007/978-3-319-77465-7_9

in our previous book 'Membrane Biophysics' (Ashrafuzzaman and Tuszynski 2012) and references (see, e.g., Ashrafuzzaman et al. 2014; Ashrafuzzaman 2015a, b). However, a quantum mechanical phenomenon, such as 'coherence,' in which the wave patterns of every part of a system stay in step, lasts less than a microsecond (μs) in the tumultuous realm of the cell. But most of the cellular events take time beyond μs. Search for quantum effects in biological cells was therefore considered unrealistic until recently. Coherent quantum processes may well be present in the natural world and perhaps specifically in biological cell-based processes. Examples range from the ability of birds to navigate using Earth's magnetic field to the inner workings of photosynthesis—the process by which plants and bacteria turn sunlight, carbon dioxide, and water into organic matter (Ball 2011).

Understanding quantum effects at nanoscale or lower-level dimension cellular structures and related calculations is slowly evolving. With the development of fast, precise, and powerful quantum computing, this trend is expected to grow fast. Although isolating pinpointed structures and events in biological system inside a liquid environment is truly an issue, we shall provide a few cases here where quantum mechanical treatments have been successfully performed.

## 9.2 Quantum Mechanical Calculations to Address Ion Channel Gating Mechanisms

Quantum coherence that sustains as long as 300 femtosecond (fs) was recently detected at physiological temperature which is long enough to impact biological energy transport (Panitchayangkoon et al. 2010). Microscopically, this coherence lifetime was attributed to correlated motions within the protein matrix encapsulating the chromophores.

About a decade ago, quantum mechanical calculations were performed on KcsA channel to address the ion channel gating mechanisms (Kariev et al. 2007). Here, the charge effects on gating mechanisms were considered. KcsA channel is formed by four identical subunits with each sub-unit containing two alpha helices, connected by a long loop of 30 amino acids proofreading into the pore region. The pore region consists of the following:

- An inner pore which starts from inside of the cell,
- A cavity near the middle of the pore,
- Selectivity filter separating the cavity from the extracellular mostly hydrophilic solution.

The inner pore region and the internal cavity region are hydrophobic in nature. The selectivity filter is lined exclusively by polar main-chain atoms belonging to the so-called signature sequence. The potassium ions coordinate at their binding sites in the pore.

A series of ab initio density functional calculations were carried out on side chains of a set of amino acids, plus water, from the (intracellular) gating region of the KcsA $K^+$ channel. X-ray structures provide channel's atomic coordinates, except hydrogen, and are known from Doyle et al. (1998), MacKinnon et al. (1998), Jiang et al. (2001), as are the coordinates of some water oxygen atoms. Quantum mechanical optimization shows four water molecules to form a 'basket' under side chains, blocking the channel. When a hydrated potassium ion approaches this 'basket,' the optimized system shows a strong set of hydrogen bonds with the potassium ion at defined positions. This prevents further approach of the $K^+$ to the basket. This optimized structure with hydrated $K^+$ added shows an ice-like 12 water molecule nanocrystals. If the water molecules exchange, unless they do it as a group, the channel will remain blocked. Here, a calculation of open and closed states of the KcsA channel leads to a proposed gating mechanism, showing how proton addition to the gating region of the channel disrupts hydrogen bonds, producing channel opening. It is suggested that a similar mechanism may apply to voltage gating. Details are presented in Figs. 9.1, 9.2, and 9.3.

Kariev et al. recently performed quantum mechanical calculations to explain the open gating mechanisms of the $K_v1.2$ channel (Kariev et al. 2014). Their calculations showed key role played by hydration. The open $K_v1.2$ voltage-gated potassium channel gate can just hold a hydrated $K^+$ ion. Quantum calculations showed little change from the X-ray coordinates for the protein. The calculations included the water molecules and ions which are not in the X-ray coordinates. A sequence of steps is postulated in the calculations in which the potential experienced by the ion in the pore is found to be influenced by the position of the ion. The gate structure, with and without the ion, has been optimized. The charges on the atoms and bond lengths have been calculated using natural bond orbital calculations, giving $K^+$ $\sim 0.77$ charges, rather than 1.0. The calculation provided the optimized structure of the gate region of the $K_v1.2$ channel, showing the water structure with and without an ion in the gate. The ion is well hydrated, and the hydrated ion structure provides a sort of sleeve that allows the ion to slide through the gate region. The calculations also predict that ion must have a finely tuned energy minimum at the gate to allow conductance. For details on technique, results, etc., readers may read the original articles in Kariev et al. (2014).

As a demonstration of the quantum calculations, we have used two of the studies made by Kariev and co-workers. Ion channel is perhaps one of a very few biological fields especially in cell-based areas in which a fair amount of work has been done using various quantum mechanics methods-based calculations. Usually in the case of ion channels, instead of addressing the whole channel mechanics, a part of the macroscopic channel is chosen for study. This choice helps to pinpoint microscopic nature of a macroscopic problem and deduce the calculations locally. Few examples can be mentioned here. Ion channel conductance selectivity is an important aspect which has been scrutinized quantum mechanically by Varma et al. (2011). Quantum calculations on selectivity of channels were also performed by Dudev and Lim (2009), (2010), (2012a). Varma and Rempe have addressed

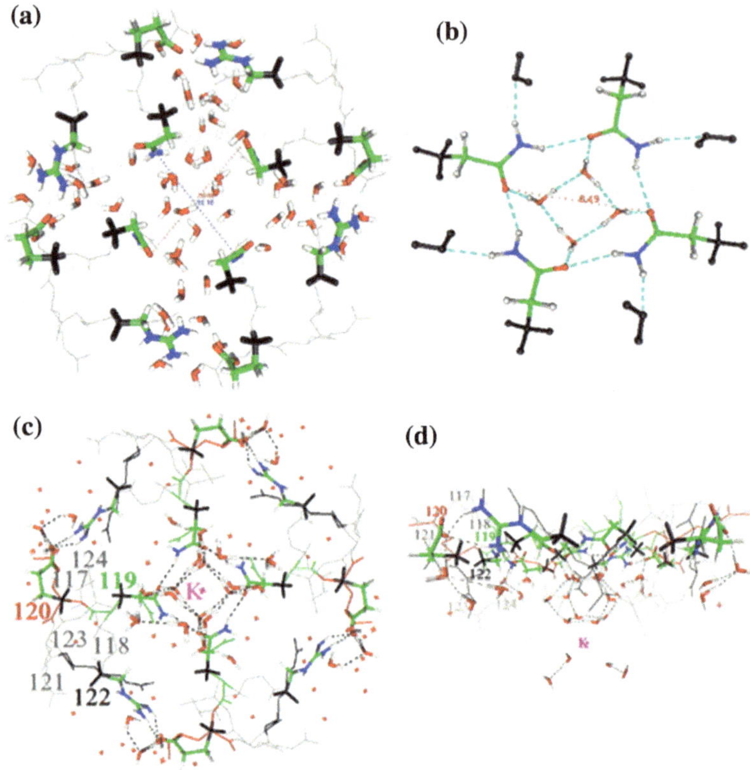


**Fig. 9.1** **a** Open (protonated) state of the KcsA channel, showing results of calculations on three amino acids from all four domains, plus water. Outer methyl groups are frozen. The distance between the two oxygen atoms of Q119 is 10.9 Å on one diagonal and 10.2 on the other. Color code: see end of caption. Outer methyl groups are frozen. The central region of the set of amino acids opens, and the distance between two oxygen atoms of Q119 is 11.0 Å: **b** the closed state, Q119 only, plus the water. Instead of 11Å, the distance across is now only approximately 6Å, not enough for a hydrated $K^+$ ion, even if the water molecules could exchange instead of blocking the channel. **c**, **d** The uncharged (closed) state of all three amino acids (two orientations), plus the remainder of the structure, not present in the computation: The system consisting of the Q119, E120, R122 amino acids (all four domains) plus 32 water molecules and a $K^+$ is optimized. These are shown as heavy lines (tubes). The amino acids adjacent to these are not included in the calculation and shown as thin gray lines; all amino acids are labeled in one domain. The small red x's are 58 water molecules (oxygen atoms) from the 1k4c X-ray structure. The heavy water molecules are the result of the optimization. As in **b**), there is a 'basket' of water molecules hydrogen bonded to Q119 that blocks the channel and also holds the domains together: The set of 12 hydrogen-bonded water molecules form a 'nanocrystal,' shown at the center of Fig. 9.1c surrounding the $K^+$ and at the bottom of Fig. 9.1d, and is a strong barrier to the approach of $K^+$ toward the channel. Hydrogen bonds are shown in Fig. 9.1a, b as dashed lines. Large amino acid labels are shown in one domain, and the corresponding amino acids are present in all four domains. Atoms: black = frozen; optimized atoms: green = carbon, blue = nitrogen, red = oxygen, white = hydrogen. For details, see Kariev et al. (2007)

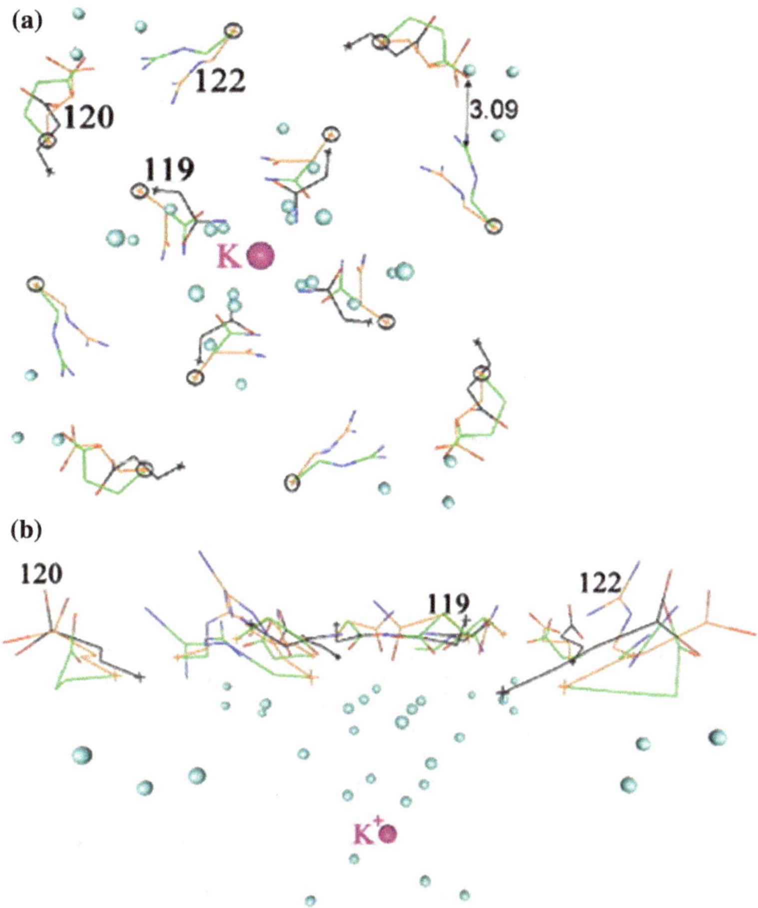


**Fig. 9.2** Two views, at 90° angles comparing the calculated and 1k4c and 1j95 X-ray structures: Carbons from the 1k4c structure are red–orange, from the calculation green, from 1j95 black. Nitrogen is blue, hydrogen not shown, and carbons frozen in the calculation that are used as methyl groups are circled in A (the 1k4c and calculated positions of those atoms are forced to coincide). Water oxygens (only calculated shown) are pale blue and can be seen to be in locations that outline the nanocrystal. R122 is missing in the 1j95 X-ray structure. One instance of each amino acid is labeled. The two oxygen atoms of the E120 carbdysfunctional without DHAoxyl to the neighboring domain R122 guanidinium C atom have optimized distances of 3.09 and 4.12 Å, compared to 4.19 and 6.17 Å in the 1k4c X-ray structure. The 3.09 Å distance is marked on the figure. For details, see Kariev et al. (2007)

extensively on the hydration of ions in relation to selectivity (Varma and Rempe 2010). Bucher et al. have applied ab initio molecular dynamics (MD), to channels, especially but not exclusively on questions of selectivity (Bucher et al. 2004; Bucher and Rothlisberger 2010). Voth group applied the empirical valence bond method to water and proton transport (Maupin et al. 2006).

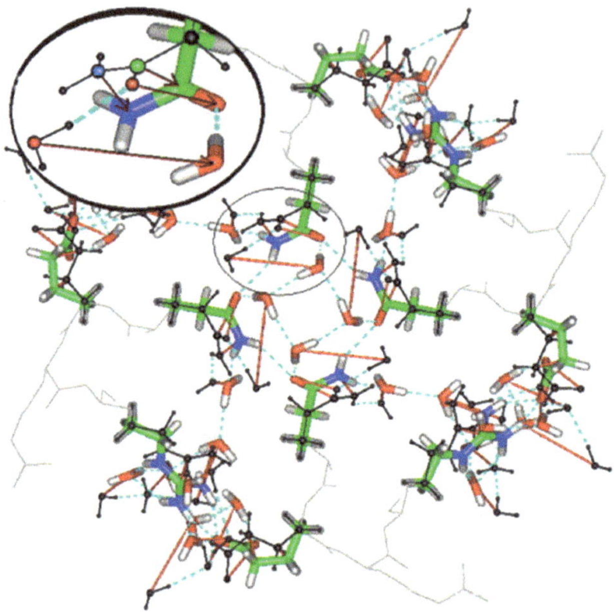

**Fig. 9.3** This figure shows how much the quantum optimization for KcsA moves atoms from the starting 1k4c positions; initial positions for all atoms in the calculation are shown in black. All thin red lines show motion of >1 Å; absence of thin red lines for atoms shown in color shows that the motion is <1 Å; these atoms essentially remained at the X-ray structure positions; most of the protein atoms are in this group. Gray atoms were not included in the calculation; they are shown in their X-ray positions. Some water moved more than 1 Å, the guanidinium nitrogen moved about 2 Å, and the water hydrogen bonded to it moves about 4 Å. The inset enlarges the key step: The glutamine amine nitrogen rotates about the single bond, and the hydrogen-bonded water accompanies it, rotating to form the 'basket.' For clarity, the water molecule to the left is removed from the inset. All four domains show the same move. Atom colors: C, green; O, red; N, blue; H (added, not from X-ray structure) white. The water O–H bond length is close to 1 Å, so that the dimensions of the figure can be seen by using water as a scale bar. Dashed blue lines are hydrogen bonds, thin black lines, initial positions. The 'basket' is in the center and only exists after the optimization. For details, see Kariev et al. (2007)

## 9.3 Quantum Mechanical Understanding of the Neural Cell Signaling

Quantum mechanical properties of cellular signaling agents may be found important to explain electrical signaling pathway. Crawford et al. recently offered a theoretical explanation based on the quantum mechanical properties of docosahexaenoic acid (DHA) regarding its role in electrical signaling membranes (Crawford et al. 2013). DHA is an omega-3 fatty acid that is a primary structural component of the human brain, cerebral cortex, skin, sperm, testicles, and retina. In regard to health benefits,

DHA is essential for the growth and functional development of an infant's brain as well as required for maintenance of normal brain function in adults (for details, see Horrocks and Yeo 1999).

Fossil Intracellular structures are found so similar to extant organisms that they were likely made with identical membrane constituents lipids and proteins, which together provided organization and specialization over time. Amino acids could be synthesized over 4 billion years ago, but only oxidative metabolism allows for the synthesis of highly unsaturated fatty acids, and thus the process might help produce novel lipid molecular species for specialized cell membranes (Crawford et al. 2013). DHA is known to have had provided the core for the development of the photoreceptor, and conversion of photons into electricity stimulated the evolution of the nervous system and brain. DHA has since been found conserved as the principal acyl component of photoreceptor synaptic and neuronal signaling membranes in the cephalopods, fish, amphibian, reptiles, birds, mammals, and humans. Despite great genomic changes, this unimaginably extreme conservation in electrical signaling membranes suggests it was perhaps nothing but DHA dictation on DNA. The quantum mechanical works of Crawford et al. predict that the unique molecular structure of DHA allows for quantum transfer and communication of π-electrons, which explains the precise depolarization of retinal membranes and the cohesive, organized neural signaling (Crawford et al. 2013).

The basis of quantum transfer and communication of π-electrons in DHA has been explained in ref. (Crawford et al. 2013). The van der Waals equation (see Crawford et al. 2013) hints that DHA has both stereochemical and electromagnetic properties. Quantum mechanics can predict the existence of energy levels inside lattices, whereby any electron in that level can be effectively spread across the whole structure, thus becoming a quasiparticle or a wave. Albert Szent-Gyorgyi postulated that common energy levels could exist in protein structures, as they contained 'a great number of atoms, closely packed with great regularity.' He considered the communication of energy between molecules in biological systems could be achieved through coherence of electrons raised to a higher energy state. The formation of a triplet state of the π-electrons around double bonds in aromatic amino acids was the basis for this mechanism (Steele and Szent-Gyorgyi 1957; Avery et al. 1961). Since Szent-Gyorgyi, the electron transfer of the energy production system in mitochondria has become well known. Bendall (1996) considered the conformal dynamics of proteins to be reliant on the long-range transfer of electrons. The method of quantum transport is quantum mechanical tunneling, a feature of proteins demonstrated by Hopfield (1974). Hackermuller et al. (2003) obtained evidence that tetraphenylporphyrin exhibits wavelike behavior, indicating quantum coherence in nature. Hammeroff and Penrose (1998) proposed a model based on quantum mechanics that can explain consciousness and is testable. In the 'Orchestrated Objective Reduction' model of consciousness, quantum coherence exists in the microtubules found in neurons. It is hypothesized that microtubules are capable of quantum computing, and quantum computations are translated to classical outputs—hence consciousness. Hammeroff (2010) then proposed the connections between neurons were linked to consciousness. Gap junctions are small enough for quantum objects to cross by quantum tunneling, allowing cohesion across regions of the brain and creating consciousness. The brain can contain numerous proteins but is absolutely dysfunctional without DHA and arachidonic acid (20:4n-6, AA). Crawford et al. (2013) hypothesize that the π-electrons in DHA could behave in similar quantum manner, explaining the unique and irreplaceable role of DHA in neuronal signaling. According to Crawford et al., we must also consider that beyond consciousness, cohesion across regions of the brain drove the evolution of symbolic thinking and behavior, which is the hallmark of humanity.

There is a planarity to the DHA molecule and a **δ (+)ve** to **δ (−)ve** polarization to the electron cloud (see Fig. 9.4). The same polarization applies to the sigma electrons around the methylene groups.

In photoreceptor membrane, the polar headgroup on the outer face is dominated by a phosphate and a strong quaternary amine (choline). On the inner side of the membrane, there is the same phosphate but a weak primary amine (ethanolamine). The resting state bilayer will be charged, with the probability of finding an electron greater in the direction of choline. Two conditions which could pull an electron out of DHA are as follows (details in Crawford et al. 2013):

(i) A sufficient electrical charge as in hyperpolarization,
(ii) The Einstein photoelectric effect-type mechanism.

Electrons from a material are dislodged by the impingement of photons when those photons reach or exceed a threshold frequency. Below the threshold

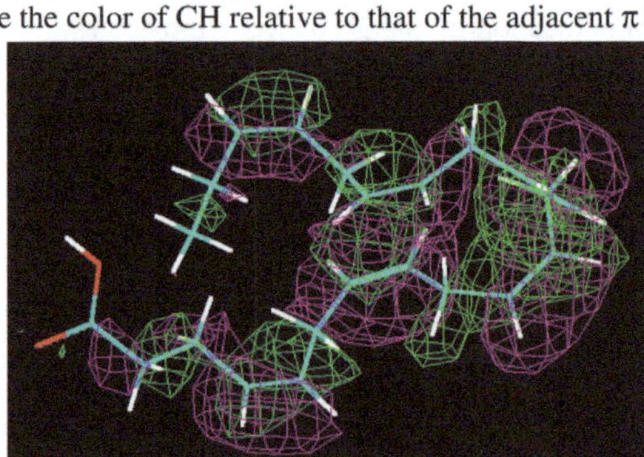


**Fig. 9.4** Molecular dynamic representation of DHA illustrating the alternative δ −ve and δ +ve distribution through the molecule. With the electrons in the methylene groups participating, a probability exists of electron cohesion across the whole molecule. The methylene participation is confirmed in the NOE analysis described later. For details, see Crawford et al. (2013)

frequency, no electrons are emitted from the metal regardless of the light intensity or the length of time of exposure to the light. Einstein's photoelectric effect explains the fact that light can eject electrons from metal even if its intensity is low, where Albert Einstein proposed that a beam of light is not a wave propagating through space, but rather a collection of discrete wave packets (photons), each with a quantum of energy. Simply, we can summarize it by saying that the photoelectric effect is caused by absorption of quanta of light. This effect is found applicable to explain the mechanism of the light-induced electron transfer in condensed phase, e.g., as found in the case of DHA in neural cell signaling (for details, see Crawford et al. 2013). Here, an important question regarding the photon energy efficiency in condensed phase has been addressed. As it is known that the energy of a photon in visible wavelengths is more than sufficient to eject an electron from retinal, could the same photoexcitation take place in the π-electrons of DHA? The isomerization of the 11-cis-retinal is achieved by a photon energy exciting a π-electron into an escape mode. DHA double bonds consist of a σ bond and a π bond. The calculations found the following: the 80 kcal/mole (the σ bond energy) and 65 kcal/mole (the π bond energy) of total 145 kcal/mole (the C = C double bond energy). The π bond energy is much less than that of σ bond. Therefore, π bonds are more reactive than σ bonds which means that all things being equal, the π bonds of DHA are more likely to undergo activation by light.

Further into the insights, based on the molecular dynamic calculations, the ground state energy of DHA is found to be 50 kJ/mol which is in the region of photoreception. Visual responses still can be detected in the electroretinograms of mice lacking critical components of all known retinal phototransduction cascades (Allen et al. 2010). Excitation of a π-electron in the DHA phosphoglycerides followed by electron tunneling, in response to hyperpolarization, could offer a theoretical solution which fits with that of Gawrisch et al. (2003) who describe low barriers to torsional rotation about the C–C bonds that link the cis-locked double bonds with the methylene carbons between them.

Electron tunneling has been explained here. If one electron is delocalized and pulled out by hyperpolarization, an immediately distal electron will take its empty place and this electron tunneling would lead to a current to flow. The Pauli Exclusion Principle suggests that no two electrons can occupy the same energy state. If one electron is pulled out, the loss leaves a hole which can only be filled by an incoming electron of the same quantum status of spin and energy. This is explained in Fig. 9.5. This process could depolarize the membrane and do so only at a quantized energy level, as is understood here, theoretically.

Crawford et al. demonstrated some remarkable conservation and irreplaceable nature of DHA in neuronal signaling and its high concentration in the photoreceptor. For details, readers should consider reading the whole paper from Crawford et al. (2013), due to lack of space we just briefed here. Here, explanations were provided through consideration of electron tunneling providing quantized signals. Membrane depolarization in phototransduction was explained using quantum

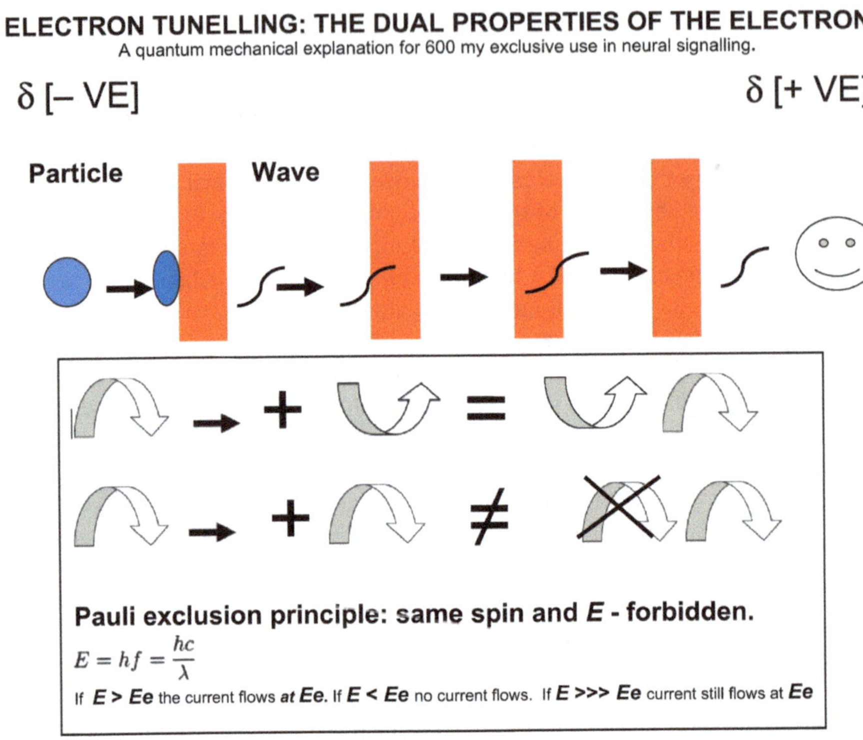


**Fig. 9.5** Dual particle/wave properties of an electron tunneling. Neural signaling is assumed to be 600 million years (my) old as the nervous tissue first arose in wormlike organisms about 550–600 my ago. Seen as particle, it is like a ball hitting a brick wall—impossible to penetrate. However, there is no such thing as a ball or brick wall; the electron's 'position' is only the sum of the molecular electromagnetic forces defining a probability function for its location. As a wave, there are probability distributions in which the electron penetrates barriers and can communicate with neighboring molecules or end up in other regions of the molecule. Removal of an electron in a neighboring orbit will reduce repulsive forces and invite the electron into a new, higher energy orbit. An electropositive charge will increase the probability of occupation. The $\pi$-electrons have opposite intrinsic angular momentums or spin. The occupation of the orbit by two electrons with the same quantum energy state is forbidden. The Pauli Exclusion Principle ensures that the energy level at which tunneling occurs is precise. For details, see Crawford et al. (2013)

mechanical treatment. This precision is essential to visual acuity and synaptic signaling. The energy-minimized structures, molecular polarization, and moment of inertia of DHA allow for its theoretical possibility of operating in the realm of supramolecular chemistry with electron quantum coherence. This is certainly a preliminary study which requires more in-depth investigations.

There are many studies in various cell-based fields already performed. Due to lack of spaces, we shall avoid going to other fields than what have been discussed for ion channels. The quantum mechanical understanding of the cell is now a reality. Before the beginning of this century, it was thought that quantum mechanics application in cell is not quite possible though predictions were made. More developments are expected to happen fast over next decades.

## 9.4 Quantum Effects in Photosynthesis

Photosynthetic complexes are naturally tuned to capture solar light efficiently. The excitation energy is then transmitted to reaction centers, where long-term energy storage is initiated. Semiclassical models are often used to describe the energy transfer mechanism. Two-dimensional (2D) Fourier transform electronic spectroscopy has been found to map the energy levels and their coupling in the Fenna–Matthews–Olson (FMO) bacteriochlorophyll complex. The FMO is a water-soluble complex. It is found to appear in green sulfur bacteria and is active in mediating the excitation energy transfer from light-harvesting antenna chlorosomes to the membrane-embedded center of bacterial reactions.

Although the spectroscopic investigations clearly document the dependence of the dominant energy transport pathways on the spatial properties of the excited-state wave functions of the whole bacteriochlorophyll complex, the intricate dynamics of quantum coherence has always been ignored. Engel and co-workers attempted first in 2007 to extend previous 2D electronic spectroscopy investigations of the FMO bacteriochlorophyll complex and obtained direct evidence for remarkably long-lived electronic quantum coherence playing an important part in energy transfer processes within this system (Engel et al. 2007). The quantum coherence was reported to manifest itself in characteristic, directly observable quantum beating signals among the excitons within the Chlorobium tepidum FMO complex at temperature 77 K.

Some photosynthetic proteins have been found to host phenomenon such as the electronic energy transfer involving quantum coherence. In photosynthesis process, sunlight is captured and transported by highly specialized proteins known as 'antenna proteins.' Structural models and absorption spectra of a few representative photosynthetic light-harvesting antennae are presented in Fig. 9.6.

In Fig. 9.6, the presented spectra were recorded at 77 K, but they are similar at ambient temperature, owing to the presence of significance inhomogeneous line broadening.

Following clockwise direction, left to right:

(a) The peridinin-chlorophyll proteins (PCPs) of dinoflagellates that contain eight blue-absorbing peridinin molecules and two chlorophyll-*a* molecules (for details, see Hofmann et al. 1996).
(b) PE545 complex, typical of varieties of cryptophyte antenna proteins, binds eight bilin chromophores with absorption that is found to vary widely among species (for details, see Wilk et al. 1999).

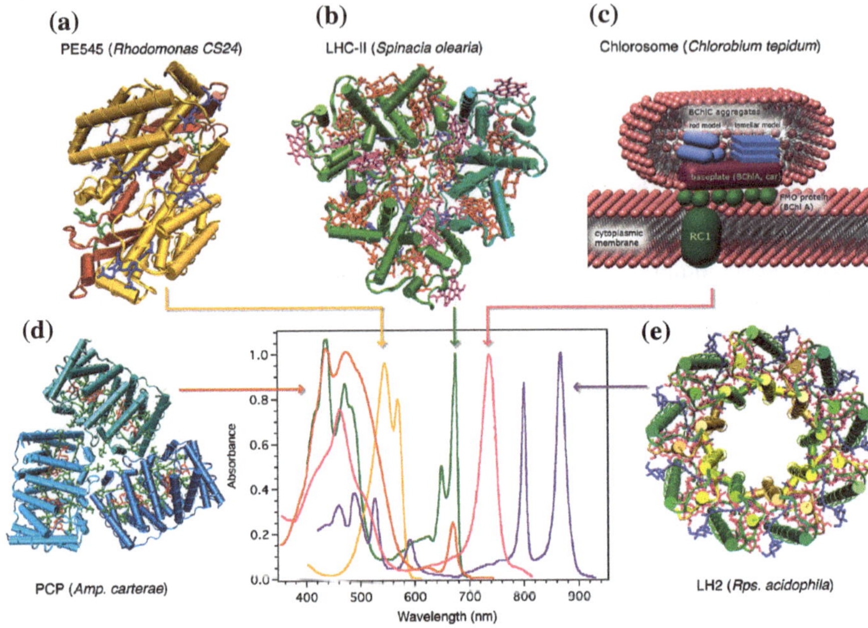


**Fig. 9.6** A few photosynthetic light-harvesting antennae, their structural models, and absorption spectra are presented here. This is copied from Scholes (2010) and Collini et al. (2009) with due permission to reproduce

(c) Primary light-harvesting complex of the higher plants, LHC-II, shown in trimeric form. Each trimer complex binds eight chlorophyll-*a* and six chlorophyll-*b* chromophores and carotenoids used for the purpose of photoprotection and regulation (for details, see Liu et al. 2004).
(d) Gigantic chlorosome antennae from green sulfur bacteria comprise about 200,000 self-assembled bacteriochlorophyll-*c* chromophores (for details, see Ganapathy et al. 2009).
(e) LH2 complex from purple bacteria is notable for its apparently beautiful ninefold symmetry (for details, see McDermott et al. 1995).It binds carotenoids and two rings of bacteriochlorophyll-*a* chromophores, 9 and 18 in the B800 ring and B850 ring, respectively.

The antenna proteins are found to act as quantum machines, use a quantum transport mechanism, guide the light, and finally store the energy in the centers of their reactions. Scientists at the Institute of Photonic Science in Barcelona, in collaboration with colleagues at the University of Glasgow, for the first time tracked the energy flow in individual proteins and discovered that the quantum coherences made the light flow in the antenna protein immune to the ubiquitous external natural effects.

Besides explaining the role of the mentioned antenna proteins in photosynthesis, the meaning of quantum-coherent energy transfer was explained in detail by Scholes in 2010 (Scholes 2010). Here, a few questions were analyzed such as why did certain photosynthetic organisms evolve to use quantum coherence in light harvesting and are these electronic excitations entangled? The analysis put forward a basis for chemists to attempt further to understand the details of the phenomenon.

The chromophores, specifically tetrapyrroles, called bilins, in cryptophyte antenna proteins are arranged with average nearest-neighbor separations of ~2 nm (Wilk et al. 1999). In comparison, nearest-neighbor separations between chlorophyll molecules in the major higher plant peripheral antenna protein LHC-II are typically about half of that distance that is ~1 nm (Barros et al. 2009). Yet, the cryptophyte antenna complexes are still able to collect incident light and transfer those excitations to specific reaction centers with a high quantum efficiency of > 95% (van der Weij-De 2006). Two-dimensional photon echo experiments, briefly explained in Fig. 9.7 (measurement of the quantum coherence), provide an answer to the question—how cryptophyte antenna complexes are able to function with the same overall efficiency as LHC-II (Scholes 2010). Quantum coherence is found to play a role, even at or near physiological temperature, to 'wire' chromophores together in the cryptophyte antennae, thereby compensating for the large interchromophore separations (Collini et al. 2010).

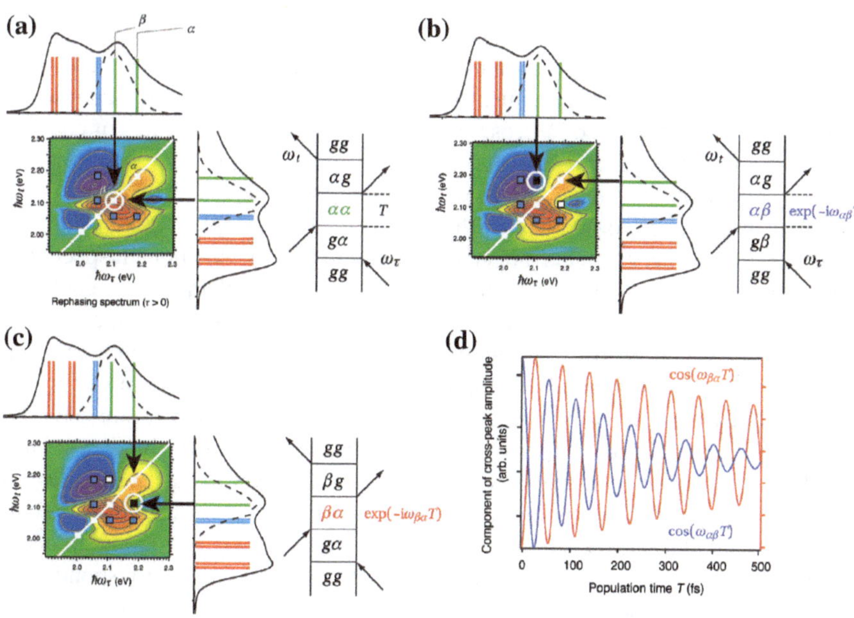


**Fig. 9.7** Measurement of the quantum coherence using 2D photon echo spectroscopy. Copied with due permission and through personal communication with Professor Gregory D. Scholes (Scholes 2010)

The 2D electronic photon echo (2DPE) spectroscopy is now a practical method for obtaining detailed insights into excited-state dynamics. The 2DPE spectra information includes the time evolution of coherent superpositions of the absorption bands, which provides a measure of quantum coherence. To illustrate this, Scholes considered a representative rephrasing 2DPE spectrum for a population waiting time $T$. The data are plotted as the photon echo 'emission' frequency $\omega_t$ versus the conjugate frequency $\omega_\tau$ that represents excitation. They are plotted here in Fig. 9.7 on an arcsinh intensity scale to accentuate the representation of weak features in the cross-peaks. The dynamics was followed by recording a series of such spectra for various population times. The data shown here (see Fig. 9.7) were recorded for an aqueous suspension of cryptophyte antenna proteins at 294 K (PC645 from Chroomonas CCMP270) (Collini et al. 2010).

In Fig. 9.7a, the linear absorption spectrum is projected along the diagonal of the 2DPE spectrum. Despite happening the strong spectral line broadening, peaks along the diagonal (white line) can be identified with absorption bands in the linear spectrum, projected above and beside each 2DPE plot for reference. For example, the excitation of the absorption band labeled $\beta$ can yield an emission at the same frequency in the photon echo, according to the double-sided Feynman diagram shown on the right-hand side of the panel. That signal (analogous to a bleach signal in pump–probe spectroscopy) yields a diagonal peak in the 2DPE spectrum, as indicated.

In Fig. 9.7b, c, off-diagonal peaks in 2DPE spectra can have a few different origins. They may mean, for example, that excitation of the absorption band $\alpha$ produces an excited-state population that relaxes incoherently to band $\beta$, perhaps by electronic energy transfer. Consequently, a cross-peak in the lower diagonal would emerge with population time $T$. Cross-peaks can also provide signatures of coherent superposition states formed by the excitation pulse sequence. Two possible coherent superposition states are presented here, and depending on the corresponding off-diagonal component of the density matrix, $|\alpha\rangle\langle\beta|$ or $|\beta\rangle\langle\alpha|$, they can produce a 2DPE signal at the upper or lower cross-peak position, respectively. According to the representative Feynman diagrams that explain these signal contributions, an oscillating phase will be acquired during the population time $T$ as like as the beating sound heard when two tuning forks of different frequency are rung simultaneously. The magnitude of the oscillation frequency is determined by the energy difference between the two absorption bands in the superposition state, but the sign of the oscillation is opposite for each peak in the pair of the cross-peaks.

In Fig. 9.7d, cross-peak intensity oscillations in 2DPE spectra reveal quantum coherence, particularly by the observation of anticorrelated oscillations as a function of $T$. As an illustration, the cross-peak intensities versus $T$ were fitted to a sum of damped cosine functions and thereby the oscillating component corresponding to the energy difference between $\alpha$ and $\beta$ was isolated. The anticipated anticorrelation was clearly indicated. The damping time of the beats indicated the decoherence time for the superposition.

Hildner et al. employed fs pulse-shaping techniques and showed that quantum coherences in single organic molecules can be created, probed, and manipulated at ambient conditions even in highly disordered solid-state environments. A broadly distributed coherence decay time for different individual molecules giving direct insight into the structural heterogeneity of the local surroundings was reported in their studies (Hildner et al. 2011). Here, they induced Rabi oscillations and could control the coherent superposition state in a single molecule. Thus, they could carry out a basic fs single-qubit operation at room temperature. Energy transport pathways of separate individual but chemically identical antenna proteins were evaluated. Each protein was shown to use a distinct pathway. The transport paths within single proteins could adapt the optical efficiency and were found to vary over time due to changes in the environmental conditions. The coherence which is a genuine quantum effect of superposition of states is thus found responsible for maintaining transport efficiency in biological systems.

Sarovar et al. investigated the entanglement in multichromophoric light-harvesting complexes. They considered the case for the FMO protein to extract the initial state and temperature dependencies of entanglement (Sarovar et al. 2010). It was found in this study that although the FMO protein in natural conditions largely contains bipartite entanglement between dimerized chromophores, a small amount of long-range and multipartite entanglement should exist even at physiologically relevant temperatures. A rigorous quantification of entanglement in a biological system was thus analyzed here.

Quantum effects in photosynthetic complex were observed by Greg Engel and collaborators in 2007 (Engel et al. 2007). These effects were reproduced in different laboratories using varieties of approaches at a temperature of around −193 °C and subsequently at ambient temperature. These quantum effects have been observed in biological complexes, which are large, wet, and noisy systems that make the effect as also physiologically relevant. We shall avoid explaining the techniques in detail. But serious readers may go through the referred references to acquire an in-depth understanding. Although quantum effects of photosynthesis have been predicted earlier and recently demonstrated details, understandings require more studies. Future development of fine methods and techniques to address quantum effects in dynamic and noisy biological systems especially in living cells may update our understanding further.

## 9.5 Quantum Imaging

Quantum-based life science imaging is a possible future technique to generally emerge. Fluorescence-based imaging has been popular in biological systems. Whole cell imaging, cell compartment/section imaging, cell penetrating agent, particles, microbes imaging, etc., are routinely performed as there are developed imaging techniques already in place (Alivisatos 2004). The quantum properties of single spins within fluorescent particles have been explored in nanoscale magnetometry fields (Maze et al. 2008; Balasubramanian et al. 2008; Chernobrod and Berman 2005; Taylor et al. 2008; Degen et al. 2008; Cole and Hollenberg 2009; Hall et al. 2009). Such investigations are yet to be explored in biological environments. Recently, McGuinness et al. (2011) demonstrated optically detected magnetic resonance of individual fluorescent nanodiamond nitrogen-vacancy (NV) centers inside living human HeLa cells and measured their location, orientation, spin levels, and spin coherence times with nanoscale precision. They measured quantum coherence through Rabi and spin echo sequences over long time, greater than 10 h. The orientation was tracked with effective 1° angular precision over acquisition times of 89 ms. Here, we may mention that the Rabi cycle is the cyclic behavior exhibiting the Rabi flopping of a two-level quantum system in the presence of an oscillatory driving field.

Measurement of the NV quantum state within living HeLa cells in culture has been made (McGuinness et al. 2011). The NV center is an atomic-sized defect, with stable fluorescence observed in small crystals with dimensions as small as 5 nm (Bradac et al. 2010). Here, three specific investigations were made:

- Microwave spectroscopy to distinguish NV centers with otherwise identical fluorescence within the cell. Study 1,
- Intracellular quantum coherence measurement of single spin systems, including Rabi and spin echo sequences. This study provides a foundation for further intracellular magnetometry studies. Study 2,
- Orientation tracking of nanodiamonds in the intracellular medium, based on continuous ODMR monitoring. Study 3.

In a confocal microscope with microwave control, it was first demonstrated how the unique quantum properties of each NV in the absence of an externally applied magnetic field ($B_0 = 0$) could be used as a fingerprint in the intracellular environment to spectrally separate nanodiamonds (see Fig. 9.8a). Before imaging, HeLa cells cultured in media containing fluorescent nanodiamonds (see Fig. 9.8b) were washed in phosphate-buffered saline (PBS). A typical confocal image of one cell (HeLa-1), which contains two single NV centers (NV-1a and NV-1b) in the same focal plane, is shown in Fig. 9.8c. Quantum measurement and control of the NV center are shown in Fig. 9.8d.

The coherent quantum control and measurement of the spin state of the NV probe in the intracellular environment were demonstrated (see Fig. 9.9). Coherence times for both NV-1a and NV-1b were measured by means of spin echo, and this sequence was repeated for each NV as the cell timeline progressed. This *in living cell* demonstration represents the full suite of quantum control and measurement required for nanoscale intracellular magnetometry.

Continuous optically detected magnetic resonance (ODMR) monitoring in the presence of an external magnetic field allows rotational motion of an internalized

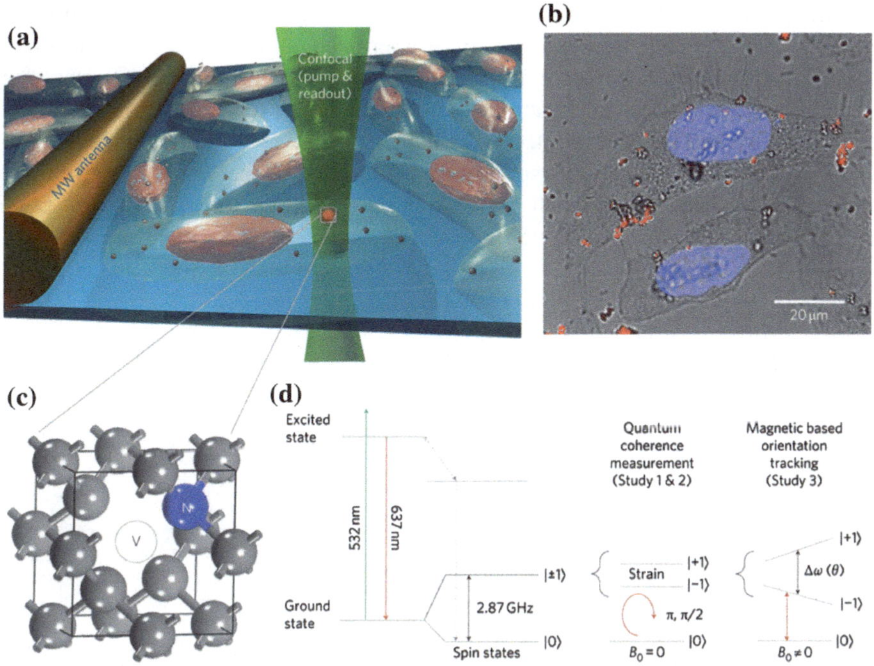


**Fig. 9.8** **a** Experimental setup with microwave (MW) control of the NV spin levels and confocal fluorescence readout. **b** Bright field and confocal fluorescence images of HeLa cells that show uptake of nanodiamonds. NV fluorescence is shown in red, and the nucleus is stained using Hoechst 33342 (blue). A Leica TCS SP2 confocal microscope was used. **c** Atomic lattice structure of the NV center. **d.** Quantum measurement and control of the NV center are shown here. Left: energy levels of the NV probe system and fluorescence dynamics. Middle: quantum control for studies 1 and 2 ($\pi$ and $\pi/2$ pulses for spin echo). Right: orientation-dependent Zeeman splitting $\Delta\omega(\theta)$ in an applied magnetic field $B_0$. For details, see McGuinness et al. (2011)

nanodiamond. This is to be measured over millisecond (ms) acquisition timescales. This was demonstrated in this study (see Fig. 9.10) (McGuinness et al. 2011).

Figure 9.10a shows two spectra taken over the life of the cell that demonstrate the timescale for rotational monitoring to be essentially unlimited, depending only on the cell lifetime. With knowledge of the applied magnetic field strength $B_0$, the nanodiamond orientation can be determined through the geometric relationship $\theta = \cos^{-1}(\hbar\, \Delta\omega/\mu_{NV}B_0)$ where $\theta$ is the angle between $B_0$ and the NV axis, $\mu_{NV}$ is the magnetic moment of the NV center, and $\Delta\omega$ is the peak separation between the ODMR transitions. Figure 9.10b tracks the value of $\theta$ that was derived from the measurement of $\Delta\omega$ over 16 h. Figure 9.10c represents the plot of the three-dimensional position and rotation of NV-2 over a 3 h period (indicated by the dashed vertical lines in Fig. 9.10b).

The quantum spin levels were found to serve as fingerprints that allow individual centers with identical fluorescence to be identified and tracked simultaneously. Monitoring decoherence rates in response to changes in the local environment may be found to provide new information about intracellular processes. The experiments performed by McGuinness et al. (2011) that are summarized here demonstrate the viability of controlled single spin probes for nanomagnetometry in biological systems. This kind of approach may be explored further to open up new possibilities for quantum-based imaging of cells and cellular compartments in life sciences.

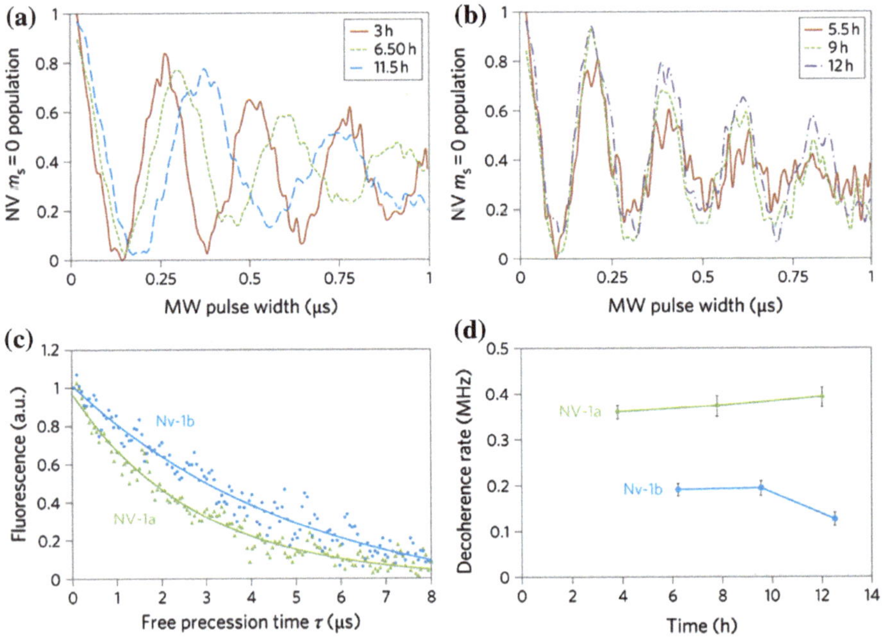


**Fig. 9.9** **a**, **b**, Rabi oscillations of NV-1a and NV-1b measured at various times during the lifetime of the cell. **c**, Initial spin echo measurements on both NV centers (statistical errors ~1% level). **d**, Time evolution of the decoherence rates extracted from the spin echo profiles for both NV centers. The uncertainties in the extracted values, quoted at the one sigma level, are determined from the fit to the data. For details, see McGuinness et al. (2011)

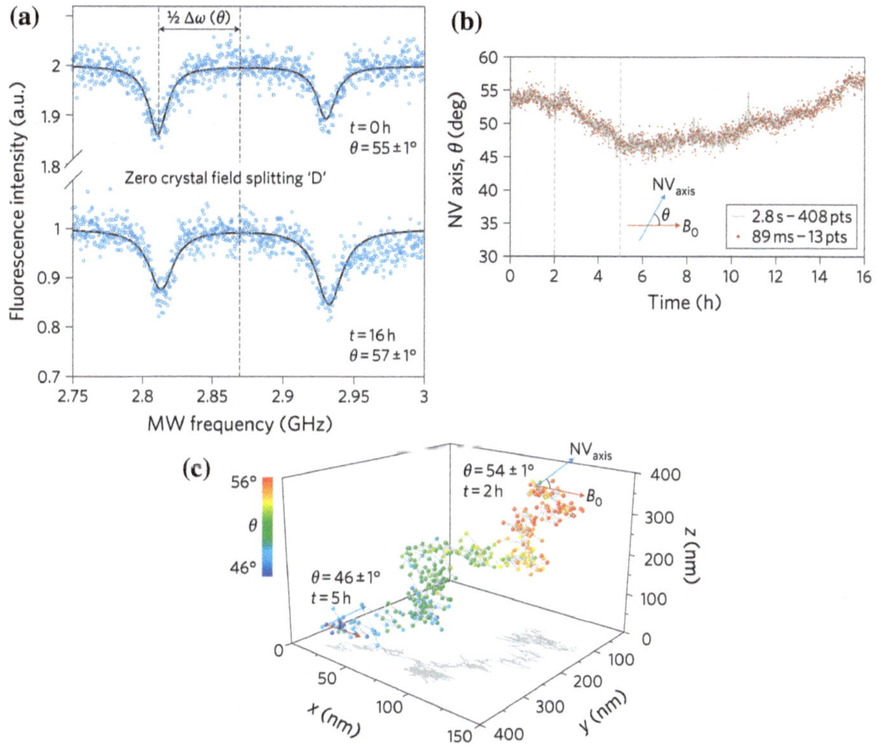


**Fig. 9.10** **a** Changes in the orientation of the NV quantization axis relative to the external magnetic field owing to nanodiamond motion are manifested in the orientation-dependent Zeeman splitting $\Delta\omega(\theta)$ observed in the ODMR spectrum shown at various times over the HeLa-2 cell lifetime. **b** Measured orientation of the nanodiamond as a function of time and comparison of different acquisition rates. **c** Four-dimensional tracking (position and orientation) of NV-2 in HeLa-2 over a 3 h period. For details, see McGuinness et al. (2011)

# References

Chance, B., Nishimura, M. 1960. The mechanism of chlorophyll-cytochrome interaction: the temperature insensitivity of light-induced cytochrome oxidation in Chromatium. Proc. US Nat. Acad. Sci., 46, 19–24.

Vredenberg, W.J., Duysens, L.N.M. 1964. Light-induced oxidation of cytochromes photosynthetic bacteria between 20 and-170°. Biochim. Biophys. Acta, 79, 456–463.

Devault, D., Parkes, J.H., Chance, B. 1967. Electron Tunnelling in Cytochromes. Nature 215, 642–644.

Ashrafuzzaman, Md., Tuszynski, J., Membrane Biophysics, Springer (Heidelberg), 2012, ISSN 1618-7210, ISBN 978-3-642-16104-9 ISBN 978-3-642-16105-6 (eBook), https://doi.org/10.1007/978-3-642-16105-6.

Ashrafuzzaman M, Tseng CY, Tuszynski JA. Regulation of channel function due to physical energetic coupling with a lipid bilayer. Biochem Biophys Res Commun. 2014 Mar 7;445 (2):463–8.

Md. Ashrafuzzaman. Phenomenology and energetics of diffusion across cell phase states. Saudi J. Biol. Sci. (2015a), 22, 666–673.

Md. Ashrafuzzaman. Diffusion across cell phase states. Biomedical Sciences Today (2015b), 1:e4.

P. Ball. Physics of life: The dawn of quantum biology. Nature 474, 272–274 (2011)

G. Panitchayangkoon, D. Hayes, K. A. Fransted, J. R. Caram, E. Harel, J. Wen, R. E. Blankenship, and G. S. Engel. Long-lived quantum coherence in photosynthetic complexes at physiological temperature. PNAS. 2010:107 (29), 12766–12770

Alisher M. Kariev, Vasiliy S. Znamenskiy, and Michael E. Green. Quantum Mechanical calculations of charge effects on gating the KcsA channel. Biochim Biophys Acta. 2007 May; 1768(5): 1218–1229.

Doyle DA, Cabral JM, Pfuetzner RA, Kuo A, Gulbis JM, Cohen SL, Chait BT, MacKinnon R. The structure of the potassium channel: molecular basis of K + conduction and selectivity. Science. 1998;280:69–77.

MacKinnon R, Cohen SL, Kuo A, Lee A, Chait BT. Structural conservation in prokaryotic and eukaryotic potassium channels. Science. 1998;280:106–109.

Jiang Y, Lee A, Chen J, Cadene M, Chait BT, MacKinnon R. The open pore conformation of potassium channels. Nature. 2001;417:523–526.

Alisher M. Kariev, Philipa Njau, and Michael E. Green. The Open Gate of the KV1.2 Channel: Quantum Calculations Show the Key Role of Hydration. Biophys J. 2014 February 4; 106(3): 548–555.

Varma S., Rogers D.M., Rempe S.B. Perspectives on: ion selectivity: design principles for K + selectivity in membrane transport. J. Gen. Physiol. 2011;137:479–488.

Dudev T., Lim C. Determinants of K + vs Na + selectivity in potassium channels. J. Am. Chem. Soc. 2009;131:8092–8101. [PubMed]

Dudev T., Lim C. Factors governing the Na(+) vs K(+) selectivity in sodium ion channels. J. Am. Chem. Soc. 2010;132:2321–2332. [PubMed]

Dudev T., Lim C. Why voltage-gated Ca2 + and bacterial Na + channels with the same EEEE motif in their selectivity filters confer opposite metal selectivity. Phys. Chem. Chem. Phys. 2012;14:12451–12456. [PubMed]

Varma S., Rempe S.B. Multibody effects in ion binding and selectivity. Biophys. J. 2010;99:3394–3401.

Bucher D., Rothlisberger U., Carloni P. QM/MM Car-Parrinello molecular dynamics study of selectivity in a potassium channel. ACS. 2004 Abstract PHYS-309.

Bucher D., Rothlisberger U. Molecular simulations of ion channels: a quantum chemist's perspective. J. Gen. Physiol. 2010;135:549–554. [PubMed]

Maupin C.M., Wong K.F., Voth G.A. A multistate empirical valence bond description of protonatable amino acids. J. Phys. Chem. A. 2006;110:631–639.

Michael A. Crawford, C. Leigh Broadhurst, Martin Guest, Atulya Nagar, Yiqun Wang, Kebreab Ghebremeskel, Walter F. Schmidt. A quantum theory for the irreplaceable role of docosahexaenoic acid in neural cell signalling throughout evolution. Prostaglandins, Leukotrienes and Essential Fatty Acids (PLEFA), Volume 88, Issue 1, January 2013, Pages 5–13

R.H. Steele, A. Szent-Gyorgyi. On excitation of biological substances. Proc. Natl. Acad. Sci., 43 (1957), pp. 478–491

J. Avery, Z. Bay, A. Szent-Gyorgi. On energy transfer in biological systems. Proc. Natl. Acad. Sci., 47 (1961), pp. 1742–1744

D. Bendall, Interprotein Electron Transfer, in: D.S. Bendall, (Ed.), Protein Electron Transfer, Bios Scientific Publishers, Oxford, UK, 1996, pp. 43–68

J.J. Hopfield. Electron transfer between biological molecules by thermally activated tunneling. Proc. Natl. Acad. Sci. USA, 71 (1974), pp. 3640–3644

L. Hackermüller, S. Uttenthaler, K. Hornberger, E. Reiger, B. Brezger, A. Zeilinger, M. Arndt, M. Wave. Nature of biomolecules and fluorofullerenes. Phys. Rev. Lett., 91 (2003), p. 090408

S. Hameroff, R. Penrose. Quantum computation in brain microtubules the Penrose-Hameroff Orch OR model of consciousness. Philos. Trans. R. Soc. London A, 356 (1998), pp. 1869–1896

S. Hameroff. The conscious pilot-dendritic synchrony moves through the brain to mediate consciousness. J. Biol. Phys., 36 (1) (2010), pp. 71–93

A.E. Allen, M.A. Cameron, T.M. Brown, A.A. Vugler, R.J. Lucas. Visual responses in mice lacking critical components of all known retinal phototransduction cascades. PLoS One, 5 (11) (2010), p. e15063

K. Gawrisch, N.V. Eldho, L.L. Holte. The structure of DHA in phospholipid membranes. Lipids, 38 (4) (2003), pp. 445–452

Horrocks LA, Yeo YK. Health benefits of docosahexaenoic acid (DHA). Pharmacol Res. 1999 Sep;40(3):211–25.

Gregory S. Engel, Tessa R. Calhoun, Elizabeth L. Read, Tae-Kyu Ahn, Tomáš Mančal, Yuan-Chung Cheng, Robert E. Blankenship, Graham R. Fleming. Evidence for wavelike energy transfer through quantum coherence in photosynthetic systems. Nature 446, 782–786 (2007).

Gregory D. Scholes. Quantum-Coherent Electronic Energy Transfer: Did Nature Think of It First? J. Phys. Chem. Lett., 2010, 1 (1), pp 2–8

Collini, E.; Curutchet, C.; Mirkovic, T.; Scholes, G. D. Electronic Energy Transfer in Photosynthetic Antenna Systems. In Energy Transfer Dynamics in Biomaterial Systems; Burghardt, I., May, V., Micha, D. A., Bittner, E. R., Eds.; Springer Verlag: Heidelberg/Berlin, Germany, 2009; Vol. 93.

Hofmann, E.; Wrench, P. M.; Sharpies, F. P.; Hiller, R. G.; Welte, W.; Diederichs, K. Structural Basis of Light Harvesting by Carotenoids: Peridinin-Chlorophyll-Protein from Amphidinium Carterae. Science 1996, 272, 1788–1791.

Wilk, K. E.; Harrop, S. J.; Jankova, L.; Edler, D.; Keenan, G.; Sharpes, F.; Hiller, R. G.; Curmi, P. M. G. Evolution of a Light-Harvesting Protein by Addition of New Subunits and Rearrangement of Conserved Elements: Crystal Structure of a Cryptophyte Phycoerythrin at 1.63-Å Resolution. Proc. Natl. Acad. Sci. U.S.A. 1999, 96, 8901–8906.

Liu, Z. F.; Yan, H. C.;Wang, K. B.; Kuang, T. Y.; Zhang, J. P.; Gui, L. L.; An, X. M.; Chang, W. R. Crystal Structure of Spinach Major Light-Harvesting Complex at 2.72 Å Resolution. Nature 2004, 428, 287–292.

Ganapathy, S.; Oostergetel, G. T.; Wawrzyniak, P. K.; Reus, M.; Chew,A.G.M.;Buda, F.; Boekema, E. J.;Bryant, D. A.;Holzwarth, A. R.; de Groot, H. J. M. Alternating Syn-Anti Bacteriochlorophylls Form Concentric HelicalNanotubes in Chlorosomes. Proc. Natl. Acad. Sci. U.S.A. 2009, 106, 8525–8530.

McDermott, G.; Prince, S. M.; Freer, A. A.; Hawthornthwaite-Lawless, A. M.; Papiz,M. Z.; Cogdell,R. J.; Isaacs, N. W. Crystal Structure of an Integral Membrane Light-Harvesting Complex from Photosynthetic Bacteria. Nature 1995, 374, 517–521.

Barros, T.; K€uhlbrandt, W. Crystallisation, Structure and Function of Plant Light-Harvesting Complex II. Biochim. Biophys. Acta 2009, 1787, 753–772.

van der Weij-De Wit, C. D.; Doust, A. B.; van Stokkum, I. H. M.; Dekker, J. P.; Wilk, K. E.; Curmi, P. M. G.; Scholes, G. D.; van Grondelle, R. How. Energy Funnels from the Phycoerythrin Antenna Complex to Photosystem I and Photosystem II in Cryptophyte Rhodomonas CS24 Cells. J. Phys. Chem. B 2006, 110, 25066–25073.

Collini, E.; Wong, C. Y.; Wilk, K. E.; Curmi, P. M. G.; Brumer, P.; Scholes, G. D. Coherently Wired Light-Harvesting in Photosynthetic Marine Algae at Ambient Temperature. Nature 2010, 463, 644–647

Richard Hildner, Daan Brinks, Niek F. van Hulst. Femtosecond coherence and quantum control of single molecules at room temperature. Nature Physics 7, 172–177 (2011)

Mohan Sarovar, Akihito Ishizaki, Graham R. Fleming, K. Birgitta Whaley. Quantum entanglement in photosynthetic light-harvesting complexes. Nature Physics 6, 462–467 (2010)

Alivisatos, P. The use of nanocrystals in biological detection. Nature Biotechnol. 22, 47–52 (2004).

Maze, J. R. et al. Nanoscale magnetic sensing with an individual electronic spin in diamond. Nature 455, 644–648 (2008).

Balasubramanian, G. et al. Nanoscale imaging magnetometry with diamond spins under ambient conditions. Nature 455, 648–651 (2008).

Chernobrod, B. M. & Berman, G. P. Spin microscope based on optically detected magnetic resonance. J. Appl. Phys. 97, 014903 (2005).

Taylor, J. M. et al. High-sensitivity diamond magnetometer with nanoscale resolution. Nature Phys. 4, 810–816 (2008).

Degen, C. L. Scanning magnetic field microscope with a diamond single-spin sensor. Appl. Phys. Lett. 92, 243111 (2008).

Cole, J. H. & Hollenberg, L. C. L. Scanning quantum decoherence microscopy. Nanotechology 20, 495401 (2009).

Hall, L. T., Cole, J. H., Hill, C. D. & Hollenberg, L. C. L. Sensing of fluctuating nanoscale magnetic fields using nitrogen-vacancy centers in diamond. Phys. Rev. Lett. 103, 220802 (2009).

L. P. McGuinness, Y. Yan, A. Stacey, D. A. Simpson, L. T. Hall, D. Maclaurin, S. Prawer, P. Mulvaney, J. Wrachtrup, F. Caruso, R. E. Scholten, L. C. L. Hollenberg. Quantum measurement and orientation tracking of fluorescent nanodiamonds inside living cells. Nature Nanotechnology 6, 358–363 (2011)

C. Bradac, T. Gaebel, N. Naidool, M. J. Sellars, J. Twamley, L. J. Brown, A. S. Barnard, T. Plakhotnik, A. V. Zvyagin & J. R. Rabeau. Observation and control of blinking nitrogen-vacancy centres in discrete nanodiamonds. Nature Nanotech. 5, 345–349 (2010).

Dudev T., Lim C. Competition among Ca2 +, Mg2 +, and Na + for model ion channel selectivity filters: determinants of ion selectivity. J. Phys. Chem. B. 2012a; 116:10703–10714.

# Epilogue

Application of biophysics principles in addressing cell's nanometer dimension structures and functions started not too long ago. But recent advances in this area are not limited too. This is why I took an initiative with Springer-Verlag to publish a book with the title 'Nanoscale Biophysics of the Cell.' I have presented quite a lot of information about biological and biophysical aspects of various cellular events. To be in line with the title of the book, my focus has always been to limit the discussion in the specific nanoscale world of cell and the surrounding cell biological fields by presenting the analysis of existing and novel experimental and theoretical data along with the techniques of their production, methods of interpretations, and understanding of their applications. The book has addressed on structural and functional mechanism perspectives of both normal cell conditions and perturbed conditions under the influence of diseases. A lot of specific cell-based structural components have been taken into consideration as examples for explaining the biophysics aspects of mainly the nanoscale structure cell components.

The book started with the search of a descriptive definition of life. With the development of technology over centuries, we have been able to revise our knowledge on cell by going deep into its finest structures. Consequently, life's definition, once used to be made up using biological perspectives, now also requires considerations of physical principles and phenomena discovered in cells for a descriptive understanding. We have attempted to do so. The physics of cellular dynamics, diffusion, and transport has been explained considering existing rules and models as well as novel principles and analytical expressions. Application of biophysics in nanoscale sections of cell is a pretty new field. A lot of new physics-based models have to emerge to help understand the unexplored areas. I have produced a few conceptual approaches what readers may find as their mind opening and thought creating for further explorations in microscopic cell's low-dimension interior word.

Biological cells are enriched with molecular machines which account for most of the molecular-level solid state structures that regulate cellular functions. Their pre-creation molecular mechanisms, states, and energetics of physical existence and their physiological functional aspects have been inspected quite in detail. The book contains rigorous discussion where molecular machines have been scrutinized in



M. Ashrafuzzaman, *Nanoscale Biophysics of the Cell*,
https://doi.org/10.1007/978-3-319-77465-7

regard to their physical roles in creating cell-based natural nanotechnology. The principles of these natural nanotechnology along with artificial biophysics techniques can be utilized for regulating the functions of cellular machines and work toward repairing the cellular disorders by designing drugs. I provided quite a detailed explanation on this issue too.

The molecular machines in cancer cells appear with perturbed physiological properties. Special emphasis has been given in understanding ion channels in cancer cells. This medically important area will certainly be found attractive due to the importance in applied medical research.

Although quantum mechanics application in cell is quite new, it offers some promising tools to understand crucial cellular processes. A brief address on quantum mechanical calculations that are especially related to the quantum effects in cell dynamics has been provided.

Low-dimension cell structure-based nanotechnology has been naturally observed. Physics principles besides biological and chemical properties are also naturally in built to drive physical structures into functioning. Application of physics principles related to interactions, forces, and energies are mainly responsible to cause and regulate the statics, dynamics, and technological architectures of cells. Biological processes as detected in observational experiments are better understood using the physics laws. Disorders due to disease conditions can also be addressed through understanding the perturbations in structures and functions using related physics principles. 'Nanoscale Biophysics of the Cell' will help the readers to get a summary of huge information on all of these. We hope that this monograph will be a catalyst for future studies of cell in regard to its treatment at the nanoscale components where various technologies are available. It is my great hope that this book will become a source of empirical information and conceptual inspiration to students of biophysics, biology, biomedical sciences, and engineering disciplines, and expert researchers including those involved in pharmaceutical sciences and industries.

# Index





M. Ashrafuzzaman, *Nanoscale Biophysics of the Cell*,
https://doi.org/10.1007/978-3-319-77465-7

FSC
www.fsc.org
MIX
Papier | Fördert
gute Waldnutzung
FSC® C083411

Kolibri 360 GmbH
Ferdinand-Jühlke-Straße 7
99095 Erfurt, Deutschland
produktsicherheit@kolibri360.de